American Seacoast Forts

A Directory with Period Military Maps
1890-1950

Volume 4

Alaska and the Overseas Bases including Hawaii, the Philippines, the Panama Canal Zone, the Caribbean, Newfoundland and Bermuda

PREPARED BY

TERRANCE C. MCGOVERN
MARK A. BERHOW
GLEN M.WILLIFORD

Published by the Coast Defense Study Group Press
2025

IBSN 978-0-9748167-9-1 (Hardcover B&W)
LIBRARY OF CONGRESS CATALOG CARD NUMBER 2025941598

Library of Congress Cataloging-in-Publication data
American Seacoast Forts: A Directory / Terry McGovern, Mark Berhow and Glen Williford
p.   cm.
Includes bibliographical references and index.
Library of Congress Control Number: 2025941598
ISBN 978-0-9748267-9-1 (h.c.)
1. Military History, 2. Artillery I. Terry McGovern, Mark Berhow and Glen Williford

First Edition: August 2025
Printed in the USA by Ingram Spark

Cover Photographs:
Front cover: Battery #296 Fort Tidball Kodiak, Alaska (Terry McGovern)

Rear cover (clockwise from upper left): General map of Honolulu and Pearl Harbors, 1934 (National Archives);
Fire control stations Aleutian World War II National Historical Area Dutch Harbor, AK (Terry McGovern);
Perico Island and Flamenco Island Panama (Terry McGovern); Fort Drum in Manila Bay, the Philippines, 1930s
(National Archives)

THE COAST DEFENSE STUDY GROUP, INC.
CDSG.ORG

The Coast Defense Study Group, Inc. (CDSG) is a tax-exempt corporation dedicated to the study of seacoast fortifications. The purposes of the CDSG include educational research and documentation, preservation and interpretation of historic sites, and assistance to other organizations dedicated to the preservation and interpretation of coast defense sites. Membership is open to any person or organization interested in the study or history of coast defenses and fortifications. Membership in the CDSG will allow you to attend annual conferences, special tours, and receive quarterly newsletter and journal. To find our more about the CDSG, please visit the CDSG website at **cdsg.org.**

Acknowledgments

This book is dedicated to three key members of the CDSG who began research on this subject in the 1970s and provided much of the initial information for those that followed in this study. Robert Zink (member number 1) and Glen Williford (member number 2) were both dedicated in their studies and unfailing supportive to those that sought more information on American seacoast defenses. The third key CDSG member was Bolling W. Smith who did a large amount of research work in obtaining copies of documents from the National Archives of which many were used in this compilation.

Sighting a 10-inch gun at Fort Standish, Massachusetts
(Leslie Jones, Digital Commonwealth, Massachusetts Online Collection, Boston Public Library)

The CDSG Press

Coast Defense Study Group Press is a division of Coast Defense Study Group (cdsg.org), which publishes books of historical interest, especially concerning seacoast fortifications. The CDSG Press also offers an ever-expanding number of key reprints reports and manuals in electronic PDF format on compact disks. To order these books and other **CDSG Press** publications, please access the **CDSG Press** pages on the **CDSG web site** at **cdsg.org.**

CDSG Press is interested in new titles, especially those dealing with fortifications, please contact Terry McGovern at 703/538-5403 or at tcmcgovern@att. net if you have a title that you are seeking to have published. Visit www.cdsg.org/press.

Under the CDSG Press label, our organization has published:

Notes on Seacoast Fortification Construction by Col. Eben E. Winslow, 1920, 428 pp. 1994 reprint HC with bound drawings
Seacoast Artillery Weapons Technical Manual (TM) 9-210 by U.S. War Dept. 1944, 202 pp. 1995 reprint PB
The Service of Coast Artillery by F. Hines & F. Ward, 1910, 736 pp. 1997 reprint HC
Permanent Fortifications & Sea-Coast Defenses by U.S. Congress, 1862, 544 pp. 1998 reprint HC
American Coast Artillery Material Ordnance Dept. Doc#2042 by U.S. War Dept., 1922, 528 pp., 2001 Reprint HC
American Seacoast Defenses: A Reference Guide (3rd Edition) by Mark A. Berhow, (2015) 732 pp. HC
The Endicott & Taft Board Reports, reprint of original reports of 1886 and 1905 by U.S. Congress, 525 pp. 2007 HC
Artillerists and Engineers: The Beginnings of US Fortifications 1794-1815 by Col. Wade, U.S. Army, 226 pp. 2011 PB
World War II Harbor Defenses of San Diego by Commander (Ret.) Everett, U.S. Navy, 226 pp. 2020 HC

The CDSG Presss offers these Hole in the Head Press Books

Artillery at the Golden Gate by Brian B. Chin (Hole-in-the-Head Press), 176 pp. 2019 PB
Fort Baker Through the Years by Kristin L Baron and John A. Martini (Hole-in-the-Head Press), 99 pp. 2013 PB
Rings of Supersonic Steel (3rd Edition) by Mark Morgan and Mark Berhow (Hole-in-the-Head Press), 358 pp. 2010 PB
The Last Missile Site by Stephen Hailer and John A. Martini (Hole-in-the-Head Press), 158 pp. 2010 PB
To Defend and Deter by John Lonnquest and David Winker (Hole-in-the-Head Press), 432 pp. 2014 PB

The CDSG Press
1700 Oak Lane
McLean, VA 22101-3322 USA

AMERICAN SEACOAST DEFENSES
A DIRECTORY OF AMERICAN SEACOAST DEFENSES 1890-1950

Table of Contents

Introduction   ... 13

Chronology   ... 15
• Key Events

Design and Function of American Seacoast Forts   ... 16
• Historic Development of Coast Defenses during the Modern Era   ... 16
    The Endicott Program (1885-1904)   ... 16
    The Taft Program (1905-1916)   ... 19
    The World War I Program and the Inrerwar Period (1917-1939)   ... 20
    The 1940 Modernization Program and World War II (1940-1950)   ... 21
• Coastal Defense Objective   ... 23
• Coast Defense Armaments and Equipment   ... 29
• Coast Defense Organization and Fire Control   ... 34
• A Typical U.S. Coast Artillery Fort   ... 40
• Coast Artillery Garrison - Daily Life   ... 42
• Aftermath and Today   ... 45

Period Military Maps   ... 45
• Confidential Blueprint Series Maps (1915-37)   ... 45
• WW II-Era Maps (1933-46)   ... 46
• Glossary of Terms used in Maps   ... 47
• Symbols and Abbreviations (1921)   ... 49
• Symbols and Abbreviations (1940)   ... 52

Comments on what is included in this Directory   ... 58

U.S. Coast Artillery 1890-1945 Harbor Defense locations

Directory to American Seacoast Forts 1885-1950 in 4 volumes
    (21 Continental US Harbor Defenses and Overseas Harbor Defenses)
• Volume 1 North Atlantic Coast: New England – Long Island Sound - New York
• Volume 2 Mid-Atlantic, South Atlantic, and Gulf Coasts: Cheasapeake Bay to Galveston
• Volume 3 Pacific Coast San Diego to Puget Sound
• Volume 4 Alaska and Overseas Bases: Hawaii - Philippines - Panama - Caribbean - Newfoundland
    - Bermuda

Volume 4: Alaska and the Overseas Possessions

Sitka, Alaska—WWII Program sites   ... 59
    FORT BABCOCK   ... 61
        Battery #290
    FORT PIERCE   ... 63
        Battery   #291
    FORT ROUSSEAU   ... 63
        Battery   #292
    Fort Ray   ... 64
        Batteries   AMTB Watson Point; AMTB Whale Island

Sitka Fire Control Sites   ... 66
    Hill 800
    Sitka Point
    St. Lazaria Island
    Shoal's Point, Kruzof Is FORT BABCOCK
    Lava Point
    Sound Island
    Lisianski Peninsula
    Clam Island
    Abalone Island
    Makhnati Island FORT ROUSSEAU
    Kayak Island
    Kita Island
    Biorka Island FORT PIERCE
    Little Biorka Island
    Akaku Sound
    Golf Island
    Watson Point
    Whale Island

Seward, Alaska—WWII Program sites   ... 85
    FORT McGILVARY   ... 85
        Battery   #293; AMTB Lowell Point
    FORT BULKLEY   ... 86
        Battery   #294
    Fort Raymond   ... 87

Seward Fire Control Sites   ... 91
    Caines Head
    Rocky Point
    Caines Head FORT McGILVRAY
    Rugged Island FORT BULKLEY Alma Point, Carol Cove
    Topeka Point
    Barwell Island
    Chamberlin Point
    Lowell Point

Kodiak, Alaska—WWII Program sites ... 105
    FORT J.H. SMITH ... 105
        Batteries   #403; #295 (planned)
    FORT TIDBALL ... 107
        Battery   #296
    FORT ABERCROMBIE ... 107
        Batteries   #404; #297 (planned); AMTB Spruce Cape
    Fort Greeley ... 108
        Battery   AMTB Puffin Island

    Kodiak Fire Control ... 112
        Soquel Point Site # 1
        Narrow Cape Site 1A
        Cape Greville site 2
        Cape Chiniak Site 3
        Bald Hill site 3A
        Round Top Site 3A
        St. Peter's Head site 4 FORT J.H. SMITH
        Midway Point site 6
        Mansfield Ridge site 7
        Artillery Hill site 8
        Puffin Island site 9
        Gibson Cove site 9
        Long Island Sites 10-14 FORT TIDBALL
        Spruce Cape site 15
        Miller Point Piedmont Point site 16 & 16A FORT ABERCROMBIE
        Spruce Island SE Site 17
        Spruce Island NW Site 18
        Kizhuyak Point Site 18

Dutch Harbor, Alaska—WWII Program sites ... 147
    FORT SCHWATKA ... 148
        Batteries   #402; #299 (planned); AMTB Amaknak Spit
    FORT LEARNARD ... 149
        Battery   #298; AMTB Eider Point
    FORT BRUMBACK ... 149
    Fort Mears ... 149

    Dutch Harbor Fire Control Sites ... 155
        Ugadaga Bay Site 1
        Erskine Point Site 2
        Constatine Point Site 3
        Summer Bay North Site 4 FORT BRUMBACK
        Summer Bay South Site 5
        Amaknak Spit site 6
        Hill 400 site 7

Ulakta Island site 8 FORT SCHWATKA
Hog Island site 9
Eider Point site 10 FORT LEARNARD
Cape Wislow site 11

Oahu Island, Hawaii

The Harbor Defenses of Honolulu—Fort and Gun Battery Descriptions　　　… 179
　　Sand Island Military Reservation　　　… 180
　　　　Batteries　　　Harbor; AMTB #4
　　FORT ARMSTRONG　　　… 184
　　　　Battery　　　Tiernon
　　Punchbowl Military Reservation　　　… 184
　　　　Battery　　　#304
　　FORT DeRUSSY　　　… 186
　　　　Batteries　　　Randolph; Dudley; AMTB #5
　　FORT RUGER　　　… 190
　　　　Batteries　　　Harlow; Birkhimer; Dodge; Hulings; #407
　　Black Point Military Reservation　　　… 197
　　　　Batteries　　　Granger Adams; S.C. Mills
　　Wiliwilinui Ridge Military Reservation　　　… 198
　　　　Battery　　　Kirkpatrick
　　Koko Head Military Reservation　　　… 198
　　　　Battery　　　#305 (planned)

The Harbor Defenses of Pearl Harbor—Fort and Gun Battery Descriptions　　　… 199
　　FORT KAMEHAMEHA　　　… 199
　　　　Batteries　　　Hasbrouck; Selfridge; Jackson; Hawkins; Barri-Chandler;
　　　　　　　　　　　Closson; AMTB #2
　　Pearl Harbor Military Resercation (Ford Island)　　　… 205
　　　　Batteries　　　Adair; Boyd
　　Aliamanu Military Reservation　　　… 208
　　　　Battery　　　Burgess
　　FORT WEAVER　　　… 209
　　　　Battery　　　Williston
　　FORT BARRETTE　　　… 212
　　　　Battery　　　Hatch
　　Kahe Point Military Reservation　　　… 214
　　　　Battery　　　Arizona
　　Puu O Hulu Military Reservation　　　… 216
　　　　Batteries　　　Hulu; #303

The Harbor Defenses of Kaneohe Bay and the North Shore of Oahu … 218
    FORT HASE … 218
        Batteries   Pennsylvania; Demerrit (#405); French (#301); Sylvester; AMTB #7
    Kualoa Military Reservation … 221
        Batteries   Cooper (#302); AMTB #8
    Kahuku Military Reservation … 222
        Battery   Kahuku
    Opaeula Military Reservation … 225
        Battery   Riggs
    Brodie Camp Military Reservation … 225
        Battery   Ricker
    Kaena Point Military Reservation … 226
        Battery   #409

MANILA & SUBIC BAY, THE PHILIPPINES

The Harbor Defenses of Manila Bay—Fort and Gun Battery Descriptions … 228
    FORT FRANK … 230
        Batteries   Greer; Crofton; Koehler; Hoyle
    FORT DRUM … 235
        Batteries   Marshall; Wilson; McCrea; Roberts; Hoyle
    FORT HUGHES … 239
        Batteries   Gillespie; Woodruff; Craighill; Leach; Fuger
    FORT MILLS … 245
        Batteries   Cheney; Crockett; Wheeler; Geary; Way; Grubbs; Morrison; Ramsey; James; Keyes; Cushing; Hanna; Smith; Hearn; RJ 43
The Harbor Defenses of Subic Bay—Fort and Gun Battery Descriptions … 270
    FORT WINT … 270
        Batteries   Warwick; Hall; Woodruff; Flake; Jewell

PANAMA CANAL ZONE

The Harbor Defeses of Cristobal—Fort and Gun Battery Descriptions … 274
    FORT RANDOLPH … 277
        Batteries   Webb; Tidball-Zalinski; Weed; Railroad
    FORT DeLESSEPS … 282
        Batteries   Morgan; AMTB #3b
    FORT SHERMAN … 284
        Batteries   Mower; Stanley; Howard-Baird; Kilpatrick; MacKenzie; Pratt; #151 (planned)

The Harbor Defenses of Balboa—Fort and Gun Battery Descriptions ... 294
    FORT AMADOR   ... 294
        Batteries   Birney; Smith
    FORT GRANT   ... 299
        Batteries   Buell; Burnside; Parke; Newton; Warren; Carr-Merrit-Prince;
                       Railroad
    FORT KOBBE (ex-Ft. Bruja)   ... 310
        Batteries   Haan; Murray; AMTB #6
    Taboga Island Military Reservation   ... 313
        Batteries   (planned)

OTHER U.S. DEFENSES

The Harbor Defenses of Vieques Sound, Puerto Rico, Virgin Islands (Roosevelt Roads) ... 316
    FORT SEGARRA, St. Thomas Island   ... 316
        Batteries   #401; #314; AMTBs; #315 (planned)
    FORT CHARLES W. BUNDY, Puerto Rico   ... 318
        Batteries   #406; #265; #268; AMTB
                       #152 (planned), #155 (planned), #311 (planned)
    Vieques Island   ... 319
        Batteries   #153, #154, #266, #267, #285 (all planned)
    Culebra Island   ... 319
        Batteries   #312, #313 (all planned)

The Harbor Defenses of San Juan, Puerto Rico—WWII Program sites ... 320
    FORT BROOKE, El Morro   ... 320
    FORT AMEZQUITA Las Cabras Island   ... 321
        Batteries   Reed; AMTB
    FORT MASCARO Punta Salinas   ... 322
        Batteries   Buckey (#261); Pence (#262)
    Punta Escambrón   ... 322
        Battery   Schwan (#263)
    Punta Cangreios   ... 323
        Battery   Lancaster (#264)

Planned Defenses in Guantanamo Bay, Cuba   ... 324
        Batteries   Conde Bluff; Leeward Point; Cuzco Hill

Planned defenses in Trinidad   ... 328
    Batteries   2 x 12-inch, #271, #272, #273, #274, #275, AMTBs (all planned)

Planned Defenses in Jamaica   ... 329
    FORT SIMONDS
        Battery   #285

Harbor Defenses in Newfoundland, Canada—WWII Program sites ... 330

    FORT McANDREW, Argentia ... 330

        Batteries   #281; #282; AMTB Roche Point; AMTB Ship Harbor

    Harbor Defenses of St. John, Newfoundland ... 331

Defenses in Bermuda —WWII Program sites ... 333

    FORT BELL

        Batteries   #284; #285; AMTB

VOLUME 1: THE NORTH ATLANTIC COAST
PORTLAND TO NEW YORK

Portland & Kennebec River, ME-—Fort and Gun Battery Descriptions

Portsmouth, NH—Fort and Gun Battery Descriptions

Boston, MA—Fort and Gun Battery Descriptions

New Bedford, MA—Fort and Gun Battery Descriptions

Narrangansett Bay, RI—Fort and Gun Battery Descriptions

Long Island Sound, CT & NY—Fort and Gun Battery Descriptions

New York, East, NY —Fort and Gun Battery Descriptions

West Point Military Academy practice batteries

New York, South, NY & NJ —Fort and Gun Battery Descriptions

VOLUME 2: THE MID-ATLANTIC, SOUTH ATLANTIC, AND GULF COASTS
DELAWARE BAY TO GALVESTON

Delaware Bay, DE & NJ—Fort and Gun Battery Descriptions

Chesapeake Bay (Baltimore, Potomac River, James River, Hampton Roads)

Baltimore, MD—Fort and Gun Battery Descriptions

Potomac River, MD & VA—Fort and Gun Battery Descriptions

Entrance to the Chesapeake Bay—Fort and Gun Battery Descriptions

Cape Fear River, NC—Fort and Gun Battery Descriptions

Charleston, SC—Fort and Gun Battery Descriptions

Port Royal Sound, SC—Fort and Gun Battery Descriptions

Savannah, GA—Fort and Gun Battery Descriptions

St. Mary's River - Fort Clinch—Fort and Gun Battery Descriptions

St. John's River, FL—Fort and Gun Battery Descriptions

Key West, FL—Fort and Gun Battery Descriptions

Tampa Bay, FL—Fort and Gun Battery Descriptions

Pensacola, FL—Fort and Gun Battery Descriptions

Mobile Bay, AL—Fort and Gun Battery Descriptions

Mississippi River, LA—Fort and Gun Battery Descriptions

Galveston, TX—Fort and Gun Battery Descriptions

VOLUME 3: THE PACIFIC COAST
SAN DIEGO TO PUGET SOUND

San Diego, CA—Fort and Gun Battery Descriptions
Los Angeles, CA—Fort and Gun Battery Descriptions
San Francisco, CA—Fort and Gun Battery Descriptions
Columbia River, OR & WA—Fort and Gun Battery Descriptions
Willapa Bay, WA—Fort and Gun Battery Descriptions
Grays Harbor, WA—Fort and Gun Battery Descriptions
Puget Sound, WA—Fort and Gun Battery Descriptions

HECP observing stations Sitka at Fort Rousseau (Matt Hunter)

INTRODUCTION

The United States has long focused on defending its seacoasts against overseas enemies due to its geopolitical situation with its long coastlines and generally peaceful borders with Canada and Mexico. Earlier American fortification efforts resulted in the First System, the Second System, and the Third System of coastal defenses. The great brick and stone forts built or remodeled during 1820 to 1860 are well known from their military importance and use during the American Civil War. For many years after that great internal conflict most U.S. fortification efforts languished. After 1885, due to the great advances in military technology and America's increasing worldwide economic presence that the United States embarked on a new round of fortification building to protect its shores. The U.S. Army expended much of its limited manpower and resources to protecting America's coast from 1890 to 1950.

The technical development and tactical objectives of the coast defenses of the modern era (1890 to 1950) is a product of America's earlier policies and experiences. Until the advent of air power and the missile age, the defense of the United States has primarily been one of defending our shores from naval attack. Only during the nation's early years did the threat of land invasion exist. Accordingly, the United States relied on the ships of its Navy to provide the first line of defense. Its U.S. Army was called upon to provide the second line of defense by building forts at key points along its coastline to defend major harbors. This defense policy of denying an enemy fleet access to its major harbors and anchorages developed into the array of former coastal fortifications that remain today.

Based on concerns over external threats and on internal politics, the United States government has built coastal fortification in a series of construction programs. After inheriting the remains of the fortifications from the colonial era and the revolutionary war, the first in a series of national fortification construction programs began in 1794 and these programs continued into the late 1940s. For the ease of use in this book, these fortification periods have been organized into distinct groups and have been named the following: First System (1794-1801), Second System (1802-1815), Third System including the American Civil War (1816-1867), 1870s Period (1868-1879), Endicott Program (1885-1904), Taft Program (1905-1916), World War I & Interwar Period (1917-1939), and the 1940 Modernization Program including World War II (1940-1950).

The legacy of these seacoast defenses is a series of concrete structures scattered along America's coastline, many now in public shoreline parks. The nature of fortifications, the fact that they were designed to withstand the pounding of naval artillery, has allowed these massive structures to withstand the attack of both the natural elements and economic development. That these structures are still standing many years after their effective use ended draws our attention to them. They captivate us regardless of whether it's a large brick and stone multi-story structure surrounded by a dry ditch or an odd shaped, concrete structure covered with thick vines or bright graffiti and surrounded by worn fences sporting weathered warning signs. Visitors to our seashores are curious about the nature of these structures. Some of the questions that they ask are: What are these structures? Why are these structures here? When were these structures built? This directory provides answers to many of these questions. Many of these parks have visitor's centers and gift shops selling range of books and other items, but few have any books which explain the fortifications that once existed at that site. This directory will fill that void.

This directory is a companion work to the CDSG Press's *American Seacoast Defenses – A Reference Guide* by Mark Berhow. This directory is to aid students of American seacoast fortifications to locate and visit the key defenses of the "Modern Era" of coastal defense (1890 to 1950), which is defined by its use of concrete, steel, and breech-loading rifles. The following pages provide a brief review of the function and history of the development of coastal fortifications in the United States. This history reflects the politics of changing external threats to our nation and rapid advancement of military technology. The directory's focus is a guide to the "Modern Era" of fortifications along the Atlantic, Pacific, and Gulf coasts, as well as U.S. overseas bases by providing key maps, plans, photographs, and short description of their history.

The directory is organized into several sections. The first section is brief history of the modern era of American coastal defenses, including background on the U.S. coast artillery material, organization, armament, design of military reservations, and garrison life. The second section is a brief description of the history and its current status (ownership, public access, remaining assets, things to see, etc.) of the major military reservations that had seacoast artillery and a short history of each major concrete gun battery. This history includes each battery is described as to its rational, authorization, construction and transfer dates, engineering cost, naming citation, armament, service history, ultimate disarming, and current status. These battery histories do not include railway and mobile artillery sites, including those with Panama mounts. Also excluded are temporary batteries, especially those using loaned naval guns, and anti-aircraft batteries that did not also serve a seacoast role. The directory is organized by Harbor Defense around the United States clockwise from Portland Maine to the Puget Sound in Washington State, followed by the Alaskan defenses of World War II, and the defenses in Hawaii, the Philippines, Panama, the Caribbean, Newfoundland and Bermuda.

Supporting the directory of defense sites is a compilation of maps for all the harbor defense reservations utilized during the period of 1900 to 1946. The map collection includes general maps of the location of elements (sites) for each defended harbor and the individual location site maps showing buildings, gun emplacements, fire control stations and other elements. Each Harbor Defense section has an overall 1920s-30s period map of the defense sites and a selected set of Confidential Blueprint Maps for each military reservation.

The maps are arranged more or less in order from the south to the north. A series of map symbol and abbreviation keys from 1921 and 1945 are included in the introduction. Some harbors (Baltimore, Potomac River, Cape Fear, Port Royal, Savannah, Tampa, Mobile, and the Mississippi River) did not receive new defenses during World War II.

12-inch Rifle on a barbette mount, Battery Godfrey, Fort Winfield Scott, California,
(Golden Gate National Recreation Area Collection, NPS)

CHRONOLOGY

Key events during the Modern Era of American coast defenses

1875   – Funding for new construction of coast defenses is stopped by the U.S. Congress

1883   – The U.S. Navy begins the first new construction program since the Civil War

1885   – President Cleveland appoints a joint army, navy, and civilian board headed by the Secretary of War, William Endicott, to evaluate the threats and needs for U.S. coastal defenses (Endicott Board)

1886   – The Endicott Board reports on state of the U.S. defenses and recommends a $126 million construction program of breech-loading cannons and mortars, floating batteries, warships, and submarine mines in 29 locations around the nation

1888   – Congress creates the Board of Ordnance and Fortifications to test weapons and implement the Endicott Program

        – Dynamite guns were developed to fire high explosive shells using compressed air

1890   – Congress approves funding for the construction of first Endicott Program batteries

1892   – The first Endicott Program battery is completed (Gun Lift Battery Potter)

1893   – First group of controlled mine casemates are completed

1894   – Buffington-Crozier disappearing carriage for 8-inch and 10-inch guns developed

1896   – Development of the Buffington-Crozier disappearing carriage for 12-inch guns

1898   – The Spanish-American War – 150 coast artillery pieces mounted

        – U.S. adds the Philippines, Guam, Puerto Rico as colonies; establishes military bases in Cuba; annexes Hawaiian Is.

1899   – 288 heavy coast artillery guns, 154 rapid-fire guns, and 312 mortars have been mounted

1901   – Reorganization of U.S. Army artillery corps to 30 batteries of field artillery and 126 companies of coast artillery

1902   – Work begins on the fortifications of Corregidor, Philippines

1904   – Work begins on the Panama Canal

        – First specially built mine planters constructed

1905   – President Roosevelt appointed a joint army, navy, and civilian board headed by the Secretary of War William Taft, to review the Endicott Program and to bring it up to date

1906   – The Taft Board reports on state of the U.S. defenses and recommends improvement in existing defenses by adding searchlights, electrification of defenses, and a modern system of fire control, as well as new defenses for newly acquired overseas bases

1907   – Establishment of the separate U.S. Army Coast Artillery Corps

1914   –World War I begins; Panama Canal opens

1915   – Report of the Board of Review on the coast defenses of the U.S., Panama Canal, and the Insular Possessions

1916   – First Coast Artillery anti-aircraft units formed

1917   – U.S. enters the World War II

        – Construction begins on the first long-range barbette batteries using existing 12-inch gun barrels

1918   – End of the World War I

1920   – Construction of the first Panama mount for 155m GPF guns in the Canal Zone

1922   – Washington Naval Treaty limits naval construction and Pacific fortifications

        – 16-inch gun and howitzer barbette batteries are constructed

1925   – Ten U.S. Harbor Defenses on active status and 15 are on caretaker status

1937   – Construction begins on first 16-inch casemated gun battery (Battery Davis)

1939   – Outbreak of World War II; U.S. Coast Artillery Corps has 4,200 troops

1940   – Congress approves the 1940 Modernization Program for 19 harbors in the U.S.

        – U.S. draft begins, coast artillery units brought up to wartime strength, national guard units federalized

1941   – U.S. enters the World War II

        – Establishment of the Harbor Entrance Control Posts (HECP)

1942   – U.S. Coast Artillery Corps has 70,000 troops

1945   – End of the World War II

1948   – All construction efforts cease, the coast defenses are abandoned, and armament salvaged

1950   – Disestablishment of the U.S. Coast Artillery Corps; remaining units reunited with Field Artillery

DESIGN AND FUNCTION OF AMERICAN SEACOAST DEFENSES IN THE MODERN ERA 1890-1950

Historical Development of American Coast Defenses during the Modern Era

The key development that led to the new American coast defense era was the development of new heavy rifled breech-loading guns that had a longer range, were more accurate and delivered a heavier projectile than the muzzle-loading smoothbore cannons of the Civil War. These new guns were made of high-quality steel that were lighter and stronger, which took advantage of new propellants that replaced gunpowder. Equally important was the development of effective breech mechanisms that could withstand the high pressures and temperatures generated by the new guns and allow for the gun to be loaded from the rear instead of the muzzle, which increased the rate of fire and allowed for improved protection of the gun crew. The new guns and mortars could accurately fire projectiles at effective ranges that were two to three times farther than the muzzle-loading smoothbore cannons used during the American Civil War. These developments coincided with the building of the new steel naval vessels that featured these new big guns starting in 1875. However, between 1875 and 1890 the U.S. Congress did not appropriate any funds for the construction of new coastal fortifications.

Seacoast Defenses built after the Endicott Board Report 1886-1904

As U.S. coastal fortifications were allowed to deteriorate in 1870s and early 1880s, new steam powered, ocean navigating iron warships were being built by foreign navies. As the U.S. Navy embarked on its new construction program, it required protected bases for its operations. The military began to lobby to overhaul the obsolete existing defenses. In 1885, a board was created by U.S. Congress to examine and report upon the state of U.S. coastal defenses. The board headed by the U.S. Secretary of War, William C. Endicott, was comprised of four officers from U.S. Army, two officers from the U.S. Navy, and two civilians. This joint board made an extensive study of fortifications, type of armament, and defense that would be needed, by evaluating current European developments. In 1886, the Endicott Board published its recommendations for new coastal fortifications to be built at 29 key harbors, along with floating batteries, torpedo boats, and submarine mines. The board's original plan called for over 1,300 guns and mortars of 8-inch or larger of the newest design to be installed. The costs of board's recommendations were estimated to be $126 million dollars (in 1886-dollar value). While the U.S. Congress took no immediate action on the board's report, the estimates provided in the report would be cited for the next 20 years as a measure of the construction progress of this new generation of U.S. coastal fortifications.

In 1888, Congress established an U.S. Army board for ordnance and fortifications who as charged with testing new weapons and to design new coastal fortifications. In 1890 Congress made the first appropriations for the first new construction of coastal fortifications in 16 years with an initial funding of $1.2 million dollars. This funding was for the first of new defenses: a 12-inch barbette battery at San Francisco; an 8-inch disappearing battery at New York Harbor; a 12-inch gun lift battery at Sandy Hook, New Jersey; and for 12-inch mortar batteries at Sandy Hook and at the Presidio, San Francisco. The design of these new coastal batteries would set the pattern of coastal defenses that would be duplicated at all of America's major harbors, and this was the beginning of what would become known as the "Endicott Program" of American seacoast defenses.

The designs used for the Endicott Program coastal fortifications demonstrated the shift in importance from the large multi-tiered multi-gunned "fortresses" to weapons emplaced in dispersed concrete "batteries" protected by earthen embankments. The "fort" became a defined reservation of land that contained guns of a range of calibers, along with the housing for the men required to man these defenses, and supply and maintenance buildings. The weapons were grouped into batteries containing from one to sixteen guns. The batteries were located along the shoreline to maximize their range and field of fire and were designed to blend into the landscape as not to be seen from the sea. The armament of these batteries ranged from weapons to engage enemy capital ships to small-caliber rapid fire guns to knock out fast moving torpedo boats, as well as to protect the fields of electrically controlled submarine mines from minesweepers. The dominance of the armament during this period is reflected by the dramatic increase in the time and cost in constructing a gun barrel and breech mechanism along with its carriage over that of the weapons of the earlier periods.

The primary weapons of the Endicott Program were 8-inch, 10-inch and 12-inch rifled breech-loading guns, a growth in size that reflected the need to match the increase in the size of opposing naval guns. These guns were mounted on both barbette and disappearing carriages that had a maximum elevation of 15 degrees and range of about seven to eight miles. The relatively unique American "disappearing" carriage allowed the gun to be raised over the parapet by using a counterweight to fire. The energy from the recoil caused the gun to drop back down behind the parapet into the emplacement to be reloaded while being protected from direct fire from attacking warships. These heavy weapons were mounted in large concrete emplacement with thick frontal walls that were in turn protected by many feet of earthen fill. Located below or adjacent to the firing platform were support areas that included the ammunition magazines containing projectiles and powder propellants. About three hundred of these heavy guns were installed around the United States during the Endicott Program in batteries of from one to six guns. It was a less expensive alternative to the armored turret mount favored by several European nations.

10-inch disappearing guns in Battery Hale, Fort Greble, Rhode Island (C.T. Gardner Collection)

The other large caliber weapon installed during the Endicott Program was the short-barreled 12-inch mortar. The mortar was designed to fire a shell in a high arc that descended down onto the lightly armored decks of warships of that era. To increase the opportunity of making a hit, these mortars were emplaced in groups of eight to sixteen mortars in square concrete pits that were protected by earthen hills. The use of these pits would give maximum protection to the mortars and their magazines from the flat trajectory of naval gunfire of the era. About four hundred of these mortars were installed around the United States during the Endicott Program.

The secondary smaller caliber weapons were installed to protect the controlled submarine mine fields from small craft that could sweep paths through the mines for larger warships, and to protect from attack by newly developed fast torpedo boats that could potentially penetrate the harbor and torpedo the shipping within. These threats called for guns that could be aimed, loaded and fired very rapidly. While not specified by the Endicott Report, several new gun and carriage systems were developed for this role ranging from 3-inch to 6-inch is size. These guns were generally mounted on either disappearing carriages or on pedestal carriages with simple steel shields. The concrete emplacements for these guns had low parapets and magazines below the guns. A rapid-fire battery had between two and six guns per battery. Over five hundred rapid-five weapons were installed during the Endicott Program.

12-inch mortar firing at Battery Alexander, Fort Barry, California (B.W. Smith Collection)

While the use of submarine mines or "torpedoes," as well as channel obstructions or barriers, has a long history in defense of harbors, it was during the Endicott Program that a widespread and a structured use of submarine mines occurred. The U.S. Army developed a system of controlled submarine mines; stationary explosive devices located below the surface of water where ships were likely to pass. The submarine mines used from 1890 to 1930 were the buoyant type (floating but anchored to the sea bottom), though during World War II the buoyant mines were replaced with ground mines (stationed on the sea floor). The mines were only deployed during times of war or for practice, otherwise they were stored disassembled ashore—the mines and their control cables became defective after extensive exposure in the water. The controlled mines were connected to shore by undersea cables and could be exploded by electrical switches from a control board on shore by the soldiers manning the mine defenses when a warship passing over these mines or by direct contact. Controlled mines were usually laid in rows across the key shipping channels to create a group of mines, usually 19, which would cover a space of about 2,000 feet long in water up to 250 feet deep. Several groups of mines were to be deployed to create a field of mines. The U.S. Army Coast Artillery Corps had dedicated units to man the mine planting vessels, fire control stations, mine and cable storage facilities, mine casemates and switchboards, and loading wharves.

The Endicott Program roughly covered a period from 1885 to 1905, and the coast artillery function was a key mission of U.S. Army during this time (and made up a large percentage of total U.S. Army manpower). This also required a more technical trained soldiers to man them which led to the U.S. Army's Artillery branch to be reorganized in 1901 and 1907 to create the U.S. Coast Artillery Corps.

Planting a mine (Stillion Collection NPS, Gulf Shores Natl. Seashore)

Seacoast Defenses built after the Taft Board Report 1906-1916

In 1905, a new National Coast Defense Board headed by U.S. Secretary of War William Howard Taft was organized by President Theodore Roosevelt and charged with reviewing the progress of Endicott Program construction and update it. In the 20 years since the original Endicott Board report was presented, numerous technical and political developments had taken place. The Board, informally known as the "Taft Board" after its chairman, established new cost estimates and its recommendations were primarily concerned with modernization of existing coastal fortifications and adding coastal defenses to the overseas territories gained after the Spanish-American War including Hawaii and the Philippines and other locations.

The modernization of existing defenses included the electrification of lighting, communications, and ammunition handling equipment, both at the batteries and throughout the fort. The early emplacements had loading platforms widened and projectile hoists were installed to improve the rate of fire. The report recommended the use of searchlights for nighttime illumination of harbor entrances. During the Taft Program was the finalized development and implementation of a coordinated system of target information gathering and processing that greatly improved the target accuracy of the major caliber guns and mortars. Up to this time, the aiming of guns at a target had been generally done from each battery with basic sighting instruments and combination of luck and experience. The new system was based on triangulation using two observers with telescopic instruments at separate position finding stations or "base end" stations communicating with the newly developed telephones to a centralized battery plotting room that provided real-time tracking and firing coordinates on a moving target. The battery plotting room personnel would

The 14-inch gun on a disappearing carriage the Taft-era Battary Osgood, Fort MacArthur, California
(Fort MacArthur Museum)

mathematically process this sighting information and other data into aiming instructions that would then be transmitted to each gun emplacement.

While the Taft Board's recommendations on the construction of new fortifications was largely limited to existing defenses at Eastern Long Island Sound, San Diego, Puget Sound, Columbia River, and Chesapeake Bay in the continental United States, major new construction projects were planned for the Philippines, Panama, Hawaii, Cuba, Puerto Rico, Alaska, and Guam. Plans for Cuba, Puerto Rico, Alaska, and Guam were not carried out. New defenses were added for the port of Los Angeles in 1909. These "Taft Program" defenses varied little from the overall designs used during the Endicott Program. Variations from the Endicott Program were the product of advancing naval armaments and the U.S. Army's twenty years of experience of operating coast defenses. To match the increased caliber of naval guns, a new disappearing gun of 14-inch caliber was developed. Another characteristic of the Taft Program batteries was the increased dispersion of batteries. The reduce density of weapons can be seen in the construction of several one-gun, 14-inch batteries and the reduction of mortar batteries from 8 mortars to 4 mortars (4 per pit to 2 per pit).

During the Taft Program, several one of kind of projects were undertaken. The Endicott Board Report called for 16-inch guns but work on the development stalled after the construction of one gun tube in 1895. Two unique 16-inch disappearing batteries were finally built in Panama and the Long Island Sound. In the Philippines army-designed armored turrets were custom built for a very small island in Manila Bay. Four 14-inch guns were mounted in two turrets at Fort Drum, which also became known as the "concrete battleship."

World War I and the Interwar Period (1917-1939)

The march of technical improvements in naval weapons continued through improvements in naval fire control and the ability for naval turrets to elevate their guns. By 1915 the newer battleships had guns that could out range the effective range of the coast artillery emplaced during the Endicott and Taft Programs. The increased angle of fire of the newer battleships also threatened the disappearing carriage batteries which were not protected from the plunging fire of these new battleships. In 1915, a National Board of Review on the coast defenses of the U.S., Panama Canal, and the Insular Possessions recommended the construction

Firing one of the 12-inch guns of the post-WWI Battery Kingman, Fort Hancock, New Jersey (NARA)

of new batteries mounting 12-inch and 16-inch guns on higher elevation, longer range barbette carriages. While efforts to introduce these coast artillery weapons had begun, the demands of World War I placed the modernization of coast defenses on hold. Many Coast Artillery units were transformed into field and heavy artillery units for service in France. As the United States was short on long range field artillery,12-inch mortars, 10-inch, 8-inch, and 6-inch gun barrels were removed from several coast artillery batteries. These existing gun barrels, ranging from 6-inch to 14-inch in caliber were quickly mounted on railway and tractor-drawn carriages. While the United States involvement in World War I was brief, it resulted in the Coast Artillery Corps mission to be divided into three specialized areas as compared to its single mission before the war. These missions, based on armament type, were fixed coast defense weapons (including controlled mines), mobile seacoast artillery, and anti-aircraft artillery.

The development of the airplane as a ground attack weapon during the World War I added the task of defending both the mobile ground army and the shores of United States from attacks by aircraft to the Coast Artillery Corps' mission. The U.S. Army developed fixed and mobile anti-aircraft weapons, as well as accessory equipment such as aircraft sound locators, rangefinders, searchlights, specialized fuses, and mechanical fire direction calculators. The primary weapon for the defense against aircraft was the 3-inch gun on a fixed carriage (in batteries of three or four guns) located at existing coast defense posts. This weapon was later supplemented with .50 caliber machine guns and mobile 3-inch AA guns. By 1938 larger caliber anti-aircraft guns were introduced including the 90 millimeters (mm), the 105 mm, and 120 mm guns.

The mobile coastal defense mission came about because of the lack of U.S. heavy artillery for the troops in Europe. Existing gun barrels, ranging from 6-inch to 14-inch in caliber were quickly mounted on railway and tractor-drawn carriages. The construction of the new mobile carriages for guns, such as the railway mounts, took months and most of these weapons never reached European theater before the war ended. The availability of this ordnance material, especially considering the economics of using existing weapons and increased desirability of weapon mobility in the interwar period, made mobile coast artillery an attractive alternative to building new fixed coast defenses. The primary railway guns selected for coast artillery use from this large stock of World War I material were 8-inch guns and the 12-inch mortars mounted on new railway carriages. Added later was an improved version of the wartime 14-inch railway gun of which only four were constructed by 1920. The surplus mobile field artillery mounted on carriages designed for road movement included the 155-mm GPF gun (derived from a 1917 French design) of which almost a thousand were available. This powerful gun became the standard tractor drawn weapon for coast defense use against secondary targets.

While mobile coast artillery had the advantage of being able to respond to coastal areas most

a 155 mm G.P.F. mobile mount in a field position at Long Point, California (Ruhlen Collection)

threaten when enemy naval forces approached, both railway and tractor-drawn weapons lacked the accuracy and protection of fixed coast artillery. Without solid and steady firing platforms and the precision of pre-calibrated fire control networks, as well as the inability of the carriages of the mobile guns to quickly track horizontally moving targets made mobile artillery much less effective than weapons in fixed emplacements. Prepared locations with circular arcs of track were prepared at a few select locations. For the 155 mm GPF mobile artillery, simple circular concrete bases were designed. These circular bases improved stability during firing and provided for rapid azimuth adjustment for horizontal tracking. One of the most common base designs developed for the 155 mm GPF guns was a central pivot and a curved rail embedded in concrete, which the gun's split carriage would traverse. This design was first constructed in the Panama Canal Zone, so this design became known as the "Panama Mount". Given the limitations of mobile coast artillery, their use was primarily an augmentation of existing defenses or to provide protection during the construction of permanent fixed coast artillery. Due to the low level of military appropriations during the 1920s and 1930s, mobile coast artillery was the only available weapon to defend vital locations until new permanent defenses could be funded and constructed.

Given the low level of overall U.S. military funding during the 1920s and 1930s, the construction and development of new fixed coast defenses were limited. The need for economy and to allow for higher gun elevations led to the abandonment of the disappearing carriage and its complex two-level emplacements. Among the last disappearing carriages built were for two 16-inch single-gun batteries (one in Panama and the other in the Long Island Sound). A newly designed high angle barbette carriage for existing stocks of 12-inch Model 1895 guns allowed effectively doubled the range of the guns over the same 12-inch gun mounted on a disappearing carriage. Construction of fifteen long-range dual gun 12-inch batteries was started in 1917 and completed by the late 1920s. The emplacement design was a departure from those of the Endicott Program. The battery design had two guns located much further apart, each gun in the center of a large ground level concrete pad to allow for an all-around field of fire. Located between the two guns was an earth-covered reenforced concrete structure containing magazines for shells and powder, the power and plotting rooms, and storage rooms. Protection of the guns from naval fire was based on dispersion; the wide separation of the key elements of the battery. Other than camouflage and nearby anti-aircraft guns these batteries had no protection from air attack. The development of a new 16-inch gun and carriage with a range of nearly thirty miles which exceeded the range of all existing naval warships was completed in 1919. The 16-inch in emplacements that were very similar to those used for the long-range 12-inch barbette batteries, with an increased distance between the two guns of the battery and the dispersed location of magazines in simple storehouses connected by a rail system. Only a few of the U.S. Army designed barrels had been constructed when nearly sixty 16-inch barrels became available from the U.S. Navy. This windfall was due to the Washington Naval Treaty of 1922 that resulted in the cancellation of several U.S. battleships and battle cruisers then under construction. The naval 16-inch barrels were to be installed in modified U.S. Army barbette carriages after 1925. Six new twin-gun 16-inch batteries were built between 1922 and 1934. During the Interwar Period, the construction of new batteries including both long range 12-inch and 16-inch guns, amounted to little more than twenty new batteries. The coming of World War II would inject new life into building modern U.S. coast defenses.

The 1940 Modernization Program and World War II (1940 -1950)

During the 1930s the U.S. Army began discussing how to protect new coast artillery batteries from attack by aerial bombardment. The debate centered on the expense of designing and construction of turret mounts for 12-inch and 16-inch guns as compared to developing protective structures made of concrete and steel. It was practical economic and time frame requirements that resulted in the eventual selection of a concrete casemate structure design to protect the current type of barbette mounts.

The prototypes of this new type of major caliber battery were built at the San Francisco defenses, during 1937-1940. These emplacements were designed for two 16-inch guns located about six hundred feet apart with complete overhead cover. Located between the two guns along a service gallery were the ammunition magazines, power generators, and support areas. The 16-inch guns were enclosed in reinforced concrete casemates. The battery's structure was made up of eight to twelve feet of steel reinforced concrete which was topped by up to twenty feet of earth as additional protection. The entire battery structure was designed to withstand a direct hit from a naval projectile or an aerial bomb. When completed the southern San Francisco battery at Fort Funston emplacement looked like a small hill, especially when camouflage and natural ground cover was added to the structure. The only exposed portions of the battery were the casemates where the gun barrels projected out through armor shields and concrete canopies. A second casemated battery on a hilltop north of San Francisco was also undertaken. Four more casemated batteries were begun at Narragansett Bay, the Delaware River, and Chesapeake Bay in 1940-1941.

In 1940 the Harbor Defense Board was charged with developing a master plan to update the harbor defenses of the continental United States. Eighteen coastal areas in United States were selected for modernization due to their military and economic importance - Portland, Portsmouth, Boston, New Bedford, Narraganset Bay, Long Island Sound, New York, Delaware Bay, Chesapeake Bay, Charleston, Key West, Pensacola, Galveston, San Diego, Los Angeles, San Francisco, Columbia River, and Puget Sound. The Harbor Defense Board recommended the adoption existing stocks of 16-inch gun as the primary weapon and 6-inch gun as the secondary weapon for the modernization program. In all the board proposed building twenty-seven new 16-inch casemated batteries; the casemating of 23 existing primary batteries (both long-range 12-inch batteries and older 16-inch batteries; and building fifty new 6-inch two-gun barbette carriage batteries, which would provide long-range fire (15 miles maximum) against secondary warships. The new 6-inch batteries would be supported by 63 existing secondary batteries, mostly 6-inch and 3-inch barbette guns from the Endicott and Taft Programs, which would be retained. Upon completion of these

One of the 16-inch guns of Battery Steele, Peaks Island M.R., Maine (Joel Eastman Collection)

A 6-inch gun of Battery Cravens, Peaks Island M.R., Maine, with a disguised SRC 296A radar behind
(Joel Eastman Collection)

new defenses 128 existing obsolete coastal batteries would be eliminated. The board estimated that the whole program would require three years to complete and cost about $82 million during 1941-3. Formal approval of this modernization plan, which would become known as the "1940 Harbor Defense Modernization Program" or the "1940 Program," was approved in September 1940.

The 1940 Harbor Defense Modernization Program greatly simplified the task of Coast Artillery Corps by reducing the number of types of batteries as well as the overall number of batteries needed to carry out their coast defense mission. This allowed a reduction in personnel and the level of effort to maintain, training and supply the pre-1940 batteries. Some of coast artillery that was declared obsolete was shipped to Allied nations to supplement their defenses, but most were scrapped for the war effort. As the nation moved closer to war, additional coastal defense projects were added to the 1940 Program, especially at newly acquired overseas bases, such as Trinidad, Bermuda, Newfoundland, and in areas where the enemy threats seem greater, such as Alaska, Hawaii, Puerto Rico, and the Canal Zone. It also became apparent that planning, construction and emplacement of the many new batteries called for the 1940 Program was going to take a much longer time then original envisioned, especially as the program was competing with rapid expansion of the whole U.S. Army and U.S. Navy. By the middle of July 1941, only four 16-inch gun batteries were ready for action and construction work had been started on just five others. With pressure from the U.S. Army Air Corps, it was decided to limit active work to those batteries that could be completed by July 1944. As a result, all work on fourteen of the thirty-seven 16-inch batteries planned for the continental U.S. was discontinued. The expansion of overseas bases during 1941 impacted the construction of the new 6-inch gun batteries in the continental U.S. by priority assigned to the completion of twenty 6-inch batteries to guard these overseas bases.

The new batteries constructed under the 1940 Program were much more standardized that those of proceeding periods. The Army developed standardized designs for the 16-inch gun batteries and the 6-inch gun batteries which were used with only minor variations for local topography and soil conditions. Both the 16-inch and 12-inch guns, whether newly installed or retained from the Interwar Period, were emplaced within reinforced concrete casemates that limited their field of fire to about 180 degrees but gave them superior protection over the old open emplacements. The new 6-inch batteries were not casemated. A cast steel shield from four to six inches thick was placed around the gun and carriage. This shield would protect the gun and its crew from all but direct hits by heavy projectiles. Between the two 6-inch guns was an earth covered steel reinforced concrete structure contain the magazine, power generators, communications, air filtering equipment, storage, and plotting room. As these batteries were being built, they were assigned a "Battery Construction Number" for record keeping purposes. As many of the new batteries were never formally named, these construction numbers were the only designation they received. While the Army never referred to the 16-inch series of batteries as whole as the "100" series or the 6-inch series of batteries as whole as the "200" series, these terms are used by modern historians and are referred to as such in this work.

As the range of these new batteries was far greater than earlier batteries it was also necessary to update the fire control networks. The 16-inch batteries received new base end stations as far as twenty-five miles away from the gun's position to allow for gun's maximum range to be effectively used. These stations were built in wide variety of forms: houses, windmills, silos, water tanks, office buildings, or buried into hillsides. Radar was added as an early warning device and as a fire control instrument allowing the operation of coast artillery at maximum range during all weather conditions.

By the start of World War II, the Coast Artillery Corps' mobile coast artillery units had dwindled from the plans of the Interwar Period, especially the railway guns units. Several tractor-drawn 155mm GPF gun regiments were available in the continental U.S., but only part of one 8-inch railway regiment was on hand. The four 14-inch railway guns continued their role in Los Angeles and Canal Zone. The primary use of mobile coast artillery was to fill in for fixed coast artillery weapons until their completion or at secondary locations. The 155mm GPF gun units were reorganized into seventy-two 2-gun batteries along the Atlantic, Gulf and Pacific coasts. Using 12-inch railway mortars and 8-inch railway guns from storage, several CAC units were formed and sent to both domestic and overseas locations to provide temporary harbor defenses until permanent works could be constructed.

As with earlier periods, an integral part of harbor defenses was the use of controlled mines across key ship channels. These mine defenses were supplement by U.S. Navy contact mines and the use of submarine nets and booms. As the primary threat during World War II turned out to be enemy submarines at most of these ports, the U.S. Navy added detection devices in outer harbor approaches and conducted offshore patrols. Because of the need for both the U.S. Army and U.S. Navy to coordinate their coast defense activities, a centralized harbor entrance command was created in 1941. The Harbor Entrance Control Post (HECP) used both army and navy personnel to provide a link between higher command and all subordinate elements of a harbor defense. These centers were responsible for monitoring all movement of shipping in and out of the harbor. To support this effort a secondary gun battery was on duty as commercial shipping traffic was examined upon entering the harbor. One of the concerns at this time was an attack by fast moving torpedo boats combined with the lack of modern rapid-fire guns. To fill this void, the 90mm anti-aircraft gun was selected to replace the existing 3-inch pedestal guns of the Endicott and Taft Programs. In late 1942, special anti-motor torpedo boat (AMTB) batteries were installed along the Pacific and Atlantic coastlines. These batteries usually consisted of two fixed mounted 90mm guns and two mobile mounted 90mm guns, and two mobile 37mm or 40mm anti-aircraft guns. These guns would be protected by earthen revetments with protected magazines. The active harbor defenses received two, three, four, or more of these AMTB batteries beginning in 1943.

Outside the continental United States, where the threat of attack and invasion was greater, new coast defense construction proceeded with greater speed and with the use of armament on hand rather than waiting for weapons sto be provided by 1940 Program. The coast defenses of Hawaii are a good example, as the Japanese attack on Pearl Harbor made new defenses the highest priority due to concerns of an invasion attempt. A series of batteries were constructed, using excess naval guns, ranging from the 14-inch turrets from the battleship USS _Arizona_ to 8-inch gun mounts from the aircraft carriers USS _Lexington_ and USS _Saratoga_. Throughout the Pacific Islands and Alaska, surplus U.S. Navy guns (5-inch, 6-inch, 7-inch & 8-inch) were mounted on shore to defend U.S. Navy installations.

With the tide of the war shifting toward the Allies after 1942 and the demands to produce war material for the mobile army, the navy, and the army air corps, the 1940 Program was pared back. While the construction of structures could keep pace with the original plan, the manufacture of weapons and their accessories could not. In response to these pressures, the 1940 Program was scaled back even further. By the war's end, the modernization program resulted in the completion of nearly 200 new batteries in the continental United States at a cost of $220 million, or about one-half the number of installations proposed in the 1940 Program, but still the most powerful collection of coastal defenses in America's military history.

The development in military tactics and technology during World War II brought about numerous changes to the concept of coast defense. It was no longer thought necessary to defend one's seacoast using just coast artillery and controlled mines. Air power and naval forces were to replace breech loading rifles and reinforced concrete. Already at the end of World War II, all except a few 90mm AMTB batteries were placed on caretaking status. During the transition years of 1946 to 1948, some new batteries started during the war were completed while many other batteries were being disposed of and guns scrapped. By 1949, the process was completed as the last of guns were scrapped. In 1950, the remaining harbor defense commands were disbanded, and the Coast Artillery Corps was abolished as a separate U.S. Army branch with its remaining units, all anti-aircraft artillery, recombined into the Field Artillery. After 150 years of being one of America's military prime missions, the building and manning of permanent coastal fortifications was over.

The U.S. Coast Defense Objective in the Modern Era

The objective of seacoast defense is to provide protection of the coastline from invasion by an enemy, and specifically the defense of important harbors, which includes securing the anchorages and bases needed for naval operations. Coast defense is not only protective in its strength but protects the nation's ability to carry war beyond its own coastline.

It is impractical to fortify the entire extent of any nation's long coastline in such a way that an enemy in command of the sea could not land upon some portion of it. The cost of such an undertaking would be excessive, as maintenance of these defenses and number of men required would make it prohibitive expensive. An example of this type of defense was the "Atlantic Wall" built by Germany in World War II (which stretched from Norway to Spain), which failed to prevent the Allies from landing in Europe in 1944. It was essential, however, that certain selected points be permanently fortified to make invasion more difficult and to protect key naval shore installations and fleet anchorages and important commercial harbors that support the nation's economy.

The resources to defending the coastline during this era were divided into two kinds of troops. The first was the Coast Artillery troops, made up the regular U.S. Coast Artillery Corps and the U.S. Coast Artillery Reserves. These technical troops manned both the fixed and mobile seacoast artillery and controlled mines defenses. The second resources were the supporting troops of the mobile ground forces of the U.S. Army which protected the both the coast defenses and unfortified coastline from enemy landings. The second would been the local National Guard troops (formally militia), while the U.S. Navy's role in coast defense was through both offensive and defenses operations against enemy warships.

To carry out this mission, seacoast weapons were divided into classifications according to their capabilities against enemy warships. Primary armament were those weapons that could theoretically destroy the primary or capital warships of enemy naval force. Throughout most of the Modern Era primary weapons were defined as seacoast artillery of initially 8-inch and larger caliber. Controlled submarine mines were also considered part of the primary armament. The second group of seacoast weapons was the secondary armament, which were designed to counter secondary or non-capital warships, such as cruisers, destroyers, and torpedo boats.

The selection of the numbers and type of seacoast weapons was determined by such factors as the importance of the coastal area, the hydrograph profile of the approaches, the topography of area, and effectiveness of seacoast weapons in defending the coastal area. The positioning of seacoast artillery was based on the attainment of effective fire and protective factors, such as concealment, other weapons, and local defense against ground or air attacks. Attainment of effective fire refers to a position which offers the widest field of fire and greatest range over navigable water. Also considered was the need to provide coverage to all areas in which an enemy warship may operate and the placement of a suitable concentration of fire on critical areas such as harbor entrances, approaches to mine fields, and narrow portions of the channel. Consistent with these requirements, batteries were sited to provide mutual support and defense against all forms of attack. The considerations for the location of primary armament included the ability to protect friendly naval forces while entering, within or leaving the harbor, and preventing hostile naval forces from approaching within effective range of the defended coastal areas. Submarine mine fields would be placed in the seaward area of the harbor entrance and within effective range of searchlight and rapid-fire secondary armaments. Both controlled and uncontrolled submarine mines are located to prevent entry into or close approach to the harbor of enemy surface warships or submarines at all times, including during night or during conditions of heavy fog or smoke. The secondary armament would be located to provide protection for mine fields, nets, booms, and other obstacles; and the attack of hostile secondary warships engaged in raids, reconnaissance, laying of mines, and torpedo fire. Since targets of the secondary armaments were within range of visual observation and assumed to move at high speed on rapidly changing courses, these batteries were sited in direct fire positions. Protective factors in site selection included protection for the power plant, plotting room, magazines, communications, exposure of the gun crew and ammunition during the service of piece, gas protection for command post and plotting room, distances between emplacements, and concealment.

U.S. Coast Defense Armament and Equipment in the Modern Era

Few weapons of the Modern Era of coastal fortification remain today. This is the result of the advancing technology that quickly made weapons obsolete and given the economic value of high-grade steel the military sold these obsolete weapons to salvage companies. The scrapping of coast artillery material also holds true for most its supporting equipment, machinery, and instruments. As a result, today we mainly only have period images of these armaments or supporting equipment.

The development of new armament and equipment over this era usually went through cycles where the level of perceived external threats to the United States generated appropriations from Congress to allow the funding of new weapon systems. The development process for new weapons required several steps. First, was the design stage which led to the prototype and testing period and then to production and installation phase. Finally, while the weapon was in service it received modifications and improvements until it was declared obsolete. The life cycle of seacoast artillery varied from a few years to as long as fifty years.

The construction of the Endicott and Taft Programs defenses relied on the growth of heavy industry in the United States. Many of items used in coast artillery forts were invented specially for that purpose and represented the cutting edge of that technology. Early defense works relied on steam, coal, and manual energy to make things work. The use of oil and the advances in electricity brought motor driven equipment, telephones, radar, computers, and electric lighting to become key ingredients in U.S. coastal defenses.

It is also important to note that different U.S. Army branches had specialized functions that need to work together to complete a weapon system. A seacoast weapon would be designed and constructed by Ordnance Department while the emplacement was designed and constructed by the Corps of Engineers. These activities were all support by the Quartermaster Corps, Signal Corps, and so forth. The final product was then turned over to the Coast Artillery Corps for use. As you may imagine sometimes the priorities of these various organizations were not always in agreement, so delays or undesired weapons systems did occur.

A 10-inch rifle on a disappearing carriage
Battery Benson, Fort Worden, Washington (Puget Sound Coast Artillery Museum Collection)

6-inch rifles on dissapearing carriages
Possibly Battery Tolles, Fort Worden, Washington (Puget Sound Coast Artillery Museum)

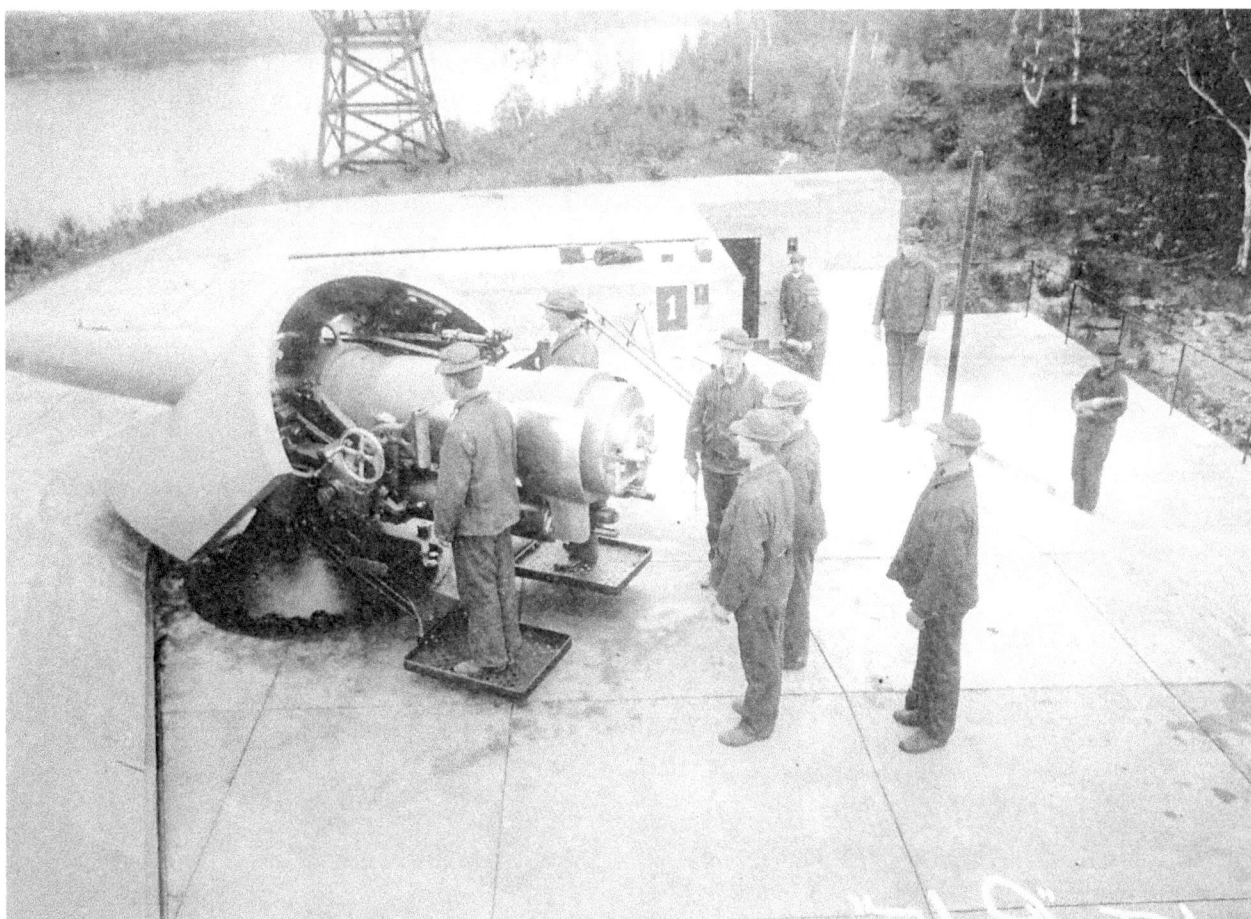

A 6-inch gun on a pedestal mount, Battery Carpenter, Fort McKinley, Maine
(Joel Eastman Collection)

3-inch guns on pedestal mounts in Battery O'Rorke, Fort Barry, California (NPS, GGNRA)

For coast artillery material, the U.S. Army insured that all items were assigned a "type" and a "model". For seacoast artillery, the type for the gun or barrel refers to the size of bore (diameter) in inches while type for carriage or mount refers to the style of operation. Associated with the type is the model which refers to year of development and any subsequent modifications until 1930 when use of the year was dropped. This nomenclature extends to projectiles, fire control instruments, searchlights, submarine mines, ammunition hoists, power generators, radar, etc.

12-inch mortars in pit A, Battery Worth, Fort Pickens, Florida (Stillions Collection, NPS)

Carriage or mount types were either fixed or mobile, they allowed the guns to elevate and provide for some horizontal movement while taking up the recoil of the discharge and return the piece to the loading position. The major caliber fixed carriages were classified as Barbette (BC) carriage, which allowed the gun to remain above the parapet for loading and firing; Mortar (MC) carriage, which allowed a short-barrel gun to fire in a high arc; the Barbette long-range (BCLR) which allowed for greater firing elevations and ranges; and the Turret (TM) mount which was a barbette carriage protected by an armored housing with ammunition supplied from below. Guns of 7-inch or lesser caliber were mounted on the Pedestal (PM) mount, which had a fixed cylindrical base on which rotated a yoke that held the gun in a cradle equipped with recoil absorbing cylinders; the Anti-aircraft (AA) mount, a pedestal mount that allowed fire at high attitude. The Fixed retractable carriages included the Gun-Lift (GLC) carriage which was a BC on an elevator platform; the Disappearing (DC) carriage where the gun is raised above the parapet for firing and retracts behind the parapet for loading. The earlier smaller caliber guns had the Balanced Pillar (BPM) mounts and the Masking Parapet (MPM) mounts, which enabled the gun to be lowered below the parapet to protect it from view. Guns on mobile carriages were used in the Interwar Period as the Railway (RY) mount cars and Tractor-drawn (TD) mounts. Other temporary coast defenses made use of available weapons with a range of carriage types, primarily former naval models.

Controlled mines were anchored to the bottom of a harbor, either sitting on the bottom itself (ground mines) or floating (buoyant mines) at depths which could vary widely, from about 20 to 250 feet. These mines were fired electrically through a vast network of underwater electrical cables at each protected harbor. Mines could be set to explode on contact or be triggered by the operator, based on reports of the position of enemy ships. The networks of cables terminated on shore in concrete bunkers called mine casemates, that were usually partly buried beneath protective coverings of earth. The mine casemate housed electrical generators, batteries, control panels, and troops that were used to test the readiness of the mines and to fire them when needed. Each protected harbor also maintained a small fleet of mine planters and tenders that were used to plant the mines in precise patterns, haul them back up periodically to check their condition (or to remove them back to the shore for maintenance), and then plant them again. Each of these harbors also had onshore facilities to store the mines and the TNT used to fill them, rail systems to load

On the deck of mine planter (Stillions Collection, NPS)

and transport the mines (which often weighed over 750 lbs.) each when loaded), and to test and repair the electrical cables. Fire control structures were also built that were used first to observe the mine-planting process and fix location of each mine and second to track attacking ships, reporting when specific mines should be detonated. The preferred method of using the mines was to set them to detonate a set period of time after they had been touched or tipped, avoiding the need for observers to spot each target ship.

Key to the successful use of coast artillery was fire control and position finding as if the guns, mortars, and controlled mines failed to strike their intended targets their mission was incomplete. Early aiming efforts relied on the skill of the gunner to hit the target, but as the weapon's range increased so did the need for specialized fire control. Using geometry, optical instruments, telephones, timing interval bells, and mechanical devices a system was devised to point weapons successfully at their targets. Key equipment included the Depression Position Finder (DPF), the Azimuth Instrument (AI), Coincidence Range Finder (CRF), Plotting Board, Range Correction Board, Fire Adjustment Board, Deflection Board, Spotting Board, Range Percentage Corrector, Data Transmission Devices, Telephone Sets, and Timing Interval Bells. Many of the devices were replaced or supplemented by the development of radar (for both surveillance and fire control duties) and gun computers (combining many of plotting room devices) during the 1940 Program.

Until the advent of radar, the use of searchlights (plus star shells and airplane flares) was used to illuminate naval targets at night. Both mobile and fixed searchlights were used for both harbor defenses and anti-aircraft defense. At first 36-inch and 60-inch searchlight were used, but the 60-inch became the standard. Searchlights were located as close as possible to the water-edge to maximize their effective range of between 8,000 and 15,000 yards. Fixed searchlights were provided with a shelter to protect the searchlight from elements, to house the electrical generator, and provide concealment of the light when it was not in use. Some positions placed the searchlight on small rail cars that allowed the searchlight to move a short distance to a more exposed operating site. Searchlights were also housed in towers, pits, and even tower that "disappeared" by pivoting. After 1940 all new searchlights assigned to the defenses were the mobile type.

By 1943, two technological advances significantly changed coast artillery fire control. The most striking was the development of radar, which, as noted, could function in any weather or visibility. The use of radar greatly reduced the need for searchlights and for fire control stations as spotting enemy warships and aircraft could now be undertaken by radar units. In addition, after decades of experimentation and development, largely stymied by inadequate funding, the coast artillery adopted gun data computers, primarily for the last generation of batteries. These replaced the plotting boards and, coupled with direct-reading observation instruments, substantially automated the fire control process, reducing the human error that had always plagued the system.

Searchlight and shelter/powerhouse, Fort Flagler, Washington (Puget Sound C.A. Musuem)

A Depression Range Finder (left) and a azimuth scope (right) in a base end station
(Al Scroeder Collection)

U.S. Coast Defense Fire Control Structures in the Modern Era

The development and changes in the optical instrument fire control system from 1900 to 1945 was a long and complicated process that changed equipment, operating procedures, and designations frequently. The reader is encouraged to consult *American Seacoast Defenses: A Reference Guide* and articles in the *Coast Defense Journal* for more detail and references to U.S. Army manuals and reports. The maps included in this guide have an extensive set of symbols indicating the locations of the various fire control structures.

By 1909, each battery was under the immediate command of the officer stationed at the battery commander's station (BC). Each battery may have had one or more additional base end stations (B) with optical spotting instruments. Small caliber batteries usually had a coincidence range finder station (CRF) nearby. Mine commanders manned their posts at the mine primary (M') station. In the defended harbor areas, called the Coast Defense Command, batteries were grouped into Fire Commands, each under the overall command of the fire commander stationed at the fire command primary station (F'). The Fire Commands were then grouped together by geographical areas under the command of the officer in command of that entire sector of the coast defense. This command was initially called the Battle Command but later was changed to the Fort Command. This officer was stationed at the primary fort command station (C'). In 1925, this chain of command was changed slightly. All forts and/or groups were under the Harbor Defense command (H). Forts (F) were also used as tactical commands. Individual gun batteries were assigned to a gun group (G). Later an additional tactical organization, the groupment (C), was added below the Harbor Defense command composed of two or more groups.

In general, batteries in each harbor defense were assigned tactical number designations, generally in numerical sequence from the south (Tactical Battery #1) to the north (Tactical Battery #2, 3, 4, etc.) on the Atlantic coast; from the north to the south along the Pacific coast; and from the east to west on the Gulf coast and along the Puget Sound during the 1940s. Note that by 1940 base end stations (B) and spotting stations (S) were often combined. This is useful in deciphering the symbols for designating the fire control observation stations on the maps: $B^1_1S^1_1$, $B^2_1S^2_1$, etc. The lower number is the tactical battery number to which the station is assigned, the upper number (or "prime" mark) is the station designation number in the series of stations assigned to that tactical battery. The number of base end stations assigned to each battery ranges from a single station to as many as 14 stations. Each station had at least one azimuth scope and/or depression range finder (DPF) scope as well as connected telephone communication equipment. 3-inch small caliber batteries had one base station with a coincidence range finder (CRF) located close to the battery.

During the period 1905 to 1940 the fire control structures were generally located on existing military reservations. The location and identities of these stations can be found on the confidential blueprint maps; in the reports of completed batteries; in the reports of completed works; and in the harbor defense engineer notebooks that are part of the CDSG ePress harbor defense document collection. After 1940 the ranges for the new guns were longer and the fire control stations were more dispersed, which resulted in the acquisition of a number of new small reservations along the coastline of each active harbor defense.

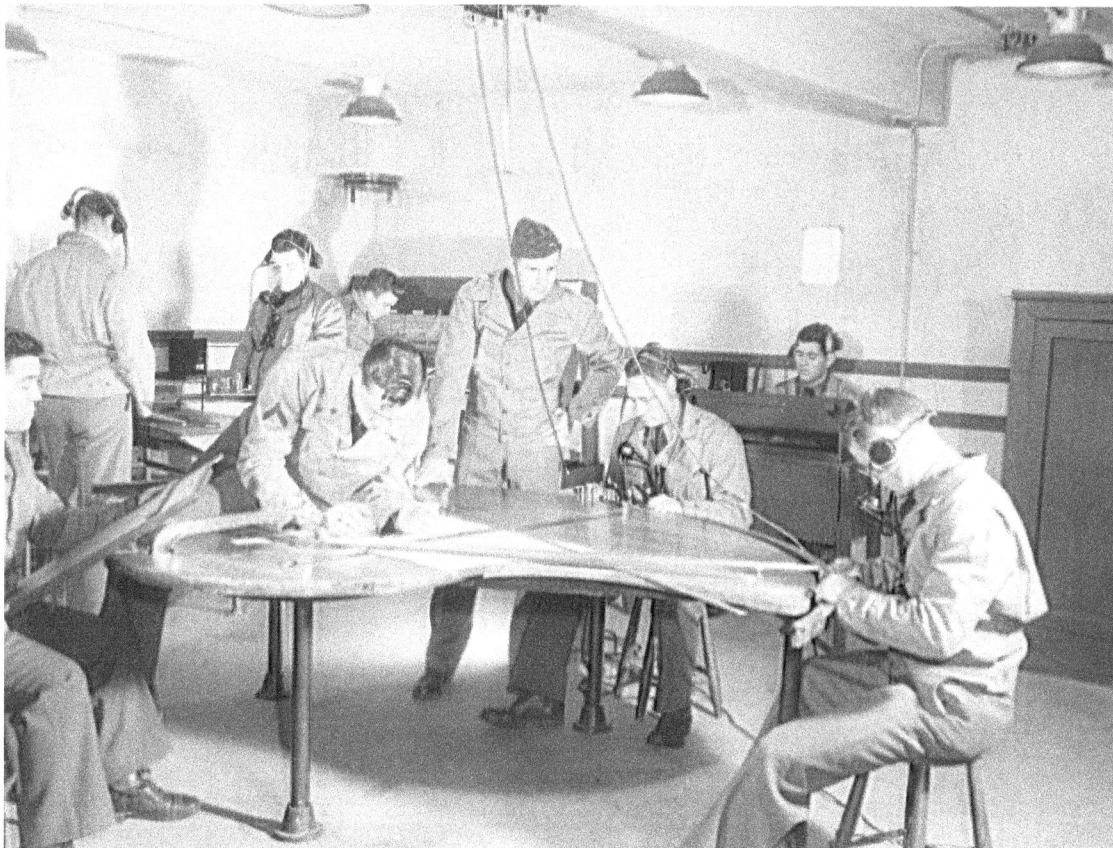

Battery Plotting Room circa 1944 (NARA)

The U.S. Coast Defense Organization Before World War II

The following organization structure of the administration and tactical command of the U.S. Coast Artillery Corps is for the 1930s period. The organization prior to 1924 was on a company basis and after 1942 on a separate battalion basis and is not discussed here.

Earlier organizations had similar purposes but used different terminology. Earlier tactical structures had Artillery Districts that were divided into Battle Commands, Fire Commands, Mine Commands, and Battery Commands, each with their own commanders, while for administrative and training purposes the CAC was divided into companies which in turn were assigned to coast artillery forts or posts.

A Harbor Defense Command is a subdivision of a Defense Command, which would cover an entire region. All elements, including materiel and personnel of a Harbor Defense Command, were located at one or more coast artillery forts. These forts consisted of defined land areas within a harbor defense in which the harbor defense elements were assigned. The forts were organized primarily to provide a centralized control over administrative and technical components of the harbor defense. The materiel provided for a harbor defense may have included various types of seacoast artillery guns, anti-aircraft guns, searchlights, controlled submarine mines, underwater listening posts, radar, observation and fire control systems, and harbor patrol boats. Harbor defenses were designated by the name the harbor or coastal area which they were

F.C. DIAGRAM OF THE COAST DEFENSES OF COLUMBIA

WAR DEPARTMENT CORPS OF ENGINEERS, U.S. ARMY.

FORTS	STEVENS				CANBY		COLUMBIA
F. & M. COMMANDS	FIRST		THIRD	FIRST MINE	FOURTH		SECOND MINE
BATTERIES	RUSSELL	MISHLER	CLARK	PRATT	GUENTHER NEW BATTERY	ALLEN	MURPHY
ARMAMENT	2-10"	2-10"	4-12"M	2-6"	4-12"M	2-6"	2-6"
CARRIAGE	D.C.-L.F.	D.C.-A.R.F.	M.C.-A.R.F.	D.C.-L.F.	M.C.-A.R.F.	D.C.-L.F.	D.C.-L.F.

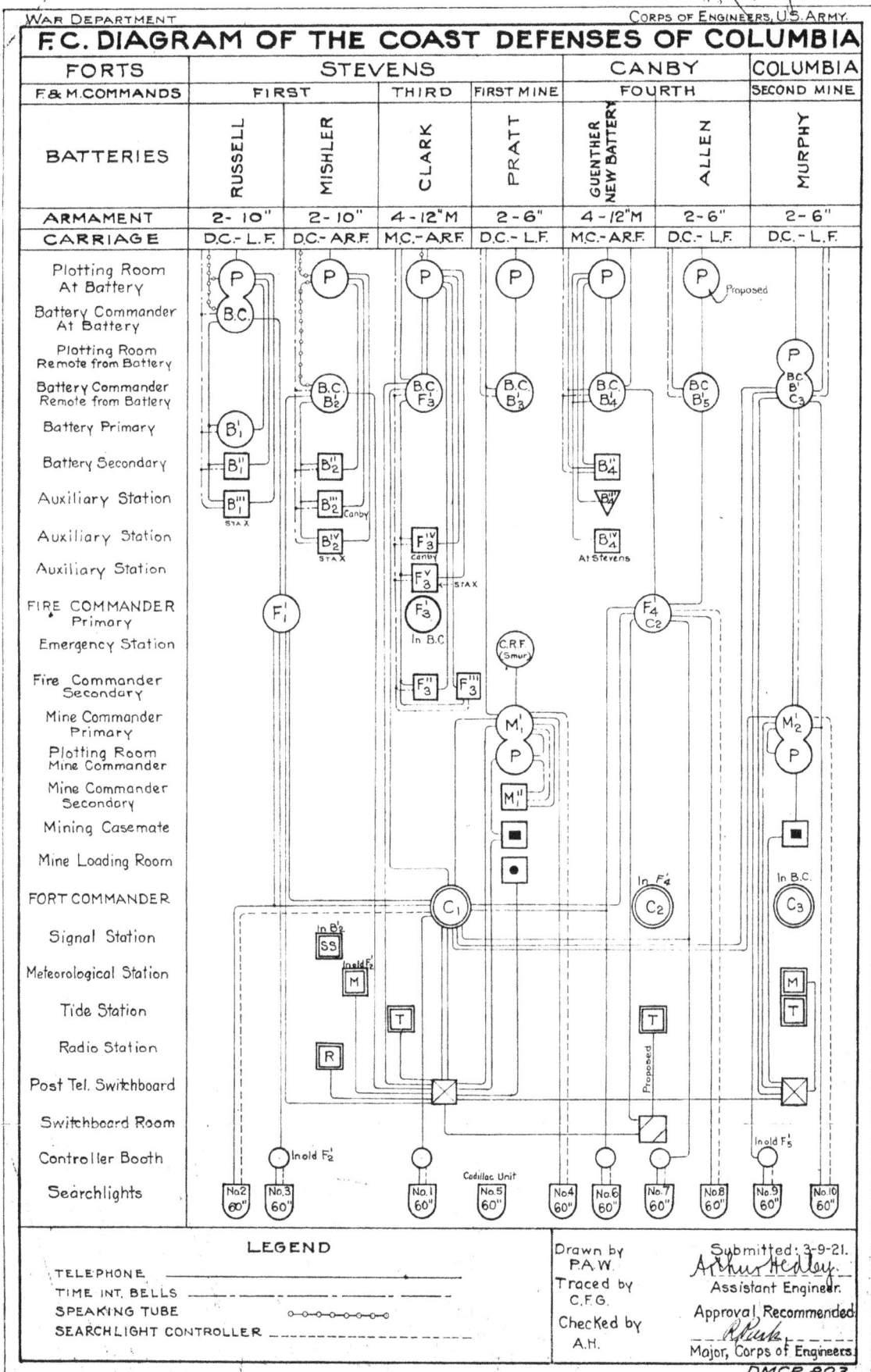

A fire control diagram showing the communications lines between the various stations.

defending, or by the name of the largest city in their immediate area. Examples are "The Harbor Defenses of San Francisco" or the "Harbor Defenses of Chesapeake Bay."

A senior U.S. Coast Artillery Corps officer was usually designated the harbor defense commander responsible for both the administration and tactical commands. He was supported by a harbor defense headquarters staff and service units from the Quartermaster, Ordnance, Medical, Signal, Engineers, and Military Police organizations. The service units usually staff the administrative headquarters and the Coast Artillery Corps the tactical headquarters. Each fort was organized with its own headquarters and fort commander, who was responsible for the administration of the post. While the fort commander was not included in the tactical chain of command, he was responsible for the training and supervision of damage control to all the fort's structures and the activities of the service units.

The basic units of the coast defense tactical command were the battery, battalion, and group. The battery was the basic combat unit of the harbor defense and contained enough men required to man one primary battery. Batteries were classified by the type according to the material with which they were equipped. The gun battery consisted of one or more fixed or mobile guns of the same caliber and characteristics to be employed against a single target and of being commanded by a single individual. It included all structures, equipment, and personnel necessary for emplacement (or mobile weapons), the conduct of fire, and the performance of service. The strength and organization of a battery depended upon the type, number, and caliber of the guns of the battery. It was divided into a battery headquarters section, a range section (containing a battery commander's detail, an observing detail, and a range detail), a maintenance section, and a gun section for each gun or mortar. Special gun batteries were the anti-motor torpedo boat (AMTB) battery and the fixed anti-aircraft battery. The mine battery consisted of the personnel, structures, and equipment other than mine planters necessary for the installation, operation, and maintenance of all or part of the controlled mine fields. It was divided into a battery headquarters section, an operations section (containing a command post detail and range detail), a casemate section, a loading and property section (consisting of loading, cable, explosive, and maintenance details), a planting section (consisting of mine planter, distribution box boat, and small boat (yawls) details) and a maintenance section. The searchlight battery consisted of the personnel, material, equipment, and structures necessary for the operation and maintenance of seacoast and anti-aircraft searchlights.

These batteries were normally administerial combined into battalions with each battery commander reporting to the battalion commander. The battalion was organized to provide administrative, training, and tactical functions. Gun battalions were composed of from two to five-gun batteries, while a mine battalion consisted of the personnel, submarine mine material, structures and vessels necessary to plant, operate, and maintain part or all the controlled mine fields. The primary purpose of the coast defense battalion was providing effective fire direction through the coordination of various types of batteries. When a harbor defense command was large, battalions will be organized into groups. A group was a tactical command containing from two to five battalions or independent batteries. As with battalions, the primary mission for the group and the group commander was to provide effective fire direction. The use of groups occurred when the number of units is greater than can be controlled by the harbor defense commander. The basis for battalions or groups was to organize batteries that covered same field of fire or water area. When large number of batteries covered the same water area then the organization was based on target selection, such as primary and secondary armaments.

For administrative and training purposes, battalions were organized into regiments up to 1942. The garrison of a harbor defense consisted of part or all of one or more regiments, and the organization of different regiments varied to conform to the special requirements of the different harbor defenses. Generally, a coast artillery regiment assigned to fixed armament consisted of a headquarters battery, a searchlight battery, a band, and three battalions. The forts were assigned to Coast Artillery Districts. The district commander commands all coast artillery troops stationed within the territorial limits of the district, including the coast artillery units of the Organized Reserves and those of the National Guard when in the service of the U.S. At the start of World War II, the headquarters for Coast Artillery Districts were in Boston, MA (1st CAC District), New York, NY (2nd CAC District), Fort Monroe, VA (3rd CAC District), Fort MacPherson, GA (4th CAC District), and Presidio, CA (9th CAC District). Overseas coast artillery units were assigned to local U.S. Army Departments, such as the Hawaii Department, etc.

View of Fort Flagler, Washington (D. Kirchner Collection)

A Typical U.S. Coast Artillery Fort in the Modern Era

While each coast artillery fort has its own unique design, it is possible to provide a general blueprint of the type and purpose of structures that you would find at a U.S. coast artillery fort built during the Modern Era. It is important to remember that each fort was like small self-contained city. All the services that were required to support the daily needs of its garrison and to operate the fort's weapon systems were included within the military reservation.

The reservation was typically surrounded by a fence. There was a main entrance gate with a guard house. While very few coast artillery forts had any land defenses, the use of security fencing was widespread. Recognizing this fencing is usually the first indication of a former U.S. military reservation. The main cantonment area contained a variety of buildings spread over a large area, not much different in appearance of a rural college campus. This support area was subdivided into functional sections surrounding by a large parade ground area. While the overall fort was under the Coast Artillery Corps, each of the support services (Quartermaster, Engineer, Medical, etc.) had their own buildings or reservations within the fort.

The main parade ground is the focal point of the post. The fort headquarters, officer's quarters, non-commission officer housing, service clubs, and enlisted barracks usually surround it. Most of the non-tactical structures at the forts constructed during the Endicott-Taft Programs were designed to be permanent structures. These wood-frame buildings were built on stone foundations with slate roofs, sided with local brick, clapboard, or stucco. The Quartermaster Corps architect's office created standard plans for all types of buildings. Those designed at the turn-of-the-century—when most Coast Artillery forts were constructed—were of Colonial Revival style with elements of Queen Anne style in the officers' quarters. As the century progressed, new styles were adopted, such as Italianate and Spanish Revival, and these styles were used when additional buildings were constructed. Store houses and pumping plants used more practical industrial or utilitarian styles.

Officer's quarters varied in size and elaborateness depending upon the rank of officer for whom the building was intended. The Commanding Officer's Quarters was usually the largest and most elaborate of the officer's quarters, and it was placed, if possible, on the highest and most prominent location on the parade ground. Other senior officers were assigned single quarters, while many of the quarters were double quarters for two families. Large forts had a Bachelor Officer's Quarters with its own mess. Non-Commissioned Officer's quarters were usually double sets.

Parade ground and officer's quarters, Fort Casewell (BW Smith Collection)

The interiors of buildings were finished with wood floors, plaster walls with wood trim, and pressed metal ceilings. All structures where officers and men lived or worked had electricity, running water and flush toilets. Each barracks was designed to house a company or battery of 100 men and was self-contained with its own kitchen, dining room, day room, barber shop, and tailor shop. Sleeping quarters were on the second floor, while the lavatory and latrine were located in the basement in northern climates. In the south, separate lavatory and latrine buildings were sometimes built. Large forts had double barracks—two 100-man barracks-built end-to-end—which functioned as two separate barracks. Forts which served as the headquarters post for a harbor defense usually had a band barracks.

Although the parade ground was used as a general athletic field, tennis and handball courts, and baseball fields were also built in open areas of the fort. A system of permanent roads served the entire fort, and the streets were usually named. Railroads and tramways were built during the construction of the forts, and these lines often continued to be used. These forts eventually had their own water, sewer, telephone, and electrical systems. If municipal water and commercial power services were available, the army used them, but at many sites the engineers built their own water and electrical plants and distribution systems. Sewer pipes ran into the ocean. Ice houses, and in northern areas, ice ponds, were also built to provide refrigeration for food in the years before electrical cooling became available. Systems for the disposal of garbage and rubbish were also created. Garbage and combustible waste were burned in crematoria, while non-combustible materials were disposed of in landfills or dumped into the ocean. The major fuel at forts was coal, and a system of unloading, transporting, and storing the fuel was developed, usually relying on mule-drawn wagons.

A large portion of fort's reservation would be devoted to the Quartermaster Corps. The Quartermaster was tasked with providing housing, supplies, and transportation for all the troops assigned to the fort. The Quartermaster oversaw the construction of most of fort's support buildings, as well as the installation of its own quartermaster wharf and tramway to transport supplies within the fort. Storehouses, commissary, workshops, and stables were usually centered near the quartermaster wharf.

Fort Terry buildings (BW Smith Collection)

The Corps of Engineers were responsible for construction the actual fortifications known as the tactical structures (emplacements, fire control stations, casemates, power houses, etc.); the Ordnance Department provided the weapons, machinery, and instruments that went into these structures; and the Signal Corps provided the technical equipment as new technology was developed. Near the shoreline were located the fire control stations along with protected telephone exchanges, command posts, meteorological stations, seacoast searchlights positions, and reserve magazines that support the fort's weapons systems.

The fort's two main coast defense weapon systems were controlled mine fields and seacoast artillery. Controlled mine fields required an extensive infrastructure within the fort. Principal structures for the mine defense included the mining casemates from which the mines were operated; the conduits connecting the casemates with the shore; the cable terminals on the shore; the cable tanks in which the mine cables were stored when not in use; the mine storehouses in which were kept the mine cases; the loading rooms in which the mines were loaded; the magazines in which the dynamite was stored; the range stations, plotting rooms, and dormitories, the mine wharves at which the mine planter used to land and receive the loaded mines; and the tramway connecting the wharves with the cable tanks, storehouses and loading rooms.

Closer to the shoreline are the emplacements of the fort's other main weapon system the large caliber gun batteries. These gun batteries consisted of both of large caliber breach loading rifles mounted on disappearing or barbette carriages and smaller rapid-fire guns on pedestal mounts during the Endicott and Taft Programs. The purpose of these gun emplacements was to provide a stable base for these guns and carriage and a convenient platform for the personnel serving the gun. The emplacement also designed to provide the armament and the personnel the maximum protection as possible, as well as providing a safe storage place of the ammunition. These thick concrete structures were covered with earthen fill on the seaward side, while they were partially buried, these batteries are easily accessible from the rear due their open back design.

The gun emplacements from the Interwar Period and 1940 Program are quite different from these earlier designs as American coast artillery responded to the progression of larger caliber naval weapons with longer firing ranges and the advent of military aviation with aerial bombs. These emplacements are usually

Fort Mott 1936 (NARA)

one-story high but usually completely buried. The gun position consists of a gun well surrounded by a circular concrete pavement. Later many of these emplacements were completely rebuilt with thick reinforced concrete casemates to protect their weapons from aerial attack and naval bombardment.

Another primary seacoast weapon of the Endicott and Taft Programs was the seacoast mortar, actually a short barrel breech loading rifle. These batteries by definition did not require direct fire, so they were often located away from the shoreline. They were located within or behind the fort's cantonment area. A typical mortar battery had a high reinforced concrete parapet with traverses that formed a series of pits. These pits were usually open to the rear, but early designs were completely surrounded with access through a tunnel. A battery had one to four pits with two or four mortars in each. Between the pits or around their sidewall were ammunition magazines, power generator rooms, shot truck areas, storerooms, and a plotting room.

The secondary gun batteries mounted rapid fire guns for the defense of the controlled mine fields from minesweepers and to repulse fast moving naval vessels and were installed after the first round of primary gun batteries. These are simple emplacements that basically provided a stable firing platform for the weapons and a protected magazine for their ammunition. Also located around some forts were groups of three or four concrete gun blocks for anti-aircraft guns that were added in later years.

A key feature of all coast artillery forts are the fire control stations which provided the target information for the mine and gun defenses. These stations come in all shapes and size. They range from a single below-grade room with observing slots to large multi-level, multi-room towers. Constructed of both wood and concrete, these stations have been disguised as non-military structures ranging from summer cottages to grain silos. Associated with World War II fire control stations were radar stations that by 1944 replaced their function. These radar stations had antennas which were mainly located on steel towers but could be mount on other structures. These antennas sent their signals to operating rooms where measurements provide location data to plotting rooms. Support these stations were power rooms and dorms for troops manning the stations.

Several to many of the structures at most of remaining U.S. coast artillery forts. However, you may only view piles of rubble and mounds of dirt as their status and condition are constantly changing. Nearly all the seacoast armament and equipment were scrapped after World War II which accounts for the lack of actual coast artillery at the forts.

Soldiers in barracks (Stillions Collection, NPS GSNS)

Garrison Life at a U.S. Coast Artillery Fort in the Modern Era

The soldiers assigned to the defenses experienced a great change in quality of life during the years from 1890 to 1950. The early years were certainly the roughest. In general military service in the U.S. armed forces was not well compensated or widely respected in some quarters. As the permanent posts were being established, physical living conditions were sometimes poor, and relationship with the local, civilian community at times strained. Officers could afford higher standards of living for themselves and their families as well as greater social involvement with the local community.

By the end of the early modernization programs in the 1910s, the living and work conditions had greatly improved. In particular the Coast Artillery was an elite assignment, with considerable prestige. The Coast Artillery Corps was relatively well funded and equipped, had a strong technical and professional dedicated career officer contingent, and was based on teamwork activity that encouraged close camaraderie. Opportunities for duty at oversea bases in exotic tropical locations like Hawaii, the Philippines, and the Panama Canal had its advantages, especially as many of the tropical diseases had been conquered. Training was emphasized, but in all the workload was reasonable. Pay was not extravagant for the enlisted man– but decent food, recreation, and athletic events were provided on post. Soldiers tended to stay in this branch of service, often re-enlisting, and were quite good at what they were taught and with what equipment they practiced on.

63rd Coast Artillery Company on parade gound at Fort Worden, Washington in 1908
(Puget Sound C.A. Museum)

The daily schedule of the Coast Artillery troops focused on drill, inspections, maintenance, meals, and recreation. The center of activity for enlisted men was their barracks which was designed to house one or two companies or batteries of 100 men each with its own kitchen, lavatories, dining room, day (recreation) room, barber shop, and tailor shop. The barrack along usually arrayed around a parade ground. The officer quarters were usually located on the opposite side of the parade ground. The day would begin with meal, roll call, and assignment of duties. This usually was training/drill in the mornings with maintenance tasks or recreation events in the afternoons. Recreation was considered important by the U.S. Army after 1900 as it was believed that it not only maintained physical fitness but promoted competitiveness which made the men more effective in combat. Most large forts were provided with a gymnasium and bowling alley, as well as athletic fields, handball and tennis courts. Other recreation activities included visiting the post theater, service clubs, libraries, chapel, and the Post Exchange, as well as leave to visit the local cities and towns. Another aspect of garrison life were weekly inspections and parades, and soldiers who failed these inspections would end up spending their weekend cleaning barracks and latrines, rather than having a weekend pass to visit the local communities.

Mess hall set for Christmas Dinner 1911, 126th Coast Artillery Company, Fort Worden Washington
(Puget Sound CA Museum)

MODERN ERA SEACOAST FORTS TODAY

In 1950, the remaining harbor defense commands were disbanded, and the U.S. Coast Artillery Corps was abolished as a separate branch with its remaining units, all anti-aircraft artillery, moved into the Field Artillery. Meanwhile, the responsibility for limited harbor defense, primarily underwater defenses, was transferred to the U.S. Navy. The U.S. Army retained several of the old coast artillery forts for other missions, while the Navy acquired several reservations for thier use including for its new role in harbor defense. Other federal agencies had an opportunity to claim all or portions of the former coast artillery sites. Those not transferred were turned over for disposition to the U.S. General Services Administration (GSA), who offered them to state, county, and local governments, and finally to private citizens. Many of the smaller, independent plots of land which had been leased or purchased for fire control and searchlight positions were returned to original owners or sold to private owners, before selling or transferring these former forts, the U.S. Army either returned to its depots all usable equipment or auctioned items in lots to the public.

Several coast defense sites had been abandoned by the U.S. Army as active defenses by 1928 including those at the Mississippi River, Mobile Bay, Tampa Bay, Savannah, Port Royal Sound, Cape Fear River, Baltimore, and the Potomac River, and the smaller inner harbor defenses at East New York, San Francisco, and Puget Sound. Many of these reservations were reclaimed for use during World War II. The next large-scale transfer of harbor defense properties from the U.S. Army began in 1947 and continued through the mid-1950s. In the early 1970s a general series of military base closures occurred throughout the U.S. Department of Defense to reduce basing costs. Several large former harbor defense sites, including military reservations around San Francisco, New York, and Pensacola, were included. Given the large size and value of these properties, Congress passed several laws that directed the ownership of these former forts to be transferred to the U.S. National Park Service (NPS). Base closure commissions in the 1990s, 2000s, and 2010s recommended the closure and transfer of other former harbor defense sites, which included the Presidio of San Francisco, Fort Wadsworth on Staten Island, Fort Monroe in Hampton, and Fort Trumbull in New London. Only a handful of old coast defense reservations remain in military hands in 2025 — Fort Story, VA, Fort Hamilton, NY, a large part of Fort Rosecrans, CA, Fort Kamehameha, HI, Fort Hase, HI and a few other sites. Other national agencies, state agencies, and local governments acquired numerous coast artillery sites for parks and recreation areas, since they inevitably had scenic river or ocean views. Depending on how diligently the GSA protected the sites, and the length of time it took to dispose of them, some sites and structures survived in excellent condition, while others suffered at the hands of salvagers and vandals.

While many of the Modern Era forts and batteries are now located within parks, they have not been accorded the same level of protection or care as the remaining brick and stone forts. Most of the old coast defenses structures are considered to be, at the worst, a legal liability or at best, an eyesore to the park. Remaining structures have been built on, fenced in, buried, or destroyed. They have been removed as interfering with the park's primary mission of providing recreation space. Vandalism has caused considerable damage over the years. Abandoned and neglected coast defense structures have suffered from freeze-thaw cycles cracking and spalling the concrete and brick, rusting metal rebar and materials has hastened deterioration. Unchecked vegetation growth has caused some structures to collapse. And rising sea levels and increasingly violent storm surges are eroding away shoreline and destroying major structures. While most gun emplacements have been constructed in such a way to resist these attacks, many other tactical structures have collapsed, and even brick structures have been damaged or destroyed by vandals and neglect. Non-tactical structures, particularly officers' quarters, have survived at many parks and government-owned sites through adaptive reuse, but at some former posts such structures have been completely removed.

However, public interest in the history of American coast defenses has grown since the publication of *Seacoast Fortifications of the United States: An Introductory History*, by E.R. Lewis in 1970. The book publication was a pivotal event, giving the public and park personnel a well-documented interpretive history of American coast defenses. A group of coast defense history enthusiasts gathered at a meeting in 1978 and organized the Coast Defense Study Group (CDSG) in 1985. The CDSG's annual conferences, Journal, Newsletter, web site, and reprints of key coast defense books have played important roles in fostering interest in the history of American coast defense and assisting both the public and park staffs in understanding the fascinating history of these defenses and to interpret their surviving elements. These massive seacoast batteries have been able to withstand both the natural climate and economic development longer than other military features from the same periods. These structures incorporated the leading edge of technology of their time and that draws interest in studying them and interpreting purpose and history. Hopefully this will translate into efforts to preserve and restore these sites for current and future generations.

Battery Winchester, Fort Armistead, Baltimore, Maryland (Terry McGovern)

PERIOD MILITARY MAPS

This book contains a compilation of maps for all the harbor defense reservations utilized during the period 1900 to 1950. The harbor defense projects show a general map of the location of elements (sites) for each harbor and the individual site maps showing the fire control elements. A series of map symbol and abbreviation keys from 1921 and 1945 are included.

The directory is organized by Harbor Defense around the United States clockwise from Portland Maine, down the Atlantic coast to Key West Florida, across to Gulf coast to Galveston Texas, then up the Pacific Coast from San Diego California to the Puget Sound in Washington, then up to the Alaskan defenses of WWII, and followed by the defenses in Hawaii, the Philippines, Panama, the Caribbean, Bermuda, and Newfoundland, Canada. While the status information is fairly comprehensive of the larger fort and military reservations, the status of many of smaller WWII-era fire control stations is not. The authors would appreciate receiving any updated information to correct or add to what has been presented here.

Notes on Coast Defense Maps

Site maps; site plans; exhibits from project plans, supplements and annexes; confidential blueprints; D-series maps—these are all terms that have been used to describe various maps which depict sites used by the U.S. Army, at one time or another, in connection with harbor defense fortifications and fire control. These maps have been keys to ferreting out the identification of the various remaining structures during site visits, yet there is some confusion over where these maps come from, what their cryptic symbols mean, and even what they are called.

Most maps of harbor defense installations are located in the Cartographic Branch of the National Archives. Many of the more frequently seen maps have come from a variety of National Archives holdings. The two concrete-era (1890-1950) map formats most frequently seen are the Confidential Blueprint map series (1900-1935 and 1940-1948) and the exhibits from the annexes/supplements to the harbor defense projects (1940s), which cover the 1940 Modernization Program (WWII-era) construction.

Confidential Blueprint Series Maps (1915-37)

As new construction finished, maps were created, revised, and updated by the Corps of Engineers. A series of maps was reproduced as negatives from a master positive in blueprint style, which meant maps were composed of white lines on a blue or dark background. As they were classified "confidential" by the War Department, they became known as "confidential blueprints."

A number of these confidential blueprints have been found in various cartographic and textual Corps of Engineers records in the National Archives. The confidential blueprint series of maps have general maps of each defended harbor, and general maps of each of the forts and military reservations in the harbor defense. If it was warranted, larger scale maps of parts of some forts were also included. These were labeled "D" for "detail" and followed in series, D-1, D-2, D-3, etc., as required. These maps show the location of batteries, various components of the fire control and communication system, mine facilities, and all the post buildings. Identification of each structure was shown by name, symbol, abbreviation, or number.

After 1900 an optical system for fire control based on trigonometric principles was developed for more precisely aiming coast artillery guns. The structures that were built to house the optical and communication elements of this system were often numerous and small in relation to the other major buildings on a military reservation, and many required a detailed description making it complicated to label them on a map, so a set of map symbols was developed to indicate the fire control structures. As these fire control structures were built in the years following 1905, they were incorporated into the maps on which the

Corps of Engineers recorded the location of all the structural elements of the fortifications in the seacoast defenses.

Keys to the fire control map symbols began appearing in coast artillery manuals, such as drill regulations, training regulations, and later field manuals. A complete update of these maps was performed during the years 1920-1922, just after the major construction projects of the Endicott and Taft programs were completed and before some of the smaller harbor defense areas were eliminated. These maps were kept as part of the records of the various Corps of Engineer district offices around the country. Copies were turned over to qualified parties in the army, such as the Coast Artillery Corps, the Quartermaster Corps, etc. On July 12, 1922, the Coast Artillery Board at Fort Monroe requested a complete list and set of these maps for their records, which were provided in August 1922. The 1922 collection contains about 290 maps of 29 harbor areas. Other versions of these maps were found in the notebooks kept by the engineer assigned to each harbor defense. In due course, the records of the Corps of Engineers and the other branches of the army have been turned over to National Archives. The map collections have been scanned and digitally "cleaned up" to remove extraneous lines and smirches from the scanning process.

WW II-Era Harbor Defense Project Maps

The 1940 "Modernization Program" brought a new set of harbor defenses, some on existing reservations, some on entirely new reservations. The fire control system was much more widespread and frequently located on newly obtained smaller reservations located around the harbor defense shoreline. Maps for these works in this guide come from the 1944-46 supplements to the various harbor defense projects published by the army.

A Harbor Defense Project was a written document which described all existing and projected harbor defense elements, including structures, first prepared in 1932-33. Supplements to the Harbor Defense Projects were prepared 1943-44 and updated during 1945-46. The supplements detailed the progress on the construction of the new 1940s modernization program defenses with descriptions and a set of maps that showed where these new structures were located, the field of fire of the guns, radar coverage, etc. The supplements provide extensive detailed information on all tactical and physical aspects of the harbor defenses on the date of the annex, both existing and proposed, and a number of exhibits detailing the locations of elements. The supplements are generally composed of 7 annexes:

A-   Armament
B-   Fire Control (including optical instruments and radar installations)
C-   Seacoast Searchlights
D-   Underwater Defenses (mines)
E-   Antiaircraft Artillery
F-   Gas Defense
G-   Equipment (usually detailing what was on hand and what was needed)
H-   Real Estate Requirements (usually detailing sites not yet obtained

These supplements and other forms of the Harbor Defense Projects have been scanned from the National Archives and are available from the CDSG ePress as electronic PDFs. These supplements contain a very comprehensive listings and exhibits of everything that was to be in place at the completion of new rearmament program and are the key references to consult for information on the final state of the American seacoast fortifications in 1945.

A few comments on the items that appear on the confidential blueprint maps and the Harbor Defense Project maps—

A **Harbor Defense** (called a "Coast Defense" before 1925) consisted of a series of land reserves (named as "Forts" and in some cases "Camps" and "Military Reservations") on which the various components of the seacoast defense fortifications were built to guard a major commercial and naval seaport. When the harbor defenses of the United States were modernized in 1890-1910, a new system of defensive works were created. The modern forts consisted of tactical and non-tactical structures spread over hundreds of acres of land. The U.S. Army Corps of Engineers selected the locations, purchased additional land, sited, designed, and constructed the tactical structures—gun batteries, mine facilities, observation stations, plotting rooms, power plants, switchboard rooms, and searchlight shelters.

Gun Batteries: The modern seacoast artillery consisted of guns, mortars and antiaircraft weapons mounted in concrete support structures varying from the simple to the quite complex. Guns were mounted on barbette, pedestal and "disappearing" carriages. Mortars were emplaced in protected pits. Antiaircraft weapons, usually the 3-inch guns, were mounted in simple concrete platforms. The term "battery" was used to describe a set of guns under a single commander together with the entire structure erected for the emplacement, protection, and service of those guns.

Fire Control Structures: The target range and azimuth for seacoast artillery guns were determined using command and equipment systems collectively referred to as fire control and position finding. The standard systems of position finding used by seacoast artillery were based on trigonometry. Components of the system included widely spaced base end stations, command stations, plotting rooms, tide stations, meteorological stations, and cable linked telephone communication systems with protected switchboards. Radar installations were deployed for the major gun batteries and as general surveillance after 1942. The radar installations included power/control buildings and antenna towers.

Searchlights: Most searchlights installed during the period 1901-1920 were fixed, located in a structure for concealment and protection during the day, with their electrical power generator. Over the years after WWI, the mobile searchlights became more reliable, durable, and rugged. By the late 1930s, the Coast Artillery switched to using mobile searchlights and replaced fixed searchlights where at all possible so after 1940 the US seacoast defenses used mostly mobile searchlights.

Controlled Mine Facilities: Throughout the modern or "concrete" era of American harbor defenses (1890-1950), mines were considered to be one of the primary harbor defense weapons. Mines were only deployed during times of war or during limited training expertises. The mines and cables were stored ashore between use. The mine shore facilities included torpedo storerooms, loading rooms, mine wharfs, explosives storage, tramway systems, cable tanks, mine casemates, and cable vaults.

Electrical Generator Power Plants: By the turn of the century, electricity had become a vital necessity for the Coast Artillery. It was used to traverse and elevate some of the large guns, to light emplacements, to operate ammunition hoists, to power searchlights, to control submarine mines, and for communications, in addition to standard garrison uses. Most large forts had a central power plant with electrical generators. The requirement that coast defenses be self-contained resulted in power rooms being included in most batteries and mining casemates, and separate searchlight powerhouses were constructed.

Protected Switchboard Rooms: As seacoast defense artillery covered increasing distances, a need for remote accurate and instant communications was required. Telephones connected by phone lines were integrated into the fire control system utilizing protected switchboard rooms after 1906. As radio communication developed in the 1930s, fixed radio sets were often integrated with the telephone communication system in their protected switchboard rooms or housed in separate protected structures.

Garrison Buildings: These are shown in the Confidential Blueprint series maps but not on Supplement series maps.

The system of numbering for buildings was the same for all Confidential Blueprint maps in the period 1915 to 1937. All buildings of the same "type" were given the same number on all the maps. For example all barracks buildings were numbered "7."

1.	Administration Building
2.	Commanding Officer's Quarters
3.	Officer's Quarters
4.	Hospital
5.	Hospital Steward's Quarters
6.	Non-commissioned Officer's Quarters
7.	Barracks
8.	Guard House
9.	Post Exchange

10 to 19 and 100 to 199	Post Buildings
20 to 29 and 200 to 299	Quartermaster Buildings
30 to 39 and 300 to 399	Ordnance Buildings
40 to 49 and 400 to 499	Engineer Department Buildings
50 to 59 and 500 to 599	Signal Corps Buildings
60 to 69 and 600 to 699	Reserved for future requirements
70 to 79 and 700 to 799	Religious and Social Buildings
80 to 89 and 800 to 899	Government Buildings not under War Dept. Control
90 to 99 and 900 to 999	All Private Buildings (Private dwellings, stores, contractor's buildings and buildings purchased with the land but not assigned to public use.)

Fort Columbia, Washington 1913 (NARA)
From left to right is the Post Exchange, a Company Barracks, the Administration Building,
a Double Officer's Quarters and the Commanding Officer's Quarters.
Just visible behind the front row of buldings is the Quartermaster's Storehouse and the Post Hospital

Symbols and Abbreviations—1921 Confidential Blueprints

Name	Abbr.	symbol	Sta. w/o roof
Fort Commander's Station	C		
Primary Station, Fire Command	F'		
Secondary Station, Fire Command	F''		
Supplementary Station, Fire Command	F'''		
Primary Station of a Battery	B'		
Secondary Station of Battery	B''		
Supplementary Station of a Battery	B'''		
Battery Commander's Station	BC		
Primary Station, Mine Command	M'		
Secondary Station, Mine Command	M''		
Supplementary Station, Mine Command	M'''		
Double Primary Station, Mine Command	M'-M'		
Double Secondary Station Station, Mine Command	M''-M''		
Separate Plotting Room	P		
Separate Observing Room	O		
Self-contained Horizontal Base	C.R.F.		
Emergency Station	E		
Spotting Station	Sp		
Meteorological Station	Met		
Tide Station	T		
Searchlight (30, 60, etc., relates to the size of the lights)	S		
Controller Booth	C.B.		
Watchers Booth	W		
Signal Station	S.S.		
Radio Station	R		
Cable Terminal	C.Ter.		
Post Telephone Switchboard	P.S.B.		
Mining Casemate	M.C.		

Name	Abbr.	symbol
Loading Room	L.R.	▣ (square with filled circle)
Switchboard Room	S.W.B.	◩ (square with diagonal)
Central Powerhouse	C.P.H.	⊡⊣ (square with circle and tab)
Powerhouse (and Searchlight Powerhouse)	P.H.	□⊣ (square with tab)
Combined Stations, in same room		⊝ (circle marked B. / B.C)
Combined Stations, in communicating rooms		⊝⊝ (F'/B' circles); ⊝⊝ (B.C/P.)
Combined C and F' Station in same room		⊝ (circle marked C./F')
Differentiation of auxiliary plants		a⊣ b⊣ c⊣ etc.

Abbreviations used on maps

Cable Gallery	C.Gal.
Cable Tank	C.T.
Cable Hut (commercial cable)	C.H.
Coast Guard Station	C.G.S.
Engineer Wharf	Engr. Whf.
Gasoline Tank	G.Tk.
Guard House	G.H.
Latrine	L.
Lighthouse	L.H.
Lighthouse Wharf	L.H.Whf.
Magazine	Mg.
Mining Boathouse	M.B.H.
Mining Derrick	M.D.
Mining Tramway	M.T.
Ordnance Machine Shop	O.M.S.
Mine Wharf	M.Whf.
Private Wharf	Pvt.Whf.
Radio (commercial station)	Rad.
Railway Wharf	Ry.Whf.
Saluting Battery	Sl.B.
Searchlight Shelter	S.Sh.
Service Dynamite Room	S.D.R.
Steamship Wharf	S.S.Whf.
Sunset Gun	S.G.
Tide Gauge (not a Tide Station)	T.G.
Torpedo Storehouse	T.S.
Tower	Tw.
Water Tank	W.Tk.
Weather Bureau	W.B.

Additional Symbols and Abbreviations

Name	Abbr.	symbol
Pumping Plant	P.P.	
Radio Powerhouse	R.P.H.	
Searchlight Powerhouse	S.P.H.	
60 inch Searchlight No. 7	$S.^{60}_{7}$	
Coincidence Rangefinder	C.R.F.	
Quartermaster Wharf	Q.M.Whf.	

Subscripts for use in both Legend and on Face of Plat are—

Imp. Improvised.	B'' imp.	
(for temporary fire control structures only.)		
p. Portable.	S^{36}_{p2}	
(Principally used for portable searchlights etc.)		
s. Superseded.	24s.	
(for abandoned buildings, etc.)		
t. Temporary.	19t.	
(For all uses except fire control structures.)		

Datum Point—location indicated by intersection of lines or by dot at end of arrow.

Triangulation Station.

Intersection Point.

Benchmark.

Lighthouse.

Such other topographic signs as were necessary were taken from the *Engineer Field Manual* (Professional Papers, Corps of Engineers, No. 29) pages 74 to 97.

Note: Maneuver buildings were classed as post buildings.

SYMBOLS and ABBREVIATIONS 1940
FM 4-155, Reference Data (Seacoast Artillery and Antiaircraft Artillery) 1940
TABLE C.-Symbols for seacoast artillery fire-control maps, diagrams, and structures

Part 1.—Basic symbols

Name	Abbreviation	Symbol
Harbor defense command post	H D C P	
Groupment command post	Gpmt C P	
Fort command post	Ft C P	
Gun group command post	G C P	
Mine group command post	M C P	
Seacoast battery command post	B C P	
Harbor defense observation station	H D O P	
Groupment observation station	Gpmt O P	
Fort observation station	Ft O P	
Gun group observation station	G O P	
Mine group observation station	M O P	
Battery observation station	B O P	
Emergency observation station	E O P	
Antiaircraft observation post	A A O P	
Battery spotting station	S O P	
Separate observation station	O P	

Name	Abbreviation	Symbol
Operations and plotting room	O P R	
Plotting room	P	
Self-contained base range-finder station	R F	
Magazine	Mg	
Shellroom	S Rm	
Temporary or improvised fire-control structures	Imp	
Mine casemate	M C	
Mine loading room	L R	
Searchlight, 60-inch seacoast	S L	
Searchlight, seacoast, other than 60-inch	S L	
Antiaircraft searchlight	A A S L	
Searchlight shelter	S Sh	
Searchlight powerhouse	S P H	
Searchlight controller booth	C B	
Data booth	Data B	
Watchers booth	W Bth	
Meteorological station	M E T	

Name	Abbreviation	Symbol
Tide station	Td	
Signal station	S S	
Fire Control switchboard room	F S B	
Post telephone switchboard room	P S B	
Combined fire-control & post telephone S B room	F S B P S B	
Cable terminal	C Ter	
Powerhouse	P H	
Radio powerhouse	R P H	
Central powerhouse	C P H	
Pumping plant	P P	
Datum point		 • OR
Triangulation station		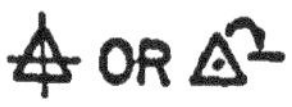 OR
Intersection point		O Black Beacon
Benchmark	B M	 BM X 1232
Lighthouse	L H	★

Other abbreviations used in this guide

BS - base end station & spotting station

HECP - harbor entrance command post

HDOP - harbor defense command observation post

HDCP- harbor defense command post

SBR -telephone system switchboard or radio room

AMTB- Anti-motor torpedo boat BC station

BC - battery commander's station

C - fort commanders station

G- group command station

M- mine station

SCR - signal corps radar

SL - searchlight

Part 2.-Numbers for harbor defense installations.—a. In harbor defense, seacoast artillery installations of each type are numbered consecutively from right to left, facing the center of the field of fire of the harbor defense. Antiaircraft installations pertaining to the harbor defense may be numbered in any convenient sequence.

b. Groupments, gun groups, mine groups, batteries, and all installations functioning directly under the harbor defense commander, such as harbor defense observation stations, searchlights, and underwater listening posts, are numbered consecutively, each type in a separate series, beginning with number 1. These numbers normally are shown as subscripts to the letter included in the appropriate symbol. Exceptions are included among the examples that follow.

Name	Abbreviation	Symbol
Harbor defense observation station	H D O P$_3$	
Fort observation station	Ft O P$_3$	
Antiaircraft observation post	A A O P 2	
Magazine or shell room	Mg 2 or S Rm 2	

c.Groupment, group, and battery observation and spotting stations assigned to a unit are numbered consecutively within the unit, each type in a separate series, beginning with number 1. These numbers are shown as superscripts to the letter included in the appropriate symbol, the unit number remaining as the subscript.

Name	Abbreviation	Symbol
Groupment observation station	Gpmt$_2$ O P$_2$	
Gun group observation station	G$_2$ O P$_1$	
Mine group observation station	M$_2$ O P$_1$	
Battery observation station	B$^1_1$ O P	
Spotting station	S$^1_3$ O P	
Emergency observation station	E$_2^1$ O P	
Temporary or improvised fire control structures	B$_3{}^2$ Imp.	

d. In certain cases it is desirable to show additional information regarding an installation, such as its size and whether fixed, portable, or mobile. Such information is placed either in the symbol or to the right thereof.

Name	Abbreviation	Symbol
60-inch seacoast searchlight; fixed, portable or mobile.	SL 2F (P or M)	2F(P or M)
Seacoast searchlight other than 60-inch	SL^{36}_{3P}	$\boxed{\frac{3}{P}}$ 36'
Antiaircraft gun battery or composite battery, fixed or mobile.	A A No. 2 (F or M)	AA 2 (F or M)

e. Where two stations are combined in one room, the symbols are superimposed one upon the other, and the letters representing each station are inclosed in the combined symbol.

Name	Abbreviation	Symbol
Combined groupment command post and fort command post.	Gpmt Ft Cp	CF
Combined battery observation and spotting station.	$B^2_1 S^2_1 \, O \, P$	$B^2_1 S^2_1$
Combined group command post and battery command post.	$G_1 B_2 C \, P$	G_1 / BC_2
Combined battery command post and battery observation station.	$B_2 C \, P \, B^2_2 O \, P$	B^2_2 / BC_2

f. Where stations are adjacent in the same structure, the symbols are tangent to each other and are arranged to show the relative location, as:

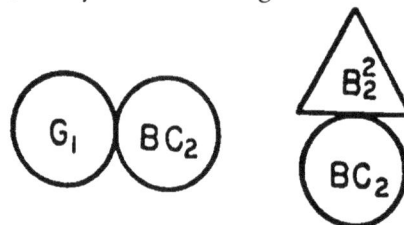

G_1 BC_2 B^2_2 / BC_2

g. Where communication may be had by voice through a passage, door, window, or voice tube, the symbols are left open at the point of contact, as:

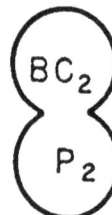

BC_2 / P_2

Part 3.—Communications symbols for use on harbor defense fire-control charts and diagrams.

Telephone cable (numerals indicate number of pairs and gage)　　26-19

Speaking tube

Mechanical data transmission line

Electrical data transmission line

Searchlight controller line

Zone signal and magazine telephone line

Firing signal line

Time interval bell line

Submarine cable (numerals indicate number of pairs and gage)　　50-19

Part 4.-Abbreviations

Cable gallery	C Gal
Cable tank	C T
Cable hut (commercial cable)	C H
Coast Guard station	C G S
Engineer wharf	Engr Whf
Gasoline tank	G Tk
Guardhouse	G H
Latrine	L
Lighthouse wharf	L H Whf
Mine boathouse	M B H
Mine derrick	M Drk
Mine tramway	M Tmy
Mine wharf	M Whf
Ordnance machine shop	O M S
Private wharf	Pvt Whf
Radio (commercial station)	Rad
Railway' wharf	Ry Whf
Saluting battery	Sl B
Service dynamite room	S D R
Steamship wharf	S S Whf
Quartermaster wharf	Q M Whf
Superseded (for abandoned buildings, etc.)	24 s
Temporary (for all uses except fire-control structures)	19 t
Sunset gun	S G
Tide gage	T G
Torpedo storehouse	T S
Tower	Tw
Water tank	W Tk
Weather bureau	W B

A DIRECTORY OF AMERICAN SEACOAST DEFENSES 1890-1950

This directory is a comprehensive guide to all the major locations and sites used for harbor defense, with maps showing what was at each site and comments on the current status of each site (extant, in ruins, destroyed, privately owned, current U.S. military use, federal, state, county, city parks, etc.) as far as information is known to the authors. While the status information is fairly comprehensive for the larger forts and military reservations, the status of many of smaller World War II-era fire control sites is not. The authors would appreciate receiving any updated information to correct or add to what has been presented here. Terms used in this reference work to describe the various periods of construction such as "Endicott-Era," "Taft-Era," "Post-World War I-era," "World War II-Era," the "100-Series" and "200-Series" batteries, etc. are terms used by modern historians and were not used by the Army to describe these programs in progress. Note that several of the planned batteries in the 1940 program were cancelled before any work was done as denoted by their *battery # in italics* and as (planned).

This directory does not cover the following artillery used for seacoast defense at various times between 1898 and 1945—the Rodman guns emplaced or re-emplaced during the Spanish American War; the Navy guns and mounts installed during the World War II years, mostly in the Pacific theater; Hawaii's World War II temporary and provisional defenses; the fixed antiaircraft gun batteries emplaced in the defenses from 1920; mobile artillery which had prepared positions including those for 12-inch railway mounted mortars and 8-inch railway guns; the Panama mount positions for the tractor drawn 155 GPF guns; and the positions on Oahu for the 240 mm howitzers.

The directory is organized by Harbor Defense around the United States clockwise from Portland Maine to the Puget Sound in Washington State, followed by the Alaskan defenses of World War II, and the defenses in Hawaii, the Philippines, Panama, the Caribbean, Newfoundland, and Bermuda.

This directory includes detailed brief histories of modern-era American coast artillery concrete gun batteries. Glen Williford created this as a personal reference guide over many years of research and study of U.S. Coast Artillery history. This battery listing includes the histories of all modern (post-1886) "fixed" or permanent concrete seacoast gun batteries emplaced by the U.S. in the country and outlying territories. The emplacements were mostly built by the Corps of Engineers, and manned by the Coast Artillery. Each battery description includes the following information where possible. The battery name in capital letters if an officially conferred name, in lower case if just an informal, local, or construction designation. The description then briefly covers the purpose for construction and the general location on the reservation, particularly in relation to other elements. In most cases the act or source of original funding (which does not include the cost of coast artillery) and date of plan submission follows. Major design features or significant variations from Mimeograph Type plans are discussed. The general dates of construction, transfer date to the Coast Artillery and engineering costs may also be included. This is followed by a description of the armament, including gun and carriage models and specific serial numbers and date of mounting if known. In general, the manufacturer of the guns and carriages are only designated only when there are multiple producers and thus duplicate serial number runs. The general order and date, that names each battery is included with a brief description of person honored. Subsequent service events, including any major alterations, accidents, armament, or name changes follows. The date of gun dismounting or at least the date of authorization for deletion is covered. A brief statement on whether the battery still exists, or when destroyed, and park or status if on public property concludes the description.

The major sources consulted were: Reports of Completed Batteries, and Reports of Completed Works, Engineer Letters of Submission, surviving Fort Record Books, Seacoast Gun Record cards and earlier Ordnance Department Seacoast Gun Ledger Books, Annexes to Seacoast Projects, General Orders naming citations, supplemented by various records in archive primary engineer correspondence files, annual reports of the Chief of Engineers, private Williford studies on emplacement accidents, temporary defenses, defenses of the Spanish American War.

ALASKA

The Alaska coastline is very long with few major harbors or population centers. The climate is harsh and access to sites very difficult. These defenses are the product of World War II and the Japanese invasion of Attu and Kiska in the Aleutian Islands. While temporary defenses were installed at many sites, only four official harbor defenses were established in Alaska— Sitka, Seward, Kodiak, and Dutch Harbor.

ALASKA PROJECTS 1941 |||| 1944

As the construction of the Sitka, Kodiak and Dutch Harbor Naval bases got underway in 1940, plans were made to provide all three with protective seacoast defense guns that would be manned by the U.S. Army's Coast Artillery Corps. As the Seward terminal of the Alaskan Railroad was the only reliable Port of Entry to the Alaskan interior, it was decided to fortify that harbor as well. Funds were authorized in late 1941 and early 1942. A few 155-mm GPF guns were rushed to the four harbors, which were the only the defenses in place on December 7, 1941. Eventually, 100 Panama mounts mounting 155mm GPF guns were authorized at 15 different locations and by 1944 some 72 emplacements had been built. Other temporary defenses included 6-inch naval guns that were emplaced during 1942 at Sitka, Annette Island, Yakutat, Cold Bay, Chernofski, Umnak Island, Nome and Adak Island.

The permanent gun batteries of the Alaskan defenses followed the standard designs for batteries already under construction on the continental United States. Kodiak and Dutch Harbor received a mix of 8-inch and 6-inch batteries, while Seward and Sitka received only 6-inch batteries. The 8-inch batteries were a "U" shaped design with a central magazine between two open gun platforms. The plotting room was a separate structure. The 6-inch batteries were a "T" shaped design. Construction began in earnest in 1942 and continued throughout 1943. As the war shifted away from the Alaskan theater, many of the coast defense construction projects were slowed and canceled in 1944, several just short of completion.

The temporary emplaced Alaskan navy guns and 155mm GPF Panama mount guns.

	# guns	Gun Type	Mount	Location
	2	6" Navy	Navy P	Shoals Point Kruzof Is. Sitka
	2	6" Navy	Navy P	Cold Bay, Mortensen Pt.
	2	6" Navy	Navy P	Chernofski
	2	6" Navy	Navy P	Umnak, Sheep Pt.
	1	6" Navy	Navy P	George Is., still emplaced
	2	6" Navy	Navy P	Yakutat, Point Carrew, still emplaced
	2	6" Navy	Navy P	Nome
	2	6" Navy	Navy P	Annette Is.
	2	6" Navy	Navy P	Adak Is.
	2	6" Navy	Navy P	Shemya Is.
	2	6" Navy	Navy P	Popof Is., Sand Point
	4	155 mm	PM	Makhnati Is. Sitka
	4	155 mm	PM	Rocky Point, Seward
	4	155 mm	PM	Chiniak, Kodiak Is.
	4	155 mm	PM	Deer Point, Kodiak Is.
	4	155 mm	PM	Bushkin Hill, Kodiak Is.
	4	155 mm	PM	Mt. Ballyhoo, Dutch Harbor
	4	155 mm	PM	Hill 400, Dutch Harbor
	4	155 mm	PM	Summer Bay, Dutch Harbor
	4	155 mm	PM	Cold Bay, Mortensen Pt.
	2	155 mm	PM	Umnak, Umnak Pass
	7	155 mm	PM	Umnak?
	4	155 mm	PM	Yakutat, Point Carrew
	4	155 mm	PM	Annette Is.
	?	155 mm	PM	Adak Is.
	?	155 mm	PM	Shemya Is.
	2	155 mm	PM	Attu Is., Murder Point
	2	155 mm	PM	Attu Is., Chichagot Pt.
	8	155 mm	PM	Amchitka Is.
AMTB	2	90 mm M1	F M3	Whittier, from Seward defenses
AMTBs	6	90 mm M1	F M3	Amchitka Is.
AMTBs	8	90 mm M1	F M3	Adak Is.
AMTB	2	90 mm M1	F M3	Shemya Is., 1 gun remains, reloacted from?

Watson, Montgomery. *The Kodiak Coastal Defense of System at Fort Greely During World War II*. U.S. Army Corps of Engineers. Anchorage, AK 2000.

Bush, James D. *Narrative Report of Alaska Construction 1941-1944,* US Army Engineer District, Alaska. Anchorage, AK, 1984

The Harbor Defenses of Sitka — Alaska

These military reservations were established as part of the major effort to defend the naval operating base and air station at Sitka. Three 6-inch batteries were built on three islands around the entrance to the harbor. As space was a premium, many U.S. Army facilities were jammed on little rock outcrops. A chain of these small islands near Japonski Island, where the Naval Air Station was built, were connected by a causeway. Fort Ray was established in 1941 as the headquarters garrison post for the harbor defenses of Sitka. The garrison were housed at Fort Ray on Alice and Charcoal Islands adjacent to Japonski Island. Battery Construction number (BCN) 290 was located on Kruzof Island designated Fort Babcock, BCN 291 was located on Biorka Island designated Fort Peirce, and BCN 292 was located at the end of the causeway from Japonski Island on Makhnati Island designated Fort Rousseau. Fire control and/or searchlight facilities were built on St. Lazaria Is., Clam Is., Kayak Is., Ataku Is., Golf Is., Kita Is., Liasianski Peninsula, Hill 800, Lava Pt., and little Biorka Island, Radar facilities were installed on St. Lazaria Is., Biorka Is., and Abalone Is., while 4-gun Anti-Motor Torpedo Boat (AMTB) batteries were located on Whale Is. and Watson Pt. In 1944, the Sitka Naval Base was closed and the work on the harbor defenses halted.

Fort Babcock (1942-1944) is located at Shoals Point on Kruzof Island which is at the entrance to Sitka Sound to the west of the city. It was named in General Order 17 of 1943 for Col. Walter C. Babcock. Babcock was the Topographical office of the 1898-99 Copper River Exploring Expedition, and in charge of construction of Trans-Alaskan Military Road. The reservation of 4070 acres was transferred for army use on August 29, 1941. It was selected to be the site of a permanent 6-inch battery of standard design in February 1942. Initially, a temporary battery was built in April 1942 using two 6-inch ex-naval guns. Several fire control stations were built and the defense's SCR-582 radar was slated to be erected on Hill 400. In common with the other Sitka construction, work was begun using private contracts but then completed by U.S. Navy Seabees. The fortification work was never fully completed when the defenses were ordered abandoned on March 13, 1944. After World War II, Fort Babcock became part of the Tongas National Forest. Access to the Fort Babcock area is only by boat, there are no developed access docks or trails, and the former U.S. Army facilities at Shoals Point are very overgrown.

Fort Babcock Gun Batteries

- **Battery #290**: A World War II battery for two 6-inch barbette guns built near Shoal Point, on the southern end of Kruzoff Island, on the western side of the sound. It followed standard plans for 200-series batteries but had the battery commander station on top of the traverse between the gun platforms, though not connected with the interior corridor. Work was started in November 1942 and suspended in May 1944 at 89% completion status. The uncompleted work was transferred on August 1, 1944 for an estimated cost to date of $378,462. It was to have been armed with two 6-inch guns Model 1903A2 on model M1 barbettes, but these were never mounted (however guns #27 and #33 were allocated to the emplacement and shipped to Sitka on March 8, 1943). These guns, without being mounted, were sent to the Black Hills Ordnance Depot on May 13, 1944. The battery was still incomplete when abandoned in 1945. The emplacement still exists. The battery is open to the public, but difficult to reach and overgrown.

- A temporary battery for two ex-navy 6-inch guns on pedestal mounts emplaced near Shoal Point at the Fort Babcock reservation on Kruzoff Island. The blocks were placed southeast of the site that would become Battery #290, relatively close to the shoreline. Work was probably done for emplacement during the first couple of months of the war, perhaps in February or March 1942. Because

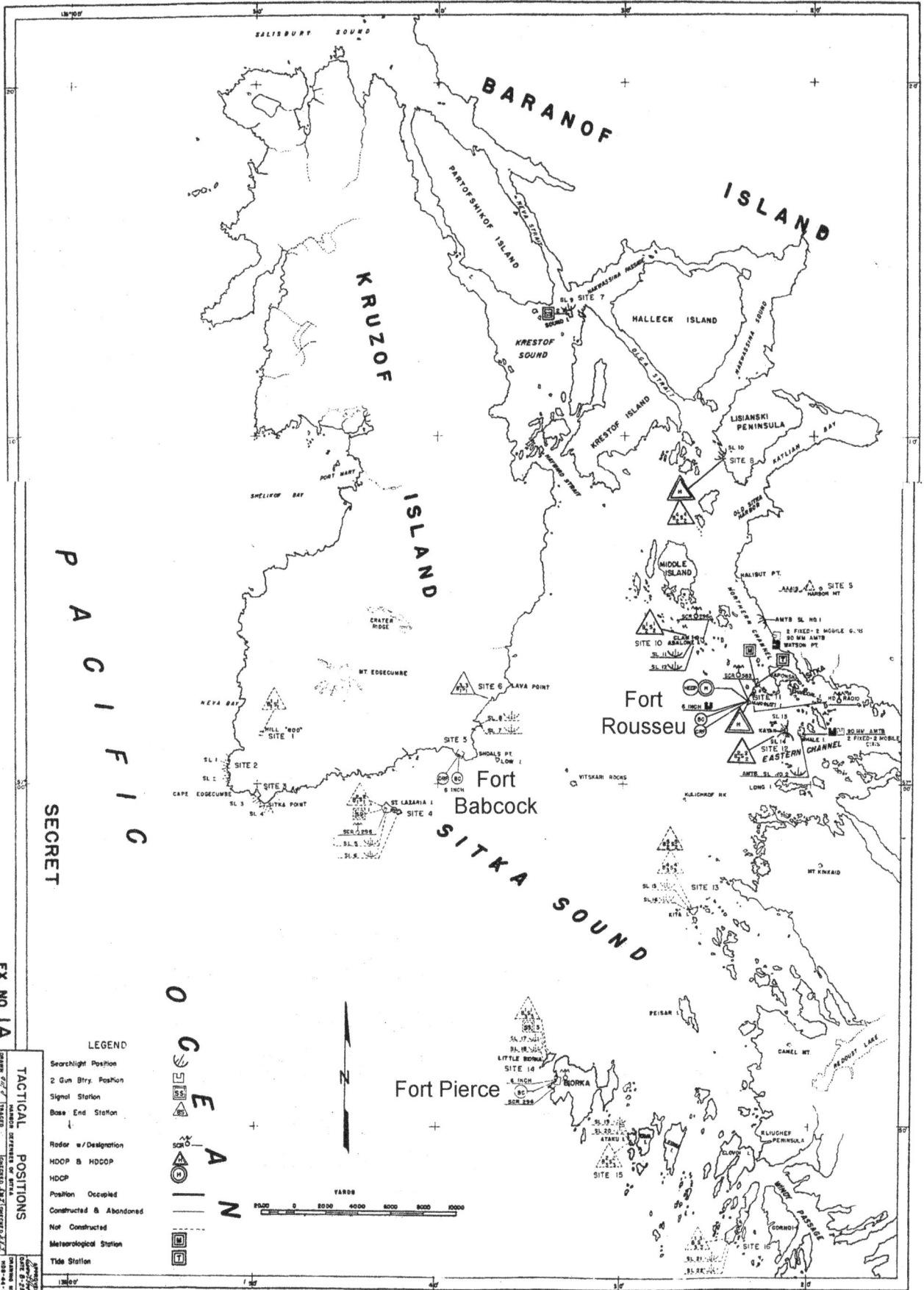

SECRET

PACIFIC OCEAN

BARANOF ISLAND

KRUZOF ISLAND

SITKA SOUND

Fort Rousseu

Fort Babcock

Fort Pierce

LEGEND

Searchlight Position
2 Gun Btry. Position
Signal Station
Base End Station

Radar w/Designation
HDCP & HDCCP
HDCP
Position Occupied
Constructed & Abandoned
Not Constructed
Meteorological Station
Tide Station

EX. No. 1A

TACTICAL POSITIONS
HARBOR DEFENSES OF SITKA

its replacement permanent battery was never completed, it probably served until the abandonment of the defenses in 1944 or 1945. Though difficult to find today, the blocks still exist though the armament is removed.

Fort Pierce (1941-1944) is located on Biorka Island which is at the southern entrance to Sitka Sound, about 13 miles from Sitka proper. The military reservation of 1660 acres was secured in late 1941. This site and the one at Fort Babcock were selected for placement of a standard 6-inch battery of the permanent system in February 1942. It was named in General Order 17 of 1943 for Capt. Charles H. Pierce. Pierce was a medal of Honor winner for actions during the Philippine Insurrection in 1899. Construction was not started on the gun battery until October 15, 1942. While most of the defensive structures were complete by early 1944, the fortification work here was ordered abandoned on March 13, 1944. The U.S. Army had built a fire control station on nearby Araku Island and had already constructed a large camp to support the battery. Several fire control and radar stations were planned on nearby islands. After World War II, Fort Pierce became the property of the United States Coast Guard and the Federal Aviation Agency.

Fort Pierce Gun Battery

- **Battery #291:** The third dual 6-inch barbette gun emplacement constructed for the defenses of Sitka. It was placed on Biorka Island, to the southwest of Sitka. Work was begun on October 15, 1942, and apparently suspended in mid-1944 when virtually complete (one report says at the 98% completion status). It was transferred in almost finished condition on August 1, 1944 at a cost estimate of $325,110.00 It mounted two 6-inch guns Model 1903A2 on Model M1 barbettes. The battery received two guns (#25 and #48) shipped here on March 8, 1943. These were at a 219-foot trunnion height. This armament was soon removed, probably in May of 1944—and sent to the Black Hills Ordnance Depot. The emplacement still exists in relatively good condition. The battery is open to the public, but access is difficult requiring a boat.

Fort Rousseau (1941-1946) is located on Makhnati Island, just to the west of Sitka at the end of a chain of small islands in the inner harbor. The land was procured for a military reservation on August 29, 1941. Before construction could begin, a new causeway of 8100-feet length joining several islands had to be built. This was not completed until mid-March 1942, and cost over $2 million dollars. The fort was named in General Order 17 of April 2, 1943 for Bvt. Maj. Gen. Lovell H. Rousseau, USV. Rousseau, while serving in the army, played a key role in the transfer of Alaska from Russia to the U.S. in 1867. While the causeway was under construction, a temporary battery was built in early 1942 using four 155mm GPF guns on Panama mounts. Also a temporary HECP was constructed at Kirushkin Island, and AA guns emplaced on Sasedni Island. The primary armament of Fort Rousseau was a permanent system standard 6-inch shielded battery, emplaced in the inner harbor to deny enemy usage of Sitka harbor or the Naval Air Station. Makhnati Island also held the joint defensive command post and an adjacent signal/observation tower. Several fire control stations were also constructed for this fort on adjacent islands. When the Sitka defenses were ordered abandoned in March 1944, the completed battery at Rousseau was kept in reduced status, After World War II, Fort Rousseau became the property of the State of Alaska. Located just west of the Sitka airport runway, it is now part of the Fort Rousseau Causeway State Historical Park. There is no land access to the park, but it is accessible by a short boat ride from Sitka.

Fort Rousseau Gun Battery

- **Battery #292:** A World War II dual 6-inch battery emplaced on Makhnati Island in Sitka harbor connected by causeway to the mainland. It was built between January 15, 1943 and June 9, 1944, for transfer on August 1, 1944. It differed from standard 200-series plans in two ways. It had a battery commander station on top of the traverse but was connected by ladder to the main corridor's interior. Also, it has a couple of small rooms in a lower, basement level beneath the entryway from the rear into the power and plotting room area. It was armed with two 6-inch Model 1903A2 guns on Model M1 barbette carriages (#15/#84 and #52/#85). Guns were held at a trunnion height of 50-feet. The battery was never named, simply referred to as Battery Construction No. 292. It served until armament removed about 1950. The emplacement still exists. The battery is open to the public, but difficult to visit requiring a boat.

Fort Ray (1941-1946) is located just to the west of Sitka, on nearby Alice and Charcoal Islands. Beginning in 1940, it was the first permanent army base built for Sitka. While not having any major fortification batteries, it became the Coast Defense HQ and main garrison cantonment for the army here. It was named in September 1941 for Brigadier General Patrick H. Ray who was stationed in Sitka in 1897. Work started in January 1941, and was done by U.S. Navy (to which the army transferred the necessary funding). In 1943 a reorganization resulted in Fort Ray being attached to the Harbor Defenses on the causeway and Japonski Island. Fort Ray also included supporting installations on nearby Galankin Island, Long Island, Whale Island, and several other islands near Sitka. Two 90mm AMTB batteries were located on Watson Point and Whale Island. Searchlights, fire control stations, and other related structures were set up on several small islands. In March 1944 the harbor defenses were authorized for abandonment, and Fort Ray was subsequently closed. After the war the property was transferred to the Alaskan Native Service. Most of the buildings were destroyed over time, and the property has been commercially redeveloped. The southwestern end of Charcoal Island has the most remaining structures, mostly warehouses which are still used for storage.

Fort Ray Gun Batteries

- **AMTB #1:** An AMTB battery for two 90mm fixed (emplacements No. 2 and 3) and two mobile 90mm (emplacements No. 1 and No. 4) emplaced on the mainland to the north of Sitka at Watson Point. It fired to the west. Probably completed and armed in 1943, it served until postwar, at least still being armed on March 10, 1945. The gun blocks of the fixed guns have been destroyed or covered over by commercial property.

- **AMTB #2:** An AMTB battery for two 90mm fixed and two 90mm mobile guns emplaced on Whale Island on the southern site of the Sitka boat harbor. Emplacements No. 3 and 4 were on concrete blocks for the two fixed guns. Probably completed and armed in 1943, it served until postwar, at least still being armed on March 10, 1945. The gun blocks for the fixed guns still exist on private property. The battery is not open to the public.

Battery Commander's Station Battery #292 (Matt Hunter 2025)

Fort Rousseau Battery #292 (Matt Hunter 2025)

Sitka Alaska World War II-era Site Locations

location	Loc#	Purpose
Hill 800	1	fire control station
Sitka Point	2	searchlight position
Makhnaki Island	11	gun position
Long Island		cable
Shoals Point	5	gun position
Cape Edgecomb		fire control station
Lava Point	6	fire control station
Harbor Mountain		gun position
Sound Island	7	signal Station Searchlight
Lisianski Peninusula	8	fire control station
South Kruzof		cable easement
St. Lazaria Island	4	fire control station
Clam and Abalone Islands	10	fire control station
Kayak Island	12	fire control station
Povorotni Point		cable landing
Kita Island	13	fire control station
Biorka & Little Biorka Islands	14	gun position fire control station
Ataku Island	15	fire control station
Golf Island	16	fire control station
Baranof to Old Sitka		cable easement
Watson Point		AMTB gun position
Whale Island		AMTB gun position

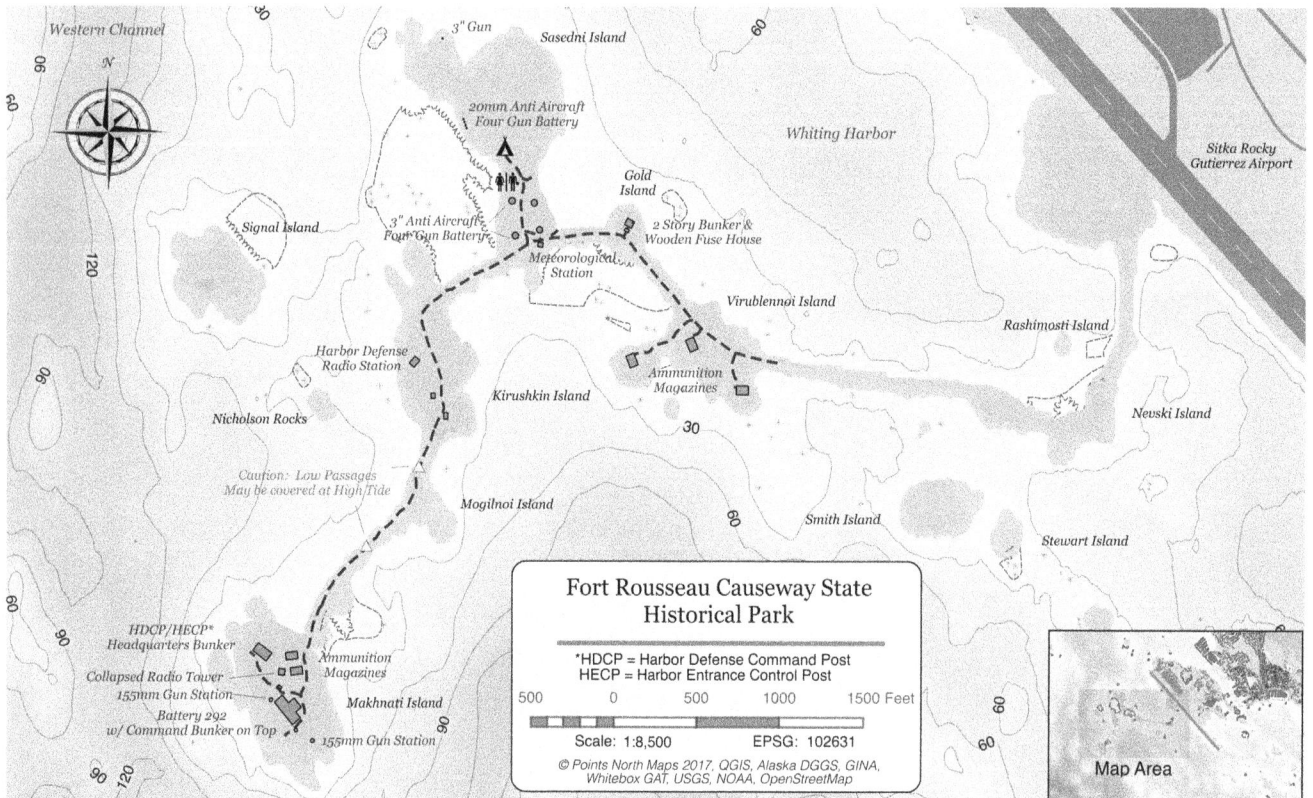

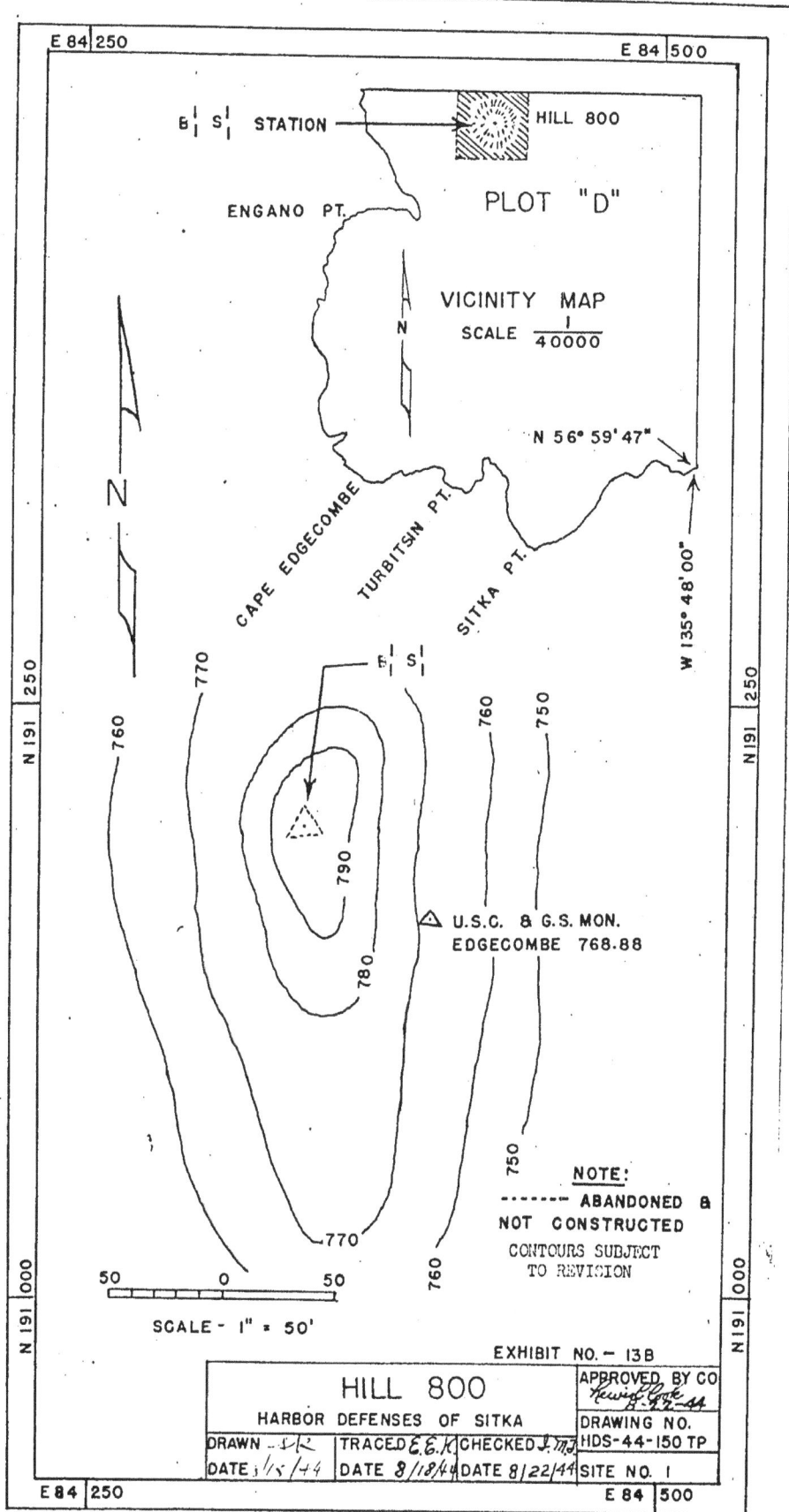

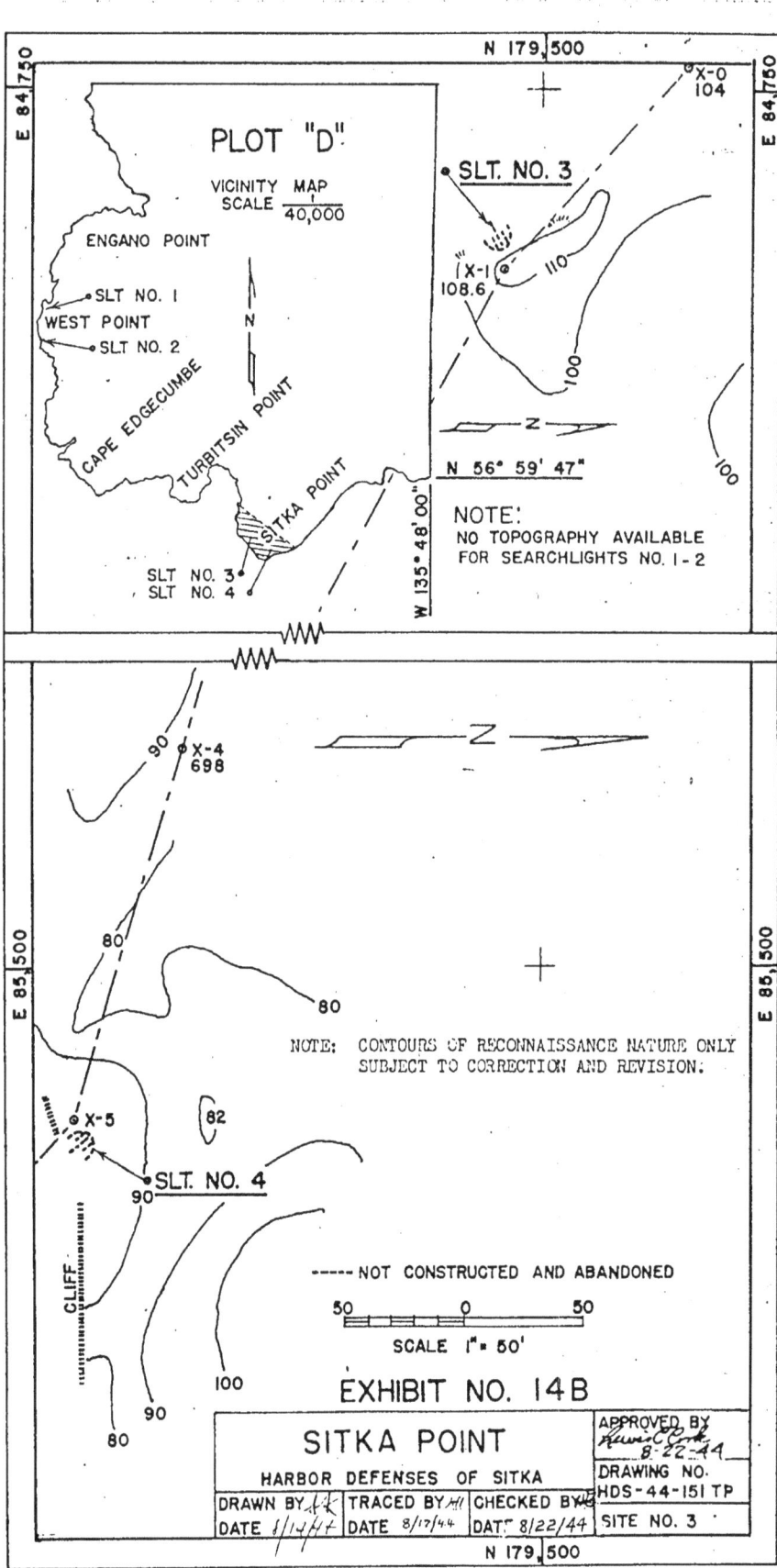

PLOT "D".

VICINITY MAP
SCALE $\frac{1}{40,000}$

ENGANO POINT

SLT NO. 1
WEST POINT
SLT NO. 2

CAPE EDGECUMBE

TURBITSIN POINT

SITKA POINT

SITKA POINT

SLT NO. 3
SLT NO. 4

N 179,500

X-0
104

SLT. NO. 3

X-1
108.6

110

100

100

N 56° 59' 47"

NOTE:
NO TOPOGRAPHY AVAILABLE
FOR SEARCHLIGHTS NO. 1-2

E 84,750

E 84,750

W 135° 48' 00"

90   X-4
698

80

80

NOTE: CONTOURS OF RECONNAISSANCE NATURE ONLY
SUBJECT TO CORRECTION AND REVISION.

82

X-5

SLT. NO. 4
90

CLIFF

100

90

80

E 85,500

E 85,500

----- NOT CONSTRUCTED AND ABANDONED

50   0   50

SCALE 1" = 50'

EXHIBIT NO. 14B

SITKA POINT	APPROVED BY
	8-22-44
HARBOR DEFENSES OF SITKA	DRAWING NO.
	HDS-44-151 TP
DRAWN BY	TRACED BY	CHECKED BY	SITE NO. 3
DATE 8/14/44	DATE 8/17/44	DATE 8/22/44	

N 179,500

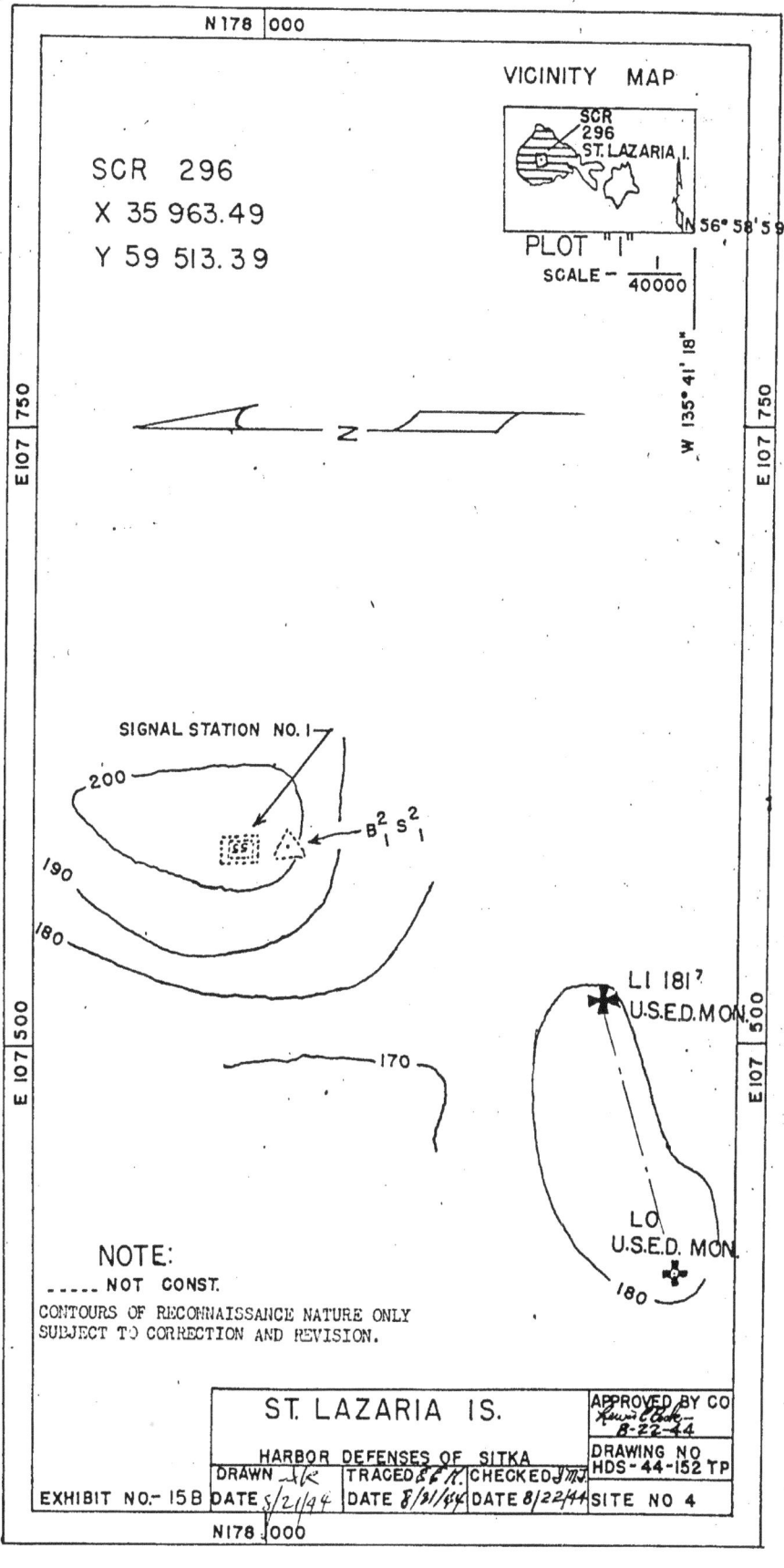

VICINITY MAP

SCR
296
ST. LAZARIA I.

PLOT "I"

SCALE $-\frac{1}{40000}$

N 56° 58' 59"

W 135° 41' 18"

N 178 000

SCR 296
X 35 963.49
Y 59 513.39

E 107 750

N

E 107 750

SIGNAL STATION NO. 1

200

$B_1^2 S_1^2$

190

180

LI 181^7
U.S.E.D. MON.

E 107 500

E 107 500

170

LO
U.S.E.D. MON.

180

NOTE:
----- NOT CONST.
CONTOURS OF RECONNAISSANCE NATURE ONLY
SUBJECT TO CORRECTION AND REVISION.

ST. LAZARIA IS.	APPROVED BY CO
	8-22-44
HARBOR DEFENSES OF SITKA	DRAWING NO
DRAWN / TRACED / CHECKED	HDS-44-152 TP
EXHIBIT NO.- 15B DATE 5/21/44 DATE 8/21/44 DATE 8/22/44	SITE NO 4

N 178 000

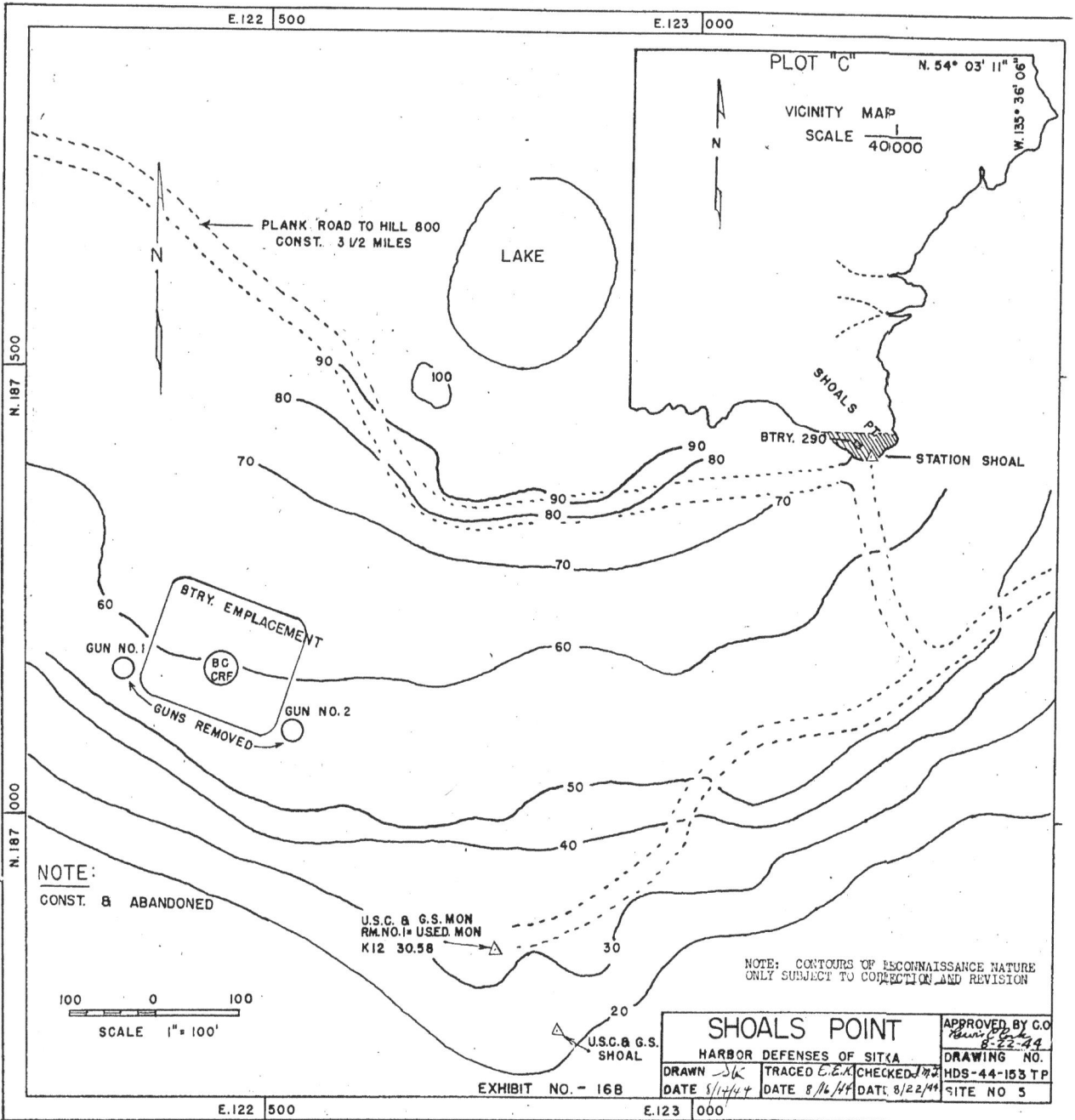

E.122 | 500   E.123 | 000

PLOT "C"   N. 54° 03' 11"

VICINITY MAP
SCALE 1/40,000

N

PLANK ROAD TO HILL 800
CONST. 3 1/2 MILES

LAKE

N

SHOALS PT.

BTRY. 290

STATION SHOAL

90   80
90
90   70
80
70

70

60

BTRY. EMPLACEMENT

60

GUN NO.1

BC
CRF

GUNS REMOVED

GUN NO. 2

50

40

NOTE:
CONST. & ABANDONED

U.S.C.& G.S. MON
RM.NO.I = USED. MON
K 12 30.58

30

NOTE: CONTOURS OF RECONNAISSANCE NATURE
ONLY SUBJECT TO CORRECTION AND REVISION

20

100   0   100

SCALE   1" = 100'

U.S.C.& G.S.
SHOAL

EXHIBIT NO. - 16 B

SHOALS POINT

HARBOR DEFENSES OF SITKA

| DRAWN | TRACED E.E.K. | CHECKED |
| DATE 5/14/44 | DATE 8/16/44 | DATE 8/22/44 |

APPROVED BY C.O.
8-22-44

DRAWING NO.
HDS-44-153 TP

SITE NO 5

E.122 | 500   E.123 | 000

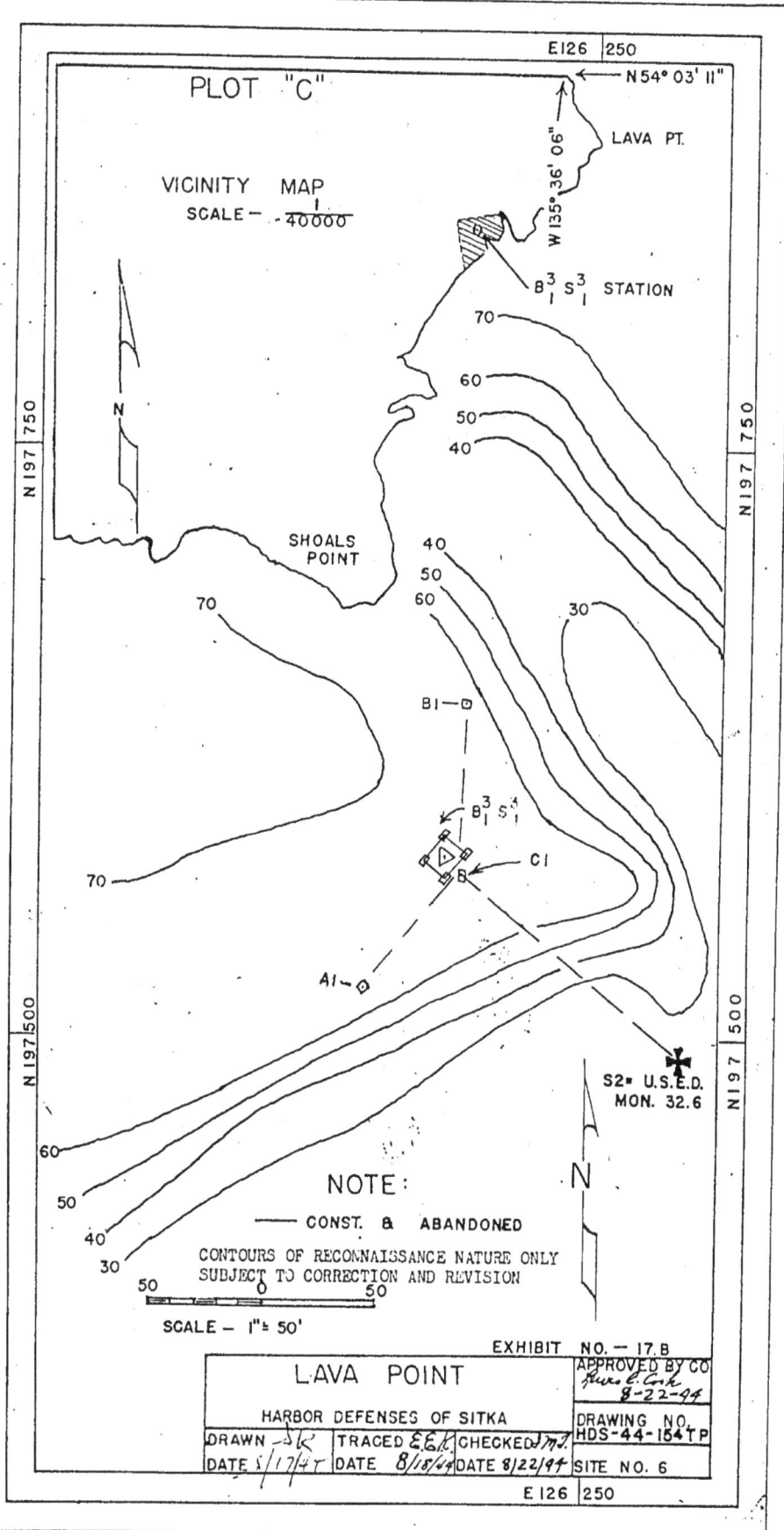

PLOT "C"

VICINITY MAP
SCALE — $\frac{1}{40000}$

N 54° 03' 11"
W 135° 36' 06"
LAVA PT.

$B_1^3 S_1^3$ STATION

70
60
50
40

SHOALS POINT

70
40
50
60
70
30

B1

$B_1^3 S_1^3$

C1
B

A1

S2= U.S.E.D.
MON. 32.6

N

NOTE:

— CONST. & ABANDONED

CONTOURS OF RECONNAISSANCE NATURE ONLY
SUBJECT TO CORRECTION AND REVISION

SCALE — I"≐ 50'

60
50
40
30
50

EXHIBIT NO. — 17.B

LAVA POINT	APPROVED BY CO
HARBOR DEFENSES OF SITKA	8-22-44
DRAWN / TRACED E.E.K. CHECKED	DRAWING NO. HDS-44-154TP
DATE 5/17/44 DATE 8/18/44 DATE 8/22/44	SITE NO. 6

E126 250

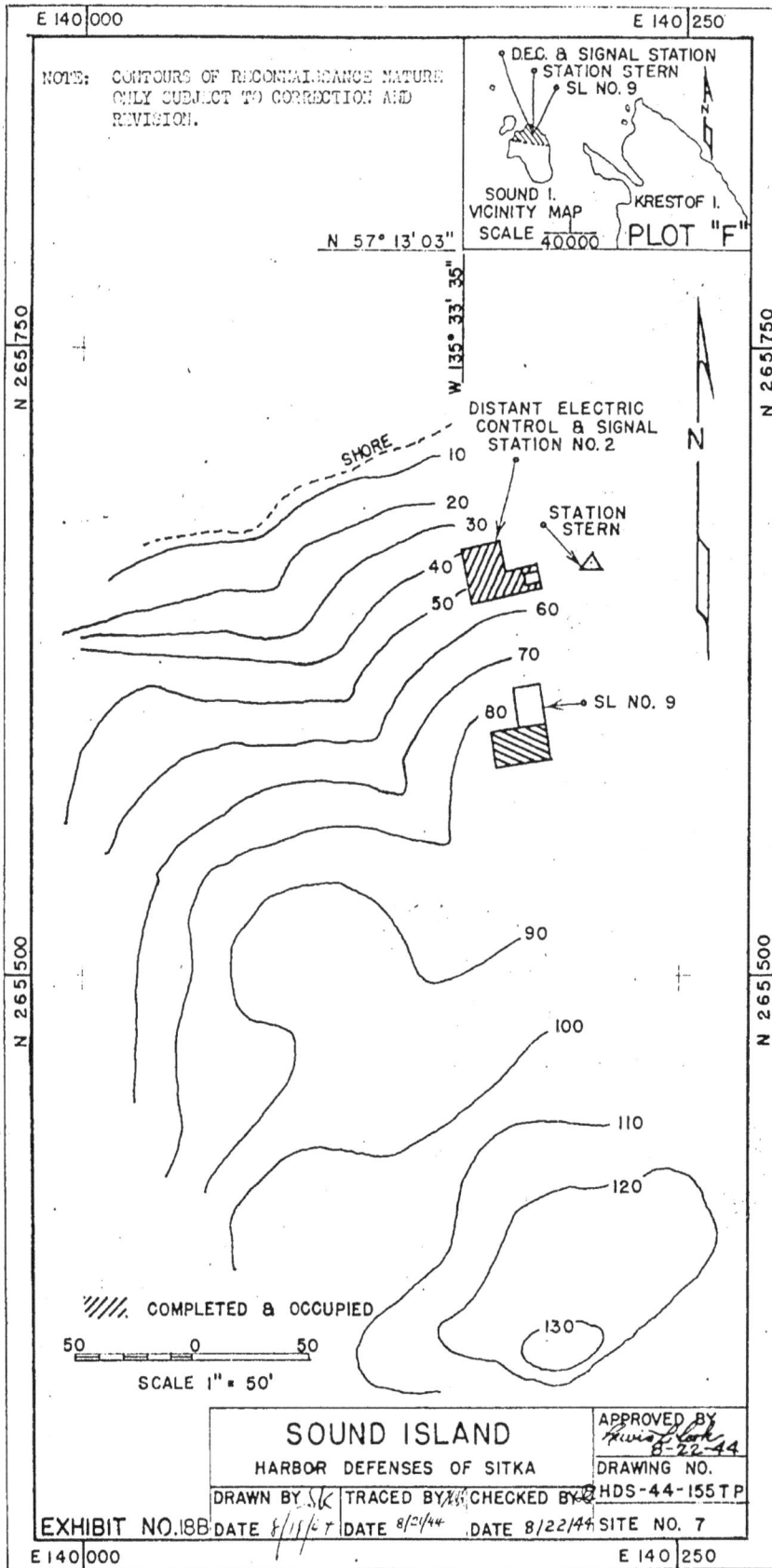

NOTE: CONTOURS OF RECONNAISSANCE NATURE ONLY SUBJECT TO CORRECTION AND REVISION.

D.E.C. & SIGNAL STATION
STATION STERN
SL NO. 9

SOUND I.
VICINITY MAP
SCALE 40,000

KRESTOF I.

PLOT "F"

N 57° 13' 03"

W 135° 33' 35"

SHORE

DISTANT ELECTRIC
CONTROL & SIGNAL
STATION NO. 2

STATION STERN

N

SL NO. 9

COMPLETED & OCCUPIED

50   0   50

SCALE 1" = 50'

SOUND ISLAND

HARBOR DEFENSES OF SITKA

APPROVED BY
8-22-44

DRAWING NO.
HDS-44-155 T P

DRAWN BY SK | TRACED BY | CHECKED BY

EXHIBIT NO. 18B DATE 8/11/44 | DATE 8/21/44 | DATE 8/22/44 | SITE NO. 7

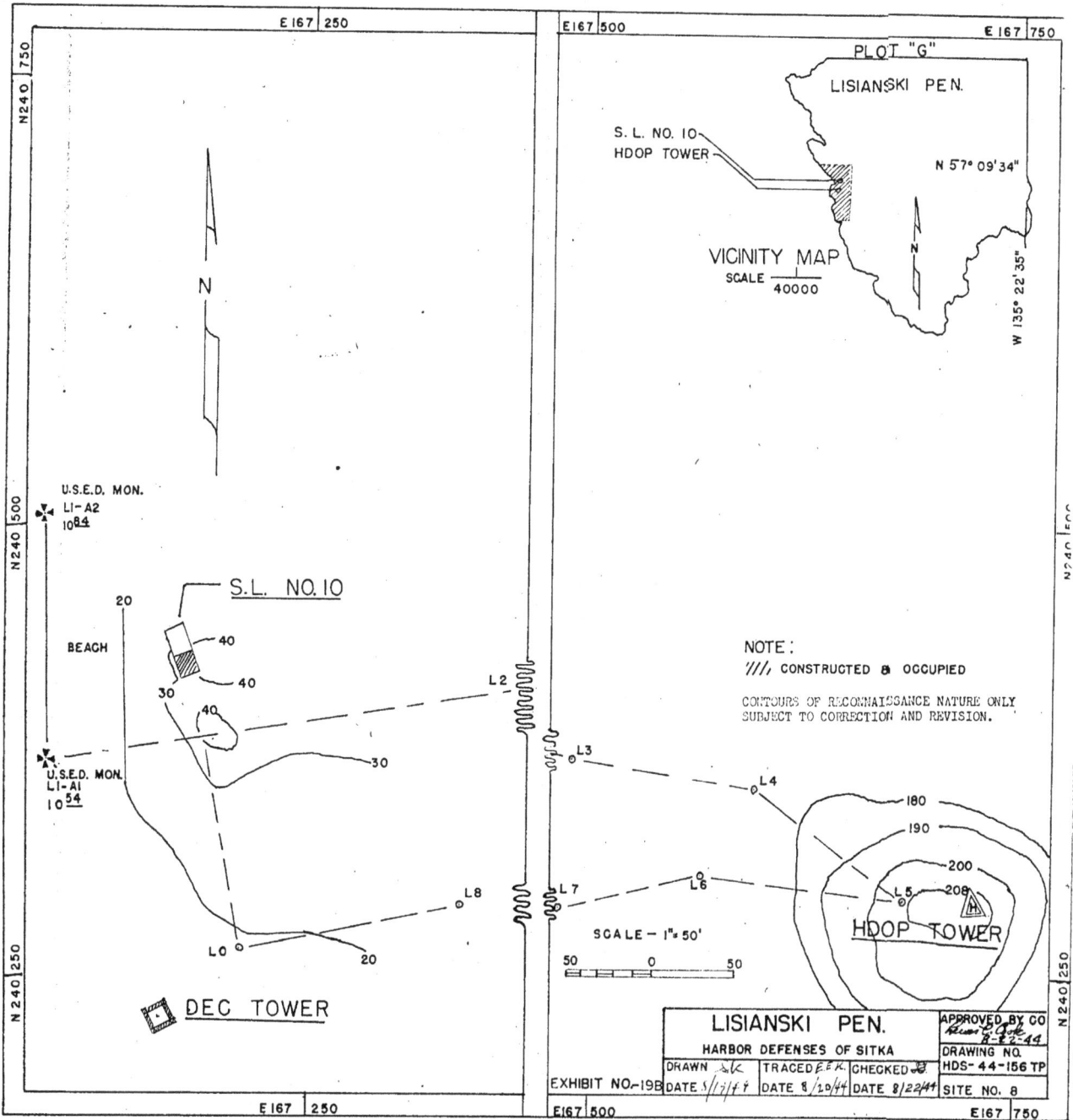

E 167 250

N 240 750

E 167 500

E 167 750

PLOT "G"

LISIANSKI PEN.

S. L. NO. 10
HDOP TOWER

N 57° 09' 34"

VICINITY MAP
SCALE —— 40000

W 135° 22' 35"

N

U.S.E.D. MON.
LI-A2
10⁸⁴

N 240 500

20

S.L. NO. 10

40

40

BEACH

30

40

30

L2

U.S.E.D. MON.
LI-AI
10⁵⁴

L3

L4

180

190

NOTE :
//// CONSTRUCTED & OCCUPIED

CONTOURS OF RECONNAISSANCE NATURE ONLY
SUBJECT TO CORRECTION AND REVISION.

200

L8

L7

L6

L5

208

HDOP TOWER

LO

20

SCALE - 1"= 50'

50   0   50

DEC TOWER

N 240 250

N 240 250

LISIANSKI PEN.

HARBOR DEFENSES OF SITKA

DRAWN	TRACED E.E.K.	CHECKED
EXHIBIT NO.-19B DATE 8/19/44	DATE 8/20/44	DATE 8/22/44

APPROVED BY CO
8-22-44

DRAWING NO.
HDS- 44-156 TP

SITE NO. 8

E 167 250

E 167 500

E 167 750

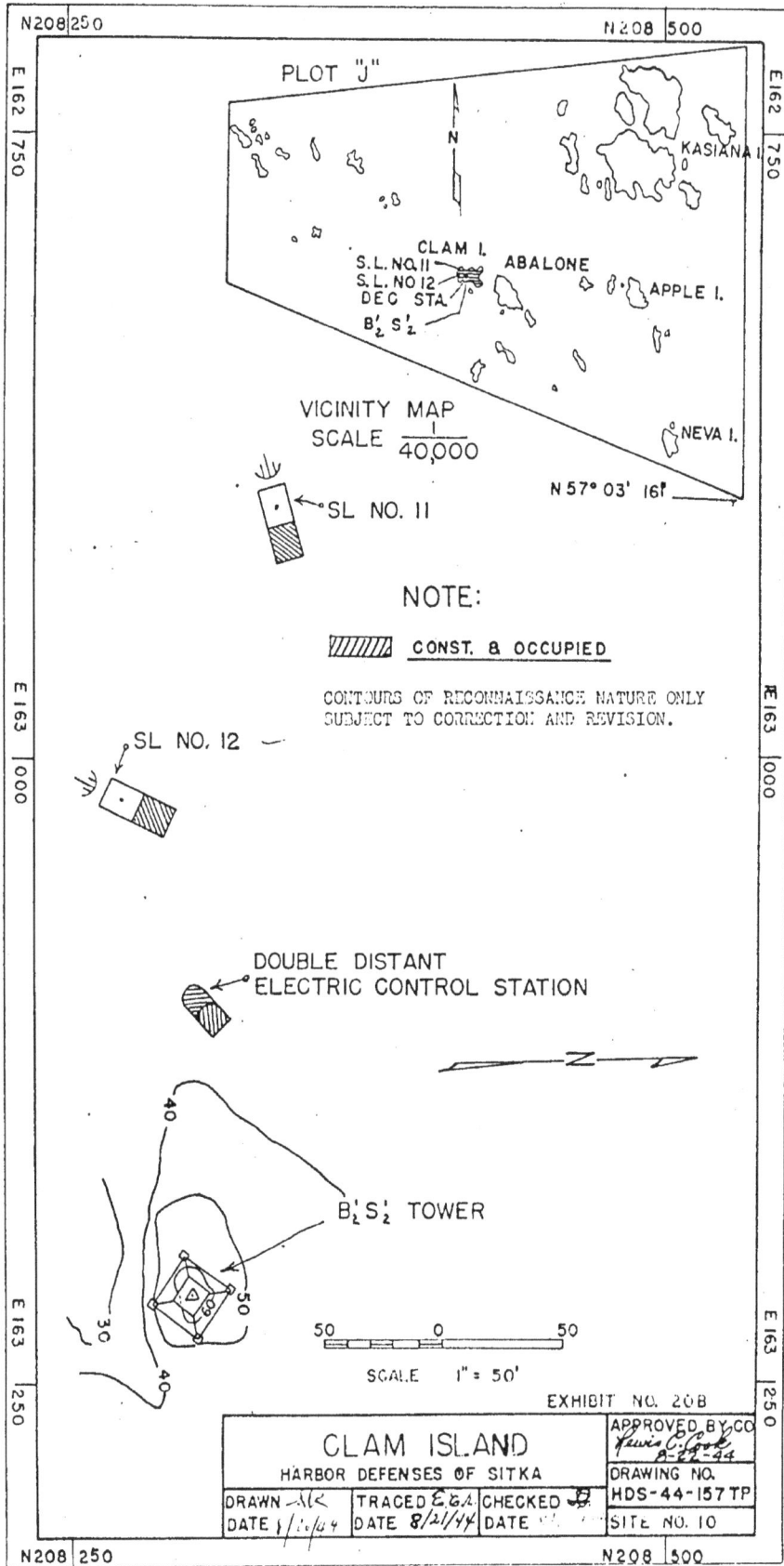

N208|250   N208|500

PLOT "J"

KASIANA I.

N

CLAM I.   ABALONE
S.L. NO. II
S.L. NO 12   APPLE I.
DEC STA
B½ S½

VICINITY MAP
SCALE $\frac{1}{40,000}$

N 57° 03' 16'

SL NO. 11

NOTE:

////// CONST. & OCCUPIED

CONTOURS OF RECONNAISSANCE NATURE ONLY
SUBJECT TO CORRECTION AND REVISION.

SL NO. 12

DOUBLE DISTANT
ELECTRIC CONTROL STATION

N

40

B½ S½ TOWER

30   50

50   0   50

40

SCALE 1" = 50'

EXHIBIT NO. 20B

CLAM ISLAND	APPROVED BY CO
HARBOR DEFENSES OF SITKA	*Lewis C. Cook* 8-22-44
DRAWN ~M<	TRACED *E.G.A*	CHECKED *J.*	DRAWING NO. HDS-44-157 TP
DATE 1/10/44	DATE 8/21/44	DATE	SITE NO. 10

N208|250   N208|500

NEVA I.

VICINITY MAP
SCALE $\frac{1}{40,000}$

KASIANA ISLAND

SCR 296 W

ABALONE I.

CLAM I.

APPLE I.

PLOT "J"

E 164,000

N 208,500

N 208,500

N 208,250

N 208,250

N 208,000

N 208,000

N 57° 03' 16"

W 135° 23' 47"

N

N

70

80

80

90

STATION ABALONE

SCR-296 W

96

70

80

70

50   0   50

SCALE 1" = 50'

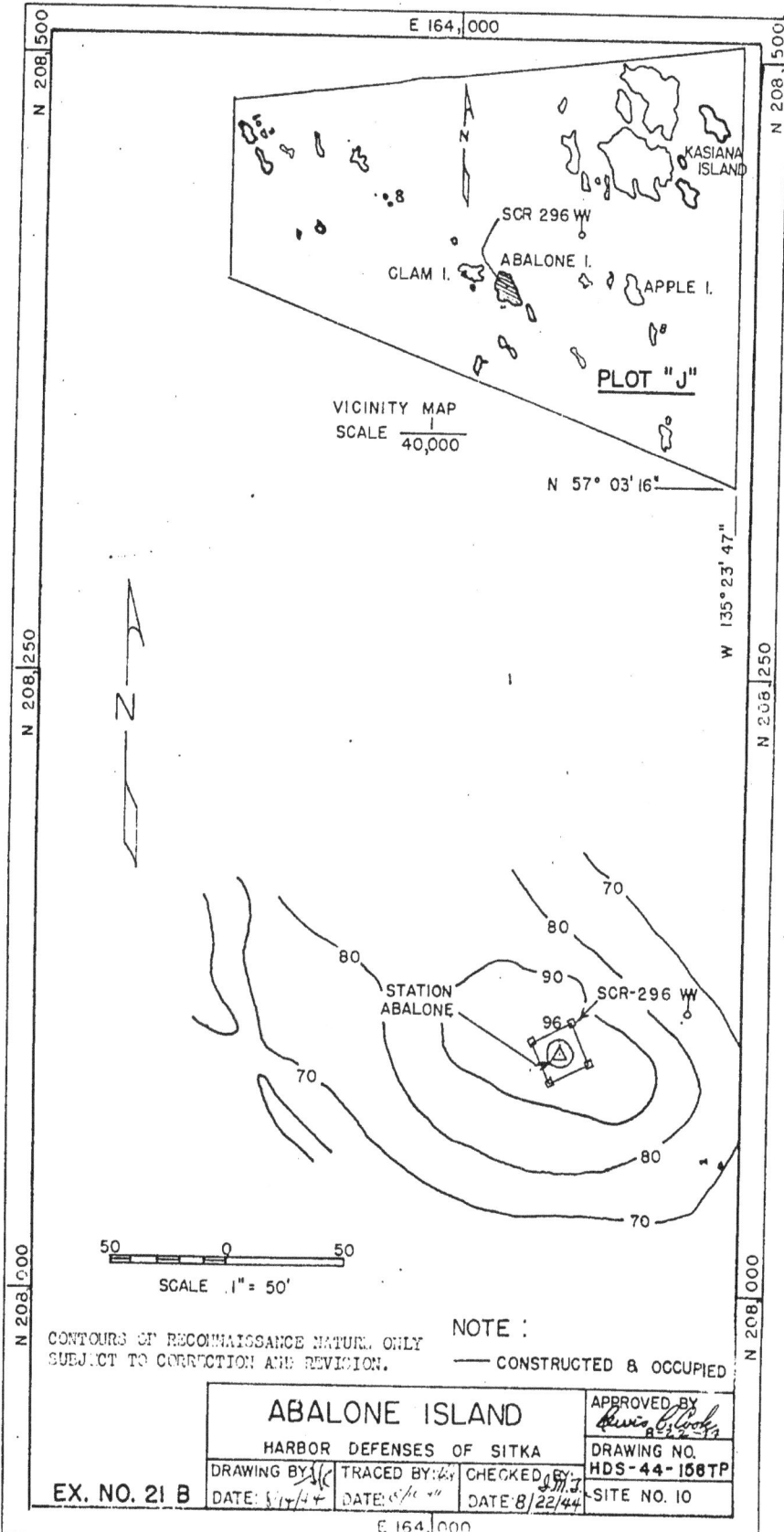

CONTOURS OF RECONNAISSANCE NATURE ONLY
SUBJECT TO CORRECTION AND REVISION.

NOTE :

———— CONSTRUCTED & OCCUPIED

ABALONE ISLAND	APPROVED BY
HARBOR DEFENSES OF SITKA	
DRAWING BY: TRACED BY: CHECKED BY:	DRAWING NO. HDS-44-158TP
DATE: DATE: DATE: 8/22/44	SITE NO. 10

EX. NO. 21 B

E 164,000

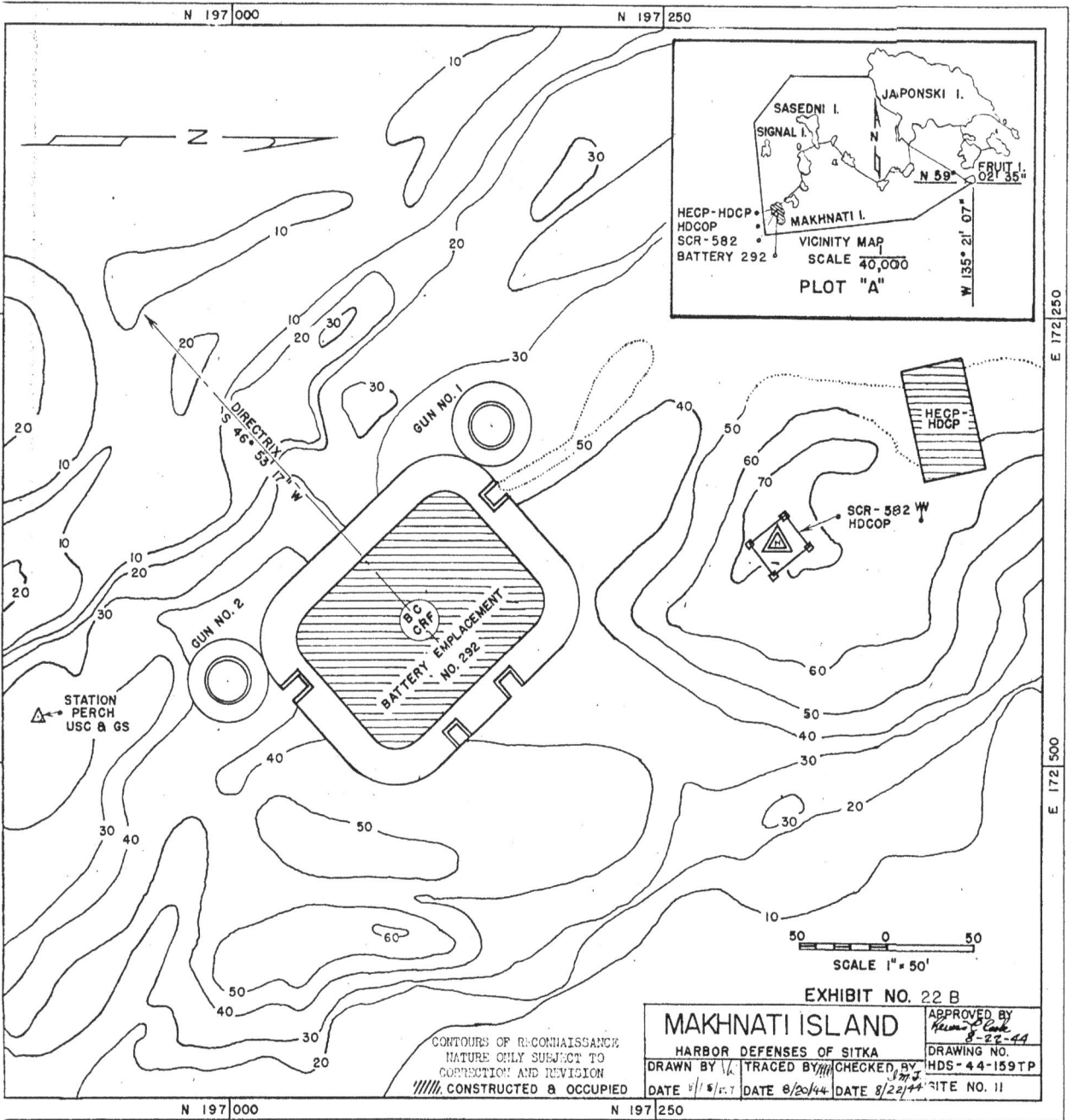

PLOT "A"

VICINITY MAP
SCALE 1/40,000

HECP-HDCP
HDCOP
SCR-582
BATTERY 292

MAKHNATI I.

SASEDNI I.
SIGNAL I.
JAPONSKI I.
FRUIT I.

N 59° 02' 35"
W 135° 21' 07"

GUN NO. 1
GUN NO. 2

DIRECTRIX S 46° 53' 17" W

B C
CRF

BATTERY EMPLACEMENT NO. 292

STATION
PERCH
USC & GS

HECP-HDCP

SCR-582
HDCOP

SCALE 1" = 50'

EXHIBIT NO. 22 B

CONTOURS OF RECONNAISSANCE
NATURE ONLY SUBJECT TO
CORRECTION AND REVISION
///// CONSTRUCTED & OCCUPIED

MAKHNATI ISLAND

HARBOR DEFENSES OF SITKA

DRAWN BY	TRACED BY	CHECKED BY	APPROVED BY
			DRAWING NO. HDS-44-159TP
DATE	DATE 8/20/44	DATE 8/22/44	SITE NO. 11

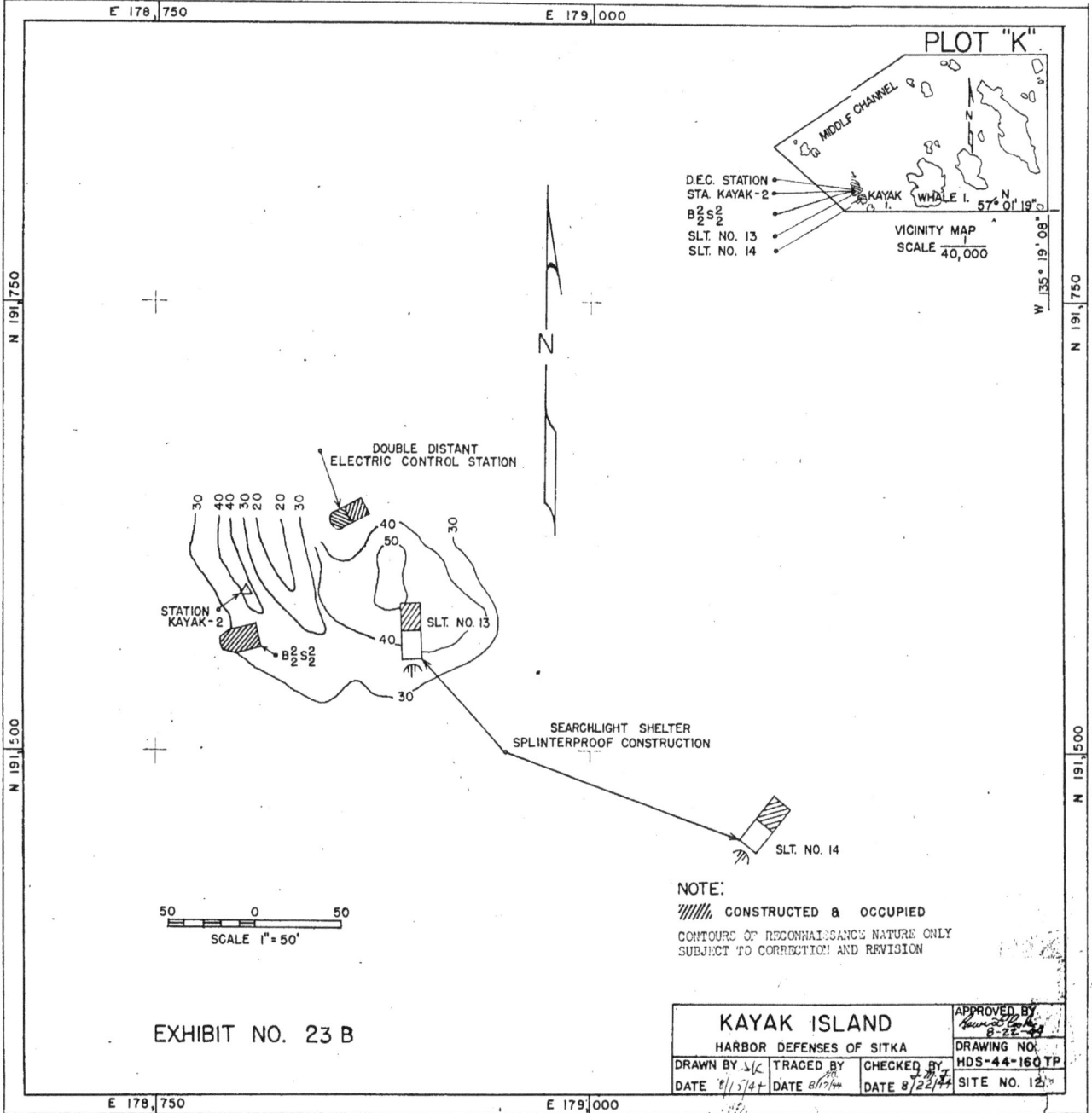

PLOT "K"

MIDDLE CHANNEL

D.E.C. STATION
STA. KAYAK-2
$B_2^2 S_2^2$
SLT. NO. 13
SLT. NO. 14

KAYAK WHALE I. 57° 01' 19"
I.

VICINITY MAP
SCALE $\frac{1}{40,000}$

DOUBLE DISTANT
ELECTRIC CONTROL STATION

STATION
KAYAK-2

$B_2^2 S_2^2$

SLT. NO. 13

SEARCHLIGHT SHELTER
SPLINTERPROOF CONSTRUCTION

SLT. NO. 14

NOTE:
////// CONSTRUCTED & OCCUPIED
CONTOURS OF RECONNAISSANCE NATURE ONLY
SUBJECT TO CORRECTION AND REVISION

SCALE 1"= 50'

EXHIBIT NO. 23 B

KAYAK ISLAND
HARBOR DEFENSES OF SITKA

DRAWN BY | TRACED BY | CHECKED BY
DATE | DATE | DATE

APPROVED BY
DRAWING NO. HDS-44-160 TP
SITE NO. 12

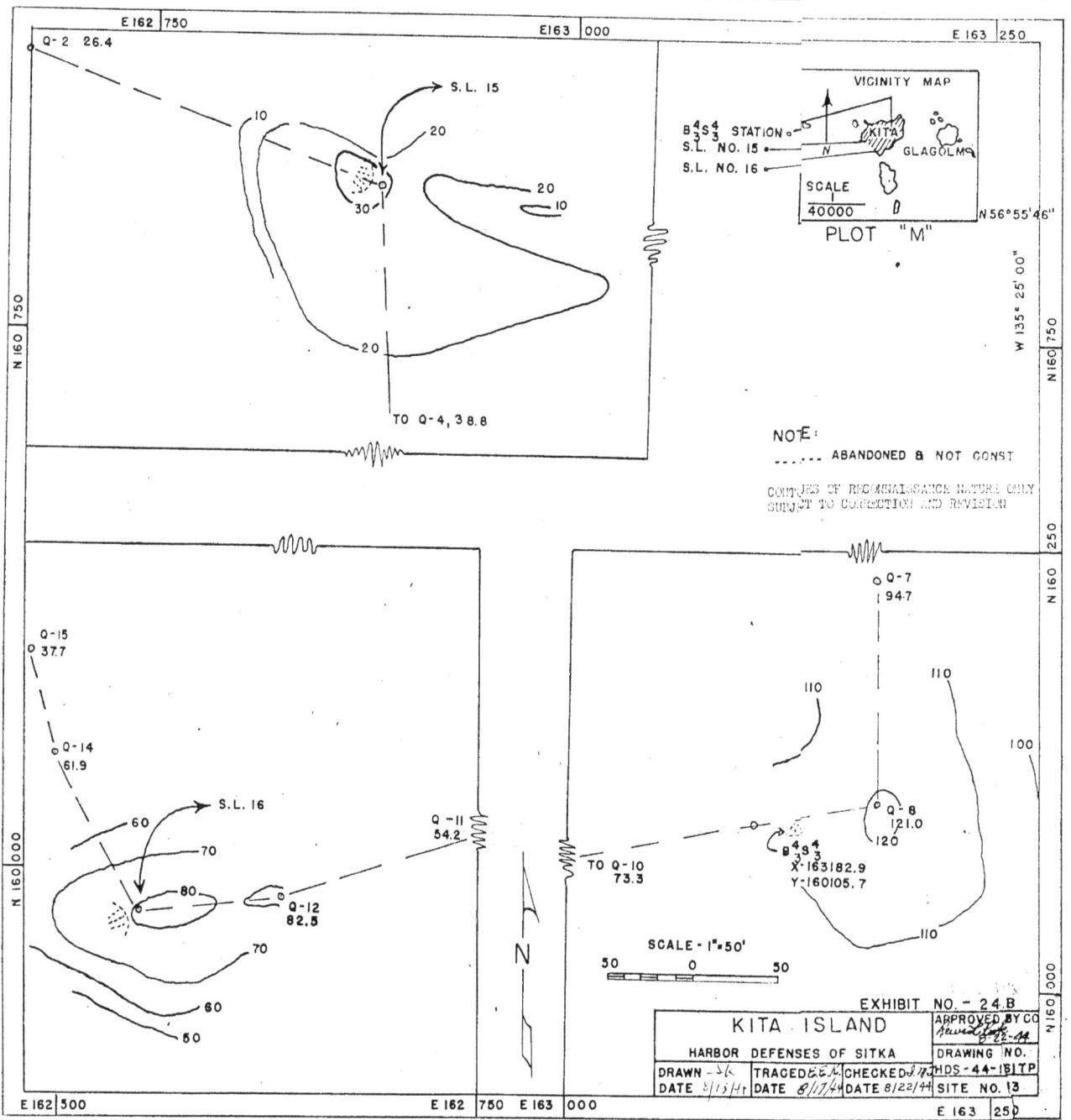

E 162 | 750

E 163 | 000

E 163 | 250

Q-2 26.4

S.L. 15

10

20

20

20

30

10

N 160 | 750

20

TO Q-4, 38.8

VICINITY MAP

B₃⁴S₃⁴ STATION
S.L. NO. 15
S.L. NO. 16

KITA

GLAGOLM

SCALE
40000

PLOT "M"

N 56° 55' 46"

W 135° 25' 00"

N 160 | 750

NOTE:
...... ABANDONED & NOT CONST

CONTOURS OF RECONNAISSANCE NATURE ONLY
SUBJECT TO CORRECTION AND REVISION

250

Q-7
94.7

Q-15
37.7

110

110

100

Q-14
61.9

S.L. 16

60

70

Q-11
54.2

Q-8
121.0
120

B₃⁴S₃⁴
X-163182.9
Y-160105.7

N 160 | 000

80

Q-12
82.5

70

TO Q-10
73.3

110

N

60

110

50

SCALE - 1"=50'

50   0   50

N 160 | 000

EXHIBIT NO. - 24.B

KITA ISLAND

APPROVED BY CO
8-22-44

HARBOR DEFENSES OF SITKA

DRAWING NO.

DRAWN -
DATE 8/15/44

TRACED
DATE 8/17/44

CHECKED
DATE 8/22/44

HDS-44-151TP
SITE NO. 13

E 162 | 500

E 162 | 750

E 163 | 000

E 163 | 250

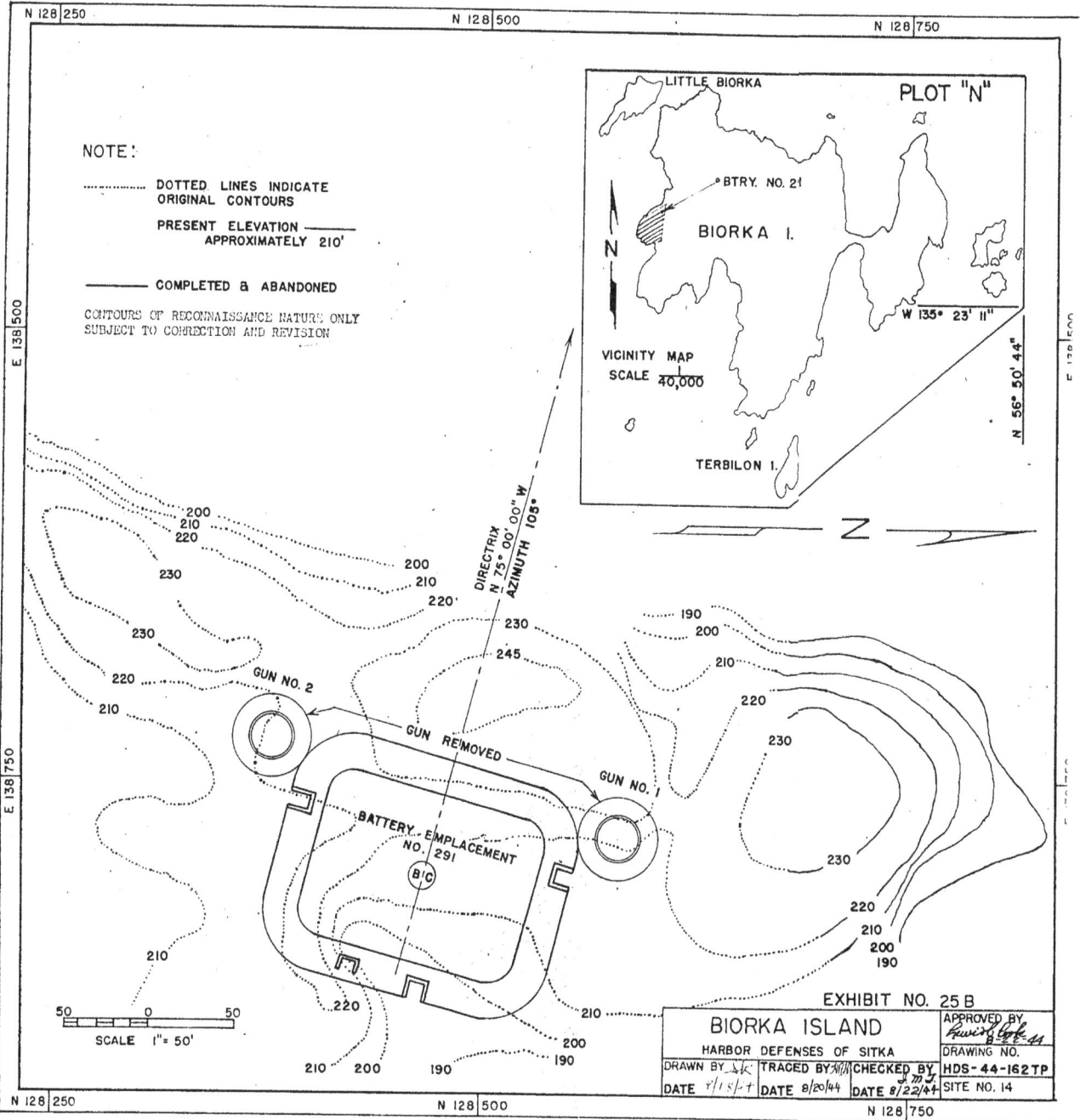

NOTE:

.... DOTTED LINES INDICATE ORIGINAL CONTOURS

PRESENT ELEVATION ——— APPROXIMATELY 210'

——— COMPLETED & ABANDONED

CONTOURS OF RECONNAISSANCE NATURE ONLY SUBJECT TO CORRECTION AND REVISION

PLOT "N"

LITTLE BIORKA

BTRY. NO. 21

BIORKA I.

VICINITY MAP SCALE 40,000

TERBILON I.

W 135° 23' 11"

N 56° 50' 44"

DIRECTRIX N 75° 00' 00" W AZIMUTH 105°

GUN NO. 2

GUN REMOVED

GUN NO. 1

BATTERY EMPLACEMENT NO. 291

B.C.

SCALE 1"= 50'

EXHIBIT NO. 25 B

BIORKA ISLAND

HARBOR DEFENSES OF SITKA

DRAWN BY   TRACED BY   CHECKED BY   DRAWING NO. HDS-44-162TP

DATE   DATE 8/20/44   DATE 8/22/44   SITE NO. 14

APPROVED BY

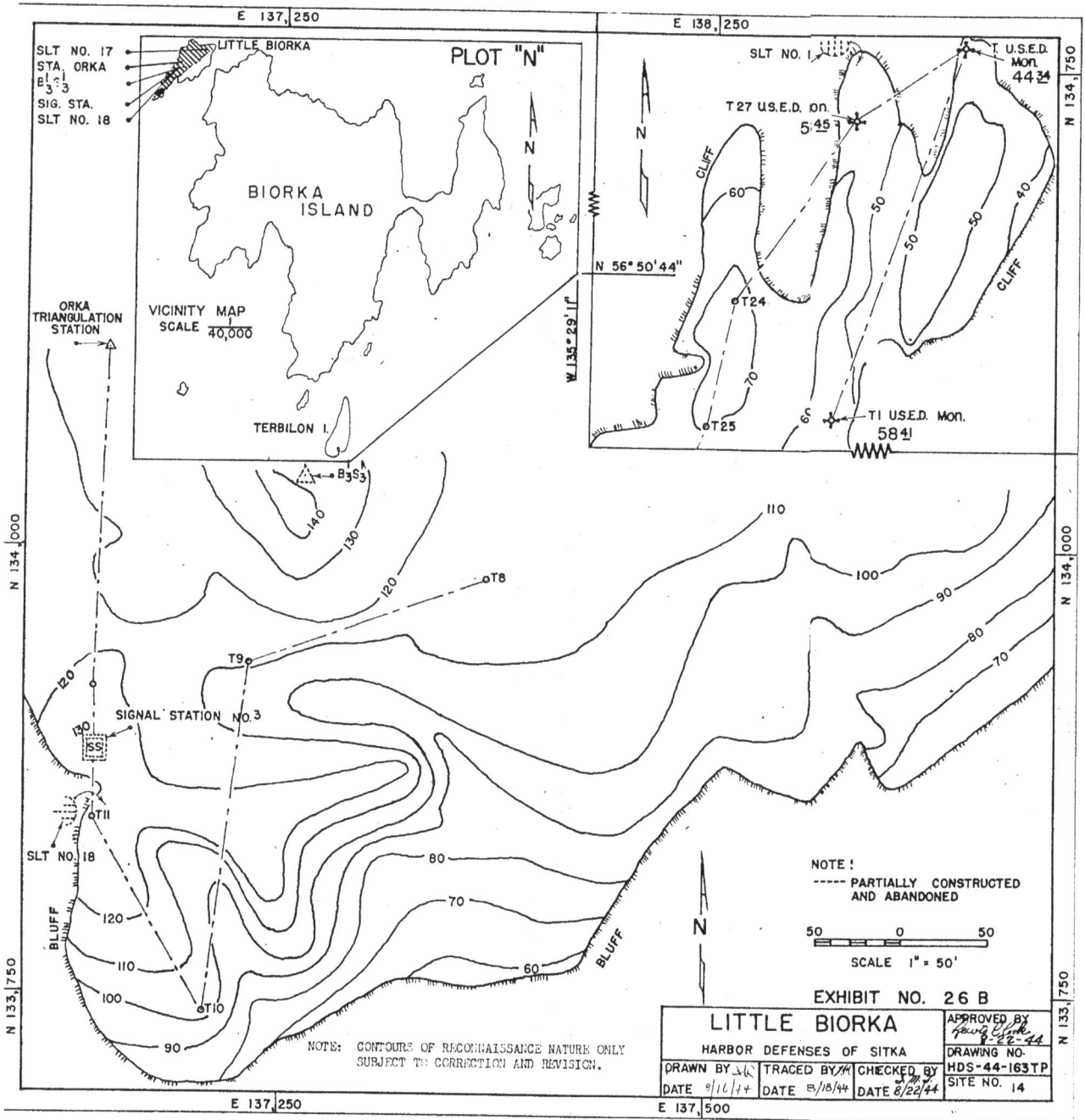

SLT NO. 17
STA. ORKA
B₃S₃
SIG. STA.
SLT NO. 18

LITTLE BIORKA

PLOT "N"

N

BIORKA
ISLAND

VICINITY MAP
SCALE 1/40,000

TERBILON I.

ORKA
TRIANGULATION
STATION

B₃S₃

SLT NO. I

N

T 27 U.S.E.D. Mon.
5.45

CLIFF

60

50

50

40

N 56° 50' 44"
W 135° 29' 11"

T24

T25

70

60

CLIFF

T1 U.S.E.D. Mon.
584!

T. U.S.E.D.
Mon.
44.34

140

130

110

120

T8

100

90

80

70

T9

120

SIGNAL STATION NO. 3

130

SS

T11

SLT NO. 18

BLUFF

120

110

100

90

T10

80

70

60

BLUFF

N

NOTE:
----- PARTIALLY CONSTRUCTED
AND ABANDONED

50   0   50

SCALE 1" = 50'

NOTE: CONTOURS OF RECONNAISSANCE NATURE ONLY
SUBJECT TO CORRECTION AND REVISION.

EXHIBIT NO. 26 B

LITTLE BIORKA

HARBOR DEFENSES OF SITKA

| DRAWN BY | TRACED BY | CHECKED BY |
| DATE 9/16/44 | DATE 8/18/44 | DATE 8/22/44 |

APPROVED BY
8-22-44
DRAWING NO.
HDS-44-163TP
SITE NO. 14

E 137,250   E 138,250   E 137,500

N 134,750
N 134,000
N 133,750

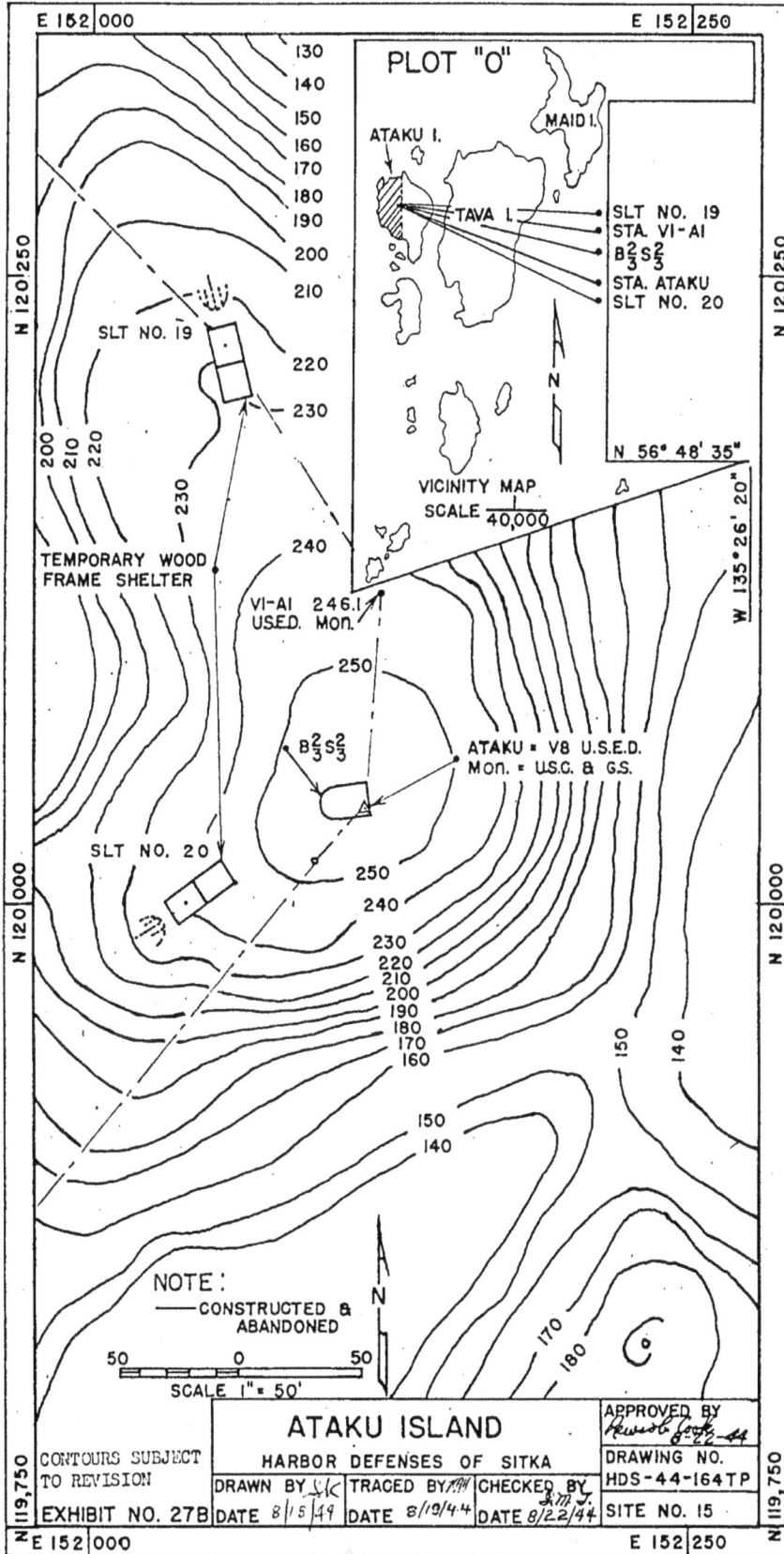

E 152 000   E 152 250

PLOT "O"

ATAKU I.

MAID I.

TAVA I.

SLT NO. 19
STA. VI-AI
$B\frac{2}{3}S\frac{2}{3}$
STA. ATAKU
SLT NO. 20

N

VICINITY MAP
SCALE $\frac{1}{40,000}$

N 56° 48' 35"

W 135° 26' 20"

N 120 250

130
140
150
160
170
180
190
200
210
220
230

SLT NO. 19

200
210
220
230

TEMPORARY WOOD
FRAME SHELTER

240

VI-AI 246.1
U.S.E.D. Mon.

250

$B\frac{2}{3}S\frac{2}{3}$

ATAKU = V8 U.S.E.D.
Mon. = U.S.C. & G.S.

SLT NO. 20

250
240
230
220
210
200
190
180
170
160

150
140

150

140

N 120 000

N 120 000

NOTE:
——— CONSTRUCTED &
ABANDONED

N

170
180

50   0   50
SCALE 1" = 50'

N 119,750

CONTOURS SUBJECT
TO REVISION

EXHIBIT NO. 27B

ATAKU ISLAND
HARBOR DEFENSES OF SITKA

DRAWN BY   TRACED BY   CHECKED BY
DATE 8/15/44   DATE 8/19/44   DATE 8/22/44

APPROVED BY

DRAWING NO.
HDS-44-164TP

SITE NO. 15

N 119,750

E 152 000   E 152 250

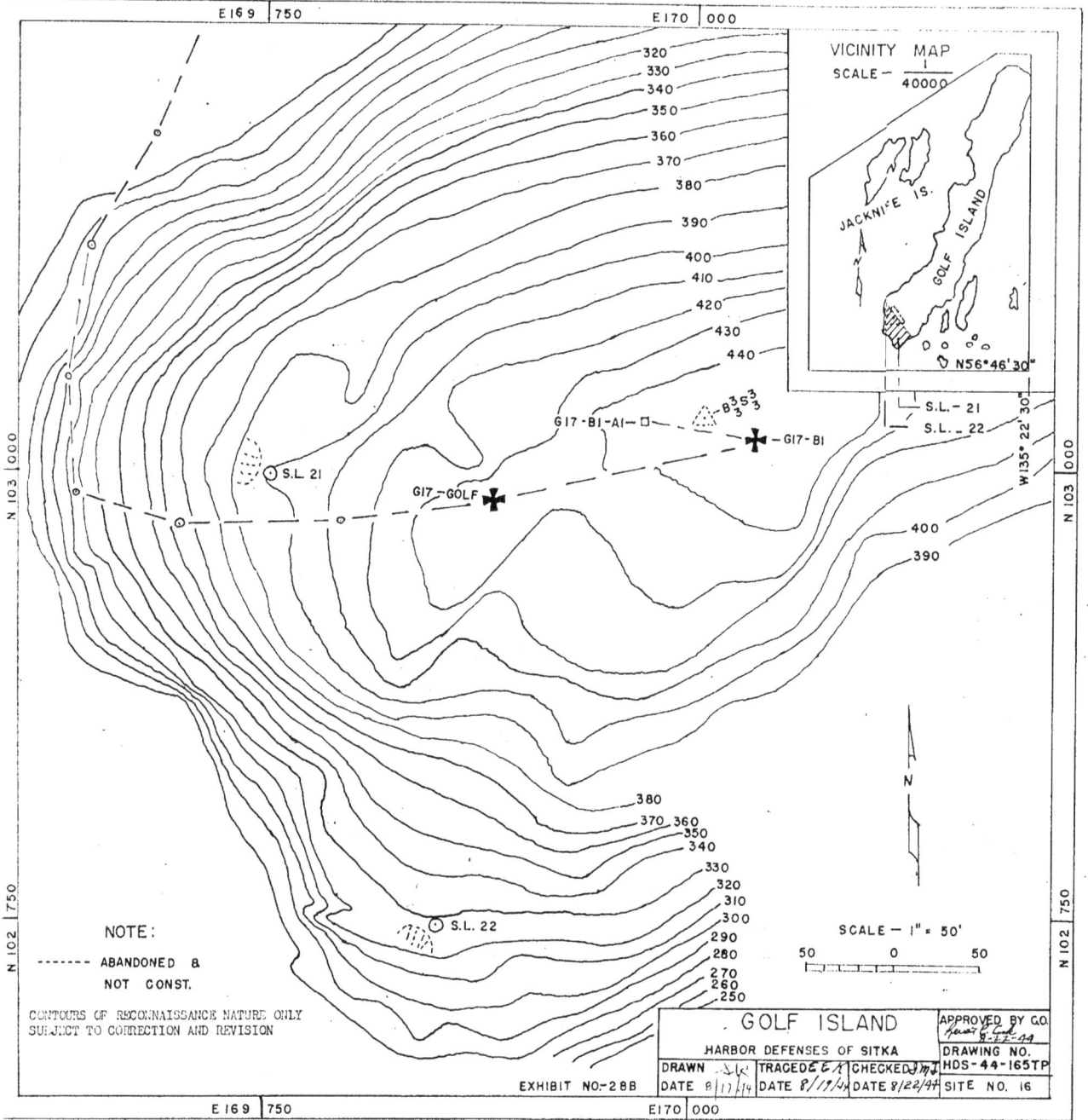

VICINITY MAP
SCALE - 1/40000

JACKNIFE IS.

GOLF ISLAND

N 56° 46' 30"

320
330
340
350
360
370
380
390
400
410
420
430
440

B 353 3

G17-B1-A1- □

G17-B1

S.L.-21
S.L.-22

400
390

G17-GOLF

S.L. 21

380
370 360
350
340
330
320
310
300
290
280
270
260
250

S.L. 22

N
N

SCALE - 1" = 50'
50   0   50

NOTE:
------ ABANDONED &
        NOT CONST.

CONTOURS OF RECONNAISSANCE NATURE ONLY
SUBJECT TO CORRECTION AND REVISION

GOLF ISLAND
HARBOR DEFENSES OF SITKA

APPROVED BY C.O.
8-12-44
DRAWING NO.
HDS-44-165TP

DRAWN | TRACED E.A | CHECKED JMJ
DATE 8/17/44 | DATE 8/19/44 | DATE 8/22/44 | SITE NO. 16

EXHIBIT NO-28B

E 169 750   E170 000

E 169 750   E170 000

N 103 000   N 103 000

N 102 750   N 102 750

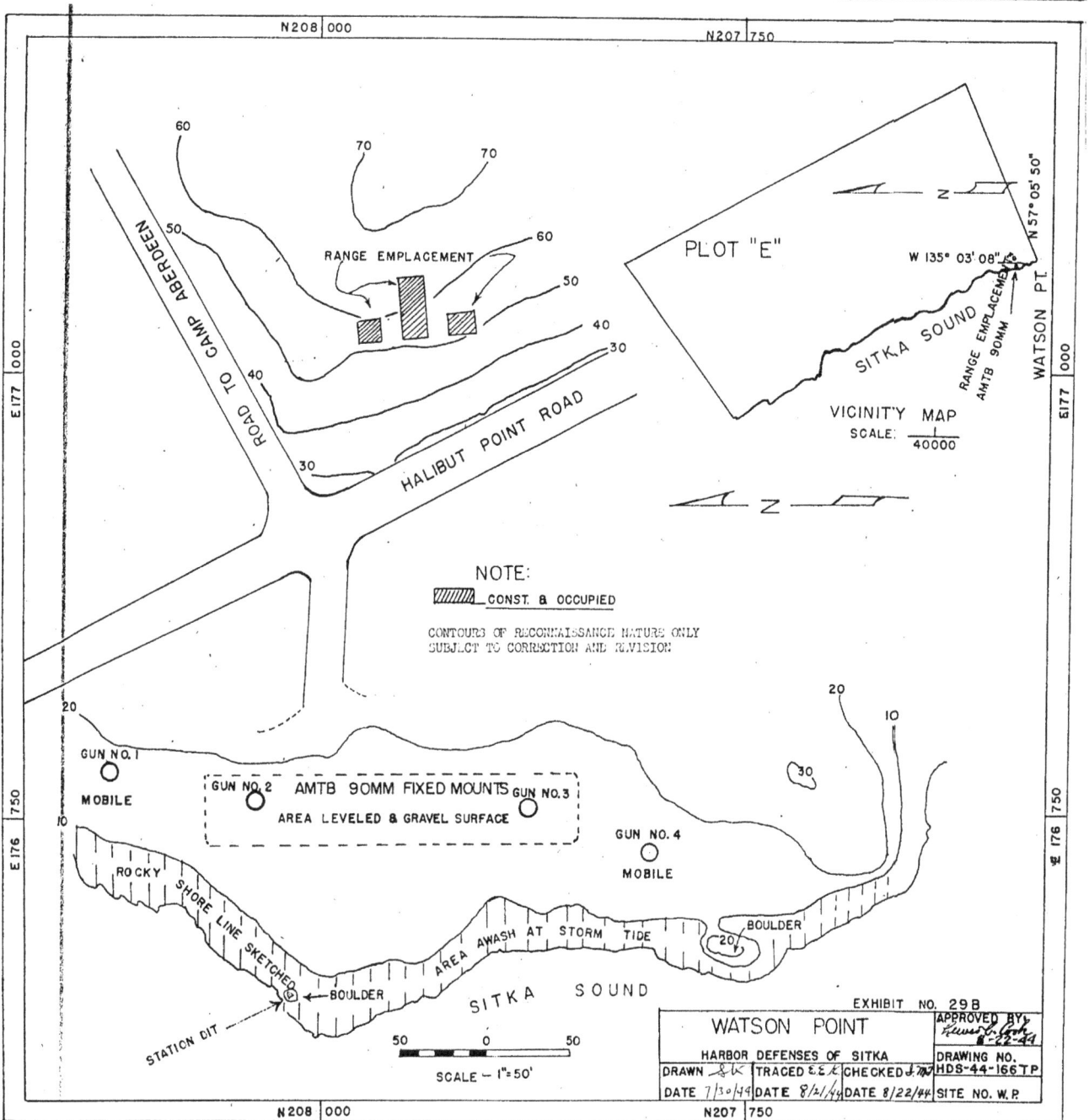

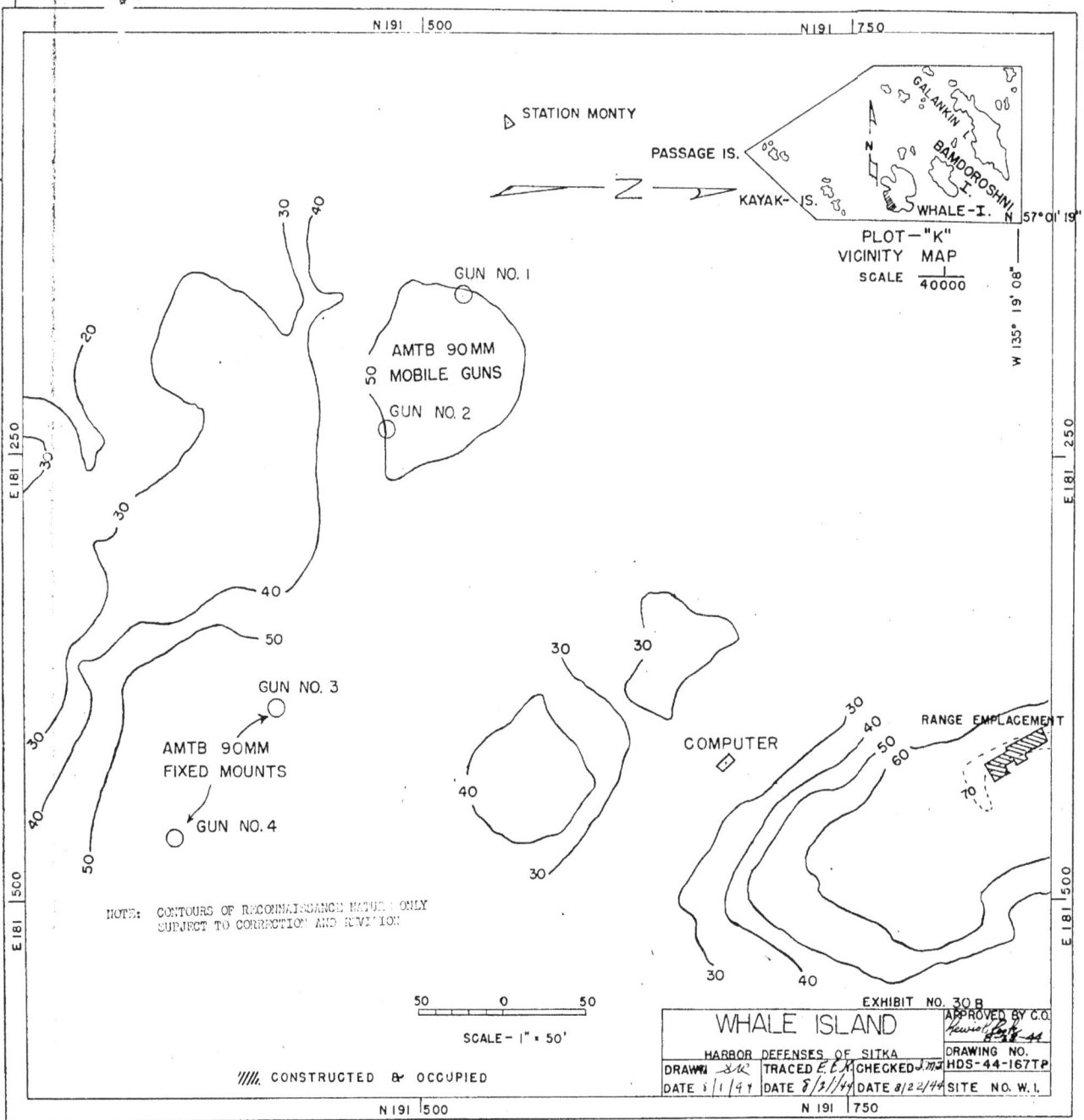

N 191 | 500   N 191 | 750

△ STATION MONTY

PLOT—"K"
VICINITY MAP

SCALE 1/40000

PASSAGE IS.

KAYAK- IS.

GALANKIN I.

BAMDOROSHNI I.

WHALE-I.   N 57° 01' 19"

W 135° 19' 08"

30   40

GUN NO. I

AMTB 90 MM
MOBILE GUNS

50

GUN NO. 2

20

30

E 181 | 250

30

30

E 181 | 250

30

40

50

GUN NO. 3

AMTB 90 MM
FIXED MOUNTS

30

40

GUN NO. 4

50

30   30

COMPUTER ◇

30
40
50
60

RANGE EMPLACEMENT

70

E 181 | 500

40

30

NOTE: CONTOURS OF RECONNAISSANCE NATURE ONLY
       SUBJECT TO CORRECTION AND REVISION

E 181 | 500

30   40

50   0   50

SCALE - 1" = 50'

EXHIBIT NO. 30 B

WHALE ISLAND

HARBOR DEFENSES OF SITKA

| DRAWN | TRACED *E.E.M.* | CHECKED *J.M.J.* |
| DATE 8/1/44 | DATE 8/21/44 | DATE 8/22/44 |

APPROVED BY C.O.

DRAWING NO.
HDS-44-167TP

SITE NO. W.I.

////. CONSTRUCTED & OCCUPIED

N 191 | 500   N 191 | 750

THE HARBOR DEFENSES OF SEWARD – ALASKA

The Seward defenses were the simplest of all built in Alaska: only two 6-inch batteries. Yet because of the rough terrain and bad weather conditions they were the among the most challenging engineering projects in Alaska and the construction experienced several delays. These military reservations were established as part of the major effort to defend the Port of Seward and railhead for the Alaska Railroad. BCN 293 was located on Caines Head designated Fort McGilvray and BCN 294 was located on the aptly named Rugged Island designated Fort Bulkley. The Rugged Island project was extremely difficult. The battery was built on the top of a granite precipice named Patsy Point and everything had to be barged out to the island including all construction equipment and cement. Most of the work was done from the barges which made it impossible to work during periods of rough weather. Fire control and searchlight facilities were located on Rugged Island, Rocky Point, Topeka Pt., Chamberlin Pt., Barwell Is., Alma Pt. and Carol Cove. Radar facilities were located at Caines Head and on Patsy Point on Rugged Island. One AMTB battery was built at Lowell Point near Seward. Fort Raymond was established as the headquarters garrison post of the harbor defense of Seward. Construction was halted on Seward's defenses after the ice-free port of Whittier became operational in 1944.

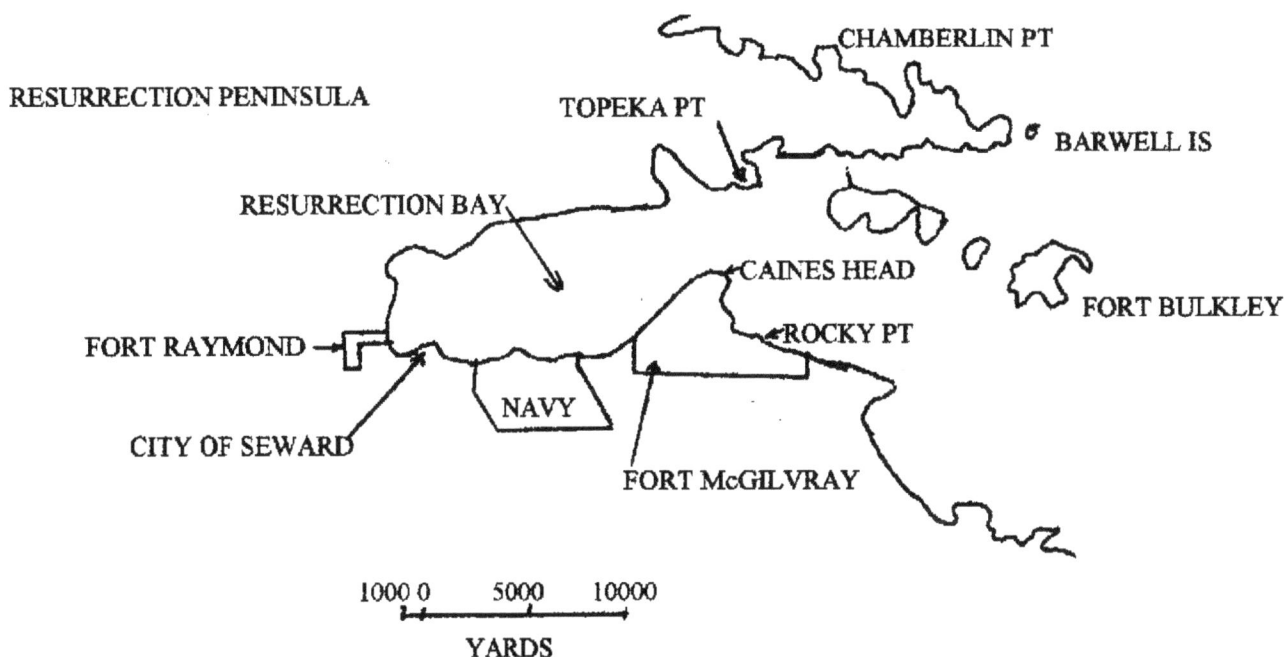

Fort McGilvray (1942-1944) is located at Caines Head which is a 650-foot high headland that rises above Resurrection Bay at the entrance to the port of Seward. Land of 4650 acres was acquired on August 19, 1941 for the Rocky Point reservation (just south of Caines Head) for construction of an emergency 155mm gun battery and an adjacent cantonment area at South Beach. Final approval to site a standard program shielded 6-inch battery at Caines Head was made on May, 1942, after a decision to re-locate the site from a much more distant Aialik Cape site to Caines Head. Work on the battery did not begin until early 1943. The post was named in General Order 17 of April 2, 1943 for Capt. John McGilvray. After service in the Civil War, McGilvray on assignment in Alaska established Fort Kenay at Cook's Inlet in 1869. Between the fort and Seward, a 90mm AMTB battery was emplaced on the beach at Lowell Point. Several fire control searchlights stations were also constructed for this fort. Work on completing the fortifications was suspended on May 3, 1944. After World War II, Fort McGilvray became transferred to the Territory of Alaska. Today it is the Caines Head State Recreation Area.

Fort McGilvray Gun Batteries

- **Battery #293:** A standard World War II dual 6-inch barbette battery emplaced at the highest level of the new Caine's Head reservation of Fort McGilvray. It had a trunnion elevation of 635-feet. The fortification was designed to provide coverage of the Ocean Terminal of the Alaska Railway. It was of standard 200-series design, with conventional magazine and gun block layout. The battery commander's station sits atop to central traverse but is not connected to the inside. Approval for construction was given on February 10, 1942. Final site selection was made on July 18, 1942. Basic concrete work was done from December 1, 1942 to December 3, 1943. It was provisionally turned over on April 11, 1944 at a status of 90% completed. It was armed with two 6-inch guns Model M1903A2 on M1 barbette carriages (guns #9 and #67). These guns were installed in early 1944 but were soon removed after authorization for dismantlement. One gun tube (#67) was shipped to San Diego for use there, the other material went to the Black Hills Ordnance Depot. The emplacement still exists at the Caine's Head State Recreation Area. The battery is open to the public.

- **AMTB #4:** An AMTB battery consisting of two 90mm fixed, two 90mm mobile, and two 37mm mobile guns emplaced at Lowell Point on Resurrection Bay as a sub post of Fort McGilvray. Two concrete gun blocks were built in mid-1943. It was turned over on November 21, 1943, the appropriate RCW not providing a cost. Guns No. 1 and No. 2 were for the fixed mounts, they had a 120-foot gun spacing and a trunnion elevation of just 13-feet. They fired to the east. The battery was abandoned on May 3, 1944, guns were probably removed shortly thereafter. The site was later sold to private interests, and parts of one gun block still exist at the commercial property of Miller's landing. The battery is not open to the public.

Fort Bulkley (1942-1944) is located at Rugged Island, which rises 700-feet above Resurrection Bay near its southern entrance. The aptly-named island is precipitous, broken granite feature with a command to the entrance of the Bay. When the Alaska Railroad neared completion to its terminus at Seward in 1915, the island was set aside for fortification use by executive order of September 15, 1916. However, at this time no project or funding for defensive works was forthcoming. It was named in General Order 17 of April 1943 for Col. Charles S. Bulkley. The fort was selected as a site for a standard, permanent 6-inch battery. The mission was to provide effective seaward defense of the Port of Seward, comprising of the docks, oil storage facilities and terminus of the Alaska Railroad. Work was very difficult as there was no dock, unpredictable seas, and very steep terrain requiring cableways. In addition to the gun battery, the defense's HDCP/HECP, SCR-582, and Met Station were to be constructed here. Considerable progress was made in 1943 as unlike other Alaskan forts, work here was supervised by the Army Corps of Engineers. In March of 1944 the fortification work here was abandoned. Among other factors, the partial replacement of the railroad terminus by a new route to Whittier was now open. After World War II, Rugged Island became part of the Alaska Maritime National Wildlife Refuge.

Fort Bulkley Gun Battery

- **Battery #294:** A World War II dual 6-inch barbette battery emplaced on the Rugged Island military reservation of Fort Bulkley. It was on the northern side of the island, high on the rugged crest firing to the north (trunnion elevation of 695-feet). The design is a modified 200-series type, with the rear power, plotting and radio room on the mirror image, opposite sites than the common layout. The BC station is unattached but sits on top of the battery's central traverse. Work was done from March 1, 1943 to March 3, 1944. It was provisionally transferred on April 19, 1944 at a state of

90% completion. While never fully completed, the armament of two 6-inch gun Model 1903A2 and Model M1 barbette carriages were delivered and emplaced. Gun tubes #20 and #68 were shipped here in November 1942. The work was suspended and then abandoned by order of May 31, 1944. The site was ordered disarmed on March 1, 1946. The emplacement still exists, though of difficult access, on the island as part of the Alaska Maritime National Wildlife Refuge. The battery is open to the public, but very difficult to reach requiring both a boat and climbing up a cliff face.

Fort Raymond (1941-1945) is located near and just north the town of Seward, Alaska. Named in General Order 10 of September 10, 1941 for Brig. Gen. Charles Walker Raymond. As an engineer, Raymond helped map the Yukon River in the new Alaskan territory in 1869. The post was established to be the headquarters for the coast defenses of Seward, and to have the major garrison and service facilities for the Army. Unlike the other Alaskan ports, Seward did not have an active U.S. Naval facility, and all fortification work for Seward was conducted by or under the supervision of the Army. The fort had no seacoast defensive works, just the usual barracks, hospital, service, and administrative facilities. When the Seward defenses were curtailed in April 1944, Fort Raymond was gradually reduced until it was closed in 1945. The fort's hospital was quickly transferred to the Territory of Alaska by the War Assets Administrator. The small landing field associated with the fort remained and became Walseth Air Force Base until it was closed in 1948. For years the former property was a campground, but the grounds are currently closed.

CAINES HEAD
STATE RECREATION AREA

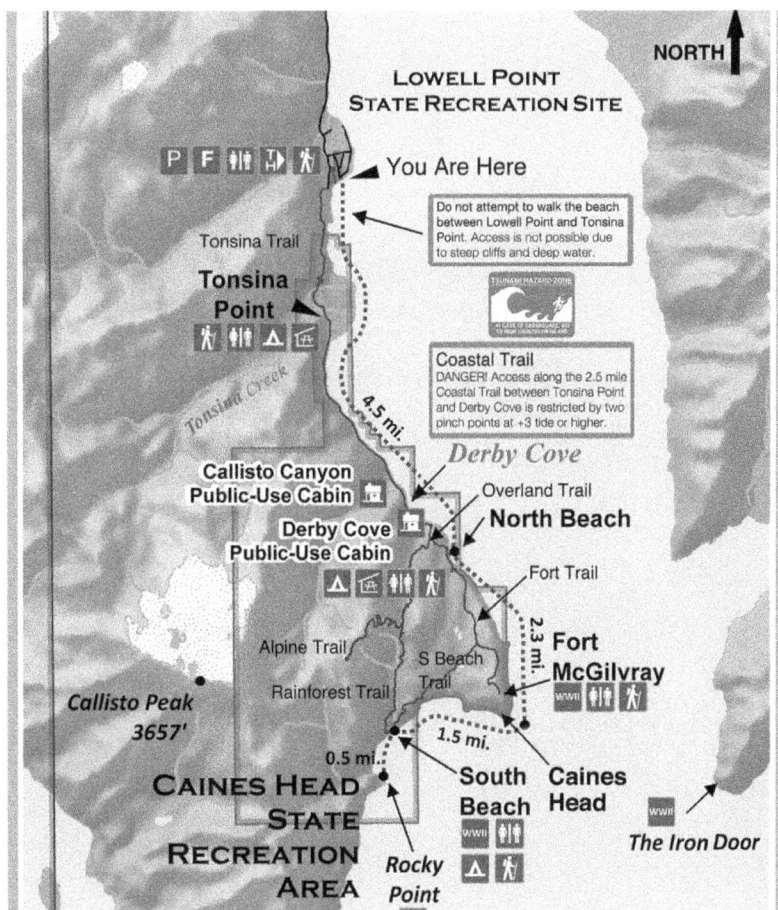

Battery #293 Caines Head Recreation Area (Glen Willford)

Battery Commander's station for Battery #294 on Rugged Island Alaska Maritime National Wildlife Refuge
(Glen Williford)

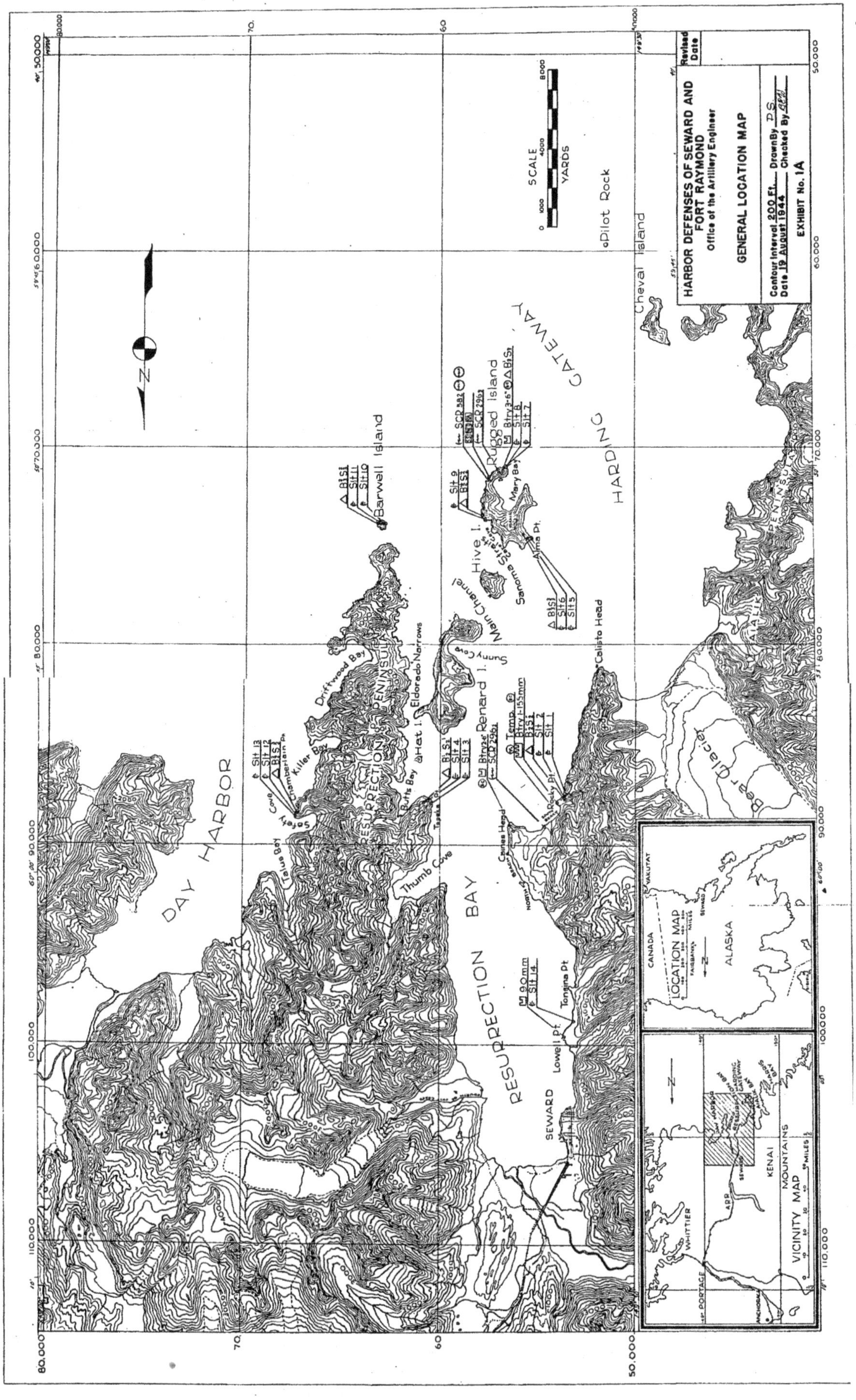

SCALE
YARDS
0 1000 2000 4000 6000 8000

HARBOR DEFENSES OF SEWARD AND
FORT RAYMOND
Office of the Artillery Engineer

GENERAL LOCATION MAP

Contour Interval 200 Ft.   Drawn By P.S.
Date 19 August 1944   Checked By

EXHIBIT No. 1A

Revised
Date

LOCATION MAP

ALASKA
CANADA

VICINITY MAP

KENAI
MOUNTAINS

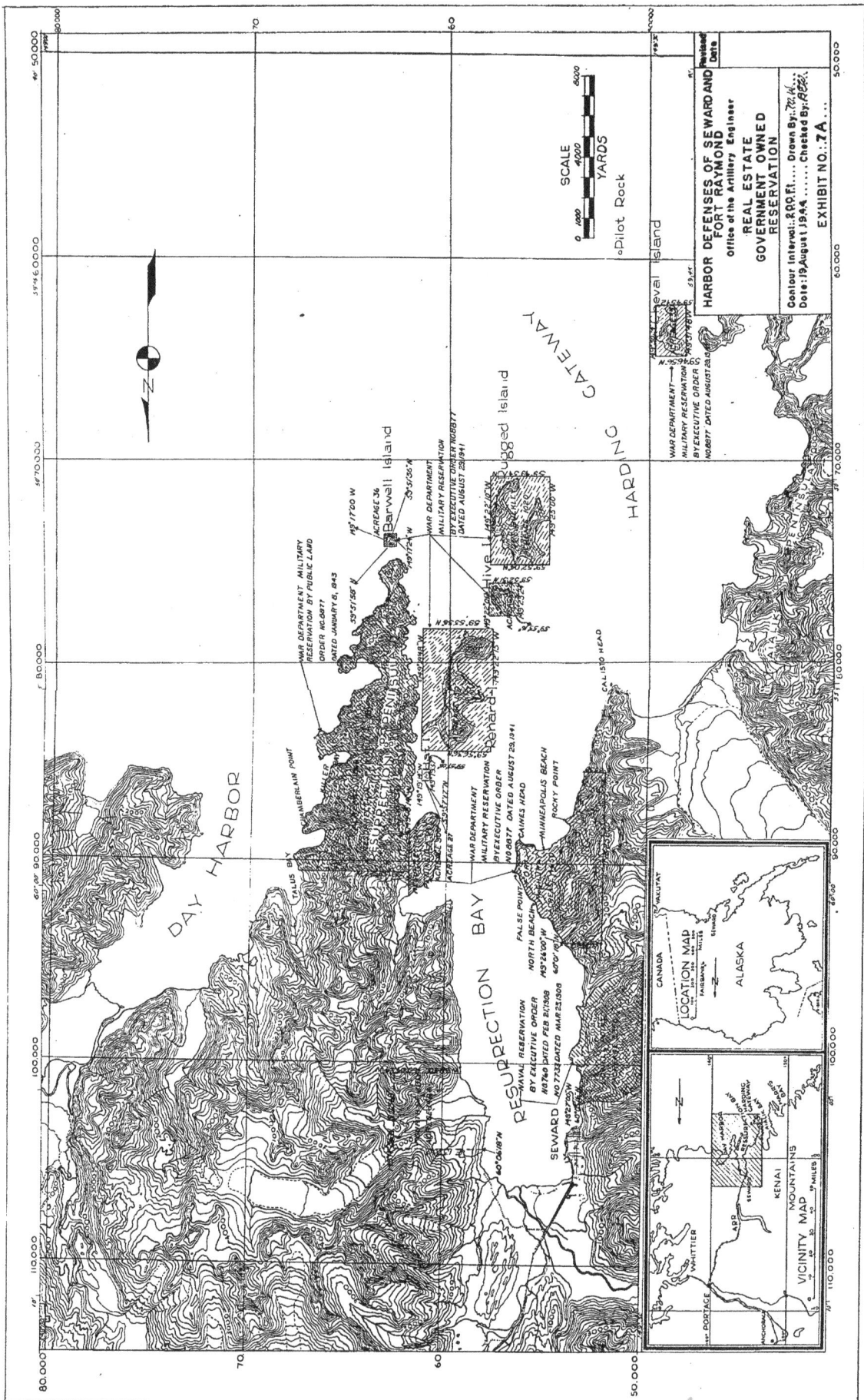

HARBOR DEFENSES OF SEWARD AND
FORT RAYMOND
Office of the Artillery Engineer
REAL ESTATE
GOVERNMENT OWNED
RESERVATION

EXHIBIT NO.:7A

Seward Alaska World War II-era Site Locations

location	Loc#	Purpose
Rocky Point		gun position fire control station
Caines Head		gun position
Resurrection Peninsula		fire control stations, cable
Reward Island		
Hive Island		
Rugged Island		gun position, fire control
Alma Point		fire control station
Chamberlin Point		fire control station
Topeka Point		fire control station
Cheval Island		
Barwell Island		fire control station
Fourth of July		
Lowell Point		AMTB battery

Welcome to

CAINES HEAD
STATE RECREATION AREA

www.alaskastateparks.org

Harbor Defenses of Seward, Alaska: Fort McGilvray 1941-1944

40x100' pile-driven North Dock:
built in1941, it is now unsafe to use

In 1941, as the United States prepared for WWII, the government recognized Seward's enormous strategic value as a vital link in the supply chain. The year-round, ice-free port and railroad connected major shipping lines to the two largest air bases in the Territory of Alaska. On June 30, 1941, the first troops arrived to defend the city, port, railroad, and Resurrection Bay.

A month later, 227 men of the 250th Coast Artillery reported to South Beach and Rocky Point. They set up two batteries, each with anti-aircraft guns and two 155mm coastal artillery guns. They built the first remote defense facilities including spotting and plotting rooms, a communication shack, and base end stations. Due to the lack of supplies, these were all made out of local timber.

In June 1942, the Japanese attacked and occupied the Aleutian Islands of Attu and Kiska, and bombed Dutch Harbor. On July 20, civilian forces employed by West Construction began building permanent, fixed harbor defenses, support facilities, and roads at Caines Head. They pioneered the 2-mile "Road of '42" from North Beach to the headland and South Beach. No roads connected to Seward so all the equipment and supplies had to be delivered by boat and unloaded at the North Beach dock or on South Beach.

In December 1942, the 267th Separate Coast Artillery Regiment joined the 250th Coast Artillery and civilian forces. The men battled time, shortages of materials and supplies, and the foes of impenetrable wilderness, the cold and unpredictable severe weather, relentless seas, darkess, and isolation.

While the fort, 6" Battery No. 293, was under construction, the 250-300 men of the Coast Artillery guarded the bay waters and sky from their garrisons at South Beach and Rocky Point. The 155mm Panama mounts were regularly fired for training purposes in preparation for the enemy that never came.

On March 25, 1943, the installations at Caines Head were named in honor of the first Army officer to serve on the Kenai Peninsula. Lieutenant John McGilvray commanded Fort Kenay in 1869, two years after the purchase of Alaska from the Russians.

By August 15, 1943, American forces had routed the enemy from the Aleutian Islands and the war effort shifted away from Alaska. By February 1944, the Caines Head project was 90% complete and had housed 2,000 troops during the course of two years. On March 29, 1944, orders were sent to dismantle the Seward Harbor Defenses. In a few short months, all the fortifications around the bay were stripped bare and abandoned to Nature.

In 1971, the newly created Alaska Division of Parks selected 1800 acres and established Caines Head State Recreation Area. In 1974, it expanded to its present size of 5,961 acres. Due to the remote location and difficult access, the historic structures remained undisturbed, reclaimed and camouflaged by Nature. In 1984, park staff began clearing a trail from North Beach to the Fort and discovered one of the best-preserved batteries in Alaska.

Caines Head State Recreation Area serves as a memorial to all the soldiers and civilians who fought for the peace, freedom, and rights we enjoy today. Reflect on their daunting challenges and many sacrifices as you walk along the historic Road of '42 and explore the Fort and other WWII structures.

North Ammunition Magazine:
a strong, concrete bunker

Elephant Shelter made of plywood and steel: fuses were stored far from ammunition and explosives

South Ammunition Magazine:
a safe distance away from the others

Fort McGilvray:
West Entrance

Fort McGilvray:
Battery Command Station

Fort McGilvray:
East 6 Inch Gun Block #2

Rusted electric conduit, hidden in the moss:
barracks and out-buildings had
electric, water, and sewer systems.

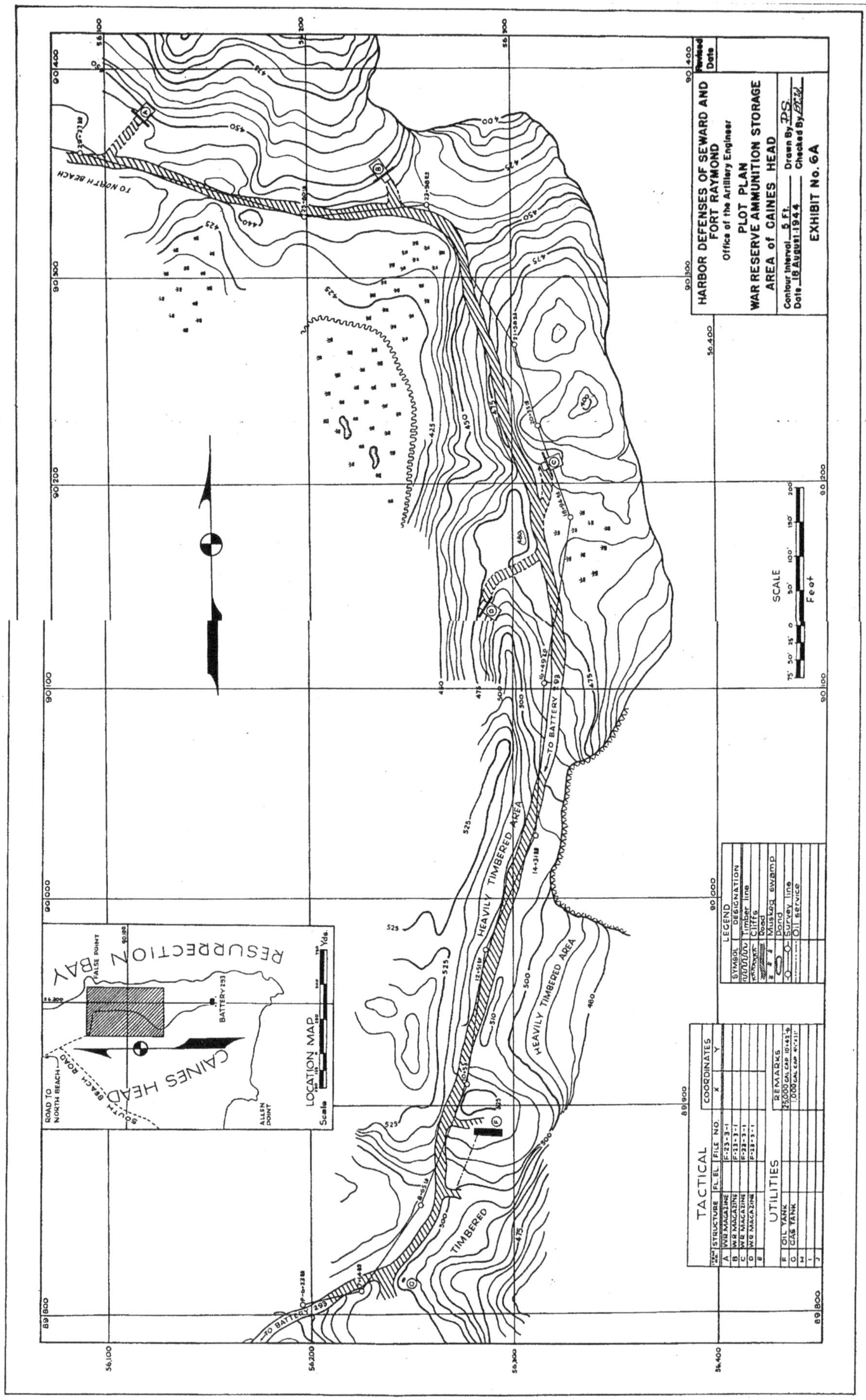

HARBOR DEFENSES OF SEWARD AND
FORT RAYMOND
Office of the Artillery Engineer
PLOT PLAN
WAR RESERVE AMMUNITION STORAGE
AREA of GAINES HEAD

Contour Interval   5 Ft.   Drawn By PS
Date 18 August 1944   Checked By

EXHIBIT No. 6A

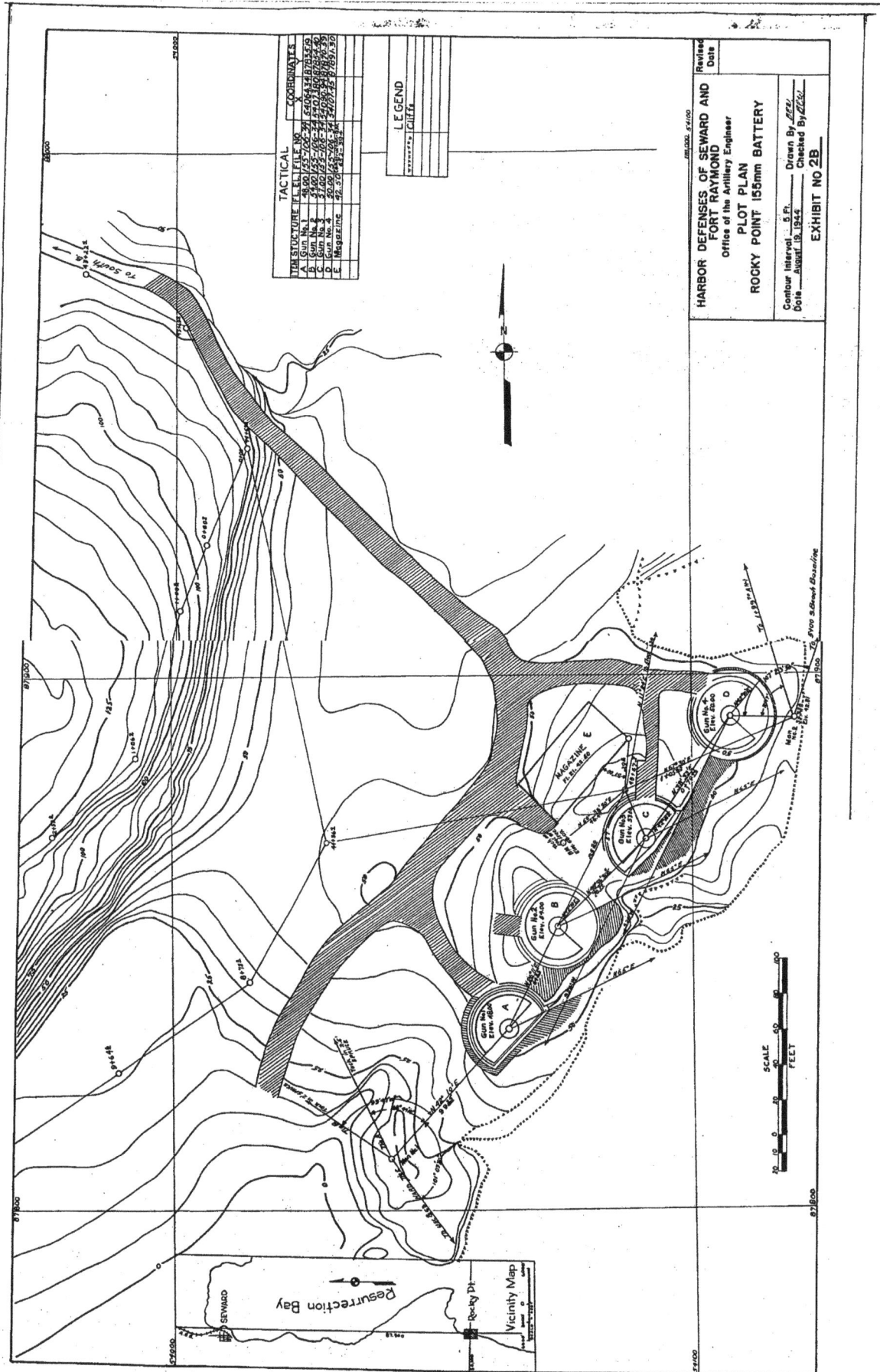

HARBOR DEFENSES OF SEWARD AND
FORT RAYMOND
Office of the Artillery Engineer
PLOT PLAN
ROCKY POINT 155mm BATTERY

Contour Interval ___5 Ft.___   Drawn By ___
Date ___August 19, 1944___   Checked By ___

EXHIBIT NO 2B

LEGEND
Cliffs

TACTICAL

ITEM	STRUCTURE	FILE/FILE NO.	COORDINATES
A	Gun No. 1		
B	Gun No. 2		
C	Gun No. 3		
D	Gun No. 4		
E	Magazine		

N

SCALE
FEET

Vicinity Map
Resurrection Bay
SEWARD
Rocky Pt.

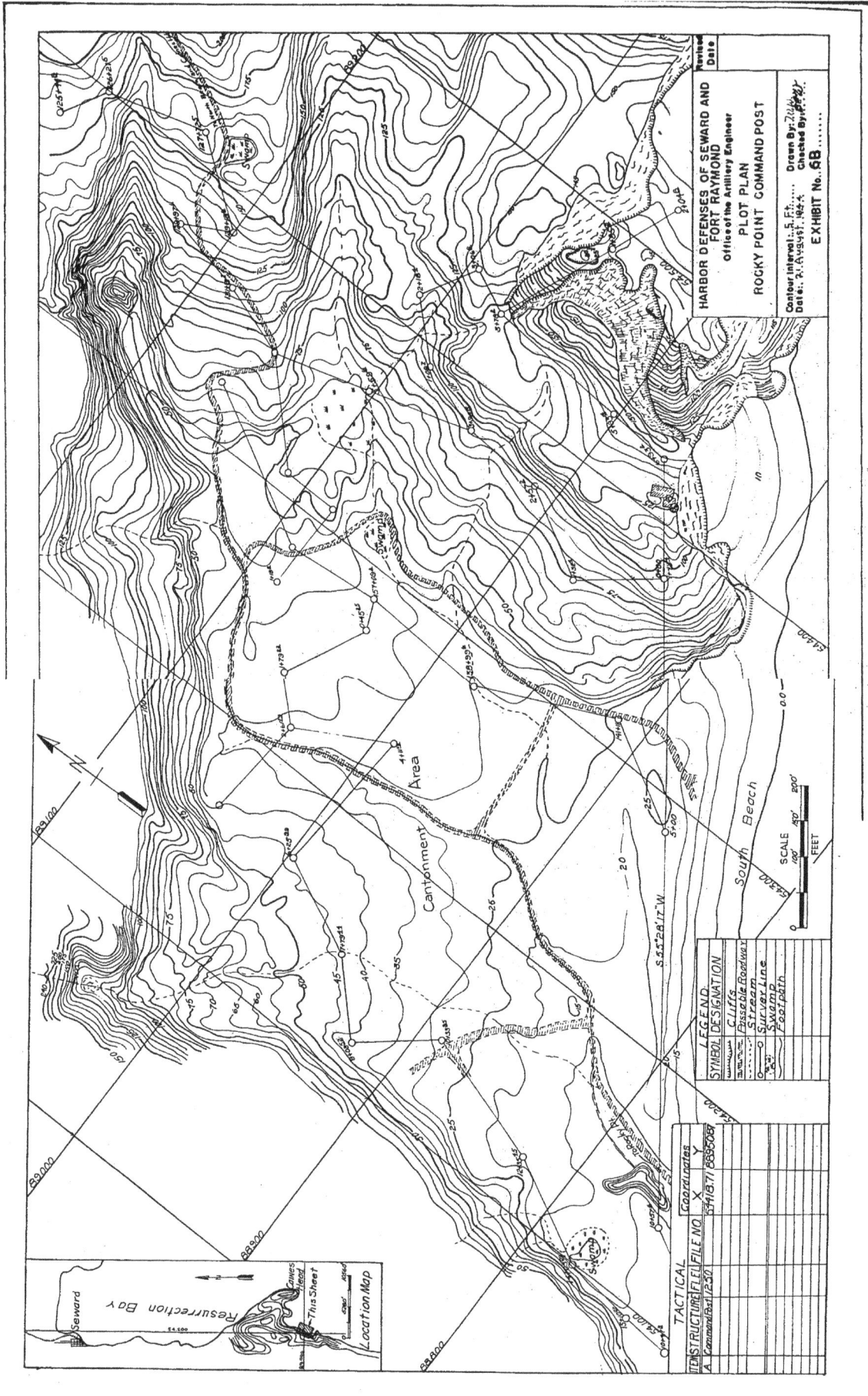

HARBOR DEFENSES OF SEWARD AND
FORT RAYMOND
Office of the Artillery Engineer
PLOT PLAN
ROCKY POINT COMMAND POST

Contour Interval...5.Ft.
Date.. 3. August. 1944
Drawn By:
Checked By:
EXHIBIT No. 6B

LEGEND
SYMBOL   DESIGNATION
Cliffs
Passable Roadway
Stream
Survey Line
Swamp
Footpath

Cantonment Area

South Beach

SCALE
100   50'   200
FEET

S.55°28'17"W.

TACTICAL
ITEM   STRUCTURE   FILE   FILE No.
A   Command Post   1:250

Coordinates
X   Y
574/8.71 / 885.00EF

Location Map
Resurrection Bay
Seward
This Sheet

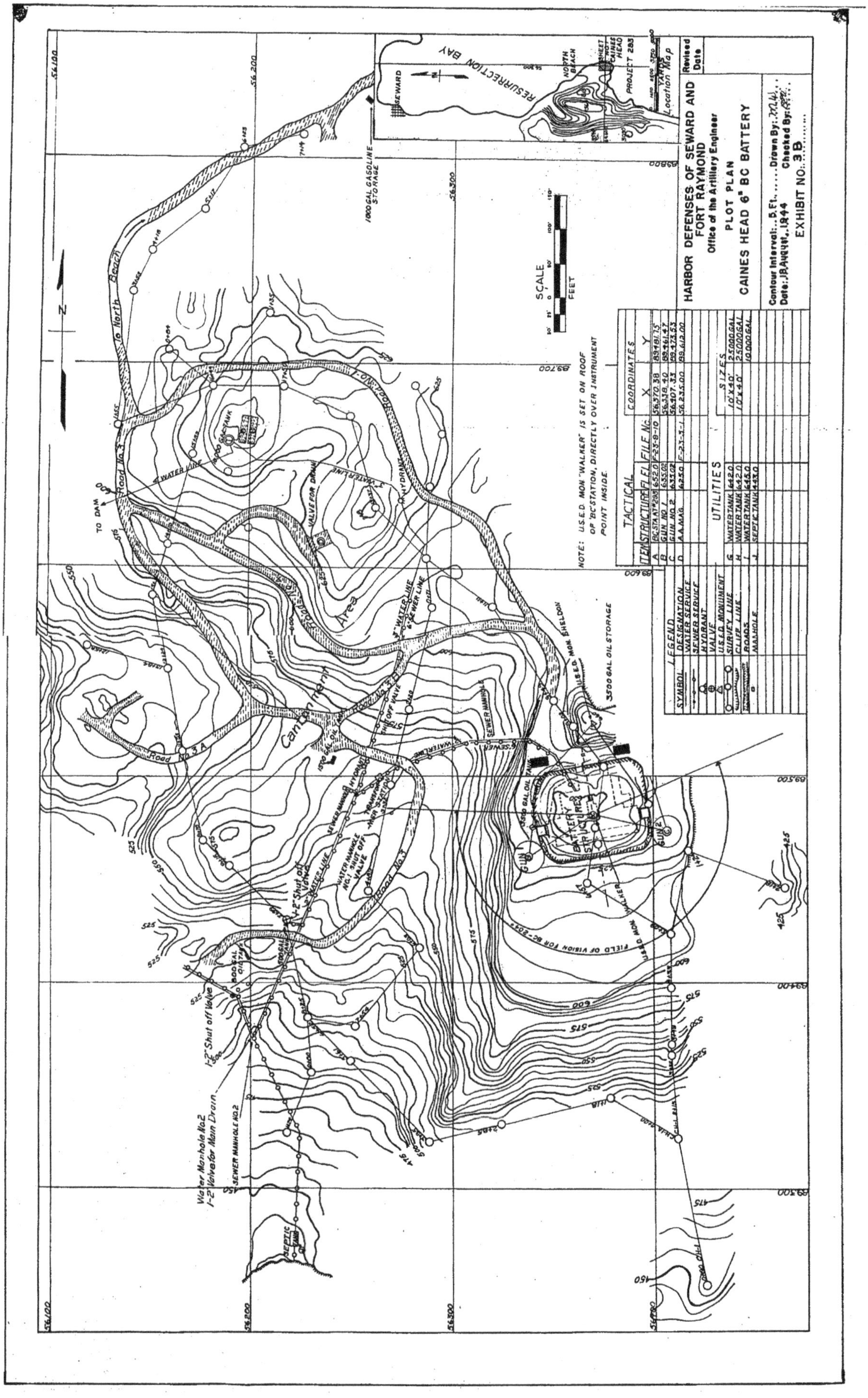
HARBOR DEFENSES OF SEWARD AND FORT RAYMOND
Office of the Artillery Engineer
PLOT PLAN
CAINES HEAD 6" BC BATTERY
EXHIBIT NO. 3B
Location Map
RESURRECTION BAY
SEWARD
NORTH BEACH
SCALE
FEET
To North Beach
Road No. 3
Road No. 3 A
Cantonment Area
1000 GAL GASOLINE STORAGE
3500 GAL OIL STORAGE
TO DAM
BATTERY STRUCTURES
GUN 1
GUN 2
FIELD OF VISION FOR BC 2022
NOTE: U.S.E.D. MON "WALKER" IS SET ON ROOF OF 'BC' STATION, DIRECTLY OVER INSTRUMENT POINT INSIDE.
TACTICAL
UTILITIES
LEGEND
COORDINATES
SIZES
SEPTIC TANK
Water Manhole No. 2
SEWER MANHOLE NO. 2

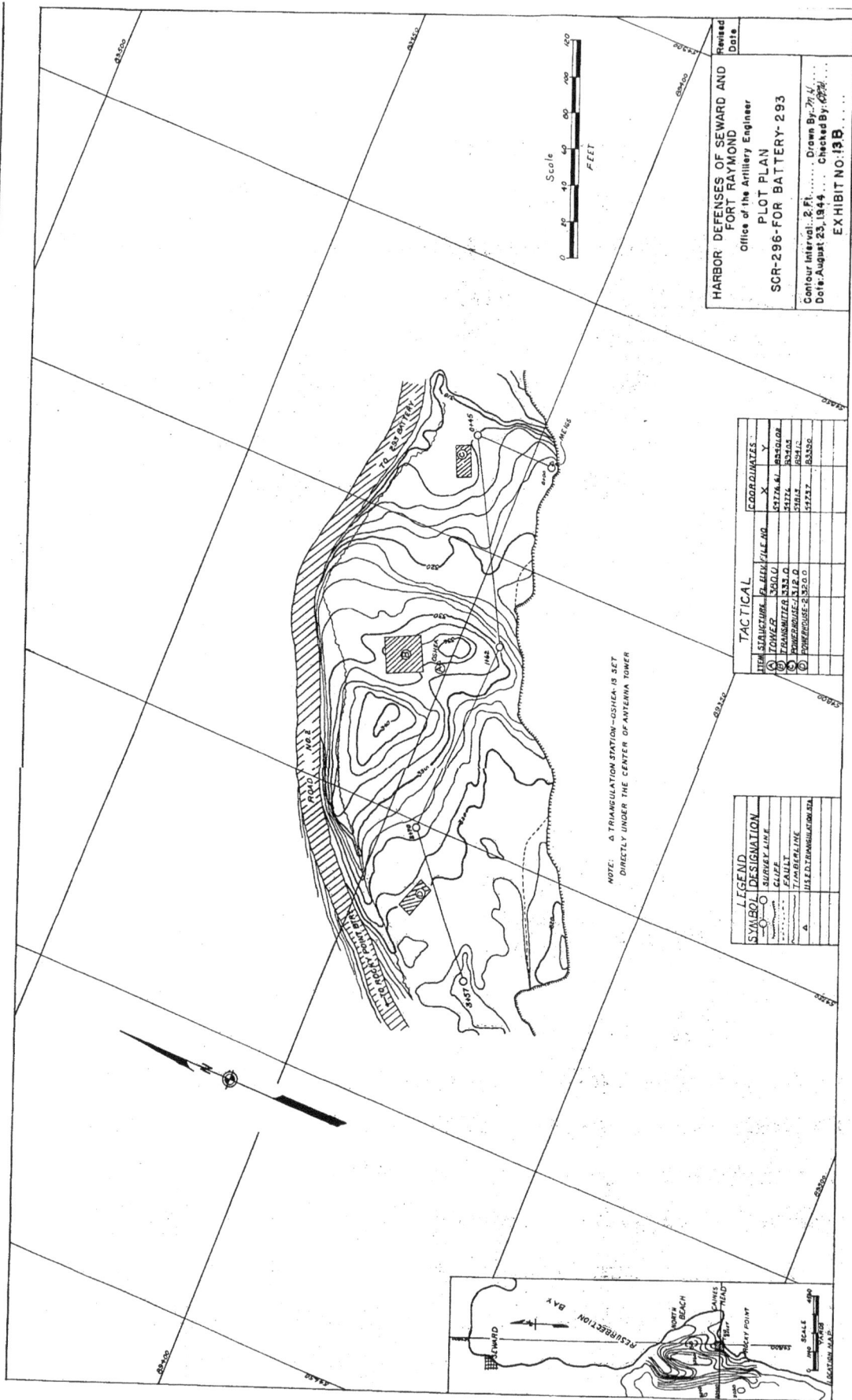

HARBOR DEFENSES OF SEWARD AND FORT RAYMOND
Office of the Artillery Engineer
PLOT PLAN
SCR-296-FOR BATTERY-293

Contour Interval: 2 Ft. ..... Drawn By: M.U. .....
Date: August 23, 1944. ..... Checked By: C.M. .....

EXHIBIT NO: 13.B.

Revised	Date

TACTICAL

ITEM	STRUCTURE	FLOOR ELEV.	FILE NO	COORDINATES X	Y
A	TOWER	360.0		51776.61	83510.02
B	TRANSMITTER	333.0		51776	83410
C	POWERHOUSE-1	312.0		51613	83410
D	POWERHOUSE-2	320.0		51717	83320

LEGEND

SYMBOL	DESIGNATION
◦—◦	SURVEY LINE
	CLIFF
	FAULT
	TIMBERLINE
△	U.S.E.D.TRIANGULATION STA.

NOTE: △ TRIANGULATION STATION—OSHEA-IS SET
DIRECTLY UNDER THE CENTER OF ANTENNA TOWER

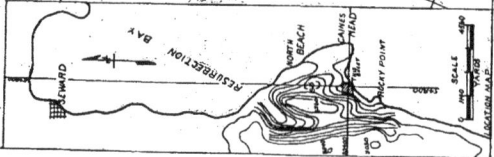

Scale
FEET
0   20   40   60   80   100   120

LOCATION MAP

RESURRECTION BAY

SEWARD

NORTH BEACH

CAINES HEAD

ROCKY POINT

SCALE
YARDS

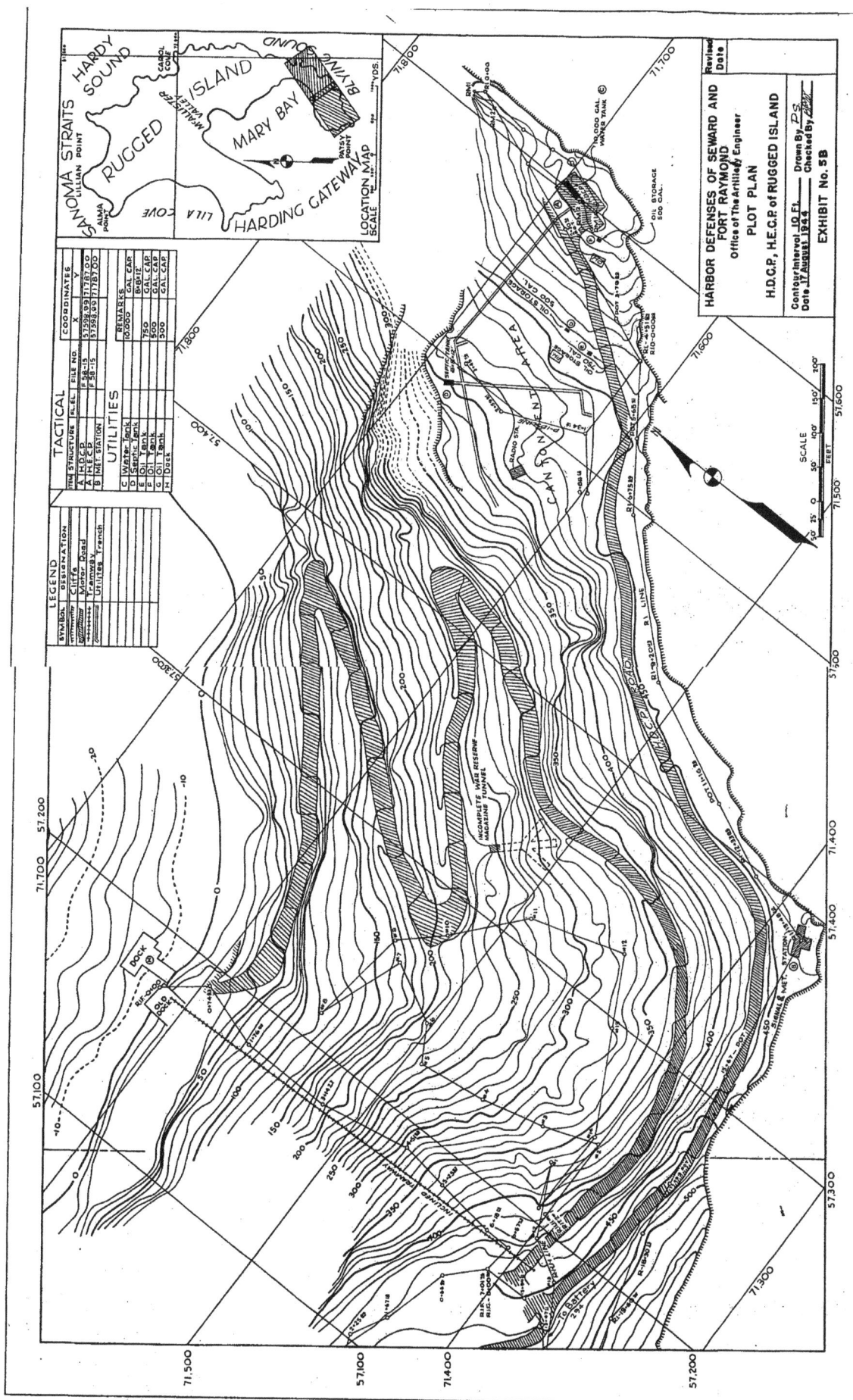

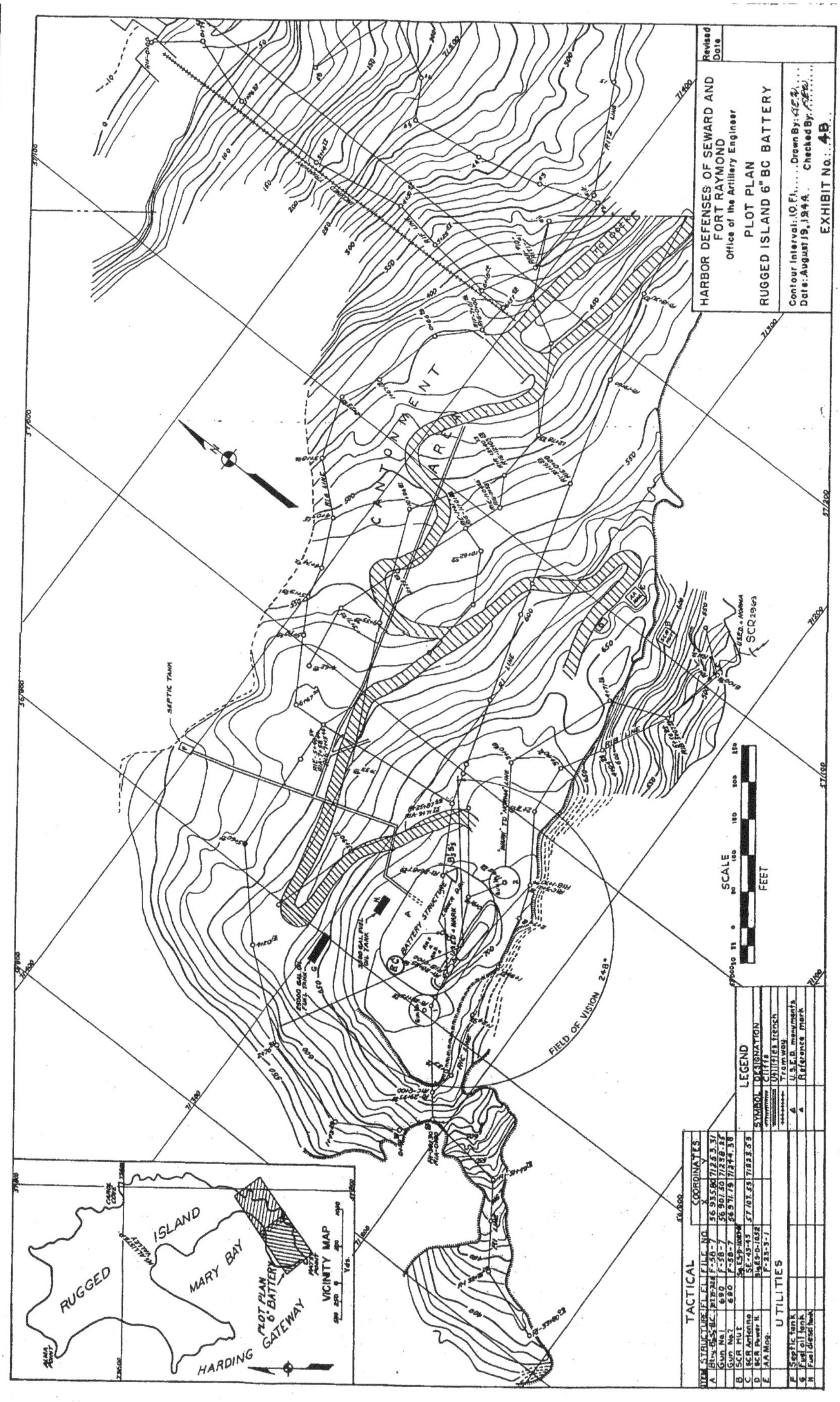

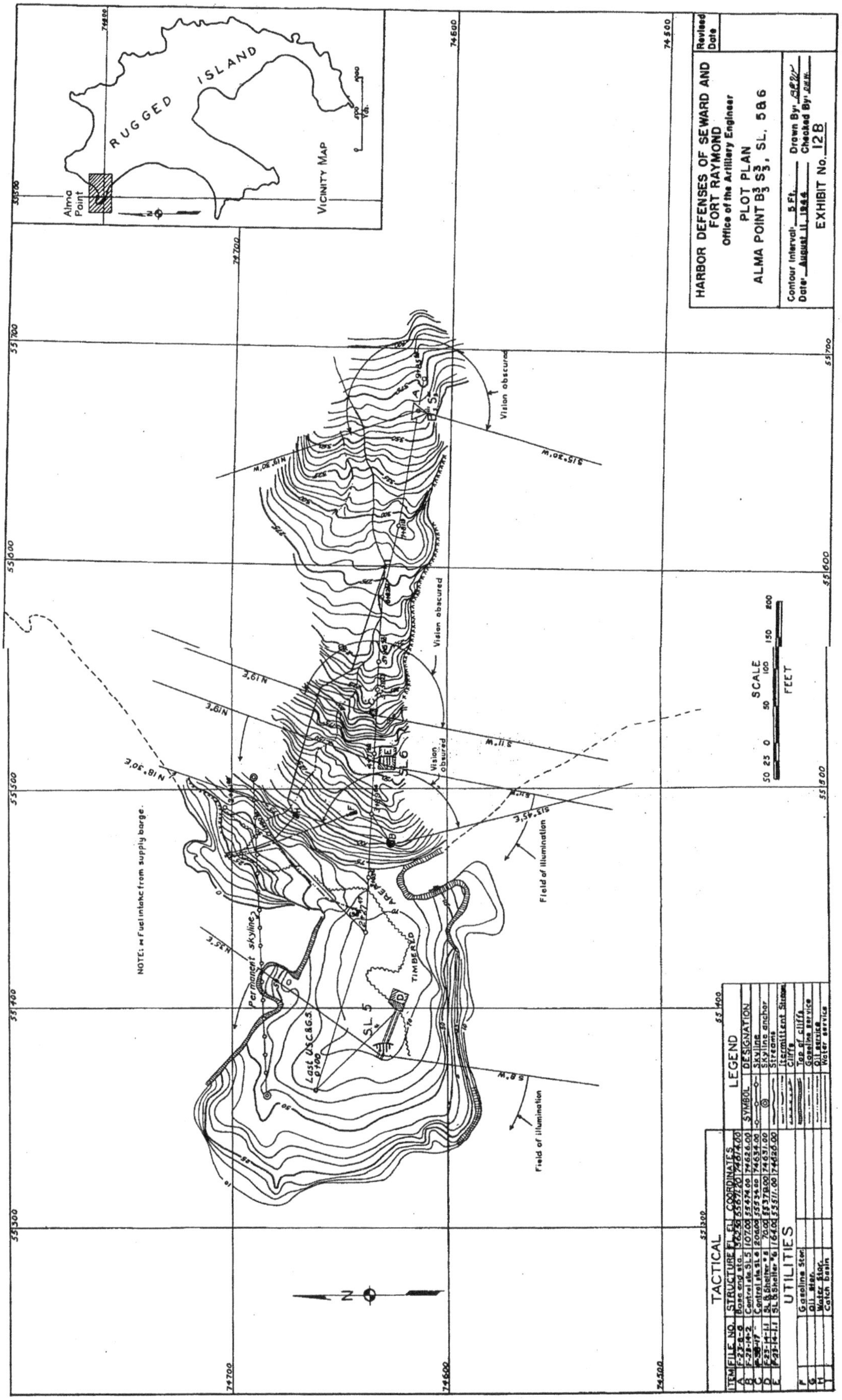

RUGGED ISLAND

Alma Point

VICINITY MAP

HARBOR DEFENSES OF SEWARD AND FORT RAYMOND
Office of the Artillery Engineer
PLOT PLAN
ALMA POINT B3 S3, SL. 5 & 6

Contour Interval: 5 Ft.
Date: August 11, 1944
Drawn By:
Checked By:
EXHIBIT No. 12B

Revised Date

Vision obscured

Field of illumination

Permanent skyline

TIMBERED

SL 5

SCALE
50 25 0   50   100   150   200
FEET

LEGEND

SYMBOL	DESIGNATION
	Skyline
	Skyline anchor
	Streams
	Intermittent Stream
	Cliffs
	Top of cliffs
	Gasoline service
	Oil service
	Water service

TACTICAL

ITEM/FILE NO.	STRUCTURE	PLAN COORDINATES

UTILITIES

	Gasoline Stor.
	Oil stor.
	Water Stor.
	Catch Basin

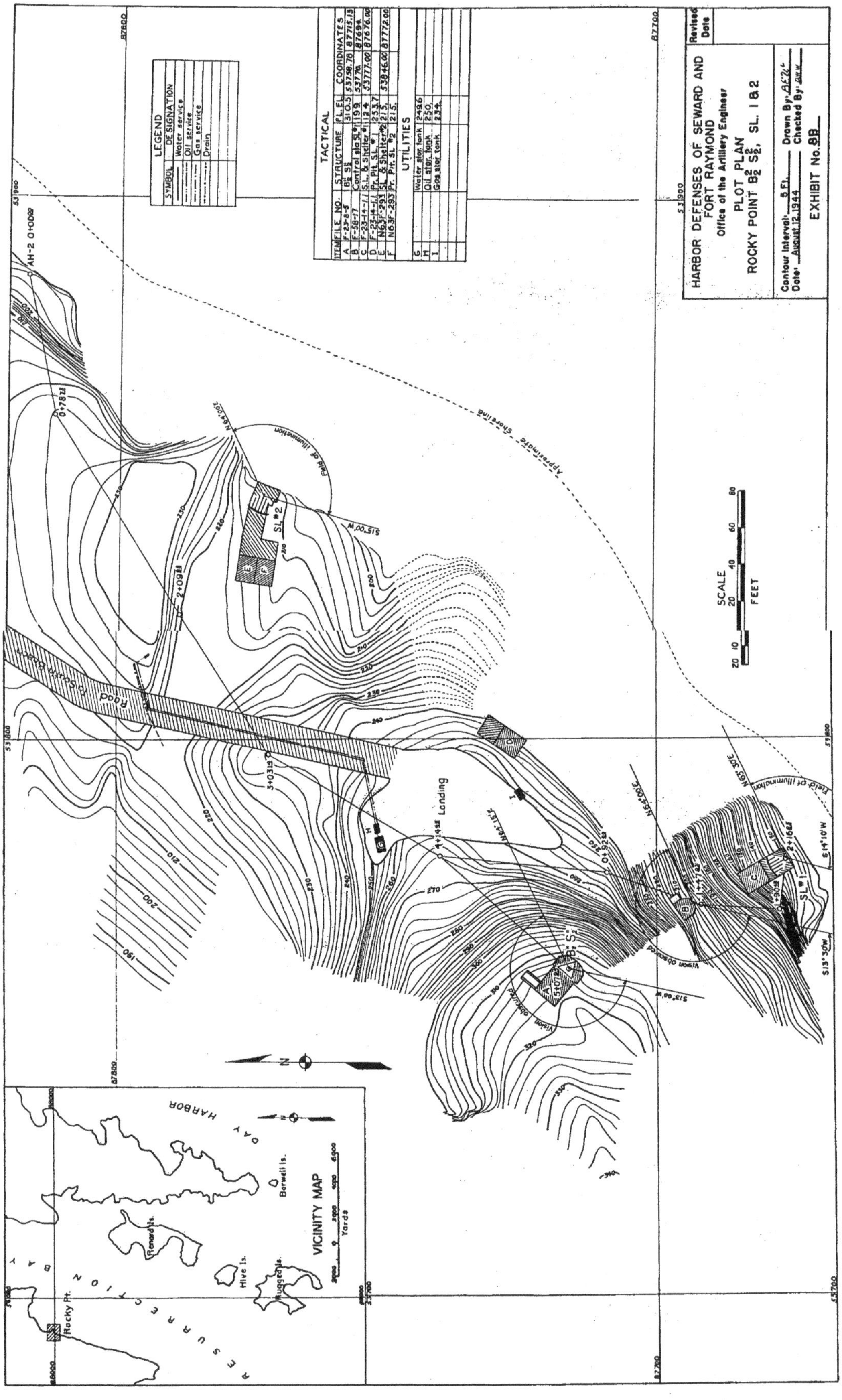
HARBOR DEFENSES OF SEWARD AND FORT RAYMOND
Office of the Artillery Engineer
PLOT PLAN
ROCKY POINT B2 S2, SL. 1 & 2
Contour Interval: 5 Ft.
Date: August 12, 1944
EXHIBIT No. 8B
LEGEND
SYMBOL
DESIGNATION
Water service
Oil service
Gas service
Drain
TACTICAL
UTILITIES
Water stor. tank
Oil stor. tank
Gas stor. tank
SCALE
FEET
Road to South Beach
Approximate Shoreline
Landing
Field of Illumination
Vision Obscured
SL. #1
SL. #2
VICINITY MAP
Yards
RESURRECTION BAY
DAY HARBOR
Rocky Pt.
Renard Is.
Hive Is.
Rugged Is.
Barwell Is.

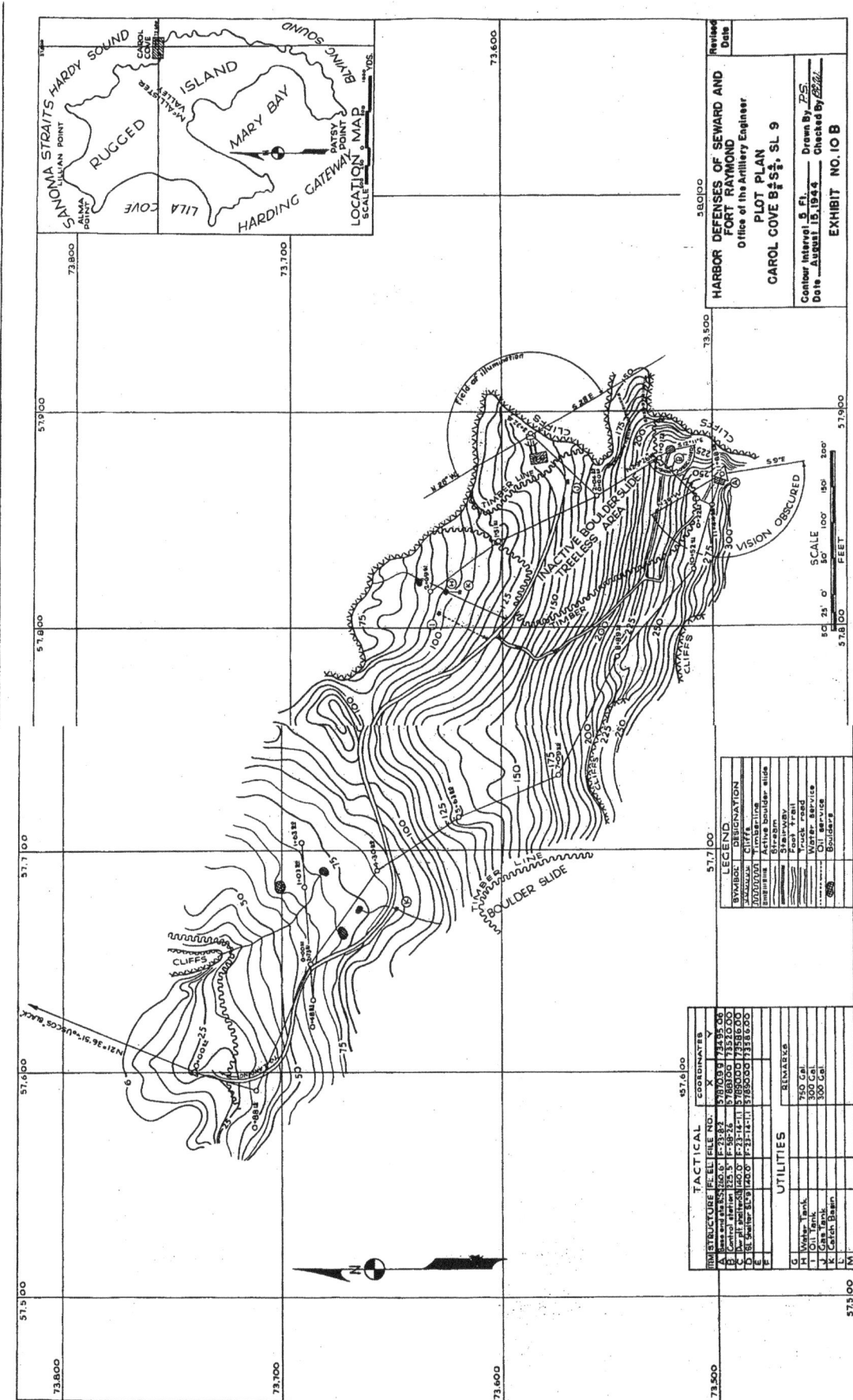

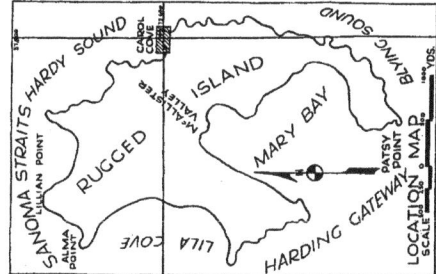

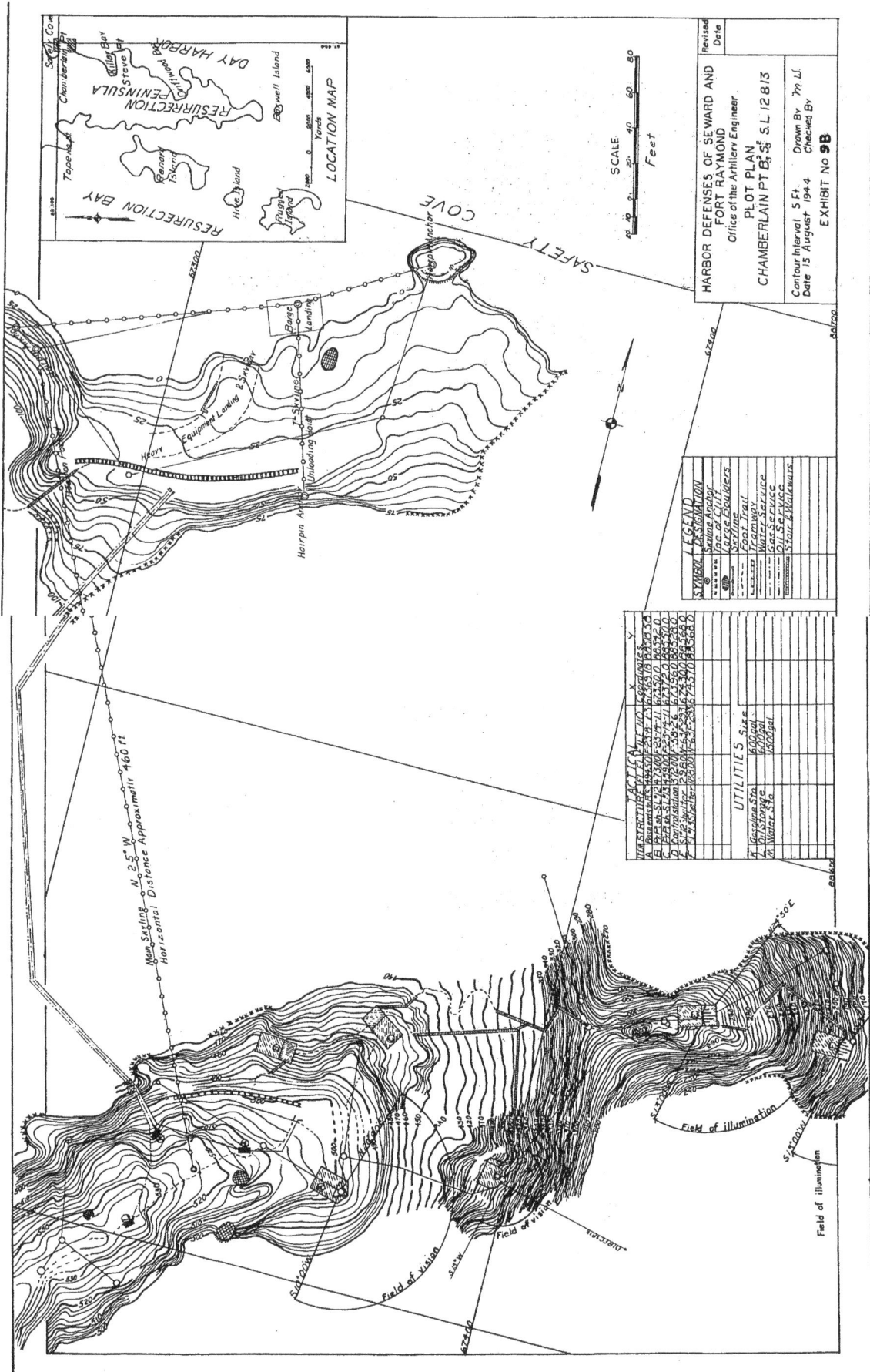

LOCATION MAP

RESURRECTION PENINSULA

RESURRECTION BAY

DAY HARBOR

SCALE

Feet

HARBOR DEFENSES OF SEWARD AND
FORT RAYMOND
Office of the Artillery Engineer
PLOT PLAN
CHAMBERLAIN PT B-2 & S.L. 12813

Contour Interval 5 Ft	Drawn By M.L.
Date 15 August 1944	Checked By
EXHIBIT No 9B

LEGEND

SYMBOL	DESIGNATION

UTILITIES	Size

TACTICAL

SAFETY COVE

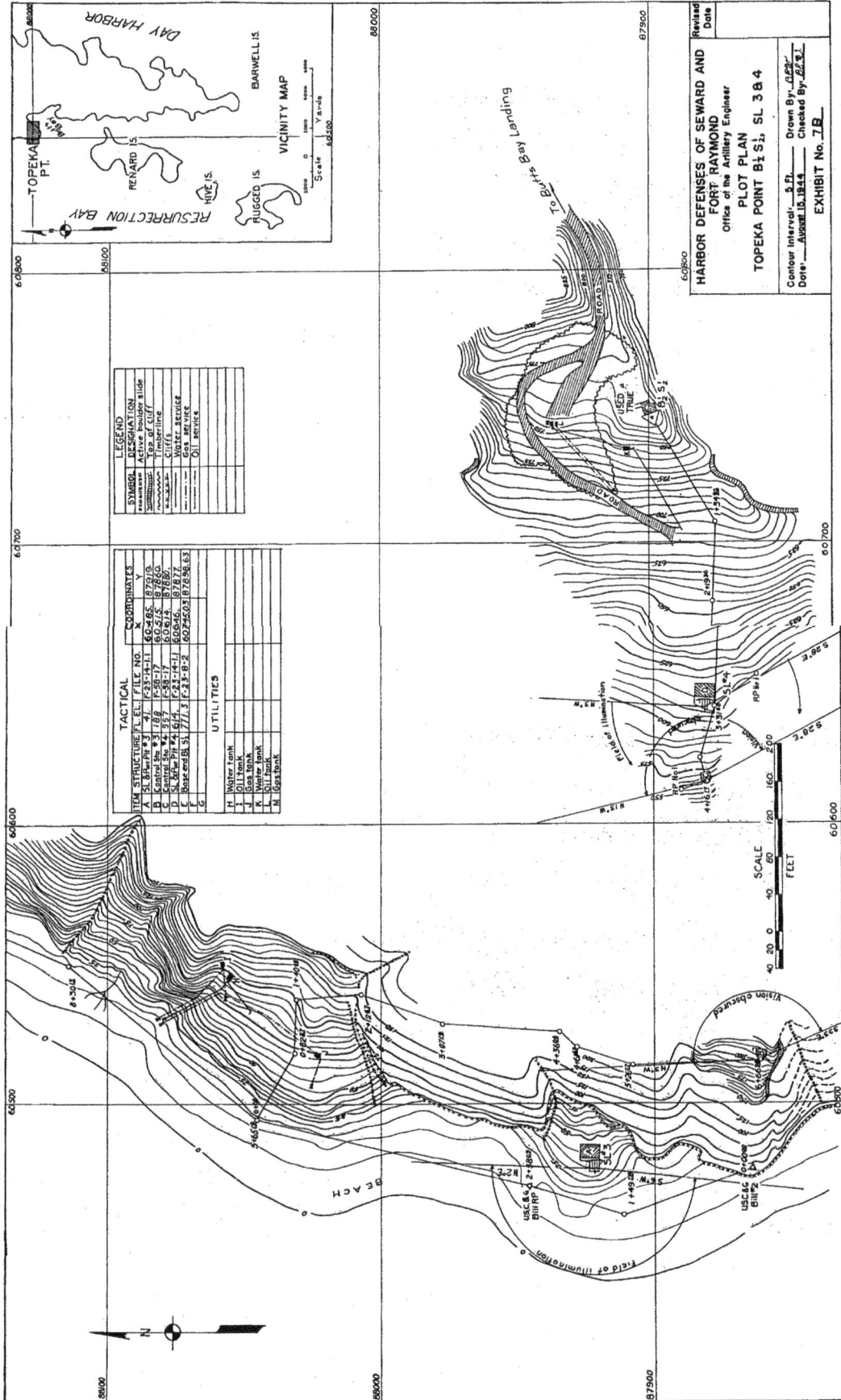

VICINITY MAP

RESURRECTION BAY

DAY HARBOR

TOPEKA PT.

BARWELL IS.

RENARD IS.

HIVE IS.

RUGGED IS.

Scale Yards

LEGEND

SYMBOL	DESIGNATION
	Active boulder slide
	Top of cliff
	Timberline
	Cliffs
	Water service
	Gas service
	Oil service

TACTICAL

ITEM	STRUCTURE	F. EL.	FILE NO.	COORDINATES	
				X	Y
A	SL B'w'r Pit *3	41	F-23-14-1.1	60.485	87910
B	Control Sta. *3	188	F-56-17	60.515	87860
C	Control Sta. *4	557	F-56-17	60614	87880
D	SL B'w'r Pit *4	614	F-23-14-1.1	60846	87877
E	Base end B. SL	771.3	F-23-8-2	607+603	87+94.63
F					
G					

UTILITIES

H	Water tank
I	Oil tank
J	Gas tank
K	Water tank
L	Oil tank
M	Gas tank

To Butts Bay Landing

ROAD

ROAD

SCALE

FEET

BEACH

Field of Illumination

HARBOR DEFENSES OF SEWARD AND FORT RAYMOND

Office of the Artillery Engineer

PLOT PLAN

TOPEKA POINT B½ S½, SL 3 & 4

Contour Interval: 5 ft.

Date: August 15, 1944

Drawn By: ___

Checked By: ___

EXHIBIT No. 7B

Revised Date

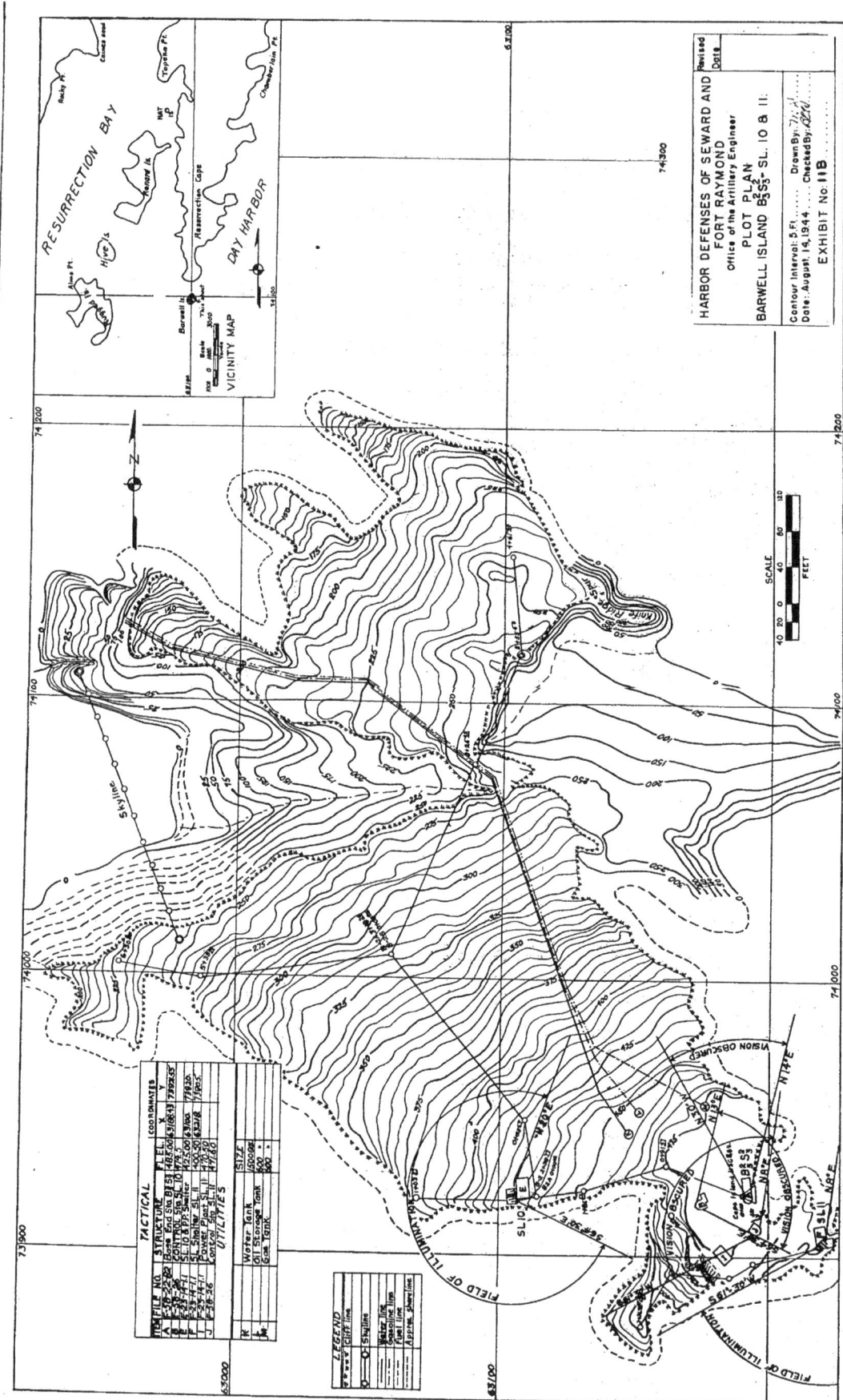

HARBOR DEFENSES OF SEWARD AND
FORT RAYMOND
Office of the Artillery Engineer
PLOT PLAN
BARWELL ISLAND B²S5"- SL. 10 & 11.

EXHIBIT No. 11B

THE HARBOR DEFENSES OF KODIAK — ALASKA

Kodiak was to have the most extensive defenses of any of the Alaskan bases: two 8-inch batteries, three 6-inch batteries, two AMTB batteries, 26 searchlight positions and 20 fire control positions. However, only three of the projected project batteries were actually built. BCN 403, a rather unique two-story 1940s-era battery, was located at St. Peter's Head on Cape Chiniak and the reservation was designated Fort John H. Smith. BCN 404 was located on Miller Point and was designated Fort Abercrombie. The lone 6-inch battery, BCN 296, was located on Long Island and designated Fort Tidball. The other two batteries, BCN 295 (Fort J. H. Smith) and BCN 297 (East Cape, Spruce Island) were cancelled. Fire control and searchlight facilities were located on Narrow Cape, Spruce Is., Kizhuyak, Artillery Hill, Soquel Pt., Midway Pt., Gibson Cove, Spruce Cape, Long Is., Cape Chiniak, Miller Point and Buskin Hill. Radar facilities were located on Long Island, Piedmont Point and Miller Point. The AMTB batteries were built on Puffin Island and Spruce Cape. The new batteries were essentially complete by the time the harbor defense was inactivated in 1945.

Fort J.H. Smith (1941-1945) is located at St. Peters Head at Cape Chiniak to the southeast of Kodiak. The reservation of 250 acres was procured on August 29, 1941 as a set-aside of public land for military purposes. A temporary battery was built in 1941 using four 155mm GPF guns on Panama mounts at Chiniak along the beach. The reservation was named in General Order 17 of April 2, 1943 for Bvt. Capt. John Hewitt Smith, an Army artilleryman of notable Civil War service. The final April, 1942 approval of permanent coast defense batteries for Kodiak selected Fort Smith for a standard 8-inch barbette battery. Originally a standard 6-inch battery was also projected for this fort, but was never authorized for construction. Work on fortifications began under U.S. Navy Seabee control in December 1942. This gun battery and separate plot and BC station was completed in 1944. Several fire control and searchlight stations were also constructed at this fort. Abandoned at the end of the war, the emplaced guns were scrapped or destroyed by 1948. After World War II, Fort Smith was transferred to the U.S. Air Force for use as a tracking station until 1975. Today, the fort is owned by the Lesnoi Corporation, a Native Alaskan holding company. Permission to visit the site is required from a local tribal council.

Fort J.H. Smith Gun Batteries

- **Battery #403:** A World War II project battery for two 8-inch guns on barbette mounts. It was built at St. Peter's Head of the Cape Chiniak reservation of Fort J. H. Smith. Work was done from December 1942 to December 24, 1943. It was of unusual design. The Alaskan 8-inch batteries generally followed a design plan of open gun blocks with surrounding aprons and exposed mounts separated by 240-feet and a central protected concrete and earth traverse containing magazines and support rooms. At this battery an entire lower, basement level was built in the traverse adding additional rooms and apparently doubling the magazine capacity. The BC station was located on top of the traverse but not connected to the battery's interior. It was transferred on December 24, 1943 (cost is not listed on the appropriate RCW). Armament was sent on October 18, 1942. The guns mounted were two 8-inch guns navy MkVI3A2 on Model M1 carriages (#120L2/#5 and #134L2/#6). It was never named, simply being known as Battery Construction No. 403. It served until the end of the war, probably being disarmed in 1946. The guns and carriages were destroyed on site by detonation—parts of them still exist scattered around the battery. The emplacement still exists on native property. A permit from the Lesnoi Corporation is required to visit the site.

- *Battery #295* **(planned):** A World War II project battery for two 6-inch guns with shielded barbettes guns planned for Fort Smith. Funds were never forthcoming to start this battery, no work was ever done.

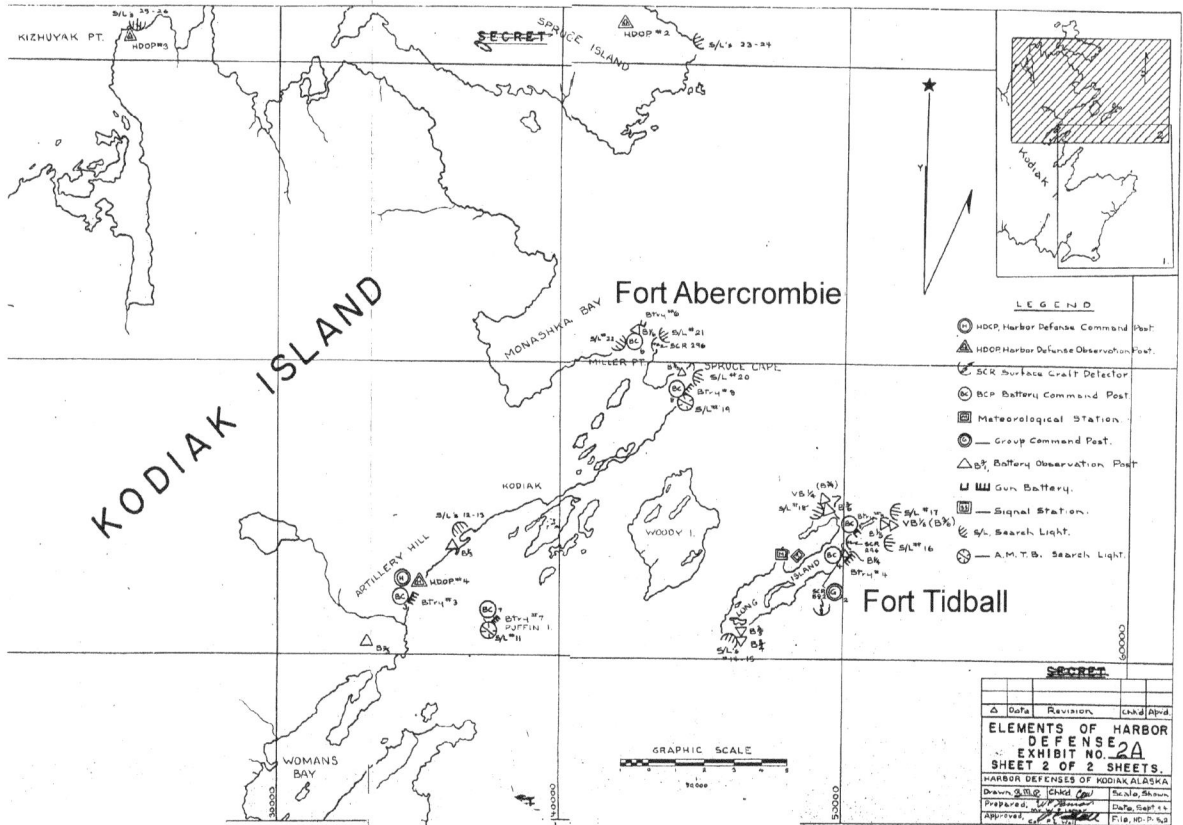

Fort Abercrombie

Fort Tidball

Fort J.H. Smith

Fort Tidball (1941-1946) is located at Castle Bluffs on Long Island in the bay to the south of Kodiak. The island of 1320 acres was obtained for military purposes on June 14, 1941. It was named in General Order 17 of April 2, 1943 for Bvt. Maj. Gen. John C. Tidball. Tidball was an Army artillery officer during the Civil War, and was the 3rd commander of the new Alaskan Department postwar. From December 1941 to March 1942 temporary batteries utilizing both a single ex-navy 6-inch gun and another using four 155mm GPF guns on Panama mounts at Deer Point were emplaced on Long Island. In April 1942 the fort was approved for location of a standard 6-inch shielded battery. Several fire control station, searchlight shelters and associated control stations and a small cantonment area of Quonset huts were also built starting in late 1942 by U.S. Navy Seabees. This battery was completed in 1944 and remained until 1950 when the barrels and carriages were destroyed on site, but the large shields remain today. Several fire control and searchlight stations were also constructed for this fort. After World War II, Fort Tidball was abandoned. Today, the fort is managed by the Great Land Trust for the Leisnoi Corperation. The island is operated as a regional nature park. Among the surviving remains are one of the only intact SCR-296 fire control radar towers. Access to the island requires permission from the Lensnoi Corperation and a boat.

Fort Tidball Gun Batteries

- **Battery #296:** A World War II project battery for two 6-inch guns with shielded barbettes emplaced at Fort Tidball. The site was on the northern end of Long Island at a site known as Castle Bluff, firing to the northwest. It was an unusual design, though with most of the features and spacing for a 1940 Program dual battery of the 200-series type. However, this battery was an unusual, split-level design. The gun blocks and magazines with connecting corridor were on one level, but the power, radio, and radiator rooms were on a whole lower story behind, with internal steps connecting the levels. Work was done following a site inspection of May 3, 1942 from September of that year until December 1943. It was turned over on December 8, 1943 (cost is not listed on the appropriate RCW). Armament was shipped here on November 26, 1942. It was armed with two 6-inch guns Model M1903A2 on model M1 barbette carriages (#37/#3 and #6/#6). It was never named, simply being known as Battery Construction No. 296. It served until abandoned around the end of 1945. While the gun tubes were removed, the carriages and shields were demolished on site, major parts still remain at the battery location. The emplacement still exists on native property. Permission from the Lesnoi Corporation is required to visit the battery and a boat to travel to the island.

Fort Abercrombie (1941-1946) is located at Miller Point to the north of Kodiak. A parcel of 317 acres was transferred from public land to a status of military reservation on June 14, 1941. The approved project for coast defense was approved by the Secretary of War on April 18, 1942. It called for a site at the fort for a standard 8-inch barbette battery. The fort was named in General Order 17 of April 2, 1943 for Lt. Col. William R. Abercrombie. Abercrombie was a professional army soldier. He participated in several of the Indian campaigns, and led an expedition to Alaska to identify a practical route to the Klondike in 1898. In December 1941 temporary positions were developed for searchlights. Heavy construction work at the fort began in late 1942 under the supervision of the U.S. Navy Seabees. In addition to the gun battery (begun in January 1943), the fort also had a major reserve magazine, power plant, several fire control and searchlight shelters and radar tower. At nearby Spruce Cape a standard 6-inch shielded battery was projected, but never built. However a 90mm AMTB battery was constructed in 1943 at this cape. After World War II, Fort Abercrombie was abandoned. A portion of the reservation was transferred to the State of Alaska as a park in 1969. Today the fort is Fort Abercrombie State Historical Park.

Fort Abercrombie Gun Batteries

- **Battery #404:** A World War II project battery for two 8-inch guns emplaced at Miller Point at the East Cape Spruce reservation of Fort Abercrombie. It was of standard late-war 8-inch battery plan, with guns spaced at 240-feet and intervening, single-level traverse magazine. Trunnion height of the site was 140-feet. Work was begun in January 1943 and finished in late December. It was transferred on December 30, 1943 (cost is not listed on the appropriate RCW). It was armed with two 8-inch, ex-navy guns Model MkVIM3A2 on Model M1 barbette carriages (#160L2/#14 and #154L2/#13). It was never named, simply being known as Battery Construction No. 404. It served until the end of the war and was disarmed probably in 1946. The guns were destroyed by detonation and throwing over the cliff in front of the emplacement but have been recovered for display purposes. Parts of one still exist at the battery site. The emplacement still exists and is on display at the Miller Point State Park. The battery is open to the public.

- *Battery #297* (planned): A projected World War II dual 6-inch barbette battery for Fort Abercrombie. The work was projected only, and no actual work was ever undertaken.

- **AMTB Spruce Cape:** A 1943 project for an AMTB battery of two 90mm fixed and two 90mm mobile guns emplaced at Spruce Cape as part of the Fort Abercrombie defenses. It consisted of two fixed gun blocks (emplacements No. 1 and No. 2), a magazine, and a concrete BC station with wooden roof. Trunnion height was 59-feet. Work was done from August 1942 to December 30, 1943. Transfer was made on December 30, 1943, no cost is given on the RCW. It served as Kodiak Tactical Battery No. 8. It was probably abandoned and disarmed in 1946-1948. The battery is located on a current Coast Guard base.

Fort Greely (1941-1944) is located near the former naval operating base on Kodiak Island. Generally known as Artillery Hill, a 175-acre piece of property was reserved for a military post on October 22, 1939. It was named Fort Greely in General Order 17, April 2, 1943 after Maj. Gen. Adolphus W. Greely. He served in the Civil War, and postwar rose to command rank. In 1881 he commanded the Lady Franklin Bay Arctic Expedition. Development of the fort began in February 1941, after Pearl Harbor it was the site for a temporary battery of 155mm guns on Panama mounts at Buskin Hill as well as a temporary 6-inch ex-navy pedestal gun. It eventually became the headquarters post for the harbor defenses and had the area HDCP. An AMTB battery (1943-1945) was located offshore on Puffin Island. For much of the war all the 155mm guns of the defenses were stored at Greely pending emergency deployment. Late in the war, the post became important as a command and staging area for the campaigns to retake Kiska and Attu. In 1953 the fort property was transferred to the Coast Guard. The Artillery Hill area is now part of the native Buskin River property.

Fort Greely Gun Battery

- **AMTB Puffin Island:** A 1943 project for an AMTB battery of two 90mm fixed and two 90mm mobile guns emplaced on Puffin Island between Kodiak and Woody Islands It consisted of two fixed gun blocks (emplacements No. 1 and No. 2), a magazine, and a concrete BC station with wooden roof. Trunnion height was 59-feet. Work was done from August 1942 to December 30, 1943. Transfer was made on December 30, 1943, no cost is given on the RCW. It served as Kodiak Tactical Battery No. 8. It was probably abandoned and disarmed in 1946-1948. The battery is on a protected island requiring permission and boat to visit the site.

Battery 296, Fort Tidball, Long Island, Kodiak, AK (Terry McGovern)

Remains of the 6 inch guns of Battery #296 at Fort Tidball on Long Island (Terry McGovern)

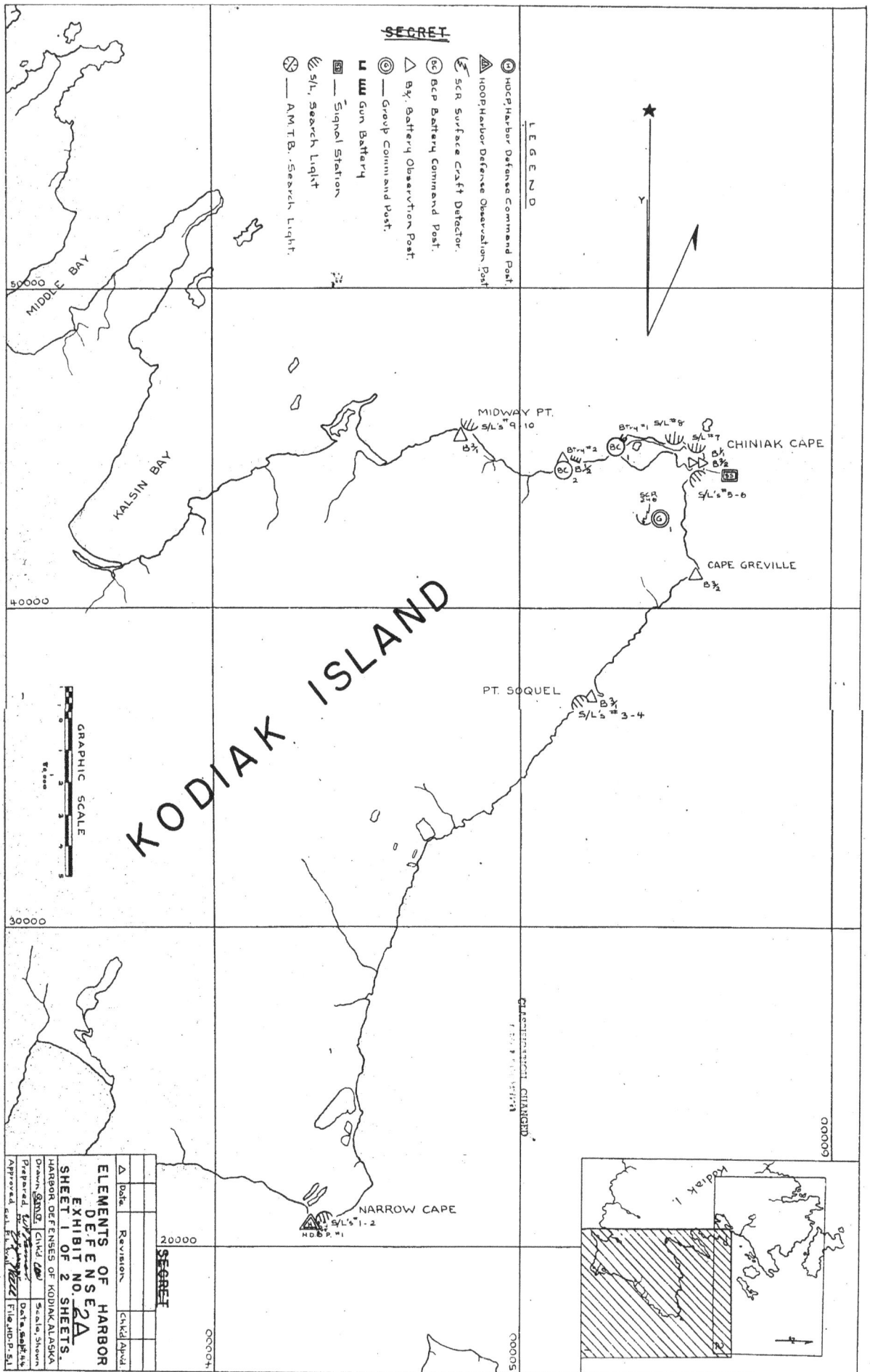

SECRET

LEGEND

HDCP Harbor Defense Command Post.
HDOP Harbor Defense Observation Post.
SCR Surface Craft Detector.
BCP Battery Command Post.
B'y. Battery Observation Post.
Group Command Post.
Gun Battery
Signal Station
S/L Search Light.
A.M.T.B. Search Light.

MIDDLE BAY

50000

KALSIN BAY

40000

30000

KODIAK ISLAND

GRAPHIC SCALE

MIDWAY PT.
S/L's #9-10
B¾

Bt'y #2
BC
B⅜
2

Bt'y #1 S/L #8
BC
S/L #7 CHINIAK CAPE
B⅛
B¾
S/L's #5-6

SCR

CAPE GREVILLE
B¾

PT. SOQUEL
B¾
S/L's #3-4

20000

NARROW CAPE
S/L's #1-2
HDCP #1

CLASSIFICATION CHANGED

60000

50000

40000

30000

ELEMENTS OF HARBOR
D E F E N S E
EXHIBIT NO. 2A
SHEET 1 OF 2 SHEETS.
HARBOR DEFENSES OF KODIAK, ALASKA

SECRET

Kodiak I.

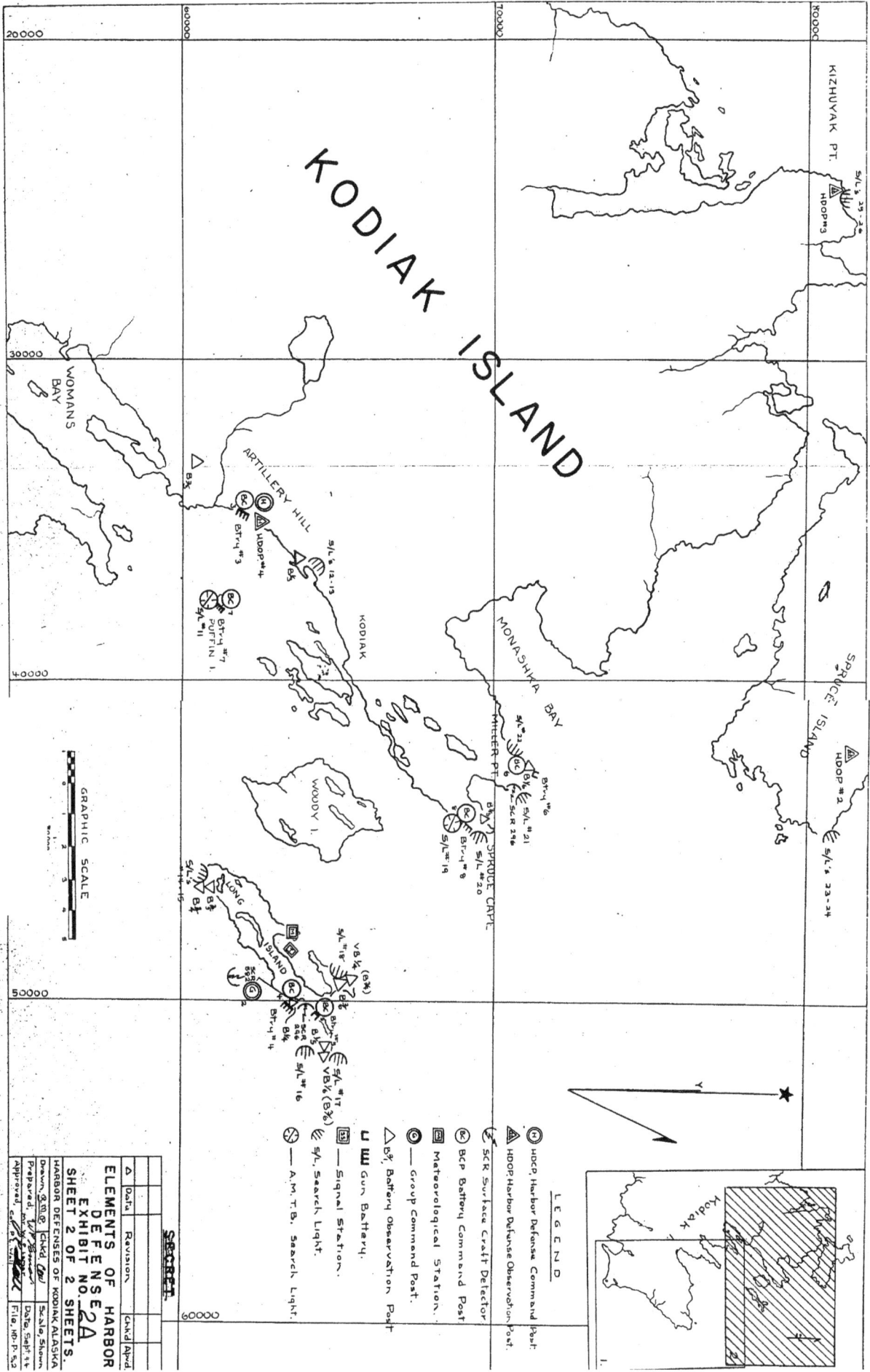

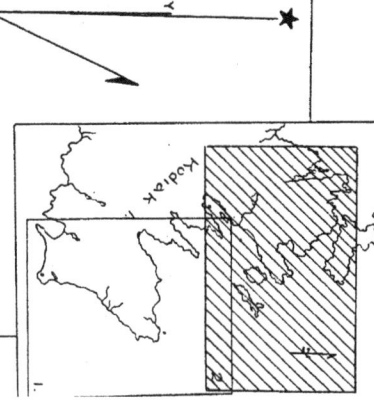

Kodiak Alaska World War II-era Site Locations

location	Loc#	Purpose
Soquel Point		fire control station searchlight position
Narrow Cape	1A	fire control station searchlight position
Cape Graville	2	fire control station
Cape Chiniak	3	fire control station searchlight position
Bald Hill	3A	searchlight position
Round Top	3A	searchlight position
St. Peters Head, Chiniak	4-5	gun position, fire control
Midway Point	6	fire control station searchlight position
Mansfield Ridge	7	fire control station
Artillery Hill	8	gun position fire control stations
Puffin Island	8A	gun position, searchlight position
Gibson Cove	9	fire control station, searchlight position
Long Island	10-14	gun position, fire control, searchlights
Spruce Cape	15	gun position, fire control, searchlights
Miller Point	16	gun position, searchlights
Spruce Island	17-18	searchlights, fire control
Kizhuyak Point	18	fire control, searchlights

Emplacment of Battery #404 at Fort Ambercrobie State Park(Terry McGovern)

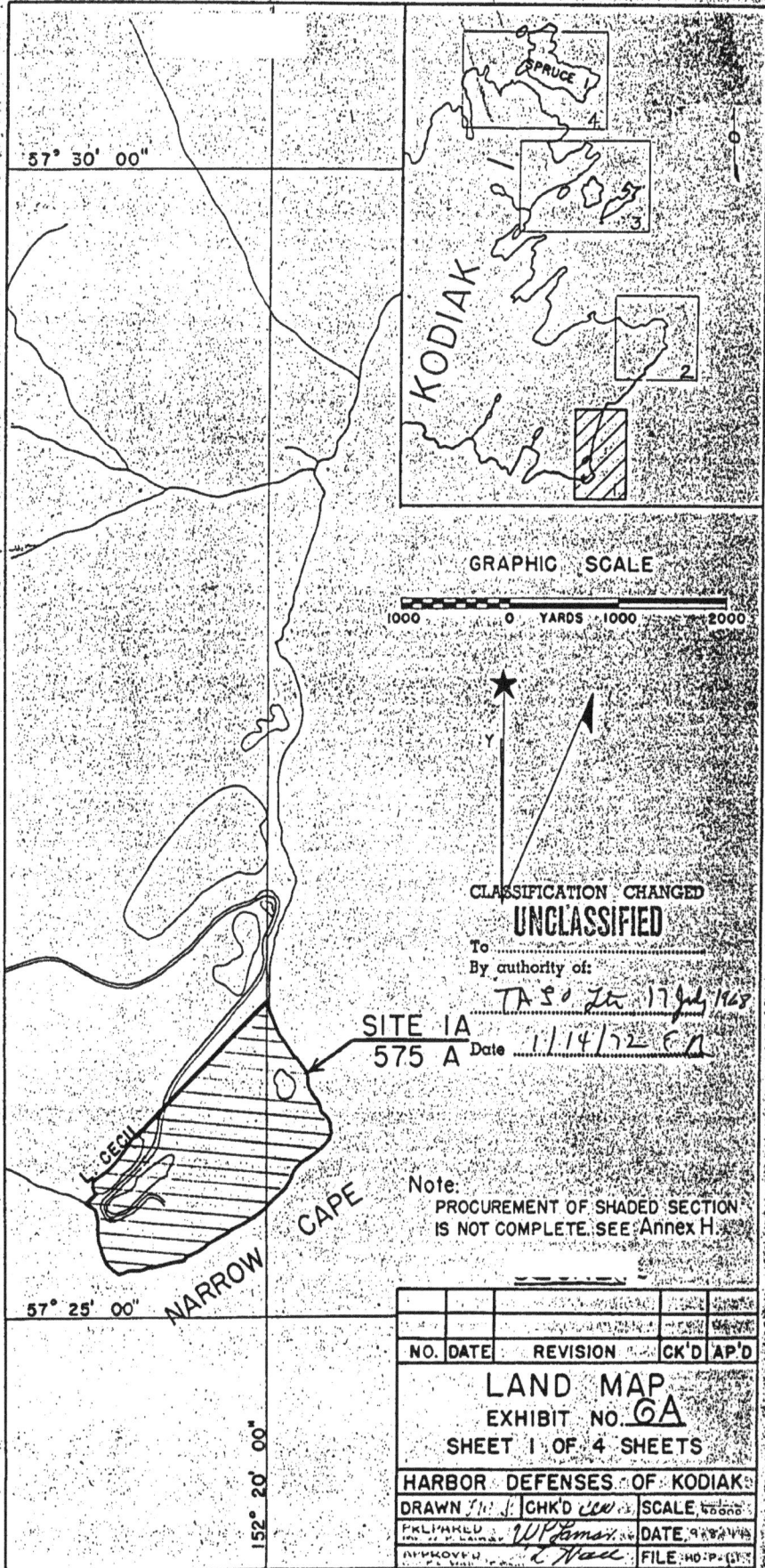

57° 30' 00"

KODIAK I.

SPRUCE I.

4

5

3

2

GRAPHIC SCALE

1000   0   YARDS   1000   2000

CLASSIFICATION CHANGED

UNCLASSIFIED

To ....

By authority of:

TASO Ltr 17 July 1968

Date ... 1/14/72 EA

SITE IA
575 A

L. CECIL

Note:
PROCUREMENT OF SHADED SECTION
IS NOT COMPLETE, SEE Annex H.

NARROW CAPE

57° 25' 00"

152° 20' 00"

NO.	DATE	REVISION	CK'D	AP'D

LAND MAP
EXHIBIT NO. 6A
SHEET 1 OF 4 SHEETS

HARBOR DEFENSES OF KODIAK

DRAWN	CHK'D CCW	SCALE 10000
PREPARED W.P.Lamar		DATE 9 8 44
APPROVED C Hall		FILE NO. P.

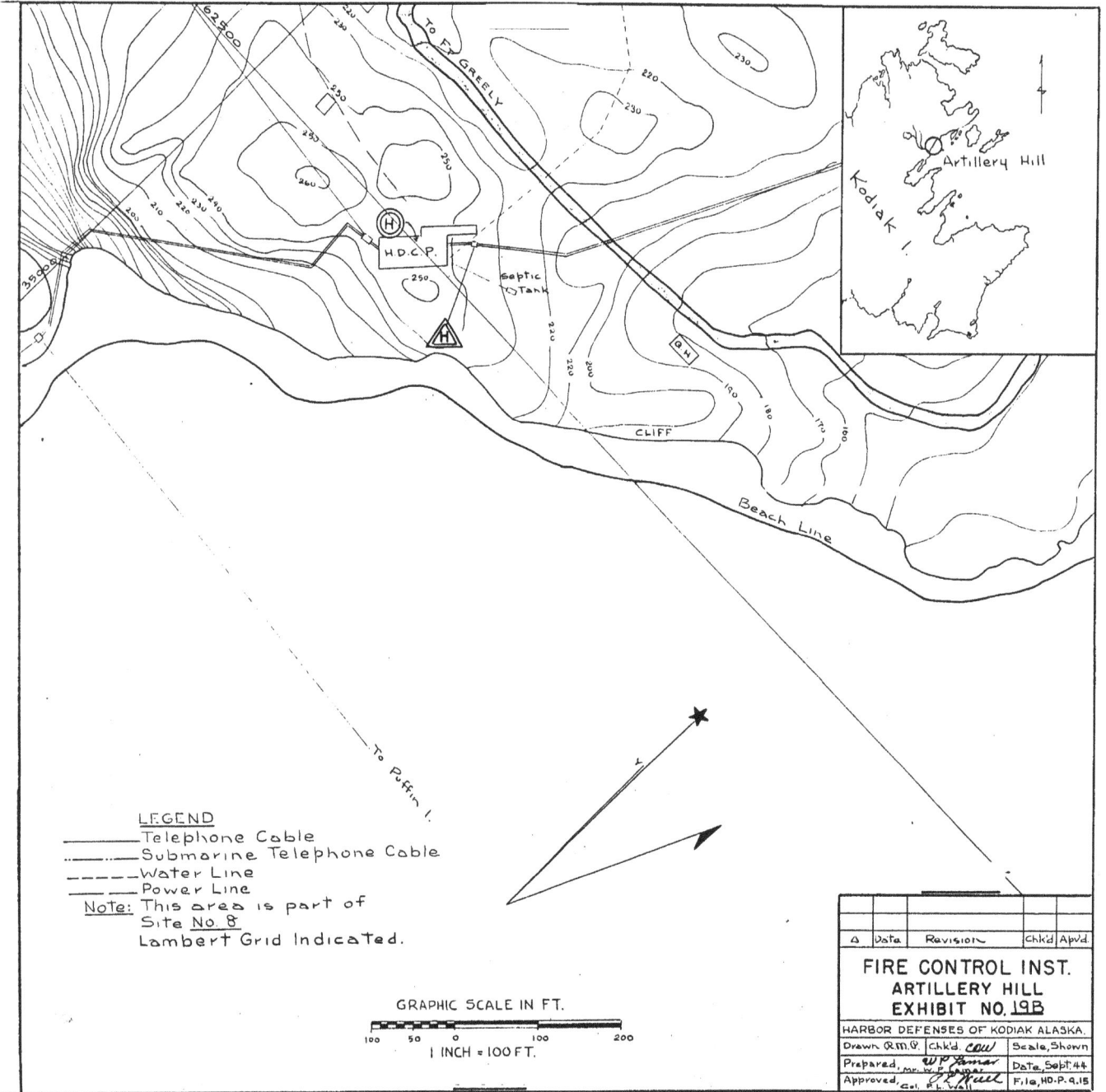

LEGEND
———————— Telephone Cable
——·——·—— Submarine Telephone Cable
— — — — Water Line
— — — — Power Line
Note: This area is part of
Site No. 8
Lambert Grid Indicated.

GRAPHIC SCALE IN FT.

100   50   0   100   200

1 INCH = 100 FT.

△	Date	Revision	Chk'd	Apvd.

FIRE CONTROL INST.
ARTILLERY HILL
EXHIBIT NO. 19B

HARBOR DEFENSES OF KODIAK ALASKA.

Drawn R.M.P.	Chk'd CDW	Scale, Shown
Prepared, Mr. W. P. Lamar		Date, Sept. 44
Approved, Col. P. L. Wall		File, HD-P-9.15

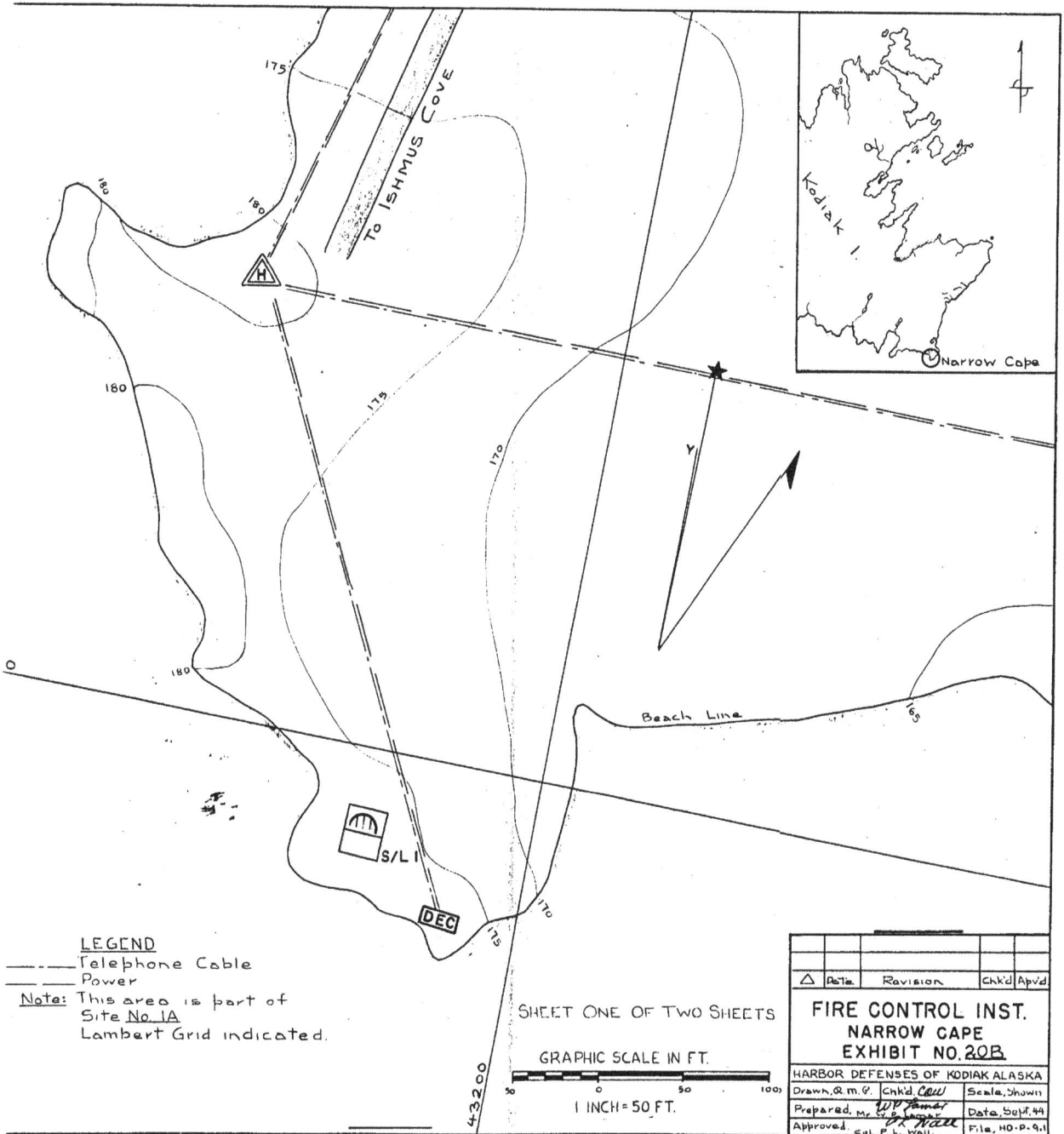

LEGEND
——·—— Telephone Cable
———— Power
Note: This area is part of
Site No. 1A
Lambert Grid indicated.

SHEET ONE OF TWO SHEETS

GRAPHIC SCALE IN FT.

1 INCH = 50 FT.

To ISHMUS COVE

Kodiak I.

Narrow Cape

Beach Line

S/L I

DEC

△	Date	Revision	Chk'd	Apv'd

FIRE CONTROL INST.
NARROW CAPE
EXHIBIT NO. 20B

HARBOR DEFENSES OF KODIAK ALASKA

Drawn. Q. m. P.	Chk'd. CRW	Scale, Shown
Prepared. Mr. W. James		Date. Sept. 44
Approved. Cal P. L. Wall		File. HO-P-9.1

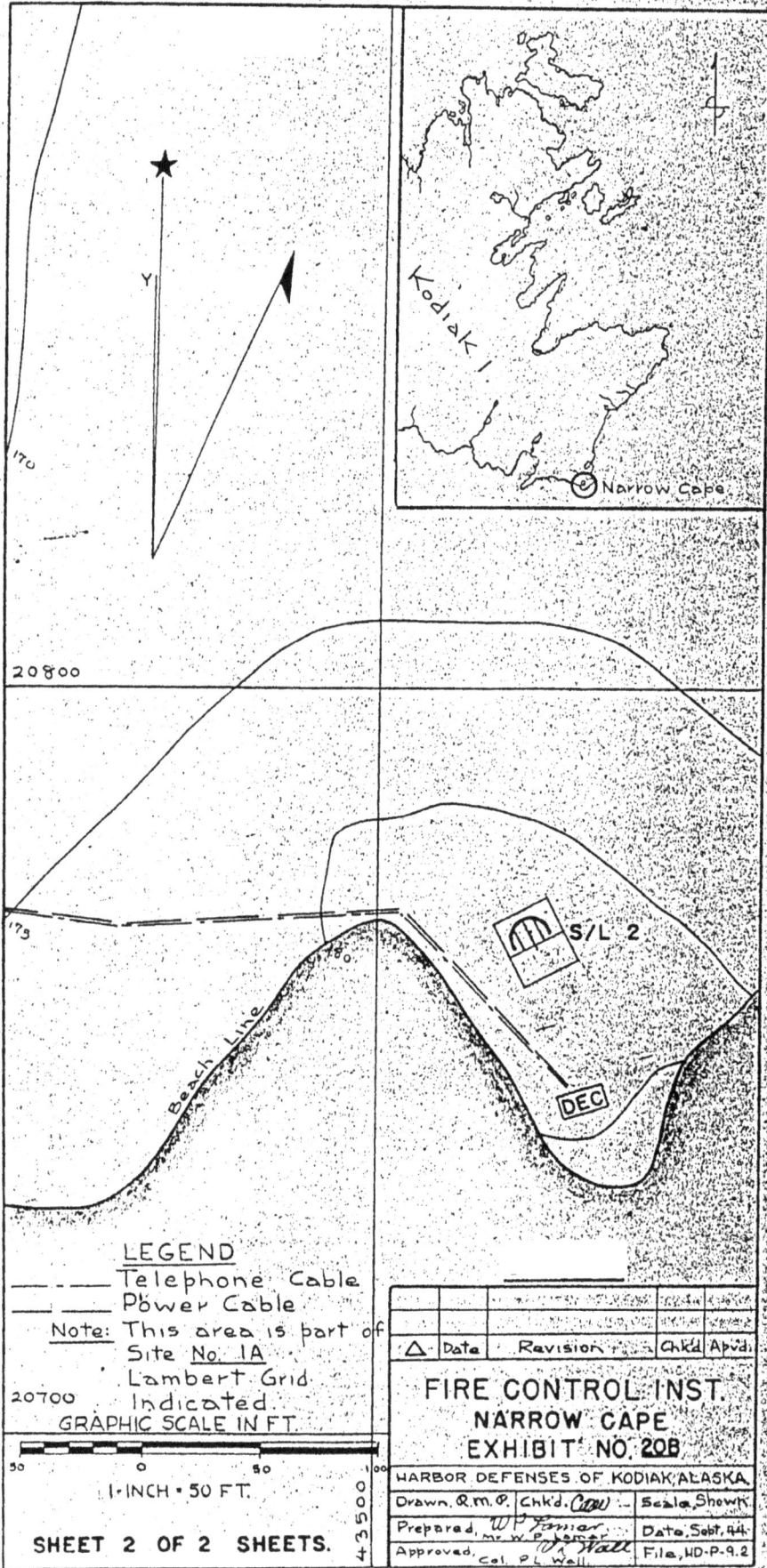

LEGEND
— · — Telephone Cable
——— Power Cable
Note: This area is part of
Site No. 1A
Lambert Grid
20700   Indicated
GRAPHIC SCALE IN FT.

50   0   50   100
1·INCH = 50 FT.

SHEET 2 OF 2 SHEETS.

△	Date	Revision	Chk'd	Ap'd

FIRE CONTROL INST.
NARROW CAPE
EXHIBIT NO. 20B

HARBOR DEFENSES OF KODIAK, ALASKA

Drawn. Q.m.P.	Chk'd. CBW	Scale Shown
Prepared, W.P. Pomior Mr. W. P. Pomier		Date Sept. 44
Approved, J.L. Wall Col. P.L. Wall		File, HD-P-9.2

Kodiak I.

Narrow Cape

S/L 2

DEC

Beach Line

170

173

20800

20700

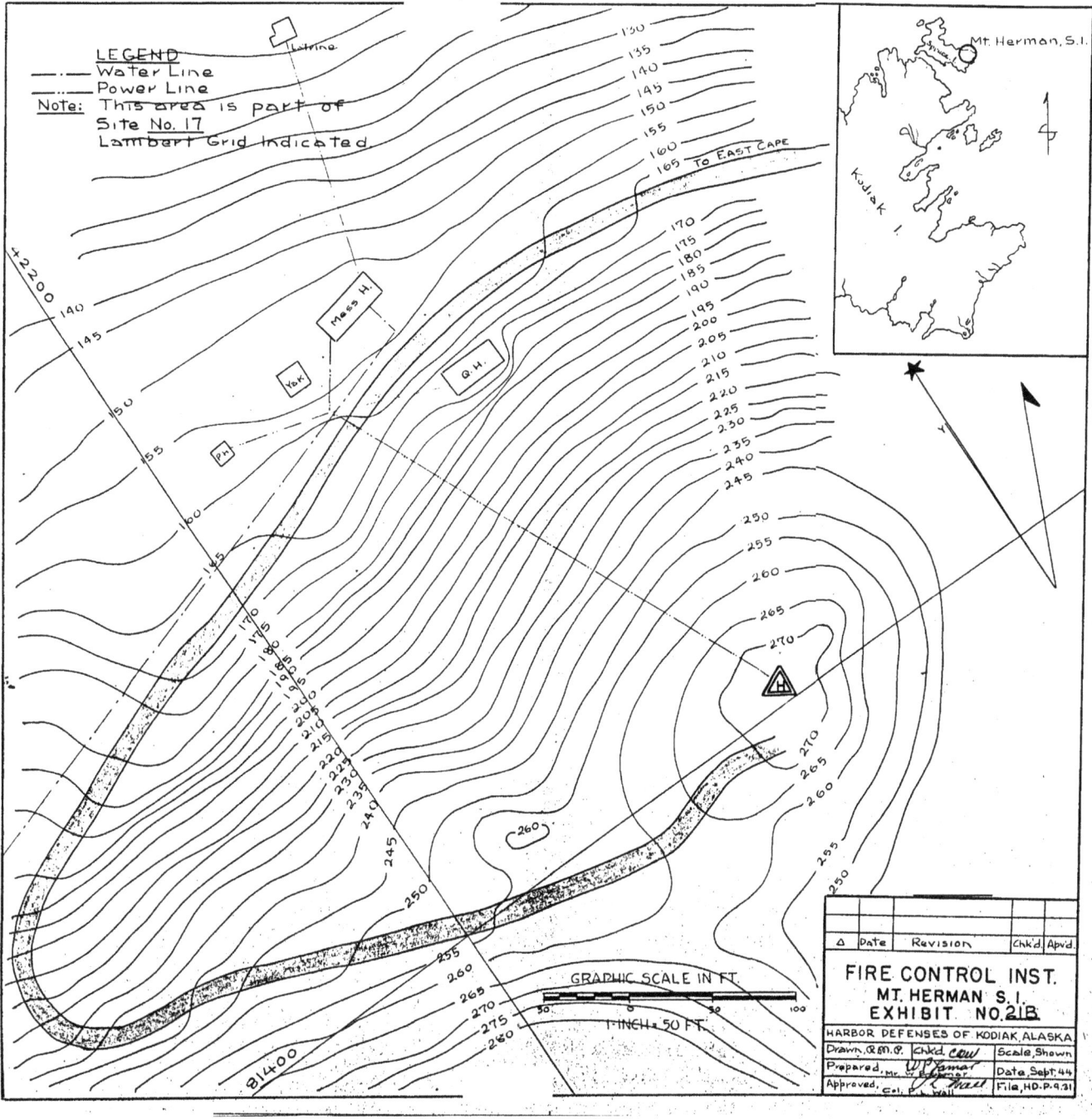

LEGEND
—·— Water Line
——— Power Line
Note: This area is part of
Site No. 17
Lambert Grid Indicated

Latrine

Mess H.

Yak

Q.H.

P.H.

TO EAST CAPE

130
135
140
145
150
155
160
165
170
175
180
185
190
195
200
205
210
215
220
225
230
235
240
245
250
255
260
265
270

+2200
140
145
150
155
160
145
170
175
180
185
190
195
200
205
210
215
220
225
230
235
240
245
250
255
260
265
270
275
280

81400

260

270
265
260
255
250

GRAPHIC SCALE IN FT.
50   0   50   100
1 INCH = 50 FT.

Mt. Herman, S.I.

Kodiak I.

△	Date	Revision	Chk'd	Apv'd

FIRE CONTROL INST.
MT. HERMAN S.I.
EXHIBIT NO. 21B

HARBOR DEFENSES OF KODIAK, ALASKA

Drawn, Q.M.P.	Chk'd CRW	Scale, Shown
Prepared, Mr. W.P.Lamar		Date, Sept. 44
Approved, Col. E.L. Hall		File, HD.P.9.31

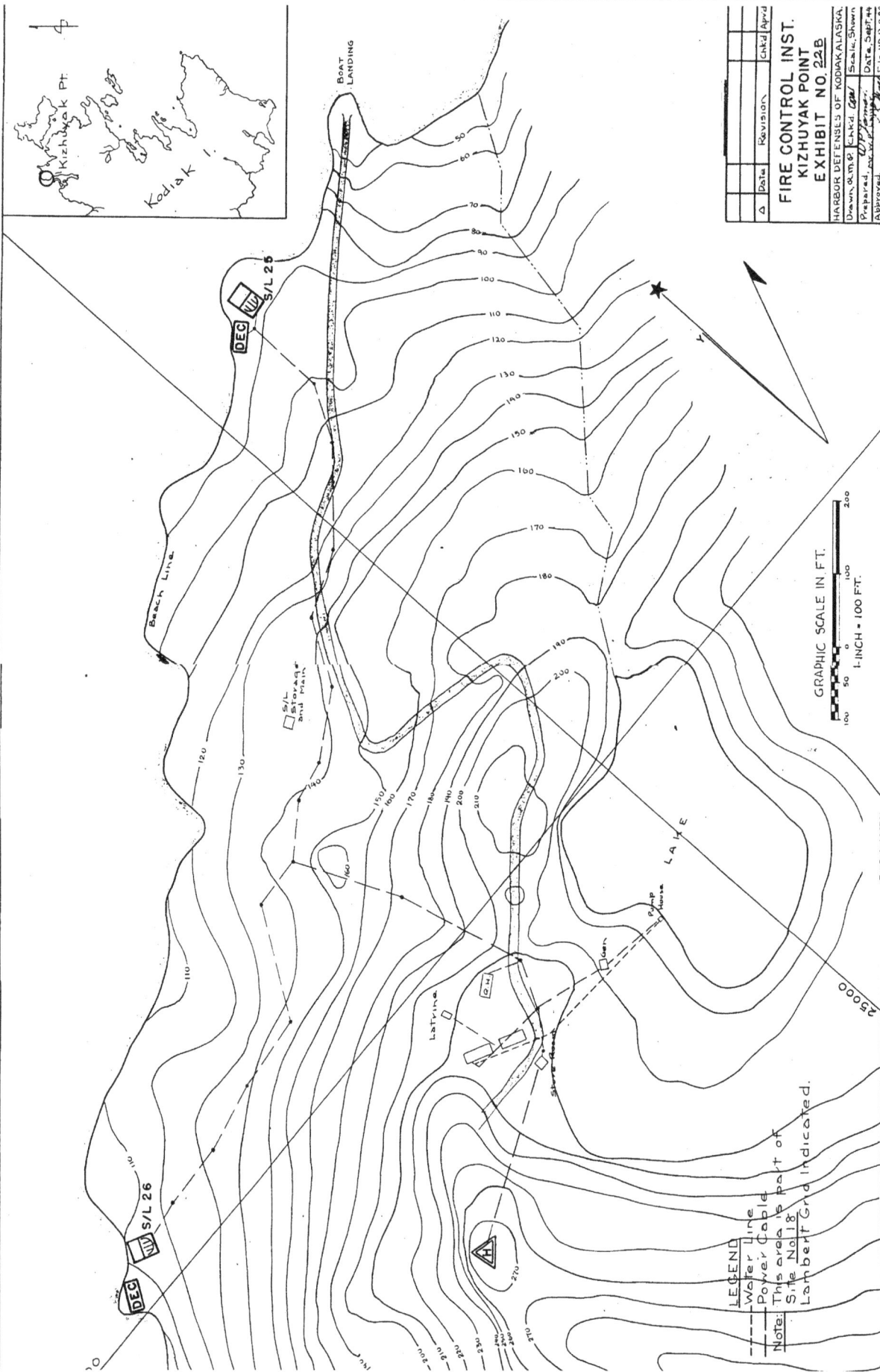

FIRE CONTROL INST.
KIZHUYAK POINT
EXHIBIT NO. 22B

HARBOR DEFENSES OF KODIAK, ALASKA.

LEGEND
—— Water Line
–––– Power Cable
Note: This area is part of
Site No. 18
Lambert Grid Indicated.

GRAPHIC SCALE IN FT.
1-INCH = 100 FT.

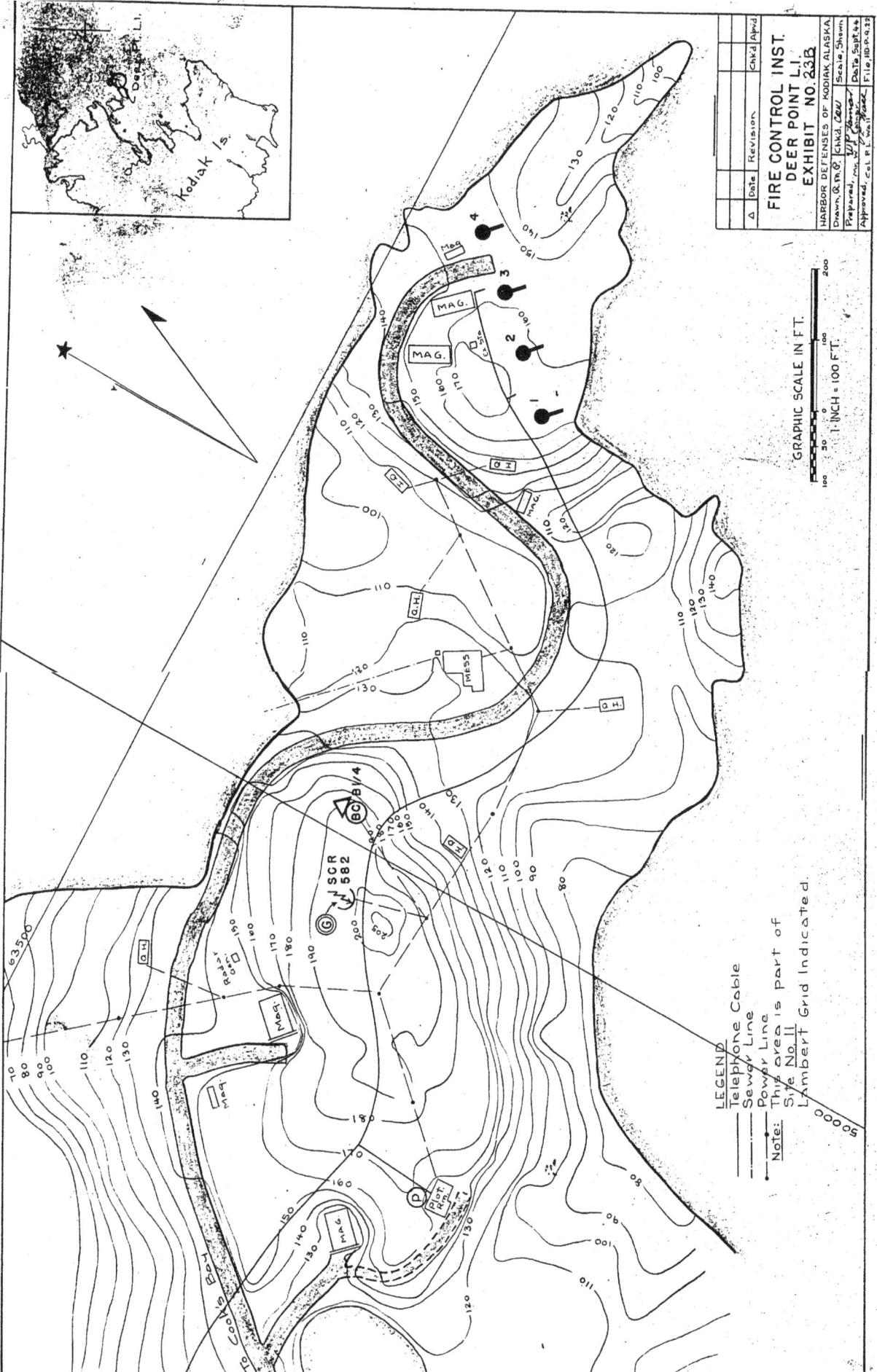

FIRE CONTROL INST.
DEER POINT L.I.
EXHIBIT NO. 23B

HARBOR DEFENSES OF KODIAK, ALASKA

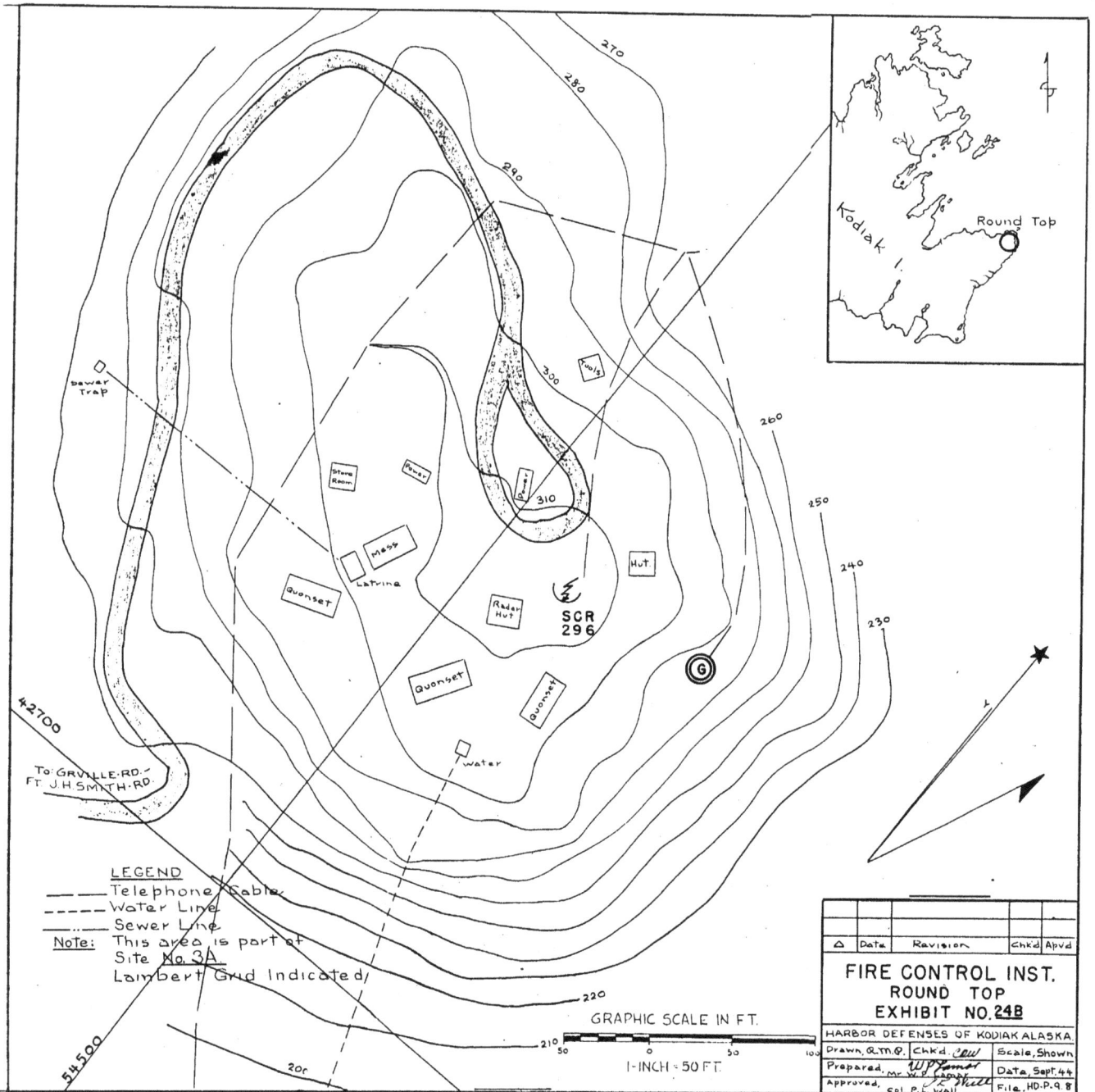

LEGEND
– – – Telephone Cable
– – – Water Line
········ Sewer Line
Note: This area is part of
Site No. 3A
(Lambert Grid Indicated)

GRAPHIC SCALE IN FT.
1-INCH = 50 FT.

FIRE CONTROL INST.
ROUND TOP
EXHIBIT NO. 24B

HARBOR DEFENSES OF KODIAK ALASKA

Drawn, Q.M.O.　Chk'd　Scale, Shown
Prepared, Mr W.P. Tomas　Date, Sept. 44
Approved, Col P.L. Wall　File, HD-P-9-8

Kodiak Is.

Curto Pt.

House

To Cooks Bay

Latrine

Hut

M

Yak

120

110

100

90

GRAPHIC SCALE IN FT.

1 INCH = 50 FT.

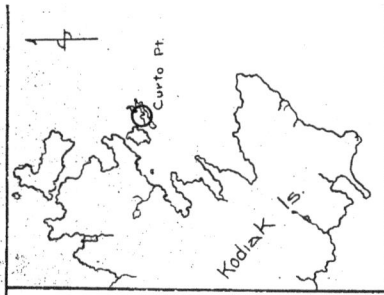

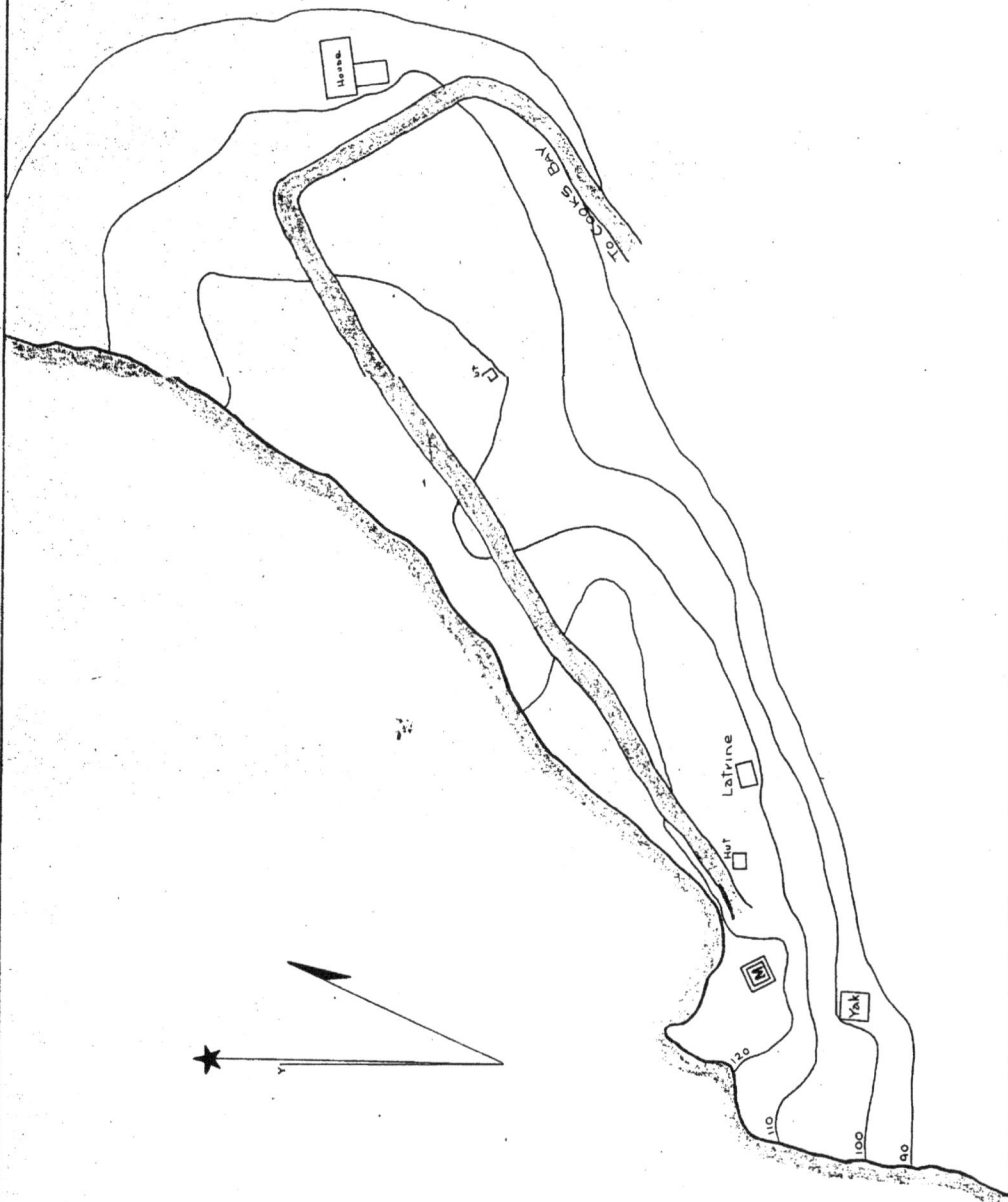

Note: This area is part of Site No.14A Lambert Grid Indicated

477100

FIRE CONTROL INST.
POINT CURTO
EXHIBIT NO. 25B

HARBOR DEFENSES OF KODIAK ALASKA

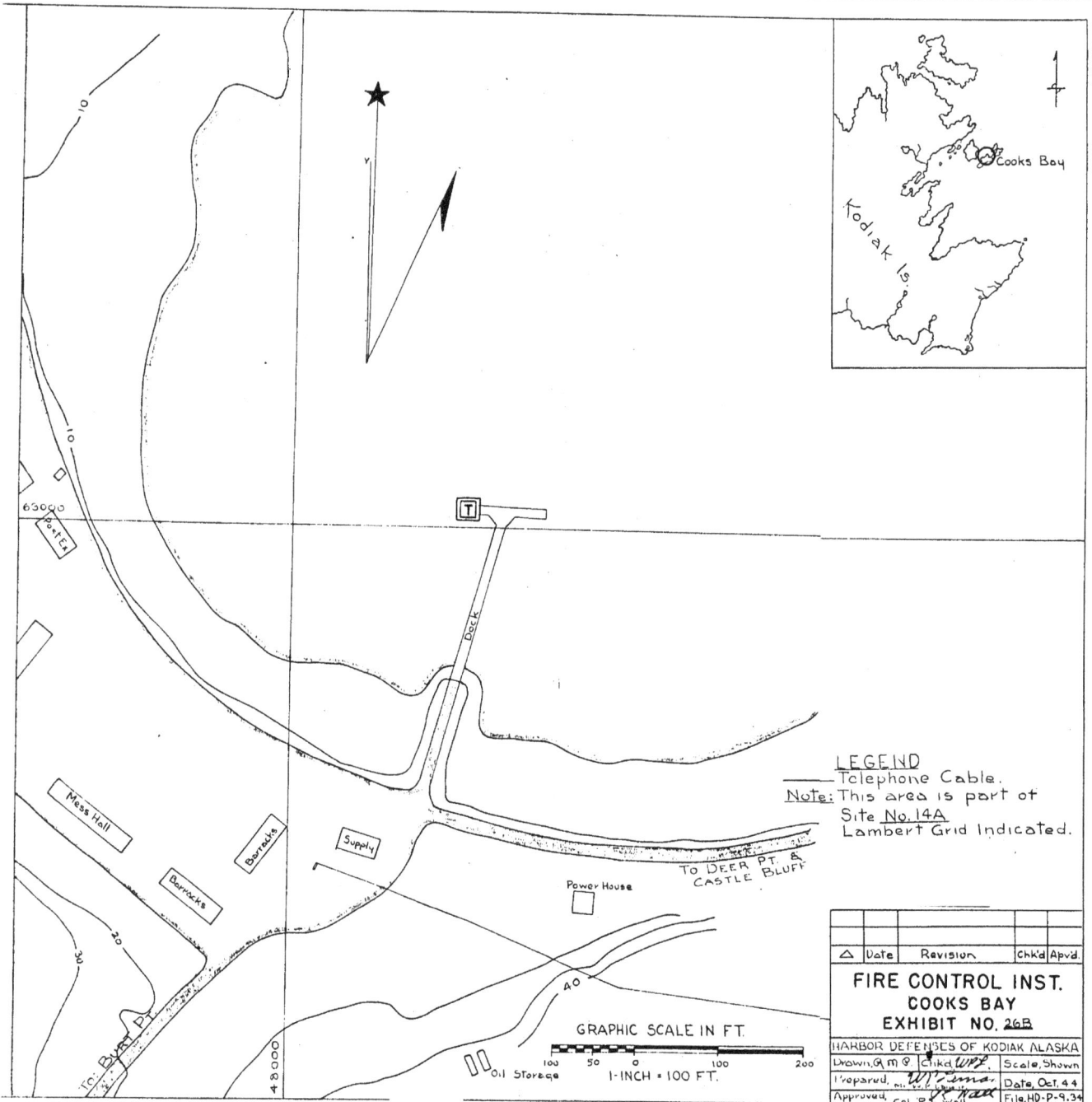

LEGEND
Telephone Cable.
Note: This area is part of Site No. 14A
Lambert Grid Indicated.

GRAPHIC SCALE IN FT.
1-INCH = 100 FT.

To DEER PT. &
CASTLE BLUFF

△	Date	Revision	Chk'd	Apv'd.

FIRE CONTROL INST.
COOKS BAY
EXHIBIT NO. 26B

HARBOR DEFENSES OF KODIAK ALASKA

Drawn, G. M. P.	Chkd WPF.	Scale, Shown
Prepared,	W. Pima,	Date, Oct. 44
Approved, Col. P. L. Wall		File, HD-P-9.34

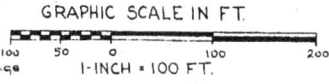

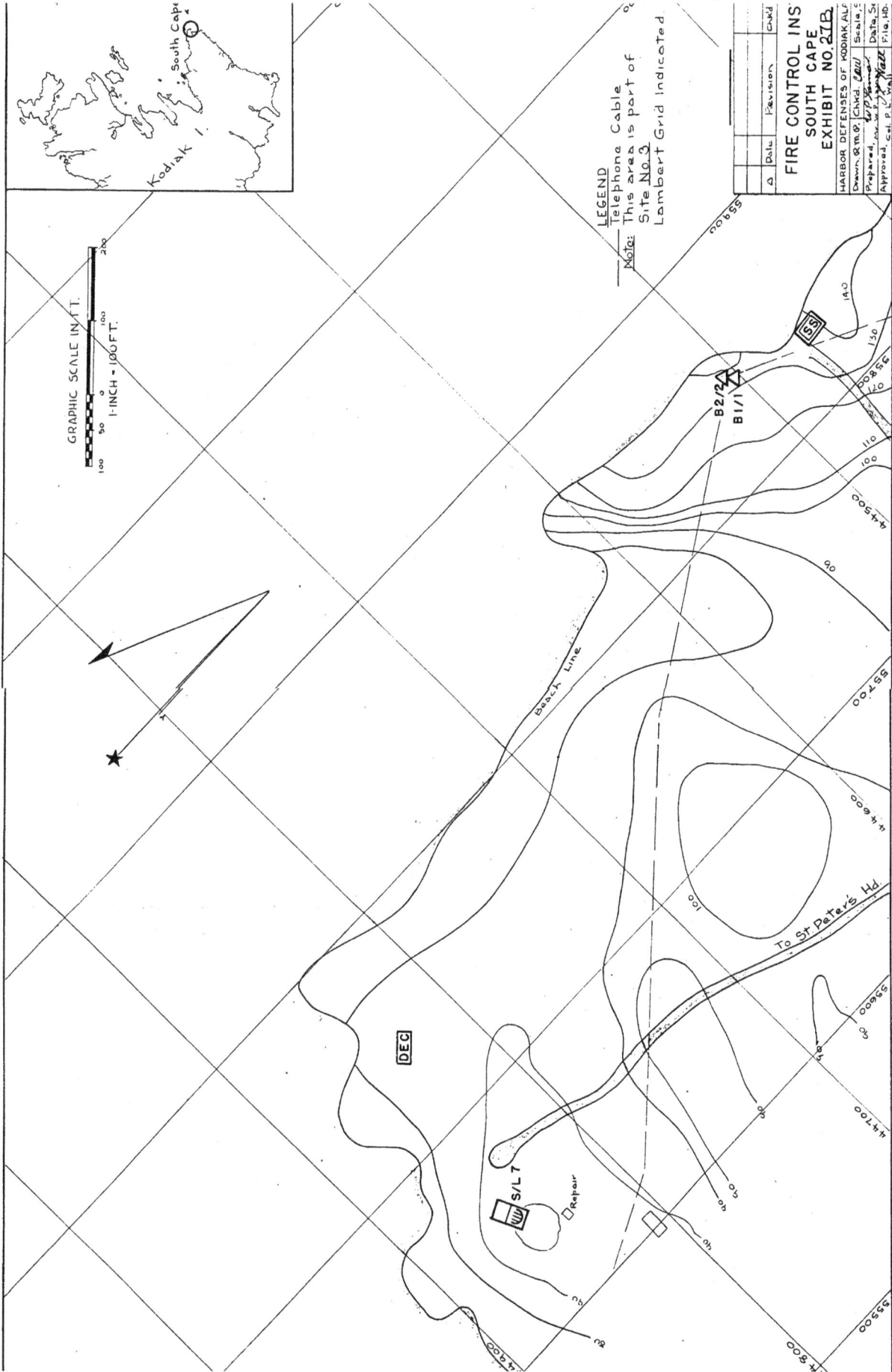

GRAPHIC SCALE IN FT.

1-INCH = 100 FT.

LEGEND
Telephone Cable
Note: This area is part of
Site No. 3
Lambert Grid indicated

FIRE CONTROL INS
SOUTH CAPE
EXHIBIT NO. 27B

HARBOR DEFENSES OF KODIAK ALE

B 2/2
B 1/1

SS

Beach Line

To St Peter's Hd

DEC

S/L 7

Repair

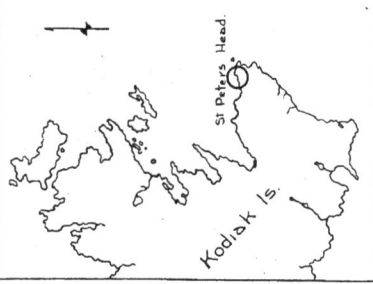

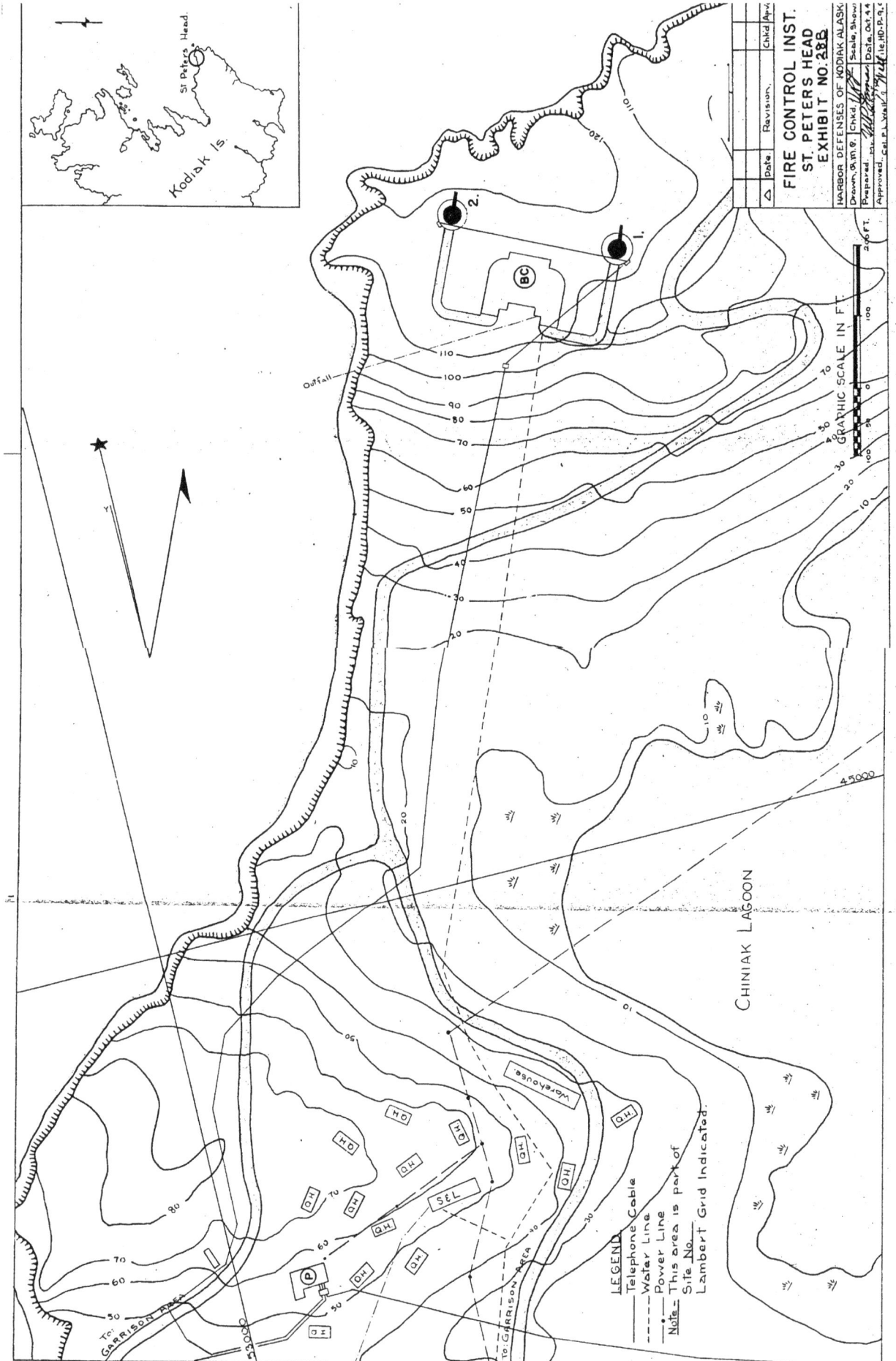

FIRE CONTROL INST.
ST. PETERS HEAD
EXHIBIT NO. 28B

HARBOR DEFENSES OF KODIAK ALASKA

Kodiak Is.

St Peters Head.

BC

CHINIAK LAGOON

Warehouse.

LEGEND
Telephone Cable
Water Line
Power Line
Note - This area is part of
Site No.
Lambert Grid Indicated.

GRAPHIC SCALE IN F

To GARRISON AREA

To GARRISON AREA

Outfall

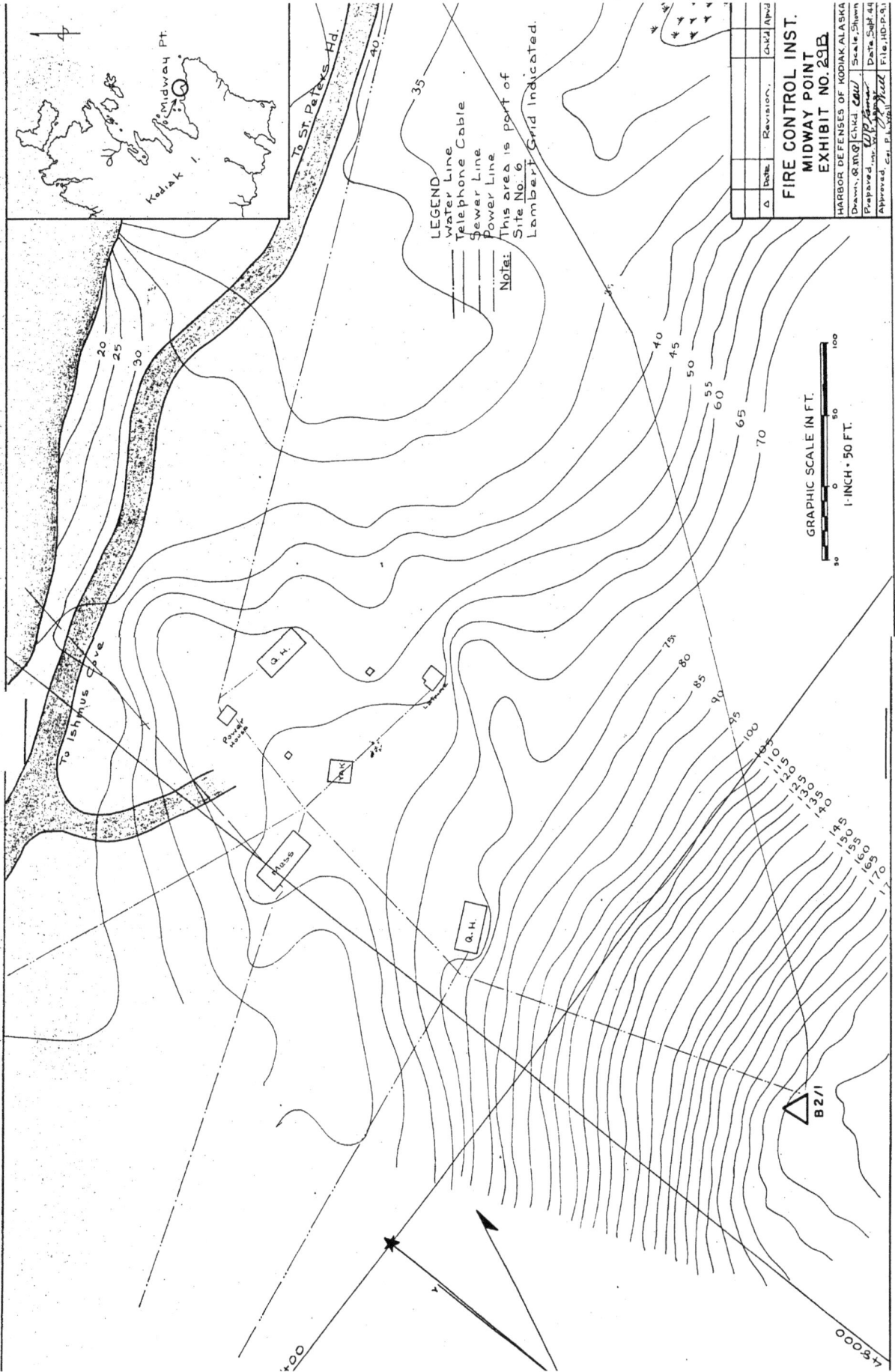

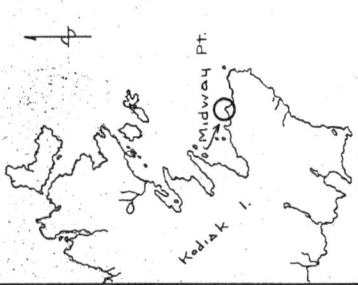

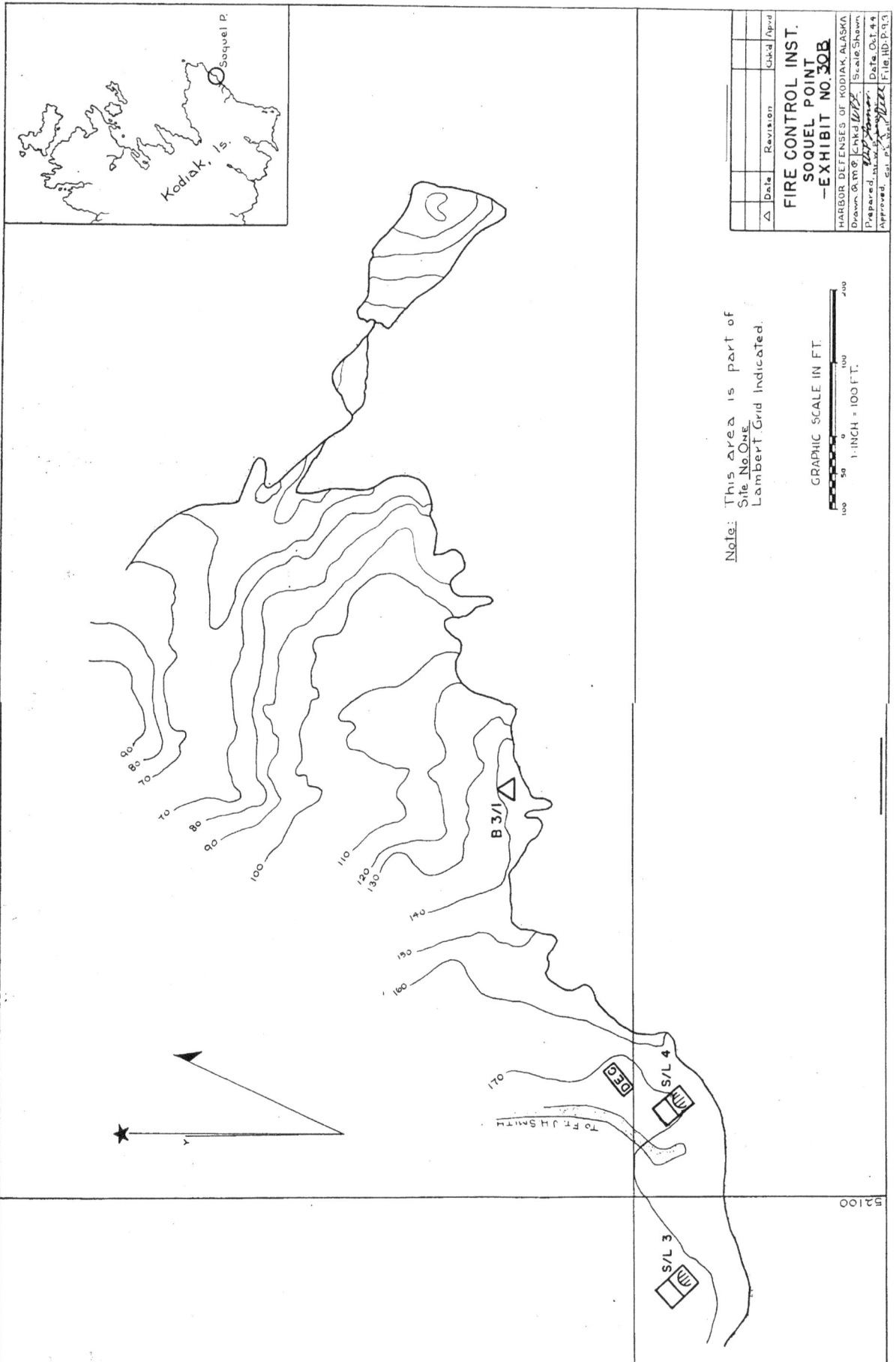

FIRE CONTROL INST.
SOQUEL POINT
—EXHIBIT NO. 30B

HARBOR DEFENSES OF KODIAK, ALASKA
Drawn R.M.G. Chkd. Scale Shown
Prepared Date Oct. 44
Approved File HD-P-9.3

Note: This area is part of Site No One. Lambert Grid Indicated.

GRAPHIC SCALE IN FT.
1-INCH = 100 FT.

B 3/1

S/L 4

DEC

To Ft. J.H.Smith

S/L 3

FIRE CONTROL INST.
CHINIAK POINT
EXHIBIT NO. 31B

HARBOR DEFENSES OF KODIAK ALASKA

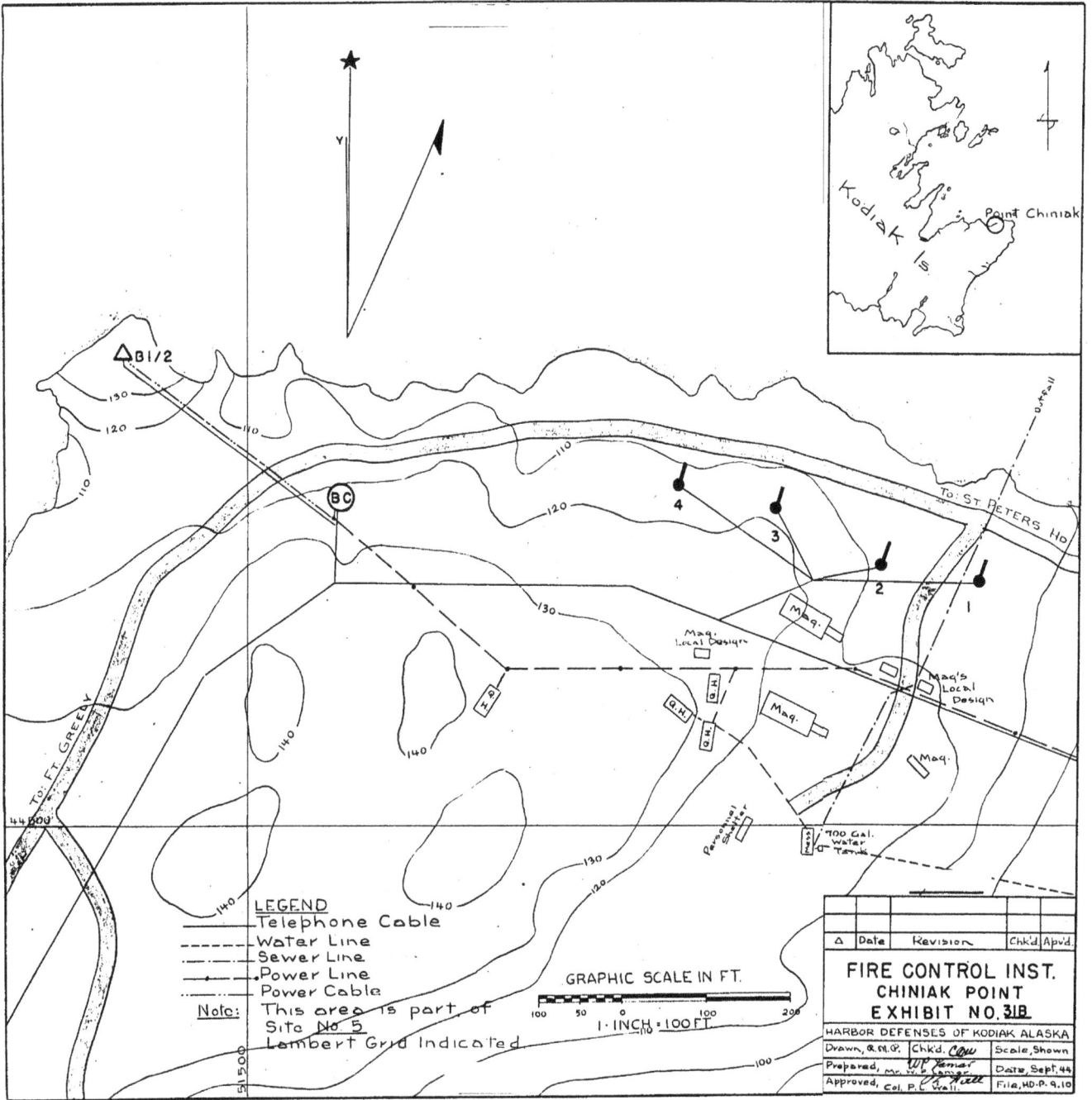

LEGEND
Telephone Cable
Water Line
Sewer Line
Power Line
Power Cable

Note: This area is part of
Site No 5
Lambert Grid Indicated

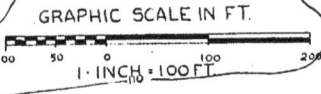

GRAPHIC SCALE IN FT.

1 INCH = 100 FT.

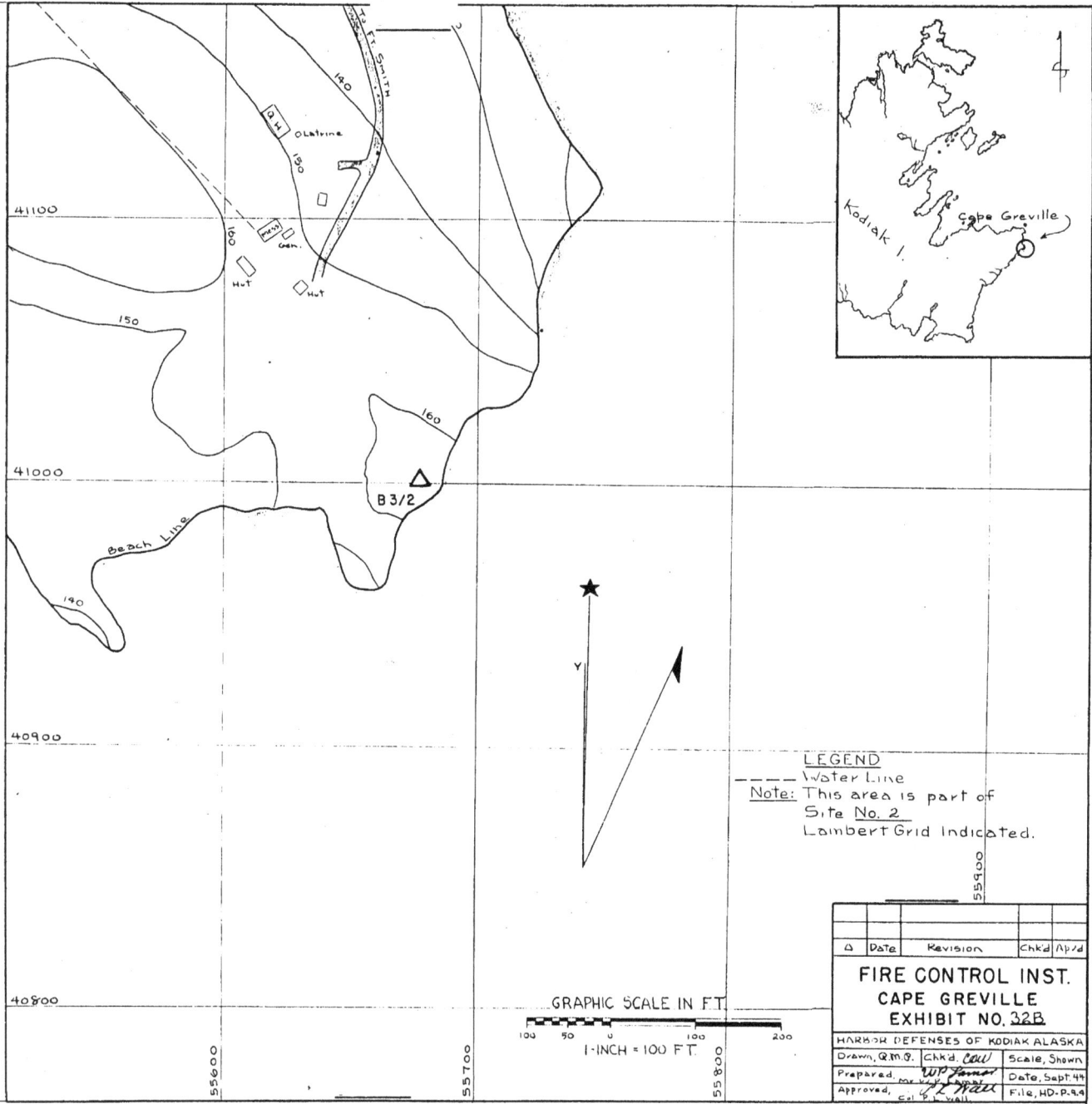

LEGEND
---- Water Line
Note: This area is part of
Site No. 2
Lambert Grid Indicated.

GRAPHIC SCALE IN F.T.
100   50   0   100   200
1-INCH = 100 FT.

△	Date	Revision	Chk'd	Ap'd

FIRE CONTROL INST.
CAPE GREVILLE
EXHIBIT NO. 32B

HARBOR DEFENSES OF KODIAK ALASKA

Drawn, Q.M.O.	Chk'd. *COW*	Scale, Shown
Prepared, *WP Jones*		Date, Sept. 44
Approved, Col P. L. Wall		File, HD-P.9.4

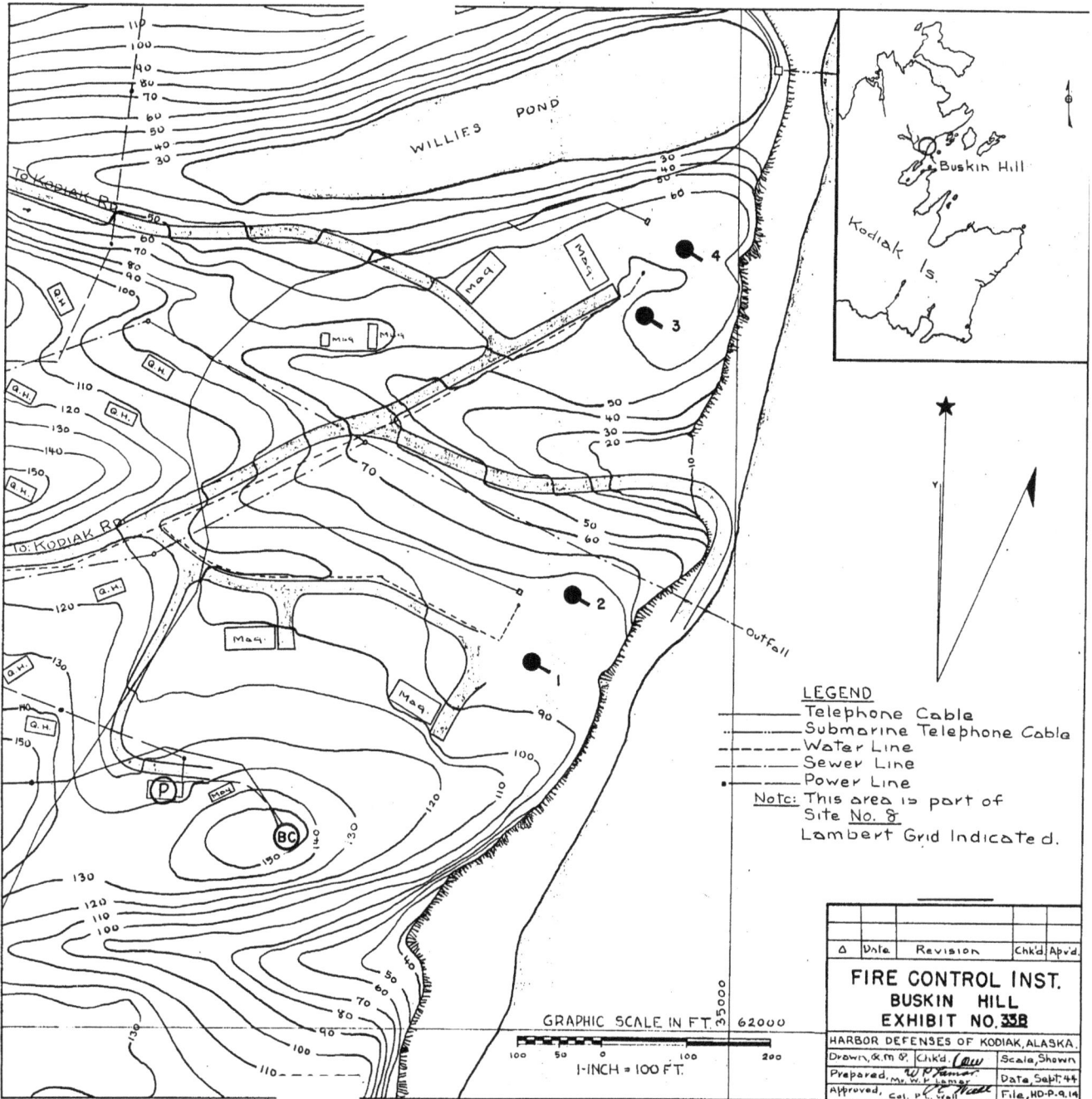

WILLIES POND

To KODIAK RD.

Mag

Mag

Mag

Mag

Mag

4

3

To KODIAK RD.

Outfall

2

1

Mag

Mag

P

Mag

BC

LEGEND
Telephone Cable
Submarine Telephone Cable
Water Line
Sewer Line
Power Line
Note: This area is part of
Site No. 8
Lambert Grid Indicated.

Buskin Hill

Kodiak Is.

GRAPHIC SCALE IN FT.

100 50 0 100 200

1-INCH = 100 FT.

Δ	Date	Revision	Chk'd	Apv'd

FIRE CONTROL INST.
BUSKIN HILL
EXHIBIT NO. 33B

HARBOR DEFENSES OF KODIAK, ALASKA.

Drawn, & m P.	Chk'd.	Scale, Shown
Prepared, Mr. W. P. Lamar		Date, Sept. 44
Approved, Col. P.C. Wall		File, HD-P-9.14

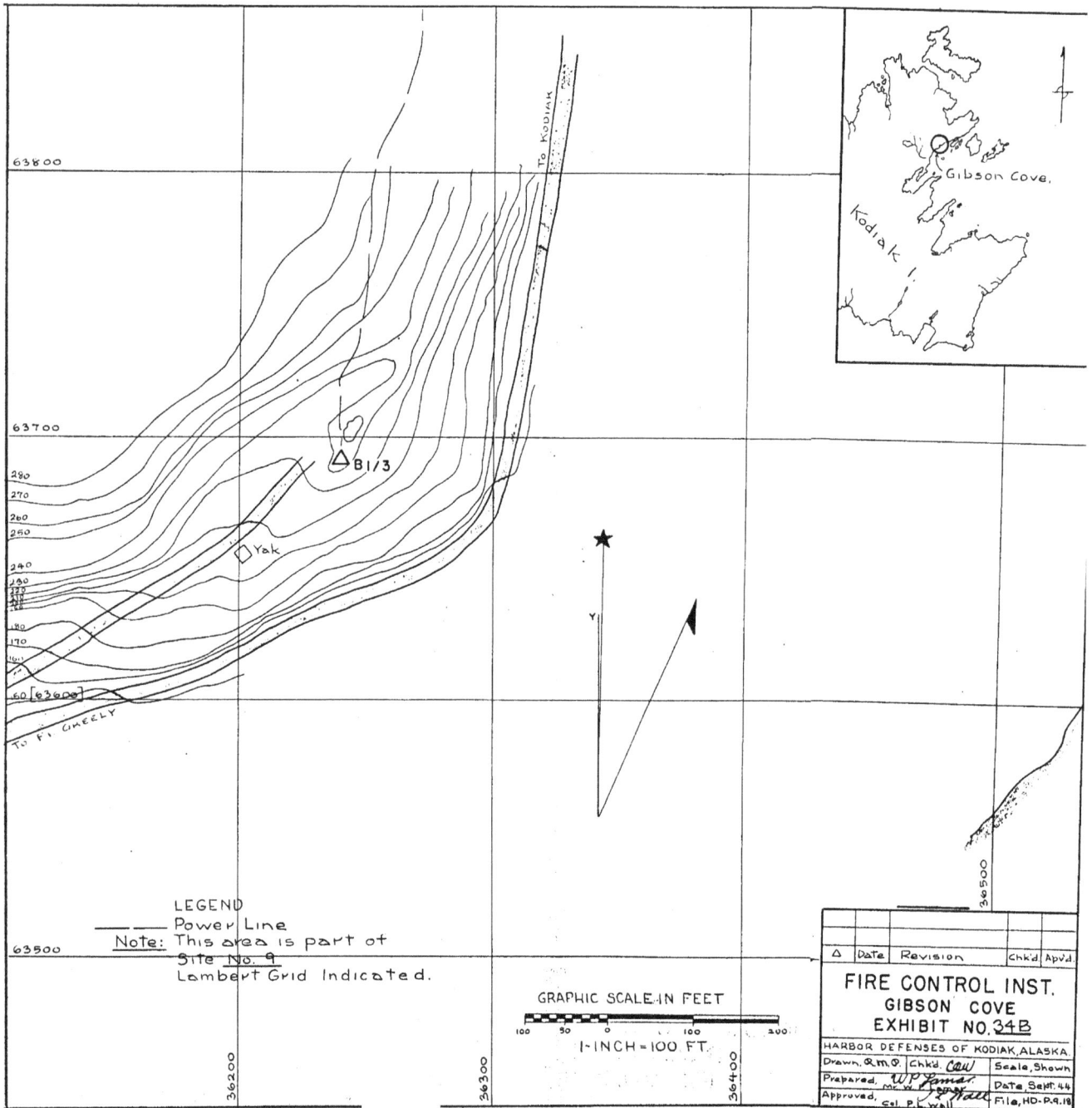

LEGEND
— — Power Line
Note: This area is part of Site No. 9
Lambert Grid Indicated.

GRAPHIC SCALE IN FEET
100 50 0   100   200
I-INCH=100 FT.

FIRE CONTROL INST.
GIBSON COVE
EXHIBIT NO. 34B

HARBOR DEFENSES OF KODIAK, ALASKA

△	Date	Revision	Chk'd	Apv'd

Drawn, R.m.O. Chk'd CAW Scale, Shown
Prepared, W.P. James Date Sept. 44
Approved, Col. P.L. Wall File, HD-P.9.18

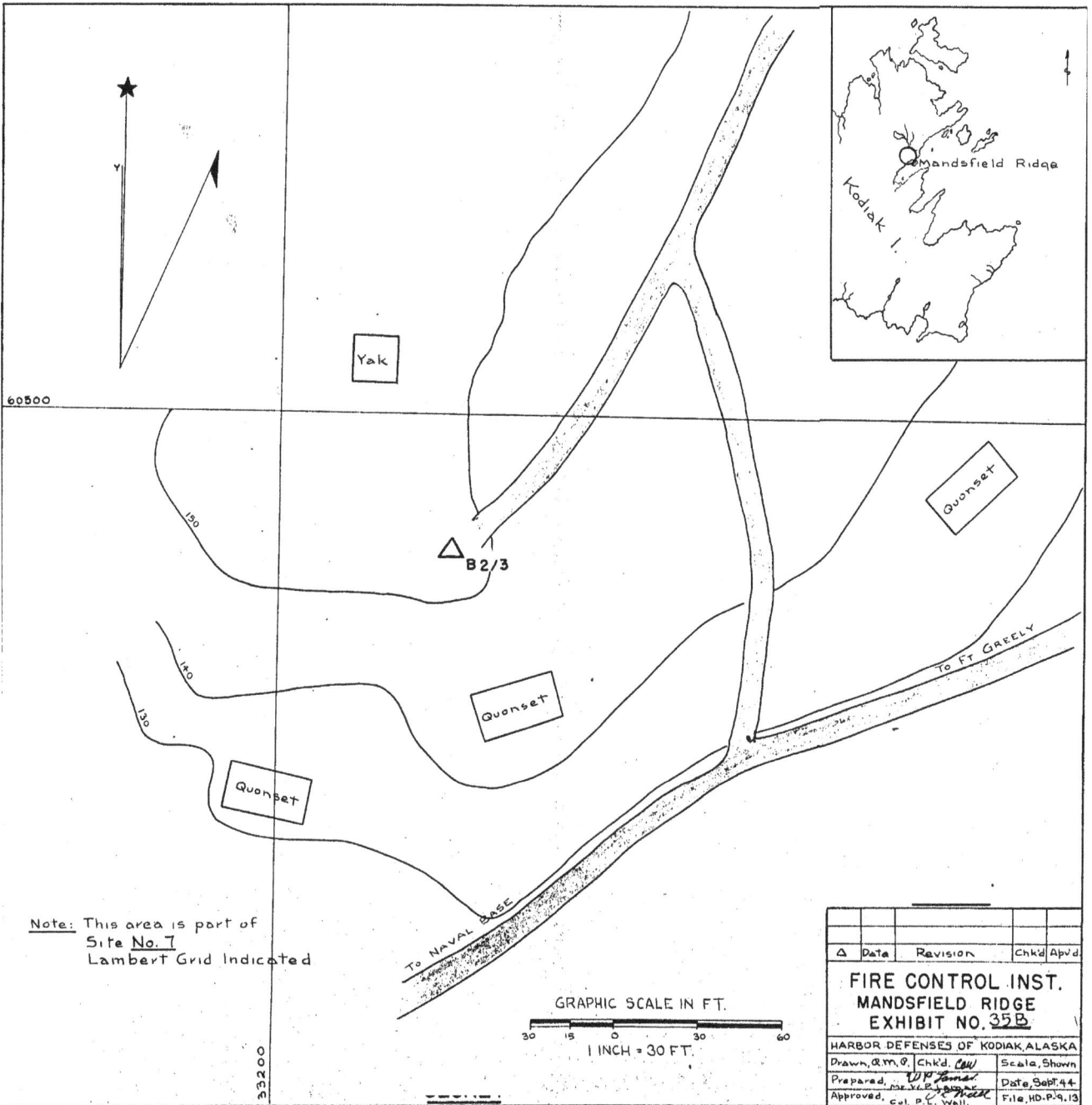

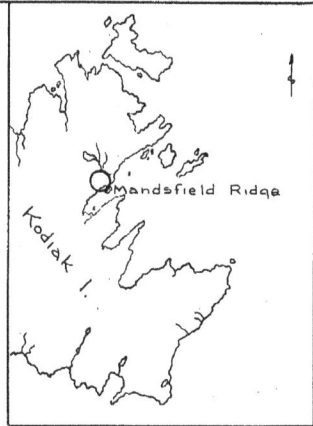

Yak

Mandsfield Ridge

Kodiak I.

Quonset

B 2/3

150

140

130

Quonset

Quonset

To FT GREELY

To NAVAL BASE

Note: This area is part of
Site No. 7
Lambert Grid Indicated

60500

33200

GRAPHIC SCALE IN FT.

30 15 0 30 60
1 INCH = 30 FT.

△	Date	Revision	Chk'd	Apv'd

FIRE CONTROL INST.
MANDSFIELD RIDGE
EXHIBIT NO. 35B

HARBOR DEFENSES OF KODIAK, ALASKA

Drawn, Q.M.O.	Chk'd. CAW	Scale, Shown
Prepared,		Date, Sept. 44
Approved, Col. P.L. Wall.		File HD-P-9.13

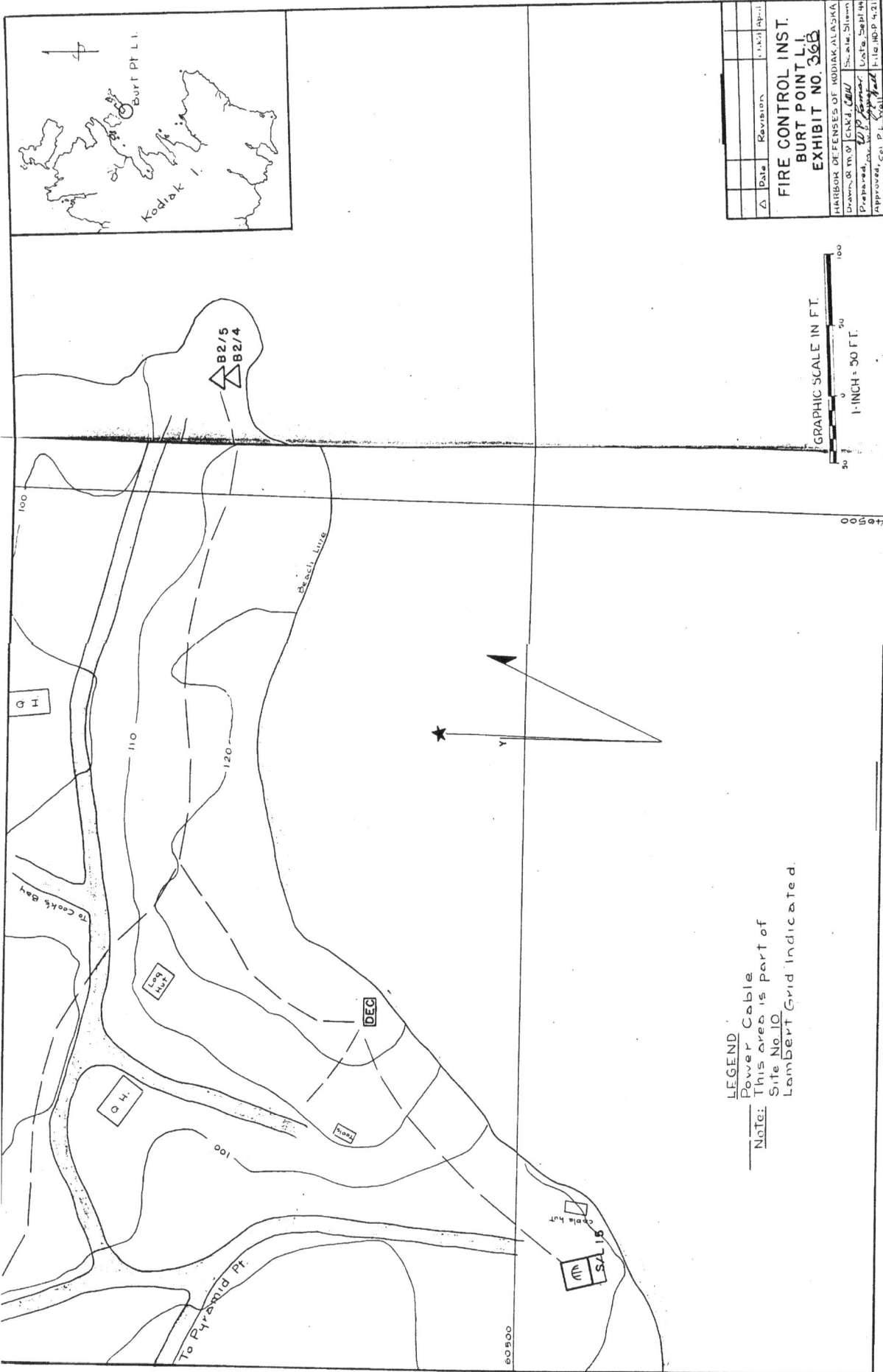

FIRE CONTROL INST.
BURT POINT L.I.
EXHIBIT NO. 36B

HARBOR DEFENSES OF KODIAK, ALASKA

GRAPHIC SCALE IN FT.

1-INCH = 50 FT.

LEGEND
Power Cable
Note: This area is part of Site No. 10
Lambert Grid indicated.

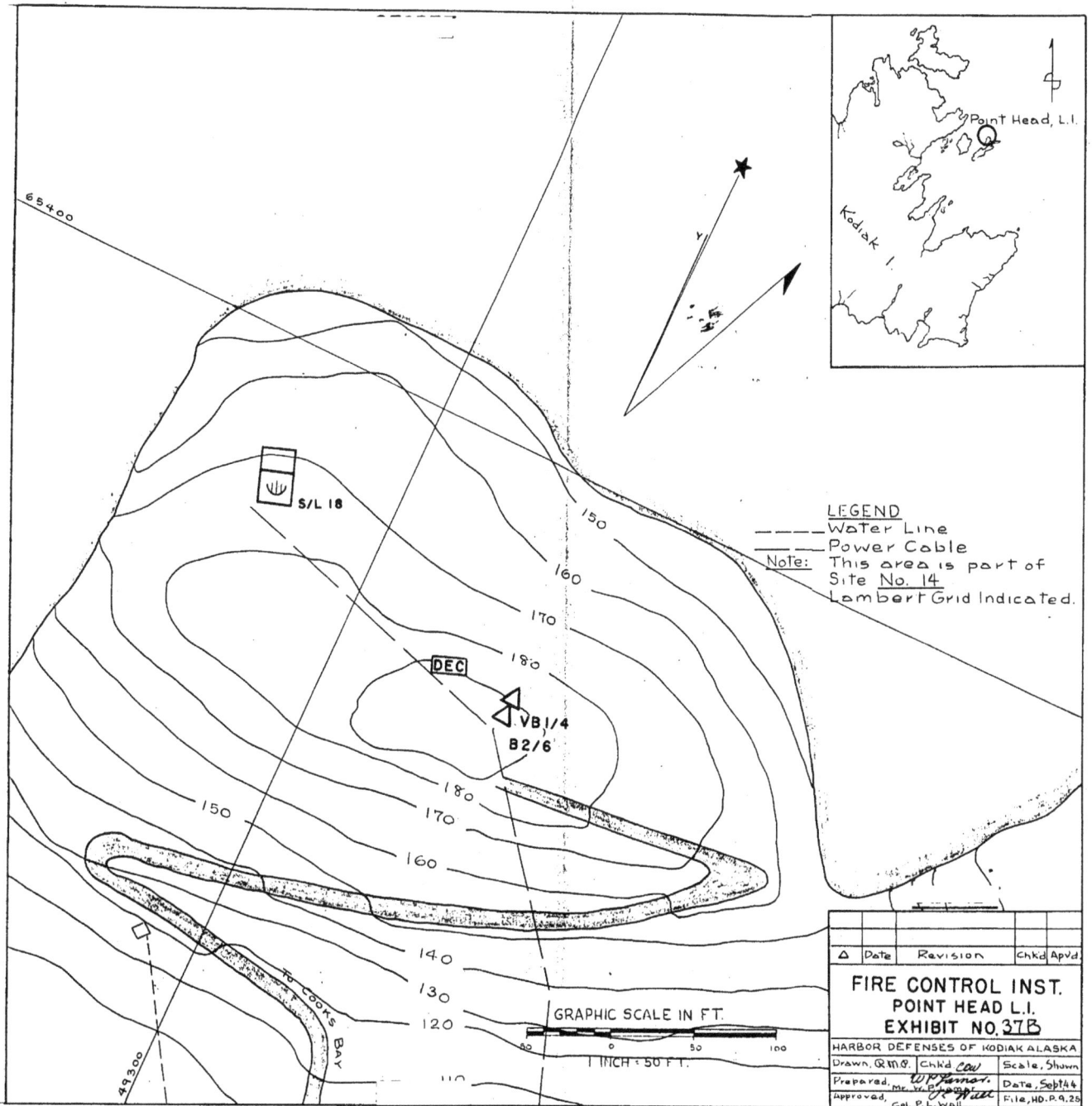

65400

Point Head, L.I.

Kodiak I.

150

160

170

180

S/L 18

DEC

VB1/4
B2/6

180

170

150

160

150

140

130

120

110

49300

TO COOKS BAY

LEGEND
Water Line
Power Cable
Note: This area is part of
Site No. 14
Lambert Grid Indicated.

GRAPHIC SCALE IN FT.
0   50   100
1 INCH = 50 FT.

△	Date	Revision	Chk'd	Apvd.

FIRE CONTROL INST.
POINT HEAD L.I.
EXHIBIT NO. 37B

HARBOR DEFENSES OF KODIAK ALASKA

Drawn, QMO.	Chk'd COW	Scale, Shown
Prepared, Mr. W. P. Jamar.		Date, Sept 44
Approved, Col. P. L. Wall		File, HD.P.9.25

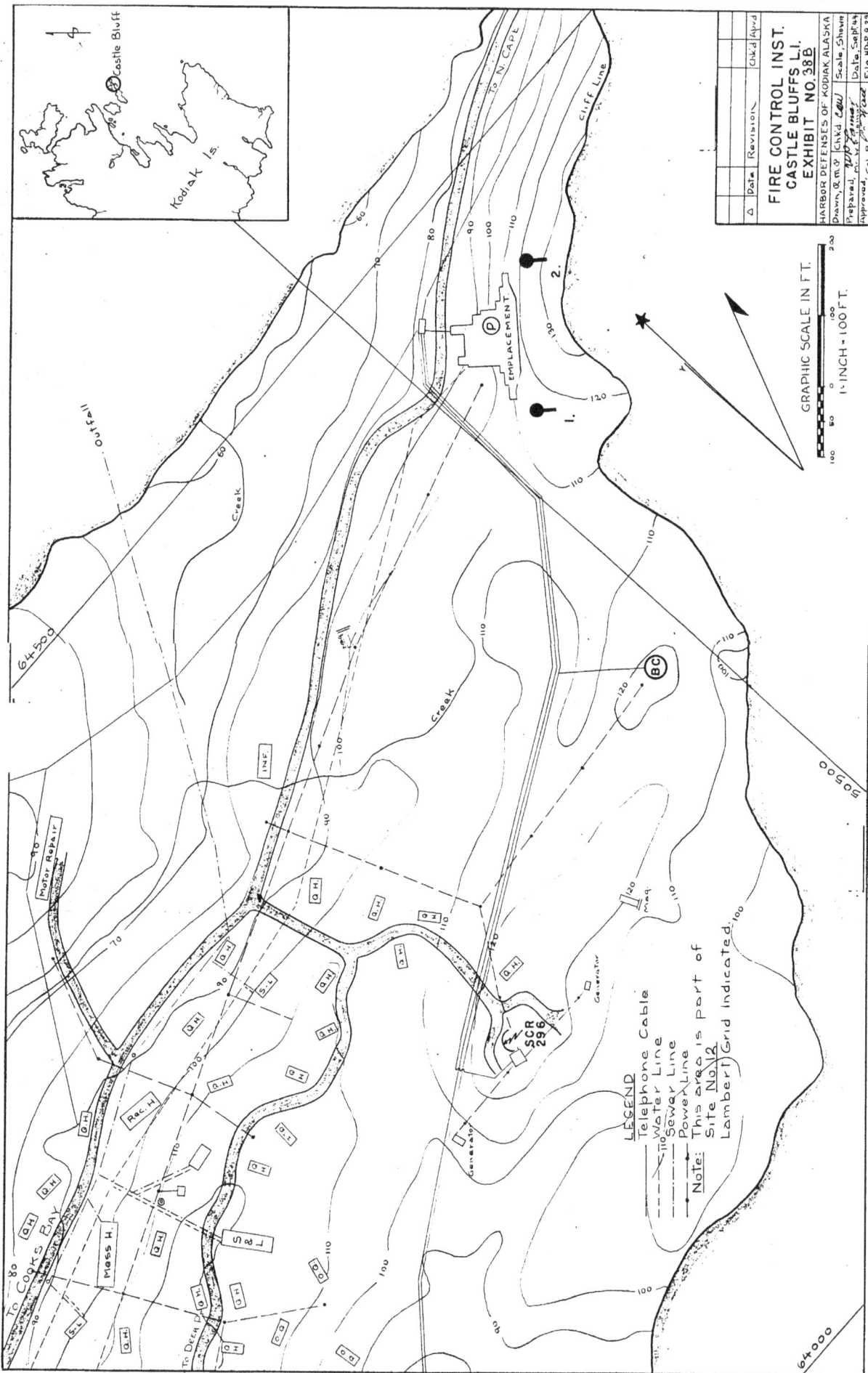

Castle Bluff

Kodiak Is.

FIRE CONTROL INST.
CASTLE BLUFFS L.I.
EXHIBIT NO. 38B

△	Date	Revision	Chk'd	Apvd

HARBOR DEFENSES OF KODIAK, ALASKA
Drawn, & m.o' Chk'd *Lew*   Scale, Shown
Prepared, *Wb James*   Date, Sept 44
Approved, col. P.C. *Bulford*   File.HD-P-929

GRAPHIC SCALE IN FT.
1 INCH = 100 FT.

TO N. CAPE

Cliff Line

EMPLACEMENT

2.

P

1.

Outfall

Creek

64,500

50,500

Creek

INF

BC

SCR 296

Generator

Motor Repair

Q.H

Q.H

Q.H

Q.H

Q.H

Q.H

Q.H

Q.H

Q.H

Q.H

Q.H

Q.H

Q.H

S.L

S.L

S.L

S.L

Rec. H.

Mess H.

Generator

TO COOKS BAY

To Dear Cove

LEGEND
Telephone Cable
Water Line
Sewer Line
Power Line
Note: This area is part of
Site No.12
Lambert Grid Indicated.

64,000

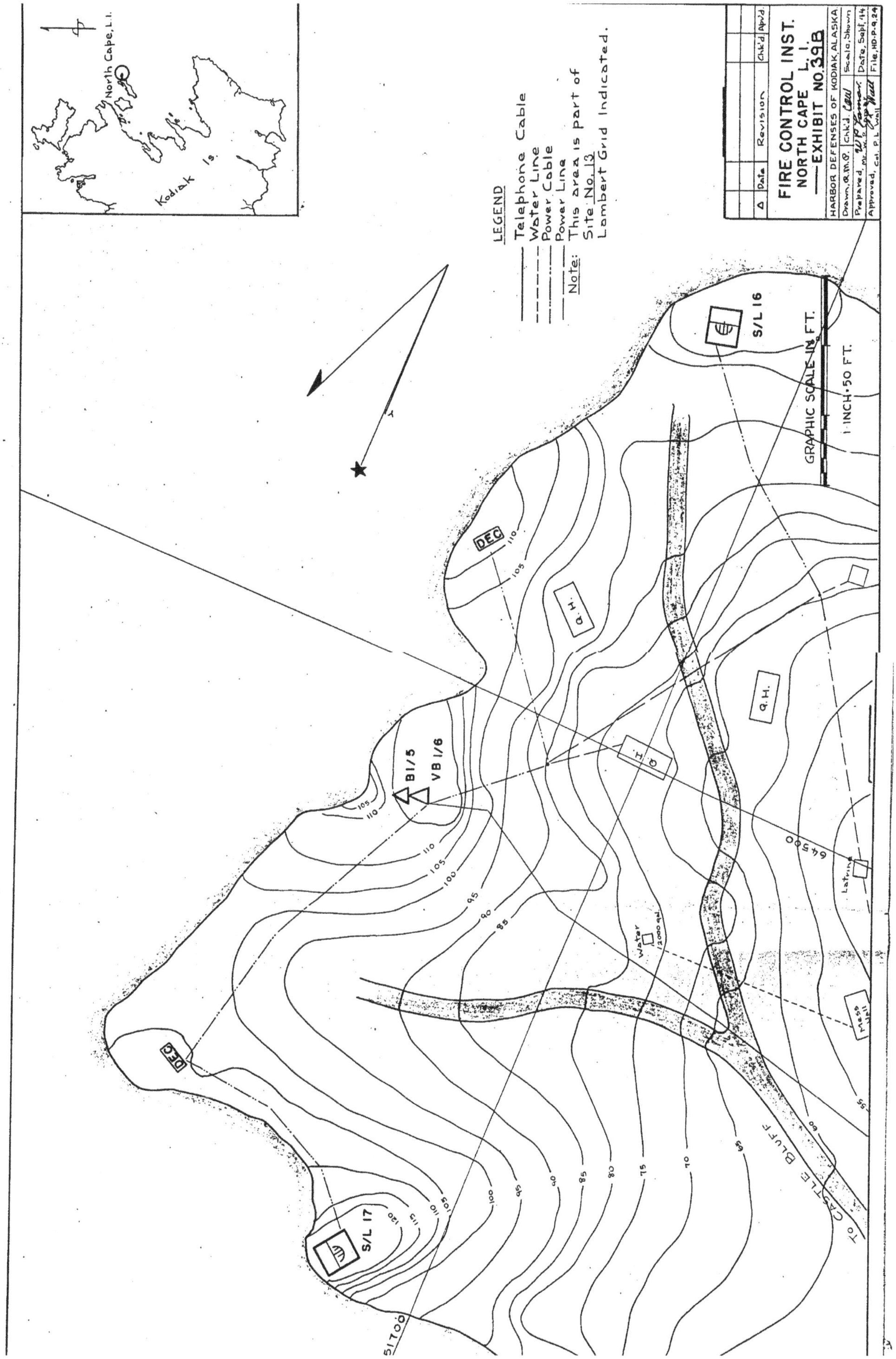
North Cape, L.I.
Kodiak Is.
FIRE CONTROL INST.
NORTH CAPE L.I.
EXHIBIT NO. 39B
HARBOR DEFENSES OF KODIAK, ALASKA
LEGEND
Telephone Cable
Water Line
Power Cable
Power Line
Note: This area is part of Site No. 13
Lambert Grid Indicated.
GRAPHIC SCALE IN FT.
1 INCH = 50 FT.
S/L 16
S/L 17
DEC
DEC
Q.H.
Q.H.
B1/5
VB 1/6
Latrine
Mess Hall
Water
TO CASTLE BLUFF

Spruce Cape

Kodiak Is.

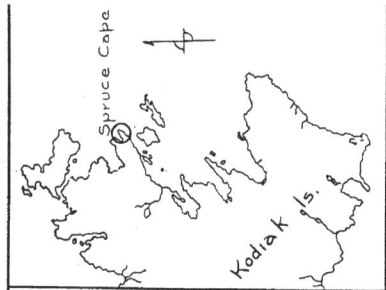

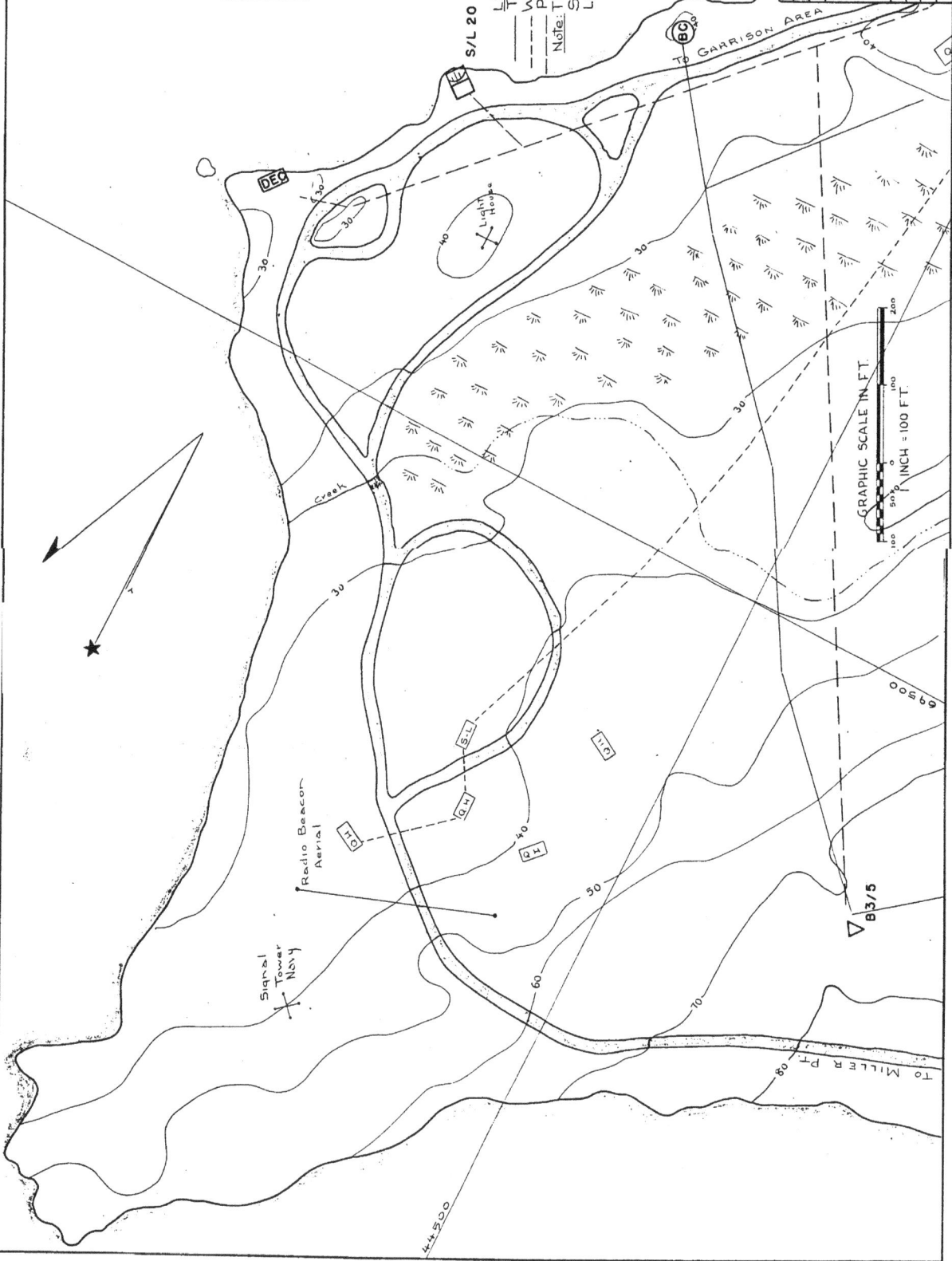

LEGEND
Telephone Cable
Water Line
Power Line
Note: This area is part of
Site No. 15
Lambert Grid Indicated.

S/L 20

DEC

Light House

BC

To Garrison Area

FIRE CONTROL INST.
SPRUCE CAPE
—EXHIBIT NO. 40B

Δ	Date	Revision	Chkd	Apvd

HARBOR DEFENSES OF KODIAK, ALASKA
Drawn, R.M.Q | Chkd | Scale, Shown
Prepared, M. | Date, Sept. 44
Approved, C.I.P. Wall | File No. HD-P 4.27

GRAPHIC SCALE IN FT.
1 INCH = 100 FT.

Creek

30

40

30

30

30

50

60

70

69,500

H-4,500

B3/5

S/L

Q/H

H1

G1

3/7

Radio Beacon
Aerial

Signal
Tower
Navy

To Miller Pt.

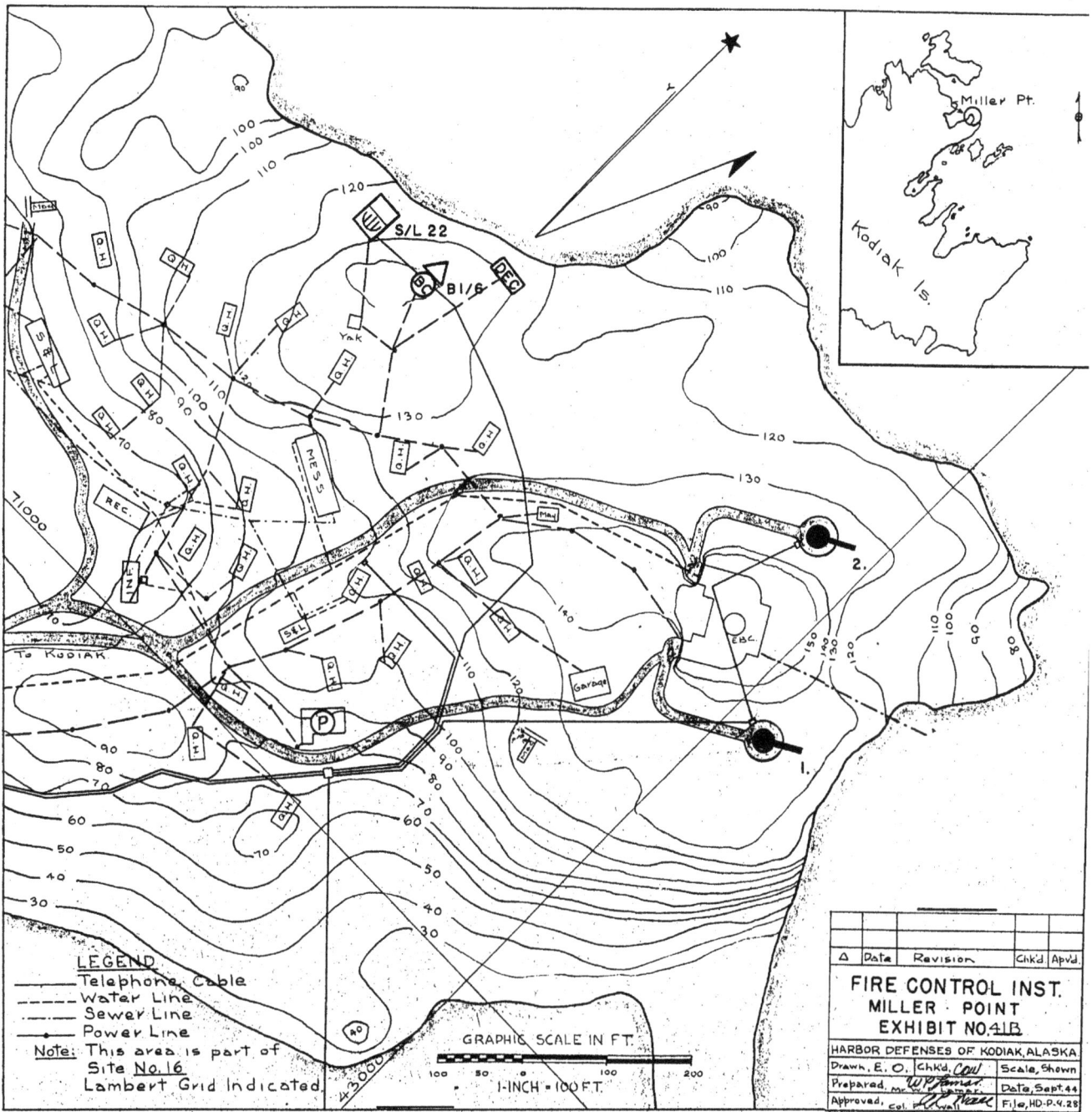

LEGEND
Telephone Cable
Water Line
Sewer Line
Power Line
Note: This area is part of Site No.16
Lambert Grid Indicated

GRAPHIC SCALE IN FT.
100 50 0 100 200
1-INCH = 100 FT.

FIRE CONTROL INST.
MILLER · POINT
EXHIBIT NO. 41B

HARBOR DEFENSES OF KODIAK, ALASKA
Drawn, E. O. Chk'd, CDW Scale, Shown
Prepared, Mr. W. P. Lamar Date, Sept. 44
Approved, Col. File, HD-P-4.28

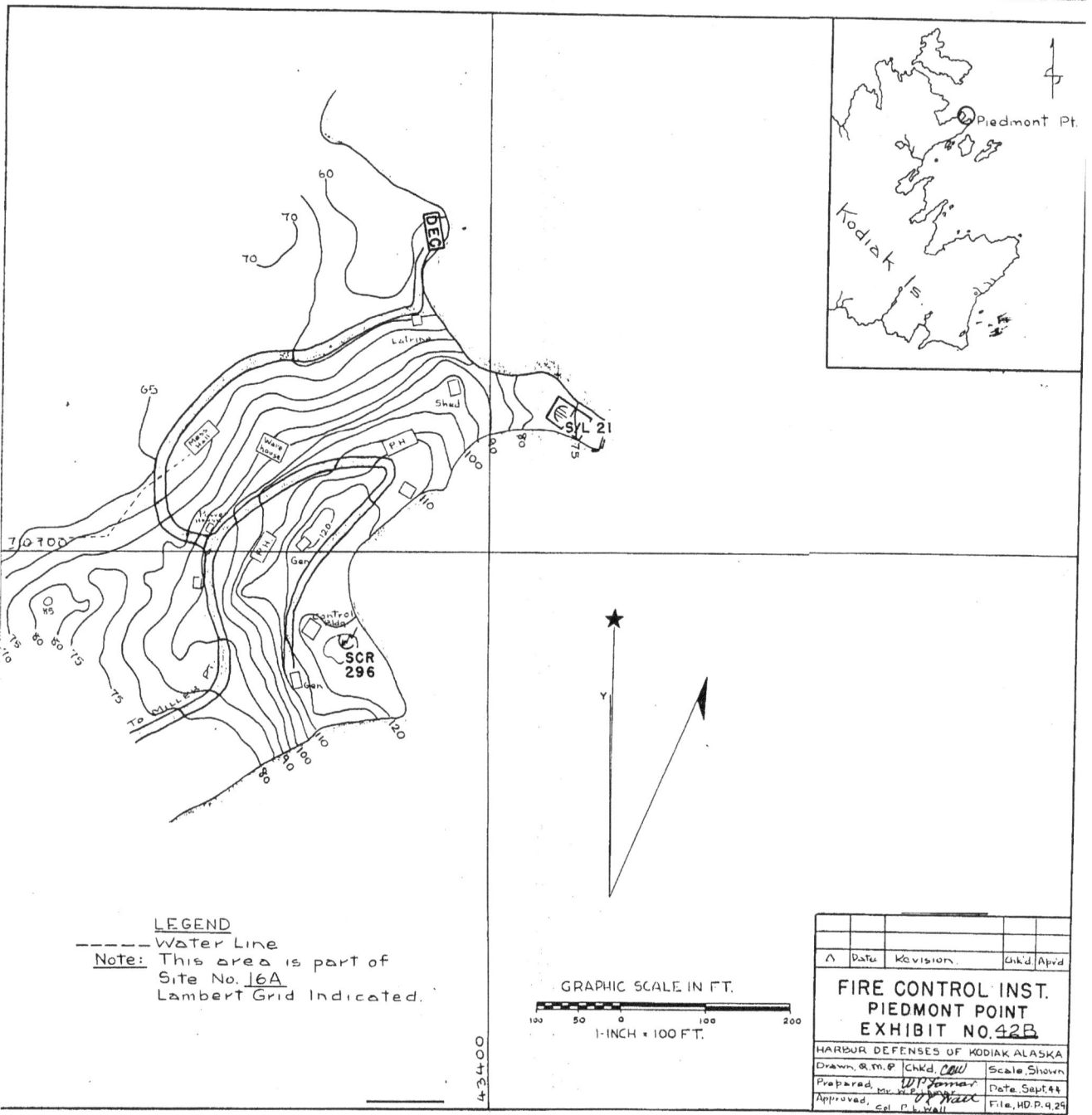

LEGEND
————— Water Line
Note: This area is part of
Site No. 16A
Lambert Grid Indicated.

GRAPHIC SCALE IN FT.
100   50   0   100   200
1-INCH = 100 FT.

Λ	Date	Revision	Chk'd	Apr'd

**FIRE CONTROL INST.
PIEDMONT POINT
EXHIBIT NO. 42B**

HARBOR DEFENSES OF KODIAK ALASKA

Drawn, Q.m.P	Chk'd. COW	Scale, Shown
Prepared, Mr. H.P. Tamar		Date, Sept. 44
Approved, Col P.L.Wall		File, HD.P.9.29

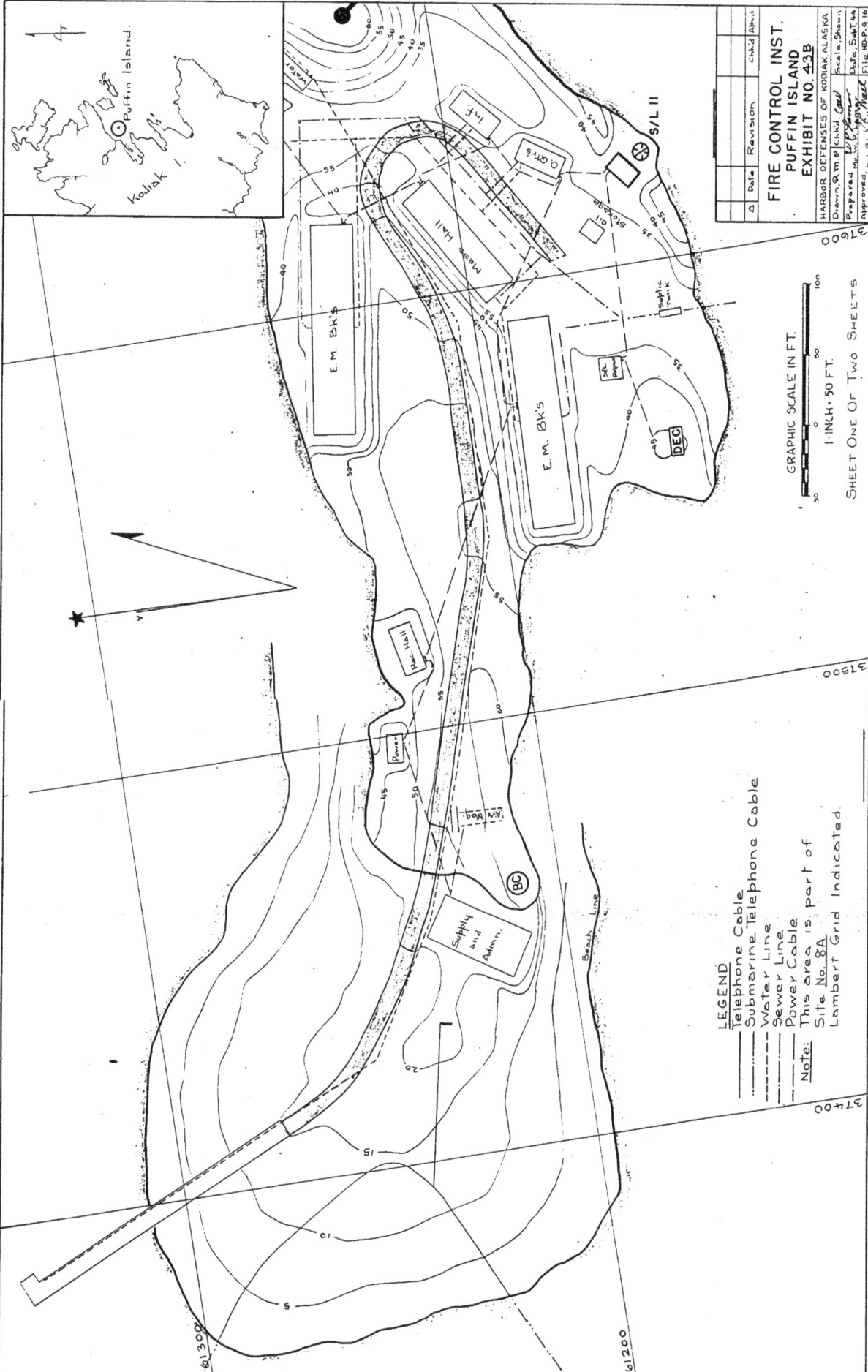

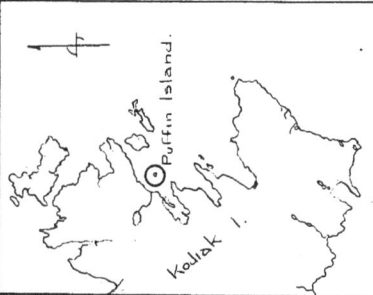

FIRE CONTROL INST.
PUFFIN ISLAND
EXHIBIT NO. 43B

GRAPHIC SCALE IN FT.
1-INCH = 50 FT.
SHEET ONE OF TWO SHEETS

LEGEND
Telephone Cable
Submarine Telephone Cable
Water Line
Sewer Line
Power Line
Power Cable
Note: This area is part of
Site No. 8A
Lambert Grid Indicated

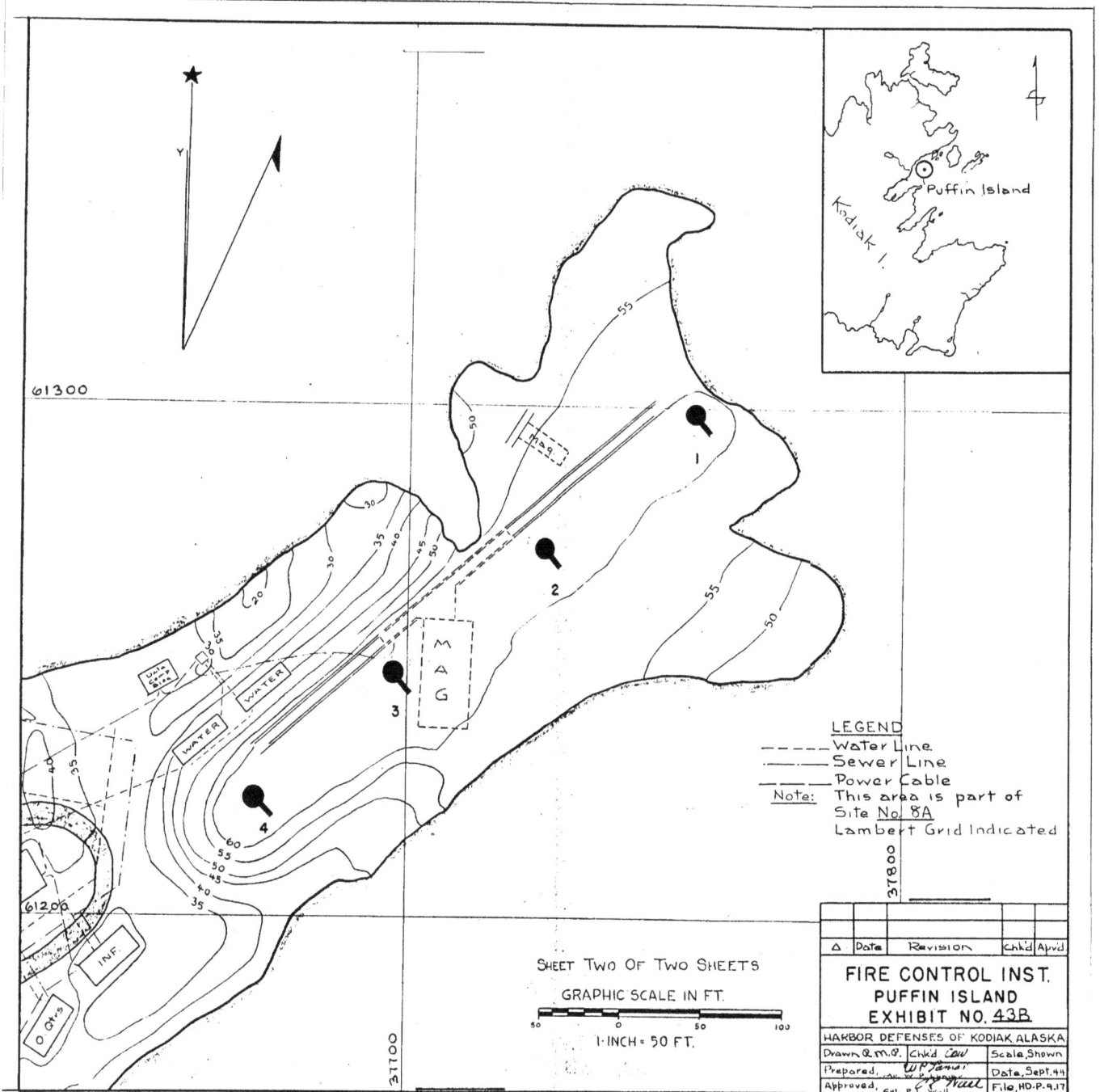

SHEET TWO OF TWO SHEETS

GRAPHIC SCALE IN FT.

1·INCH = 50 FT.

LEGEND
Water Line
Sewer Line
Power Cable
Note: This area is part of
Site No. 8A
Lambert Grid Indicated

△	Date	Revision	Chk'd	Apvd

FIRE CONTROL INST.
PUFFIN ISLAND
EXHIBIT NO. 43B

HARBOR DEFENSES OF KODIAK ALASKA

Drawn Q.M.P.	Chk'd	Scale, Shown
Prepared,		Date, Sept. 44
Approved,		File, HD-P-9.17

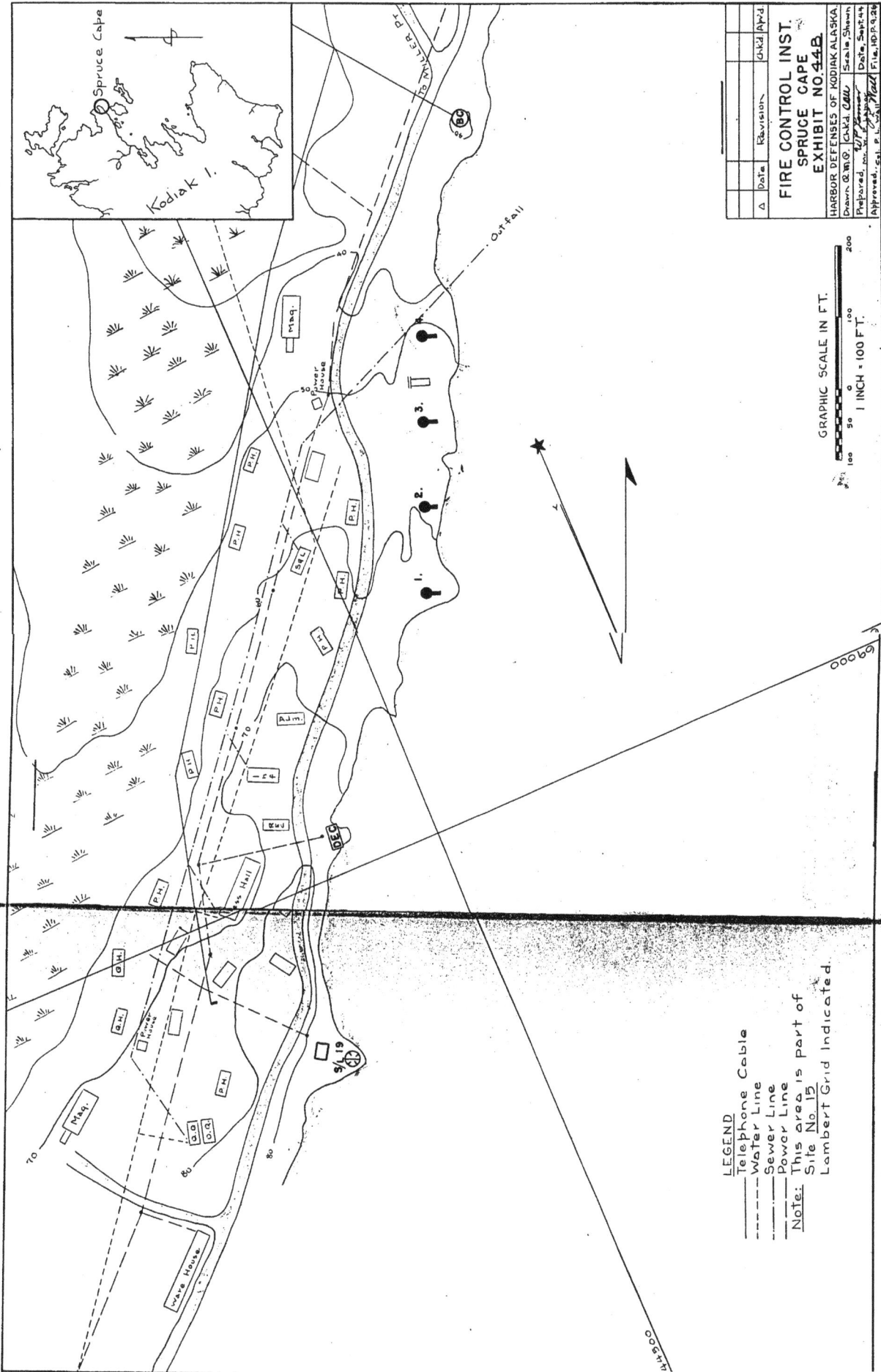

FIRE CONTROL INST.
SPRUCE CAPE
EXHIBIT NO. 44B

HARBOR DEFENSES OF KODIAK ALASKA

GRAPHIC SCALE IN FT.

1 INCH = 100 FT.

LEGEND
——— Telephone Cable
——— Water Line
——— Sewer Line
——— Power Line
Note: This area is part of Site No. 15
Lambert Grid Indicated.

Spruce Cape
Kodiak I.

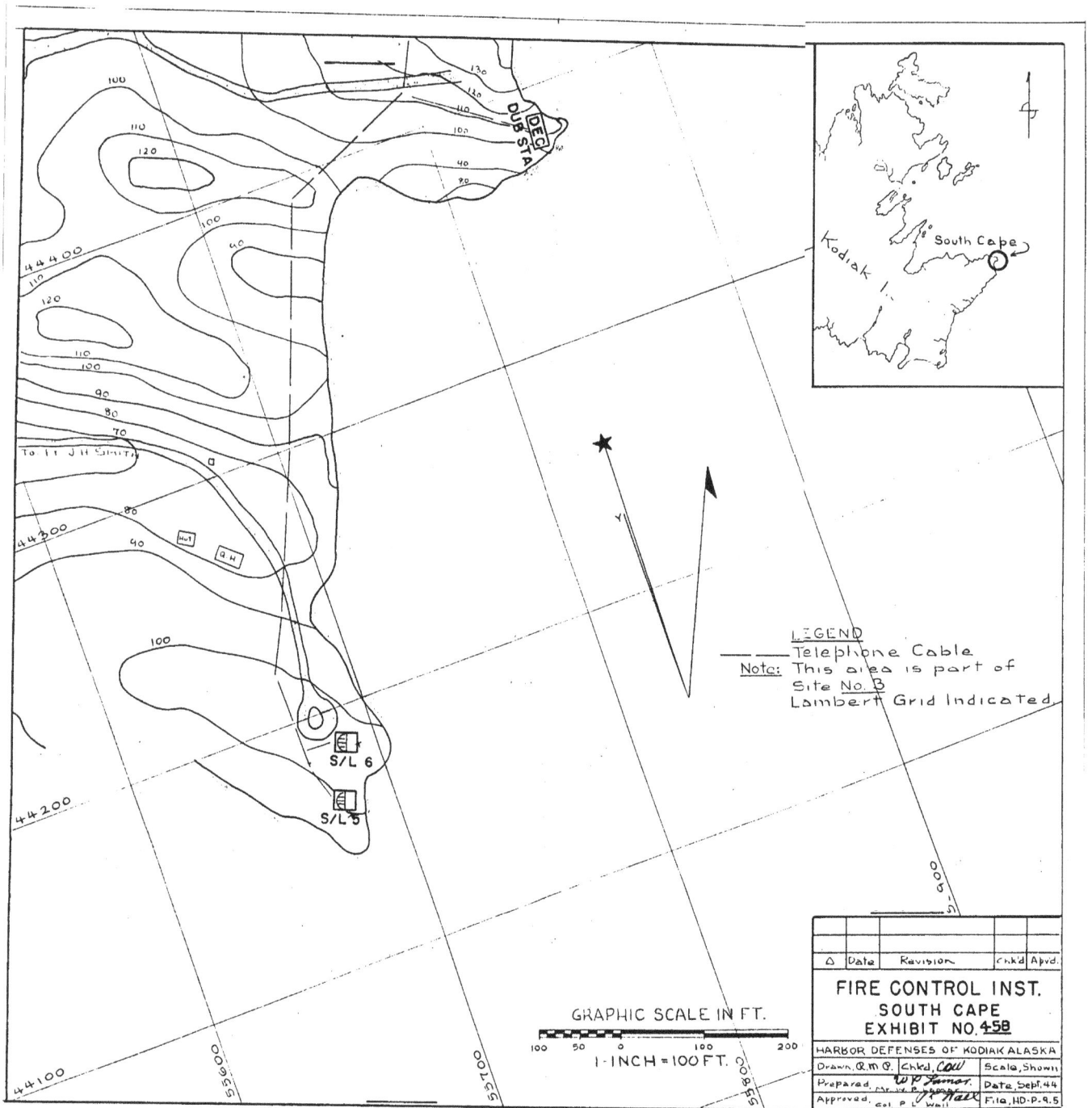

LEGEND
Telephone Cable
Note: This area is part of
Site No. 3
Lambert Grid Indicated

GRAPHIC SCALE IN FT.
100 50 0   100   200
1-INCH=100 FT.

FIRE CONTROL INST.
SOUTH CAPE
EXHIBIT NO. 458

HARBOR DEFENSES OF KODIAK ALASKA

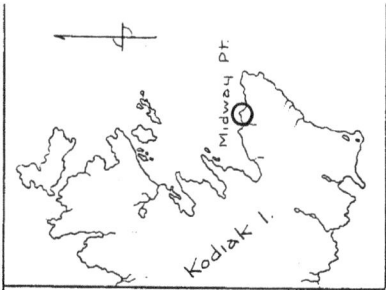

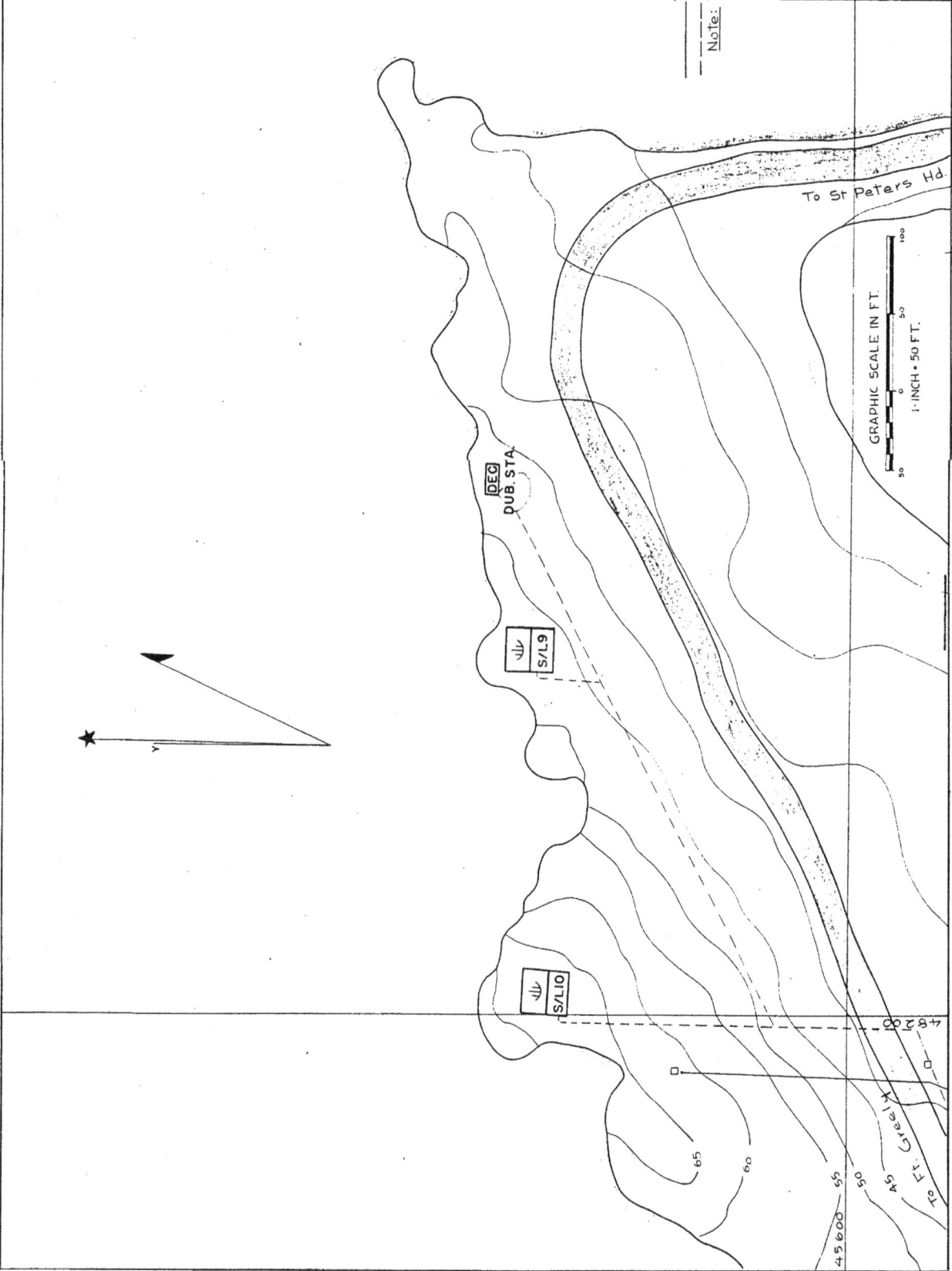

LEGEND
Telephone Cable
Power Cable
Note: This area is part of Site No. 6
Lambert Grid Indicated

FIRE CONTROL INST.
MIDWAY POINT
EXHIBIT NO. 47B

HARBOR DEFENSES OF KODIAK ALASKA

To St Peters Hd

GRAPHIC SCALE IN FT.
1-INCH = 50 FT.

DEC
DUB. STA.

S/L9

S/L10

To Ft. Greely

Midway Pt.
Kodiak I.

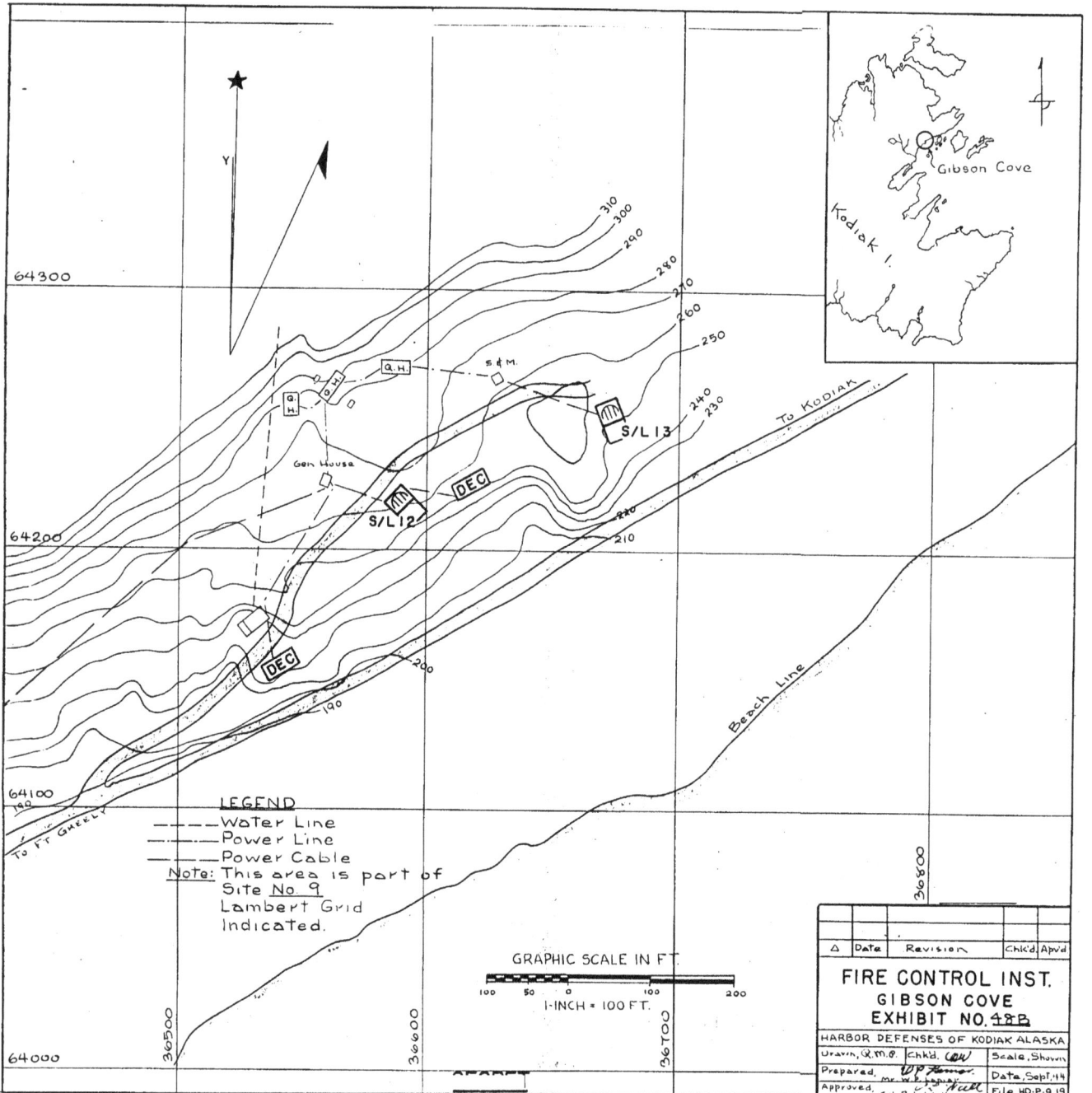

LEGEND
— — — Water Line
— · — Power Line
———— Power Cable
Note: This area is part of
Site No. 9
Lambert Grid
Indicated.

GRAPHIC SCALE IN FT.
100 50 0   100   200
1-INCH = 100 FT.

FIRE CONTROL INST.
GIBSON COVE
EXHIBIT NO. 48B

HARBOR DEFENSES OF KODIAK ALASKA

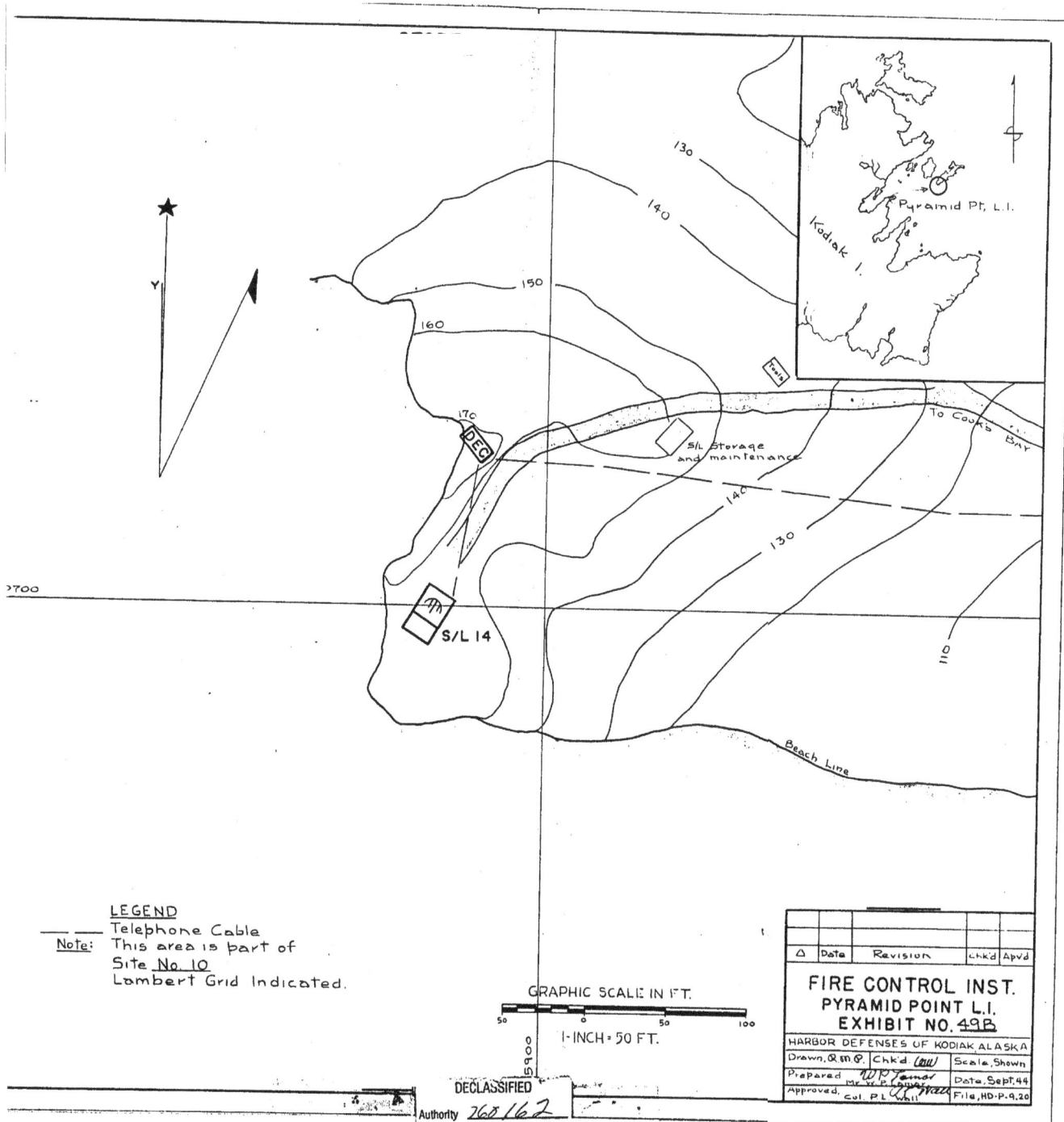

LEGEND
— — Telephone Cable
Note: This area is part of
Site No. 10
Lambert Grid Indicated.

GRAPHIC SCALE IN FT.

1-INCH = 50 FT.

DECLASSIFIED
Authority 760162

△	Date	Revision	Chk'd	Apvd

FIRE CONTROL INST.
PYRAMID POINT L.I.
EXHIBIT NO. 49B

HARBOR DEFENSES OF KODIAK, ALASKA

Drawn, Q.M.R.	Chk'd (ow)	Scale, Shown
Prepared W.P.Tamer		Date, Sept. 44
Mr. W.P.Tamer		
Approved, Col. P.L. Wall		File, HD-P-9.20

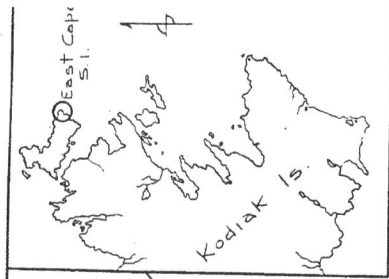

East Cape S.I.

Kodiak Is.

FIRE CONTROL INST.
EAST CAPE S.I.
—EXHIBIT NO. 50B

HARBOR DEFENSES OF KODIAK ALASKA

	Date	Revision	Chkd	Apr.

Drawn, R.M.C   Chkd, Cool   Scale, Shown
Prepared, W.P. Turner   Date, Sept. 41
Approved, M.P. Quigley Col. C.E. Wall   File, HD-P-9-31

GRAPHIC SCALE IN FT

1-INCH - 50 FT.

DEC

S/L 23

DEC

S/L 24

To Mt Herman

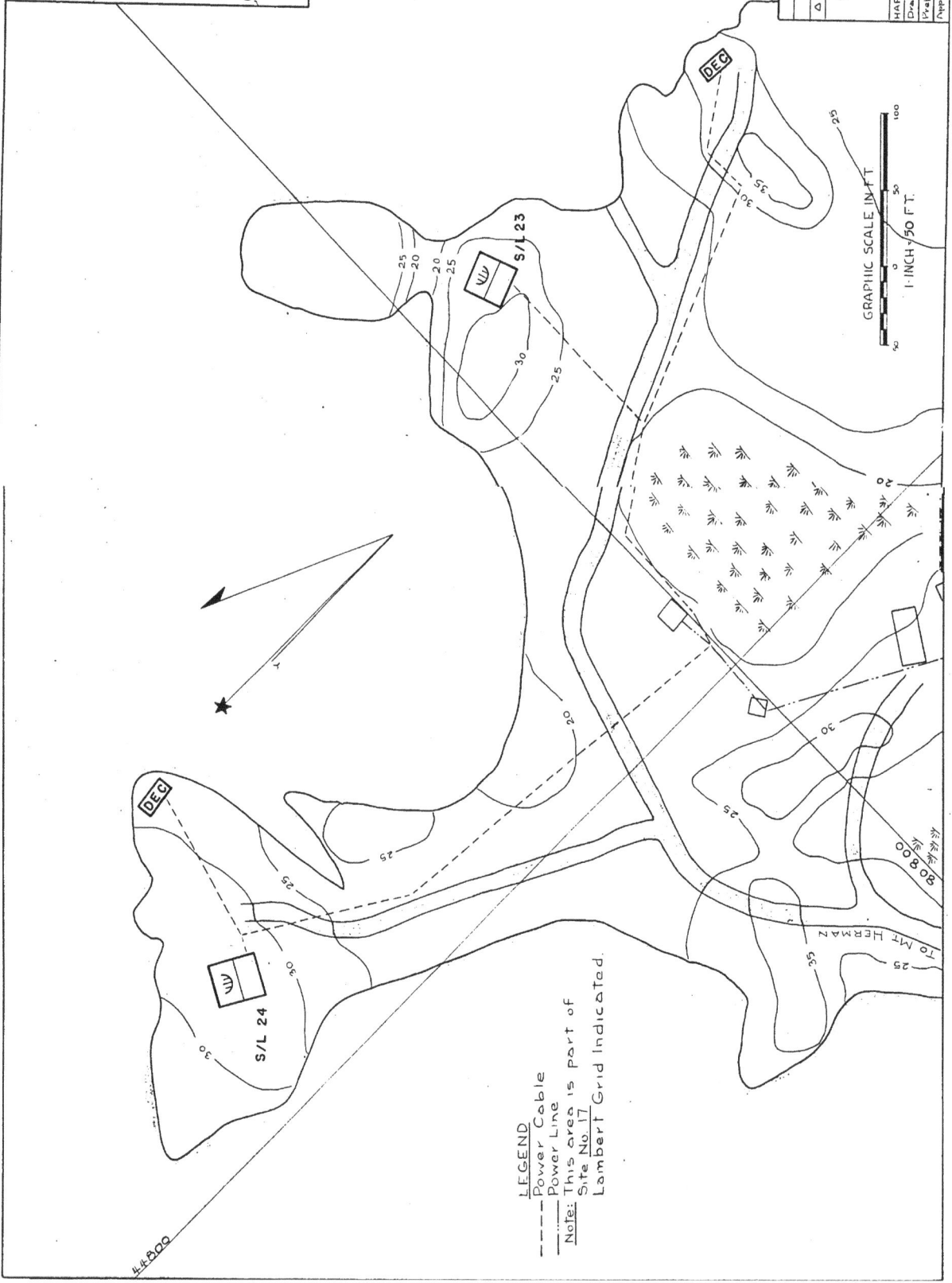

LEGEND
———— Power Cable
———— Power Line
Note: This area is part of
Site No. 17
Lambert Grid Indicated.

44,500

THE HARBOR DEFENSES OF DUTCH HARBOR – ALASKA

This military reservation was established as part of the major effort to defend the naval operating base and air station at Dutch Harbor on Unalaska Island in the Aleutian Island chain. Dutch Harbor was to receive one 8-inch and two 6-inch batteries, but one 6-inch battery was canceled. Battery #402 was built near Mt. Ballyhoo and designated Fort Schwatka. It was completed in 1944. Battery #298 was built on Eider Point and designated Fort Learnard. The reservation at Summer Bay, which had four 155mm GPF guns on Panama mounts was named Fort Brumback after Lt. Virgil J. Brumback. The cancelled battery, #299, was to have been located on Ulatka Head. Searchlight and fire control facilities were built on Ugadaga Bay, Ulatka Head, Constantine Head, the Coxcomb (Second Priest Rock), Erskine Point, Cape Wislow (aka Fisherman's Point?) and Eider Point. The AMTB batteries were built on Eider Point and Amaknak Spit below Battery #298 and Battery #402 respectively. The construction project was pretty much completed by late 1944 and probably inactivated in 1945. In 1996, the U. S. Congress designated these former military reservations a National Historic Area as a way of educating future generations both about the history of the Aleut people, and the role the Aleutian Islands played in the defense of the United States in World War II. These areas are owned and operated by the Ounalashka Corporation but as an affiliated area of the National Park Service receives funding and technical assistance for development and preservation.

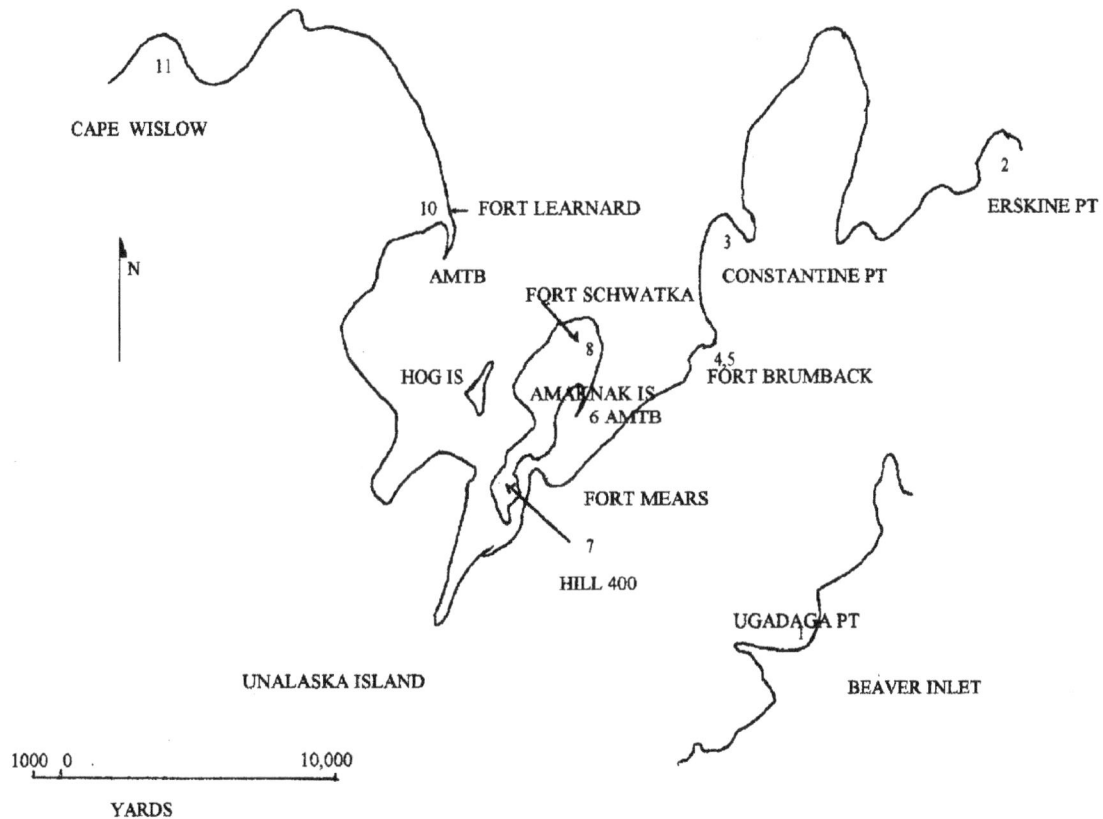

Harbor Defenses of Dutch Harbor

Fort Schwatka (1943-1945) is located at Ulakta Head on Amaknak Island in the center of Dutch Harbor. A parcel of 712 acres was acquired by the Army for a military post on June 14, 1942. It was named in General Order 17 of April 2, 1943 after Lt. Frederick Schwatka, a renowned Alaskan explorer of the 1880s. Starting in 1940 the U.S. Navy developed of an operating base with surface ship, submarine and seaplane facilities, prompting U.S. Army coast defenses. A temporary battery was built in late 1941 using four 155mm GPF guns on Panama mounts at Mount Ballyhoo. The approved project of early 1942 called for locating a standard program 8-inch barbette battery at Fort Schwatka. With a considerable engineering presence, the Navy supplied the Seabee work force to build the emplacements. The battery was completed in 1944. Along with the defenses HECP/HDCP and radar, fire control, and searchlight facilities. A 90mm AMTB was also emplaced at Amaknak Spit, close to water level. After World War II, Fort Schwatka was abandoned, the guns of the heavy battery being destroyed in place in 1948. Today the fort is part of the Aleutian World War II National Historical Area.

Fort Schwatka Gun Batteries

- **Battery #402:** A World War II standard dual 8-inch barbette battery placed at Ulatka Head on the Fort Schwatka reservation. This was the height of the hill at this end of the island, it was a very high site with trunnion elevations of 897-feet. It was of typical Alaska-type design, with concrete gun blocks separated by 240-feet and a traverse protected magazine. Work was done from November 3, 1942 until November 8, 1943. The transfer date was November 8, 1943, no cost is listed on the appropriate RCW. It was armed with two 8-inch ex-navy Mk VIM3A2 guns on Model M1 barbette carriages (#114L2/#11 and #238L2/#12). These were shipped here on March 6, 1943, and mounted at least by August 1, 1944. It was never named, just known as Battery Construction No. 402. It served in the defenses as Tactical Battery No. 3. After the war, in August 1946, the guns were destroyed by the U.S. Army with explosives, and the site abandoned. The emplacement still exists in relatively good condition on National Park Service land. The battery is open to the public, but you need to get a permit Ounalashka Corperation/NPS to drive up the road to the battery.

- *Battery #299* (planned): A proposed World War II work for the southern end of Amaknak Island. This was to be a standard dual 6-inch barbette battery. Apparently, it was either cancelled or indefinitely deferred at an early stage, for no actual construction work was ever undertaken.

- **AMTB #1A:** An AMTB battery of two 90mm fixed and two 90mm mobile guns emplaced on the Amaknak Spit on the east side of Unalaska Island to protect the harbor. It consisted of two concrete fixed gun blocks, earthen revetments for the two mobile mounts, an earth-covered magazine and an improvised frame BC station on the right flank. It was aligned along the road down the spit and fired to the east. It was destroyed in the mid-1980s when the area was rebuilt to improve the boat harbor.

Fort Learnard (1942-1946) is located at Eider Point on Unalaska Island across the bay to the northwest of Dutch Harbor. A reservation of 2542 acres was procured under authority of April 30, 1942 for a military post. It was named in General Order 17 of April 2, 1943 after Brig. Gen. Henry G, Learnard, hero of the Boxer Rebellion relief expedition of 1900. The approved coast defense project of early 1942 authorized a standard 6-inch shielded battery for this post. Under U.S. Navy Seabee auspices, the battery was begun in August 1942. Other facilities at the fort were fire control and battery commander stations, searchlights, and radar. A 90mm AMTB battery was emplaced at Eider Spit. After the end of the war the fort was abandoned, heavy equipment and the guns being destroyed on site in 1950. Today the fort reservation is owned by the Ounalashka Corporation, a Native American holding company. With permission, it is possible to visit the site, though a boat and hiking ability are required.

Fort Learnard Gun Batteries

- **Battery #298:** A World War II standard dual 6-inch barbette battery emplaced on Unalaska Island at Eider Point. It occupied a space high on the bluff of the point, having a trunnion elevation of 516-feet. It fired to the east. The plan was generally of standard 200-series type and spacing, but it did have an extended approach tunnel to the rear entry. Work was started on August 20, 1942 and completed for transfer on November 30, 1942. Guns were supplied in late 1942. It was armed with two 6-inch Model 1903A2 guns on Model M1 barbette carriages (#14/#5 and #10/#4). It was never named, just known as Battery Construction No. 298. It served actively until disarmed shortly after the end of the war. The guns and carriages were destroyed by detonation and dumping major parts over the cliff's edge. The emplacement still exists, and parts of the gun shields are still at the site. The battery is on private property controlled by the Oualashka Corporation. A permit is required to visit the site as well as boat to get there followed by a long climb to the bluff.

- **AMTB #3A:** A 1943 Program AMTB battery for two 90mm fixed and two 90mm mobile guns emplaced on the spit at the shoreline on the eastern side of the Fort Learnard reservation at Eider Point, Unalaska. Two concrete gun blocks, spaced with 90-foot centers were constructed for the two fixed guns. It was probably disarmed shortly postwar. The battery is on private property controlled by the Oualashka Corporation. A land permit is required to visit the site as well as boat.

Fort Brumback (1942-1944) is located on Constantine Point on Unalaska Island. Named on General Order, 17 of April 2, 1943 after Lt. Virgil J, Brumback. Brumback led a somewhat controversial career in the Army, his most notable contribution being with the 1884 Yukon Expedition. The post site of 604 acres was acquired on April 30, 1942. Initially its overlook of Summer Bay was selected for a battery of 3-inch RF guns, but soon this plan was replaced with the intent to use a battery of four 155mm guns. Battery Constantine Point at Summer Bay with four 155mm GPF guns on Panama mounts (although at least initially the guns were just field mounts) was located here. The battery area included Elephant Steel shelters for ready ammunition storage, a separate plotting room, a cantonment area, and a tramway to a BC station on a bluff behind the battery. Today the fort is owned by the Ounalashka Corporation, a Native American holding company. The battery is on private property but can be accessed by vehicle and a permit is required.

Fort Mears (1941-1945) is located on Unalaska Island, near the Dutch Harbor Naval Base. It was named Fort Mears on September 10, 1941 after Col. Frederick Mears, who helped develop the Alaskan Railroad system. Construction on Fort Mears began in Dutch Harbor in 1940 and was completed in 1941 with the first U.S. Army troops arriving in June 1941. The U.S. Army facilities were co-located with the Dutch Harbor Naval Operating Base and as the mission grew each expanded to fill the limited available space. It became the headquarters garrison post of the harbor defenses of Dutch Harbor. When Dutch Harbor was attacked by Japanese naval aircraft in June 1942, losses were sustained at the base. One of Fort Mears' barracks was destroyed and twenty-five soldiers killed on June 3, 1942. On 11 Aug 1942, the U.S. Army decided to turn the co-located portions of Fort Mears over to the U.S. Navy and have the Seabees construct new facilities for the U.S. Army further south on Unalaska Island. A relatively complete garrison cantonment for Army personnel was subsequently built, intended for about 4000 total personnel. A 155mm GPF battery with a combined BC-FC structure was built on Hill 400 (Loc #7/Battery #2) in 1942. A protected Joint Army-Navy Command Post and Message Center was constructed near trhe harbor and airfiueld. Abandoned after the end of the war, the army post either became dilapidated or was repurposed by the commercial fishing industry. Of interest here today is the Museum of the Aleutians, built on the foundation of a military warehouse.

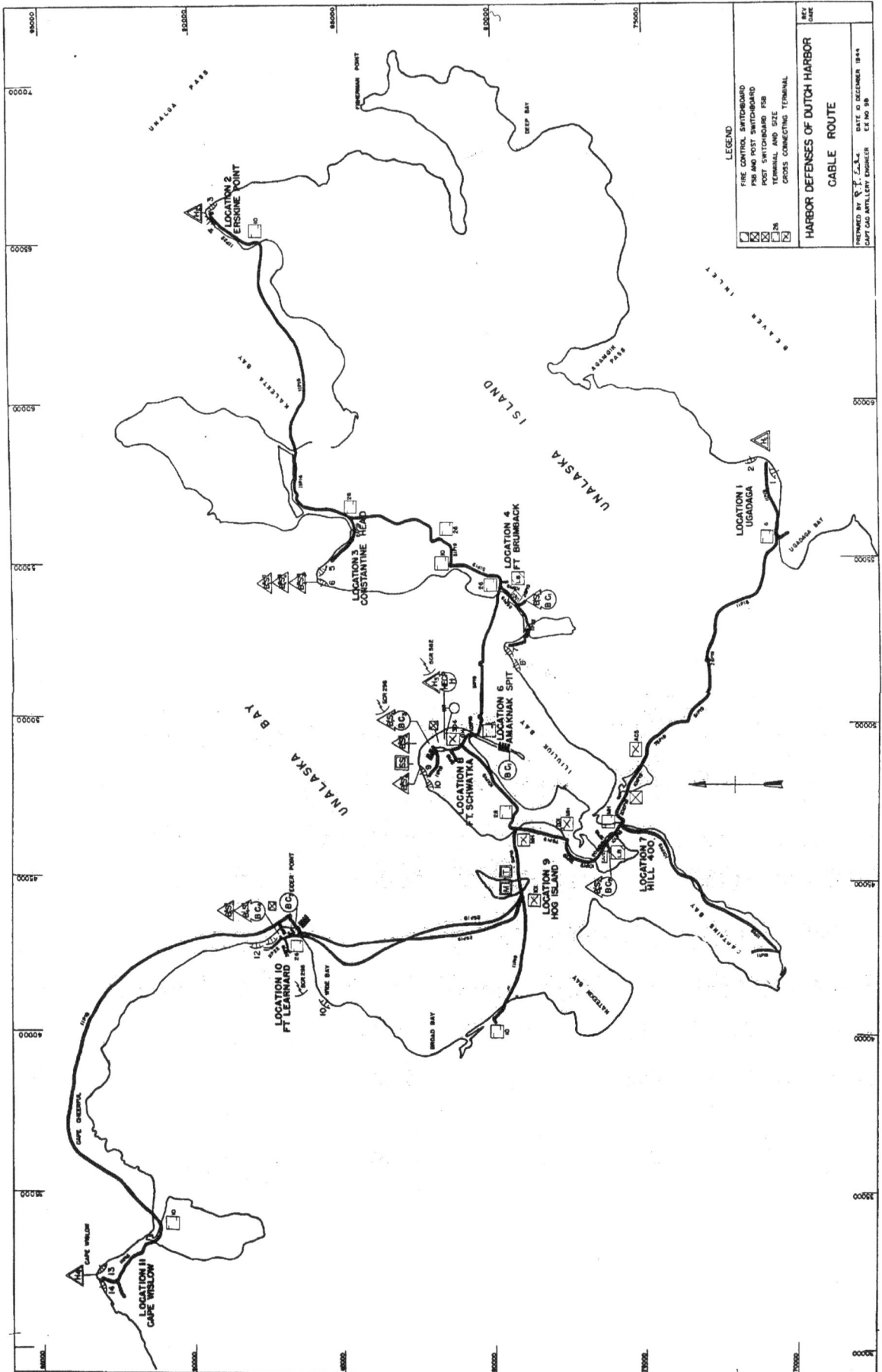

HARBOR DEFENSES OF DUTCH HARBOR

CABLE ROUTE

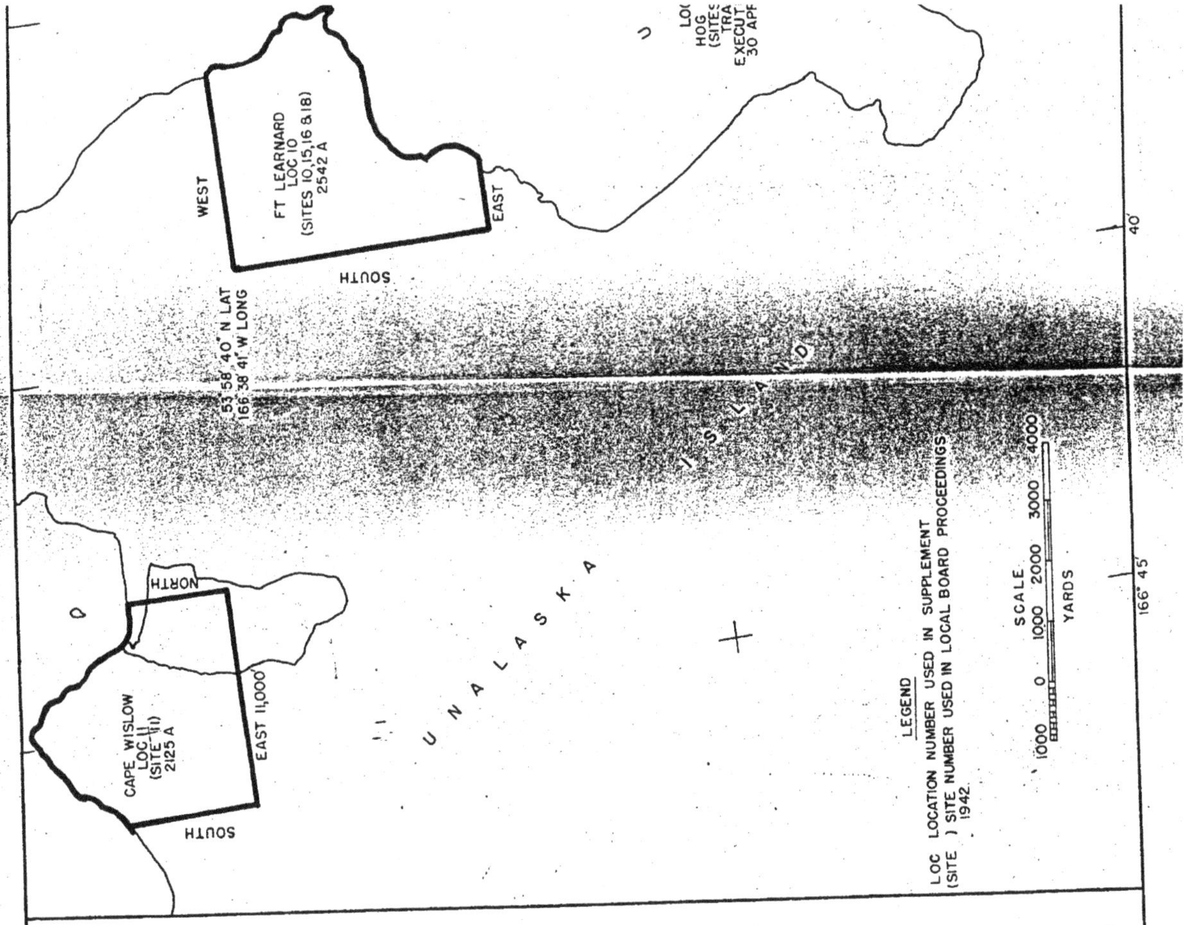
FT LEARNARD
LOC 10
(SITES 10,15,16 & 18)
2542 A
WEST
EAST
SOUTH
53°58'40" N LAT
166°38'41" W LONG
CAPE WISLOW
LOC 11
(SITE 11)
2125 A
NORTH
EAST 11,000
SOUTH
UNALASKA ISLAND
LEGEND
LOC LOCATION NUMBER USED IN SUPPLEMENT
(SITE) SITE NUMBER USED IN LOCAL BOARD PROCEEDINGS
1942
SCALE
1000 0 1000 2000 3000 4000
YARDS
40'
166°45'

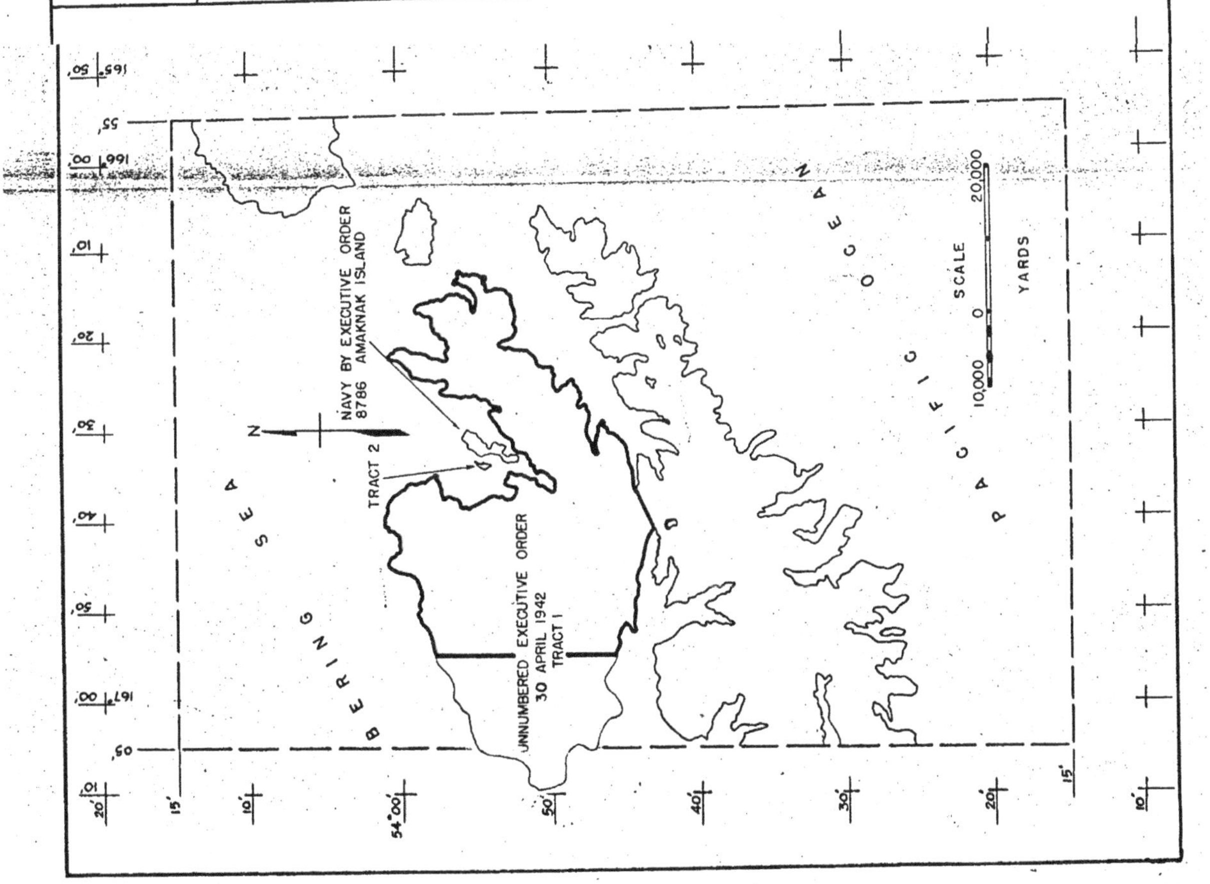
NAVY BY EXECUTIVE ORDER
8786 AMAKNAK ISLAND
TRACT 2
UNNUMBERED EXECUTIVE ORDER
30 APRIL 1942
TRACT 1
BERING SEA
PACIFIC OCEAN
N
SCALE
10,000 0 20,000
YARDS

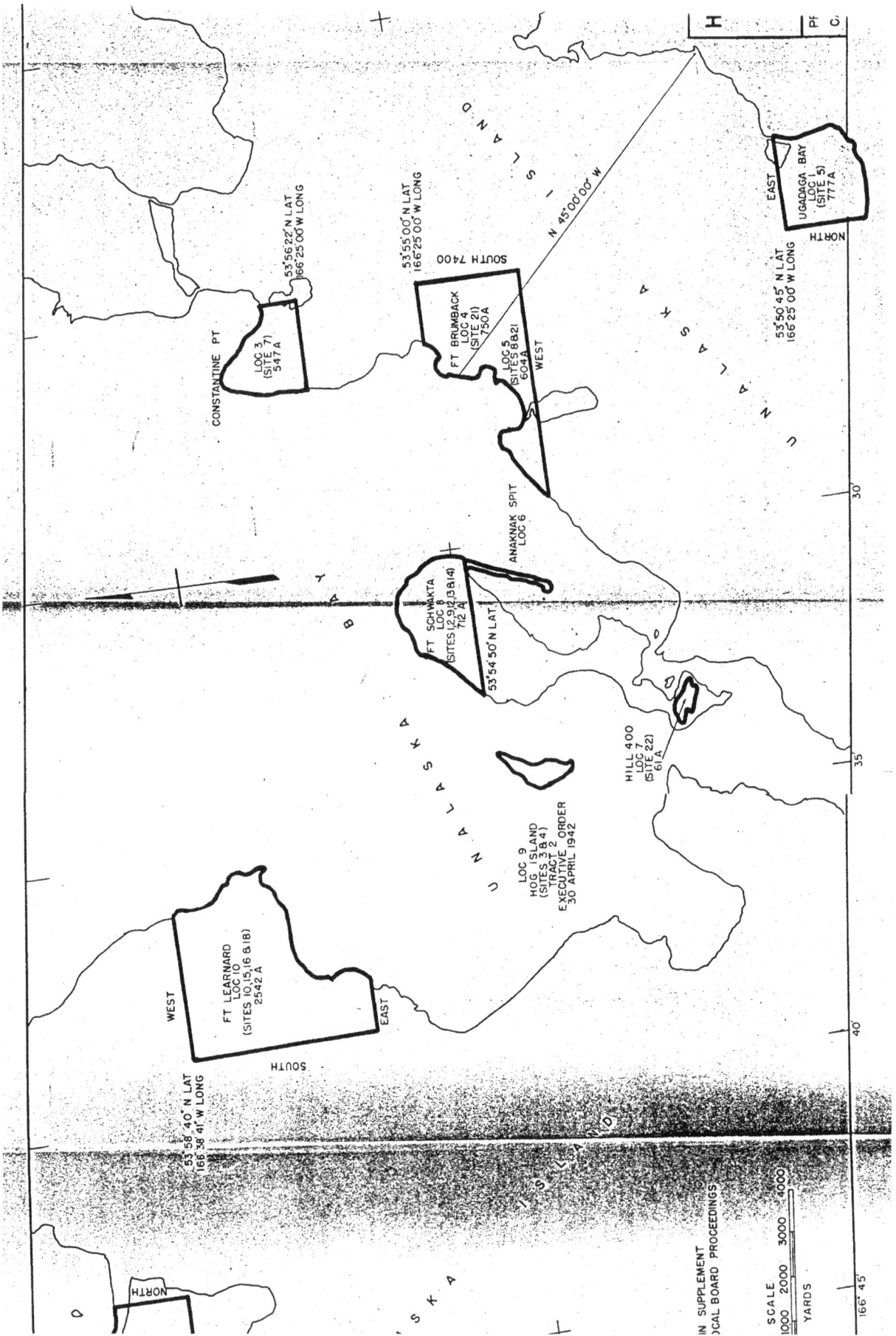

Dutch Harbor Alaska (Terry McGovern)

Fort Schwatka Aleutian World War II National Historical Area (Terry McGovern)

Fire control stations Fort Schwatka Aleutian World War II National Historical Area (Terry McGovern)

Emplacement for Battery #402 Fort Schwatka Aleutian World War II Natl. His. Area (Terry McGovern)

Dutch Harbor Alaska World War II-era Site Locations

location	Loc#	Purpose
Ugadaga Bay	1	fire control
Erskine Point	2	fire control
Constantine Point	3	fire control
North Summer Bay	4	Fort Brumback
South Summer Bay	5	155 mm battery
Amaknak Spit	6	AMTB Battery
Hill 400	7	155 mm battery garrison
Ulakta Head	8	8 inch battery Fort Schwatka
Ballyhoo		searchlights
Hog Island	9	Met & Tide
Eider Point	10	6 inch battery AMTB Fort Learnard, searchlights, fire control
Cape Wislow	11	fire control, searchlights

155 mm Panama Mount at Fort Schwatka Aleutian World War II National Historical Area (Terry McGovern)

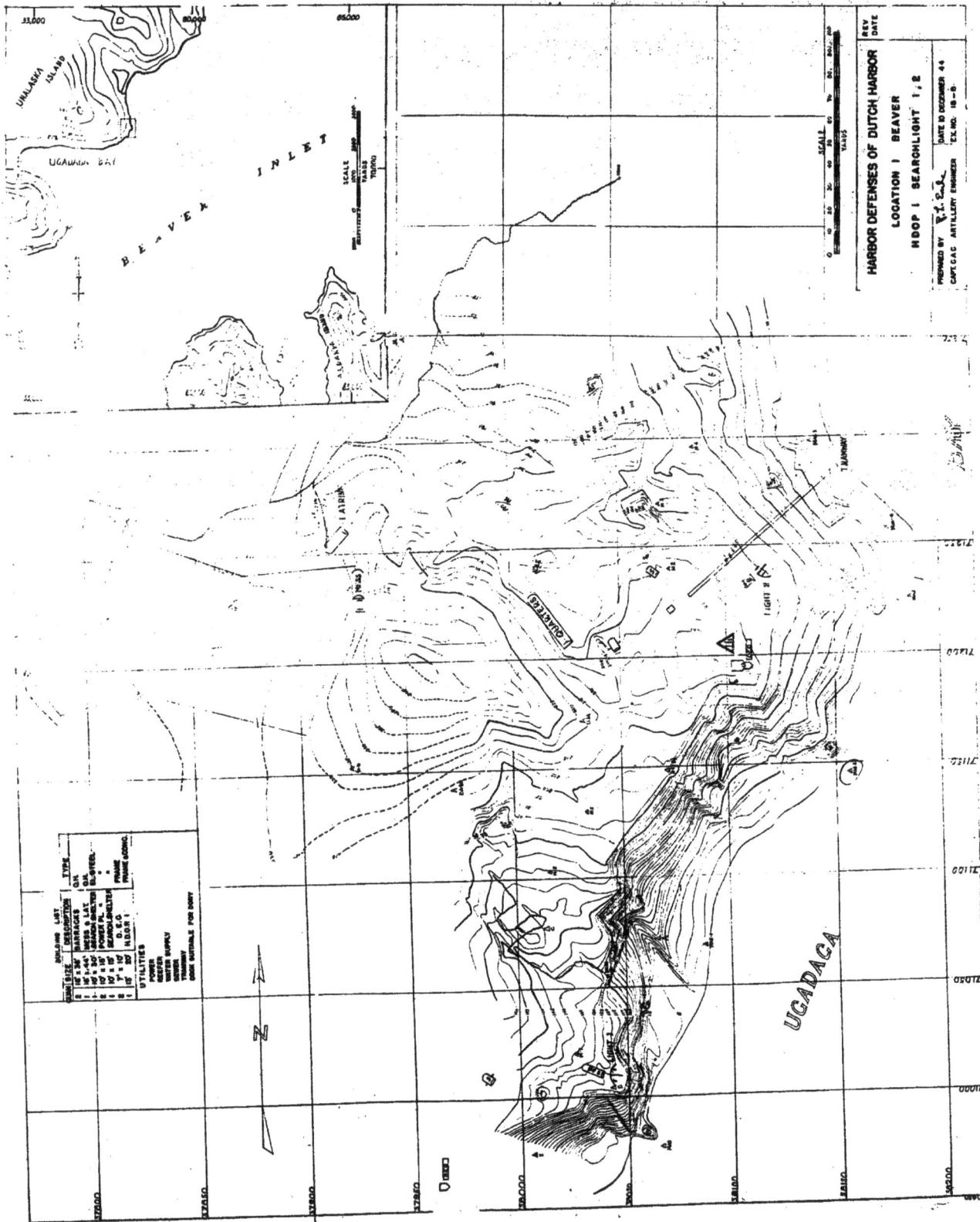

HARBOR DEFENSES OF DUTCH HARBOR
LOCATION 1 BEAVER
HDOP 1 SEARCHLIGHT 1, 2

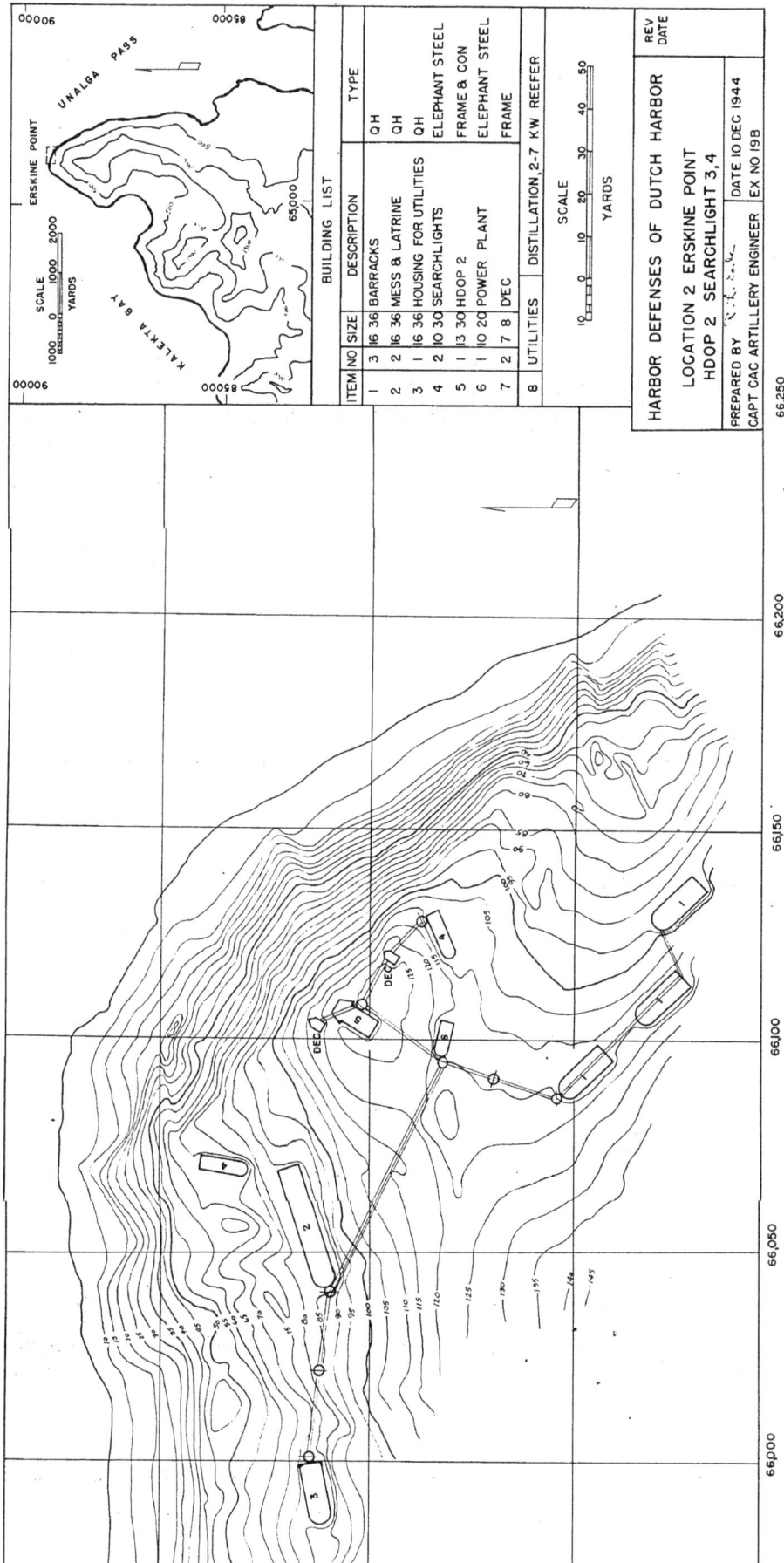

BUILDING LIST

ITEM NO	SIZE	DESCRIPTION	TYPE
1	3 16 36	BARRACKS	QH
2	2 16 36	MESS & LATRINE	QH
3	1 16 36	HOUSING FOR UTILITIES	QH
4	2 10 30	SEARCHLIGHTS	ELEPHANT STEEL
5	1 13 30	HDOP 2	FRAME & CON
6	1 10 20	POWER PLANT	ELEPHANT STEEL
7	2 7 8	DEC	FRAME
8		UTILITIES	DISTILLATION, 2-7 KW REEFER

HARBOR DEFENSES OF DUTCH HARBOR

LOCATION 2 ERSKINE POINT

HDOP 2 SEARCHLIGHT 3,4

PREPARED BY	DATE 10 DEC 1944
CAPT CAC ARTILLERY ENGINEER	EX NO 19B

SCALE

YARDS

UNALGA PASS

ERSKINE POINT

KALEKTA BAY

SCALE

YARDS

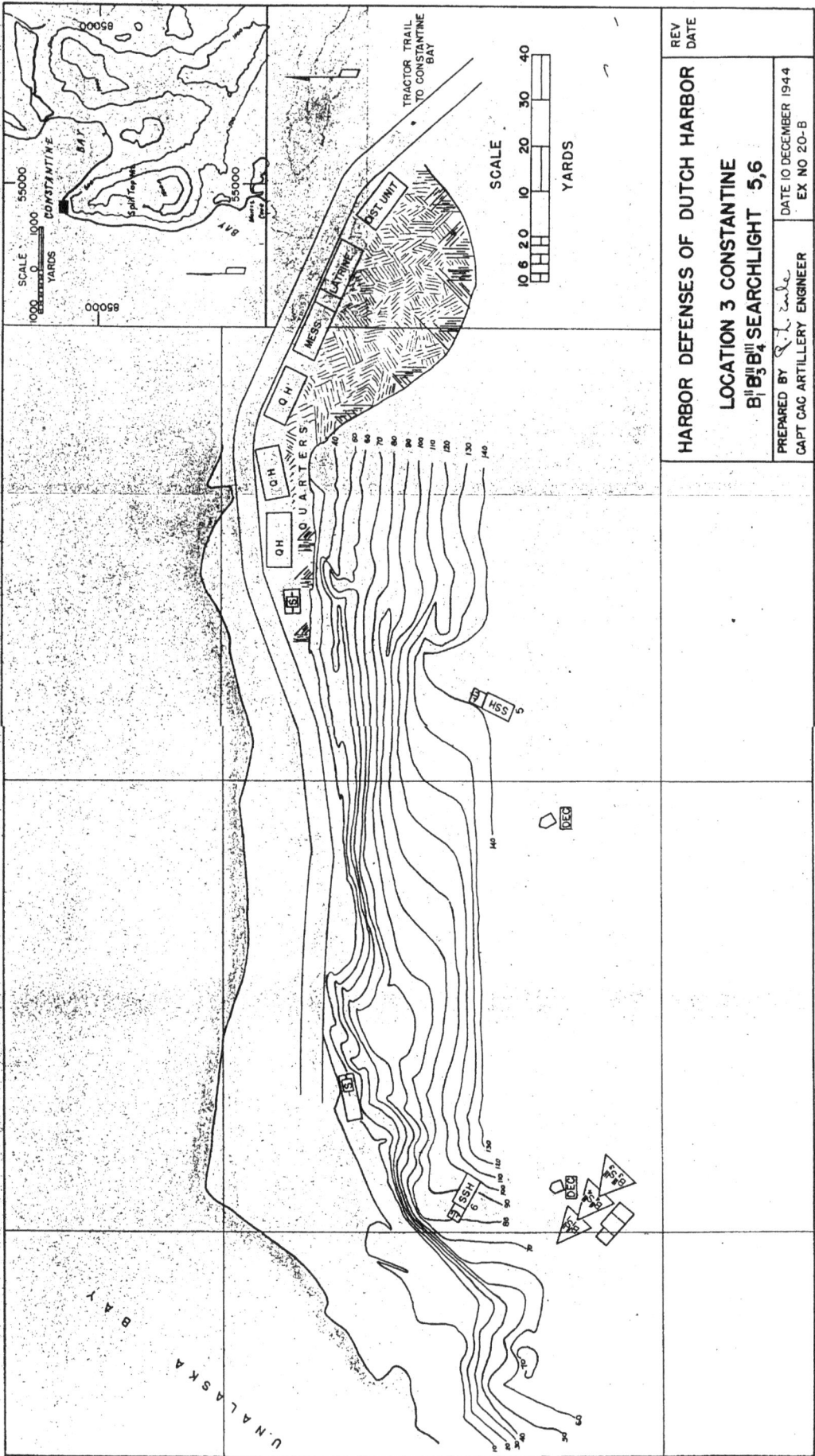

HARBOR DEFENSES OF DUTCH HARBOR

LOCATION 3 CONSTANTINE
B$_1^{II}$ B$_3^{II}$ B$_4^{III}$ SEARCHLIGHT 5,6

PREPARED BY
CAPT CAC ARTILLERY ENGINEER

DATE 10 DECEMBER 1944
EX NO 20-B

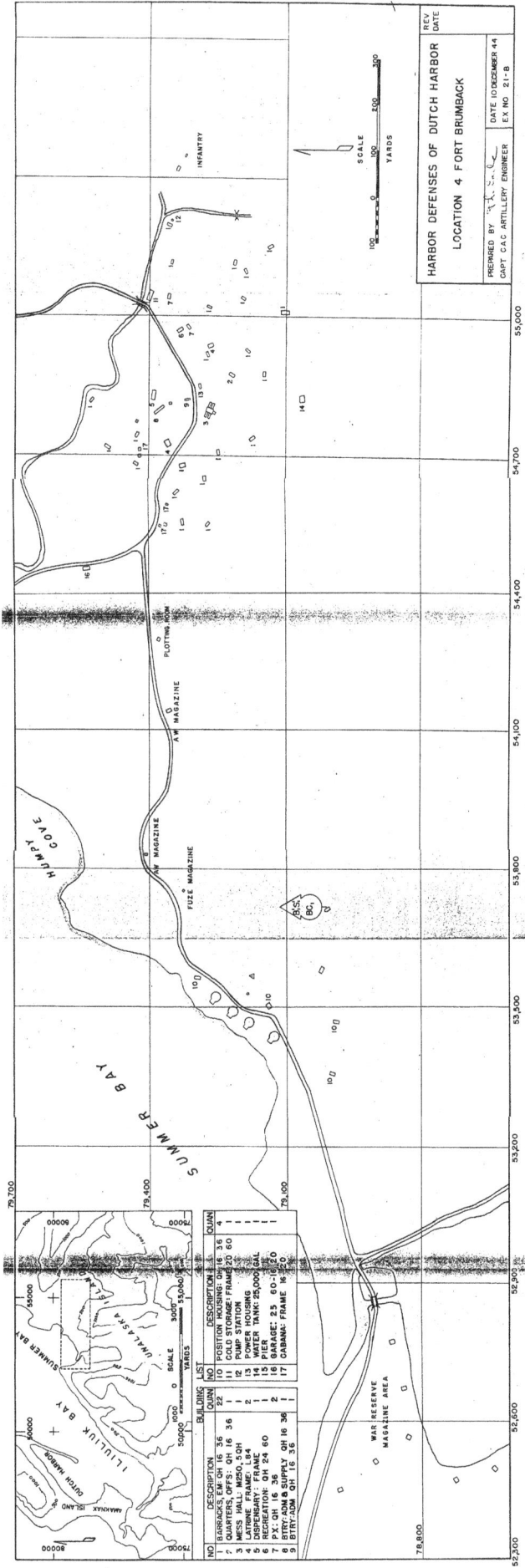

HARBOR DEFENSES OF DUTCH HARBOR
LOCATION 4 FORT BRUMBACK

PREPARED BY _____
CAPT C.A.C. ARTILLERY ENGINEER

DATE 10 DECEMBER 44
EX NO 21-B

REV
DATE

SCALE
YARDS

INFANTRY

HUMPY COVE

SUMMER BAY

PLOTTING ROOM

AW MAGAZINE

AW MAGAZINE

FUZE MAGAZINE

WAR RESERVE
MAGAZINE AREA

BUILDING LIST

NO	DESCRIPTION	QUAN
1	BARRACKS, EM QH 16 36	22
2	QUARTERS, OFFS: QH 16 36	1
3	MESS HALL: M250, 5 QH	1
4	LATRINE: FRAME: L84	2
5	DISPENSARY: FRAME	2
6	RECREATION: QH 24 60	1
7	PX: QH 16 36	2
8	BTRY: ADM B SUPPLY QH 16 36	1
9	BTRY: ADM QH 16 36	1
10	POSITION HOUSING QH 16 36	4
11	COLD STORAGE: FRAME 20 60	1
12	PUMP STATION	1
13	POWER HOUSING	1
14	WATER TANK: 25,000 GAL	1
15	PIER	1
16	GARAGE: 25 60-16 20	1
17	CABANA: FRAME 16 20	1

SCALE
YARDS

UNALASKA

ILIULIUK BAY

SUMNER BAY

DUTCH HARBOR

AMAKNAK ISLAND

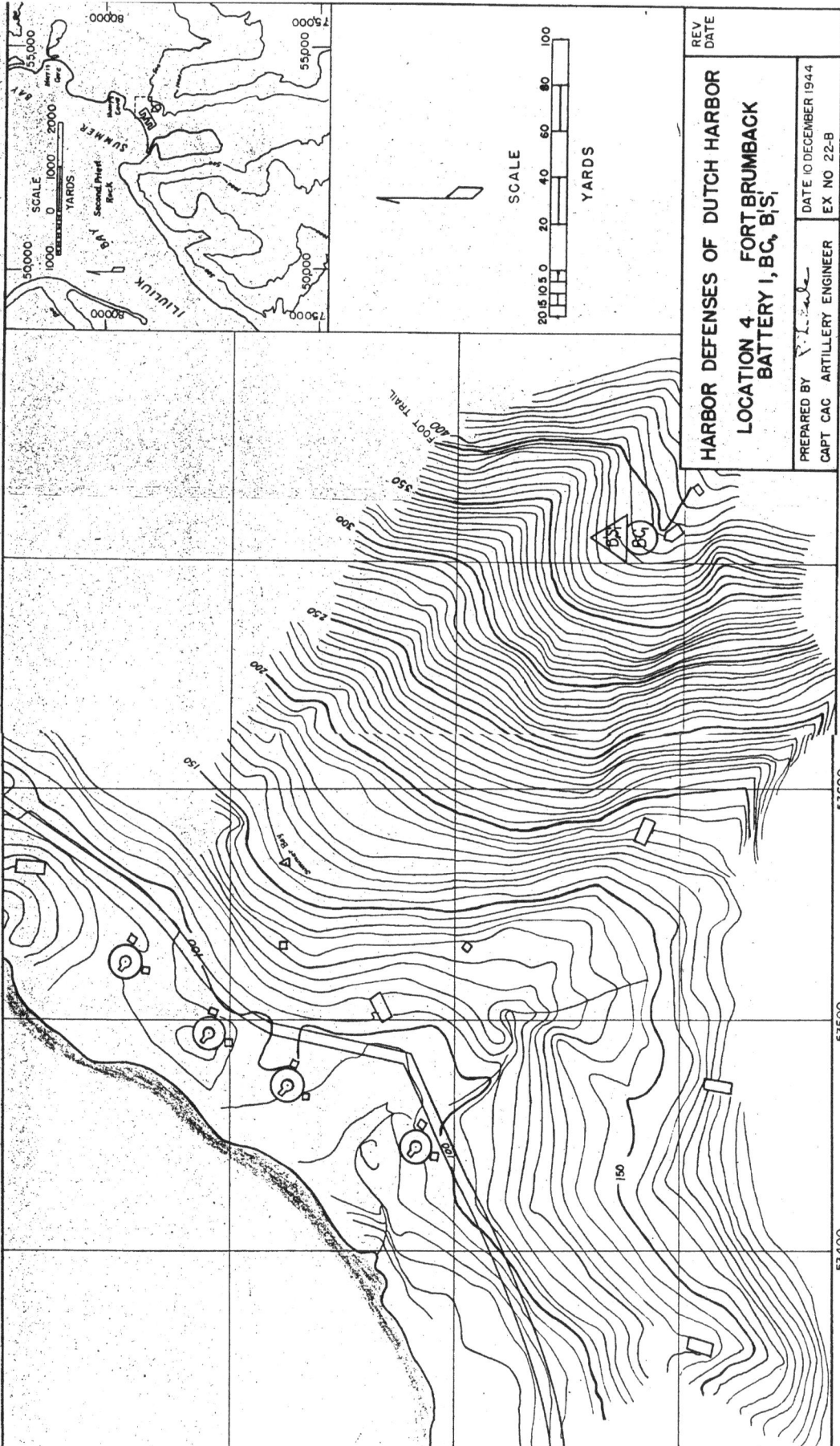

HARBOR DEFENSES OF DUTCH HARBOR

LOCATION 4   FORT BRUMBACK
BATTERY I, BC, B,S

PREPARED BY
CAPT CAC ARTILLERY ENGINEER

DATE 10 DECEMBER 1944
EX NO 22-B

REV
DATE

BUILDING LIST

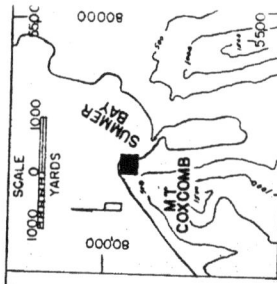

ITEM	NO	SIZE	DISCRIPTION	TYPE	REMARKS
1	2	16·36	BARRACKS	Q.H.	STANDARD Q.H.
2	1	16·36	MESS HALL	Q.H.	SEE DWG CB 341-5
3	2	10·30	SLP.P. SHELTER	STEEL	SEE DWG CB 401-2
4	2	7·10	D. E.C.	FRAME	SEE DWG CB 442-1
5	1	16·24	POWER PLANT 2·2½ KVA		¾ OF PACIFIC HUT
6	1		TRAMWAY		SEE DWG D 10-14-2

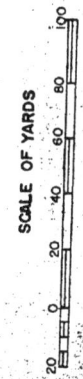

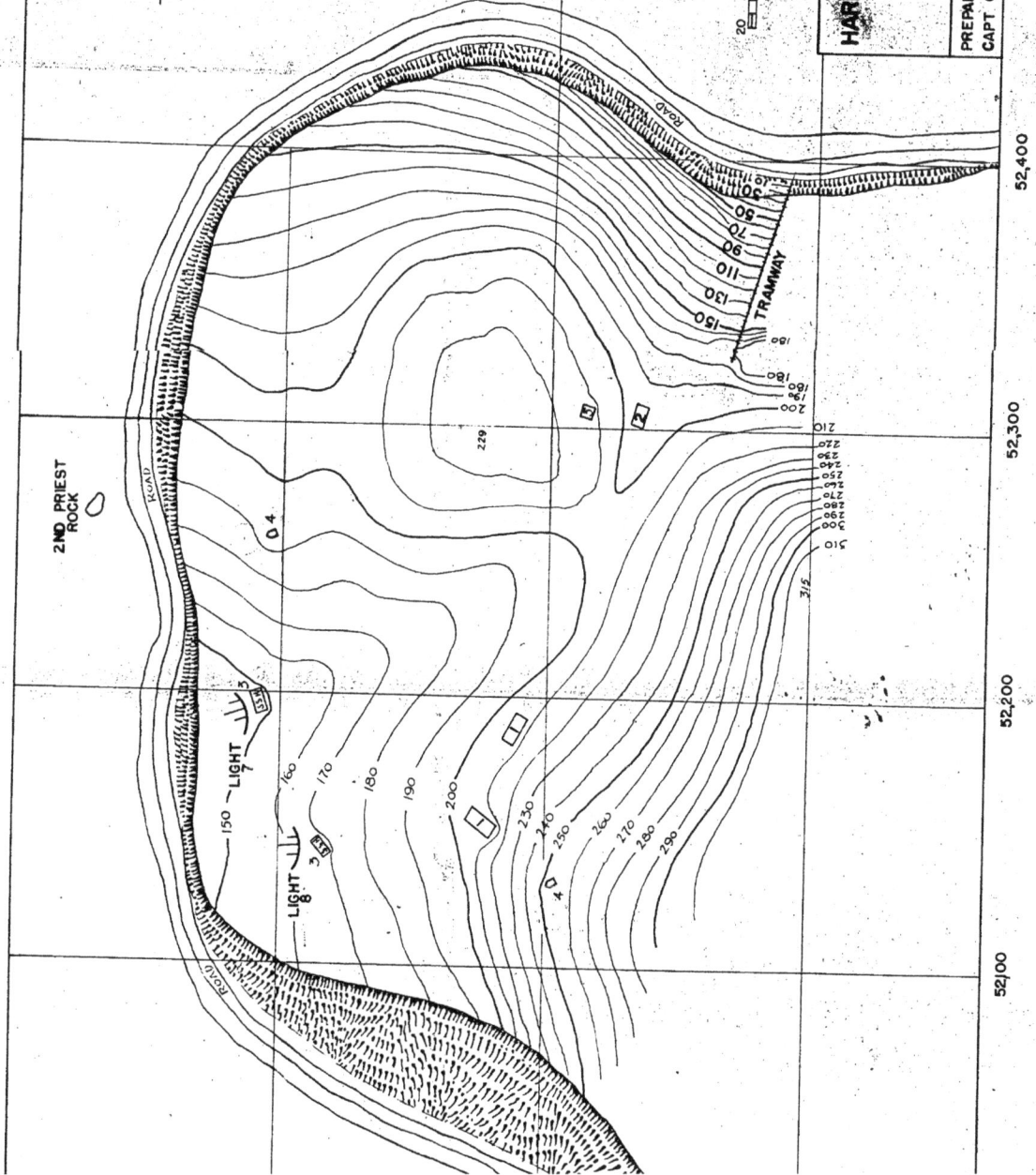

SCALE OF YARDS

HARBOR DEFENSES OF DUTCH HARBOR
LOCATION 5
COXCOMB, SECOND PRIEST ROCK
SEARCHLIGHTS 7, 8

PREPARED BY	DATE 10 DECEMBER 1944
CAPT CAC ARTILLERY ENGINEER	EX NO 23-B

REV
DATE

SUMMER BAY

SCALE
1000   1000
YARDS

MT COXCOMB

2ND PRIEST ROCK

ROAD

TRAMWAY

LIGHT 7

LIGHT 8

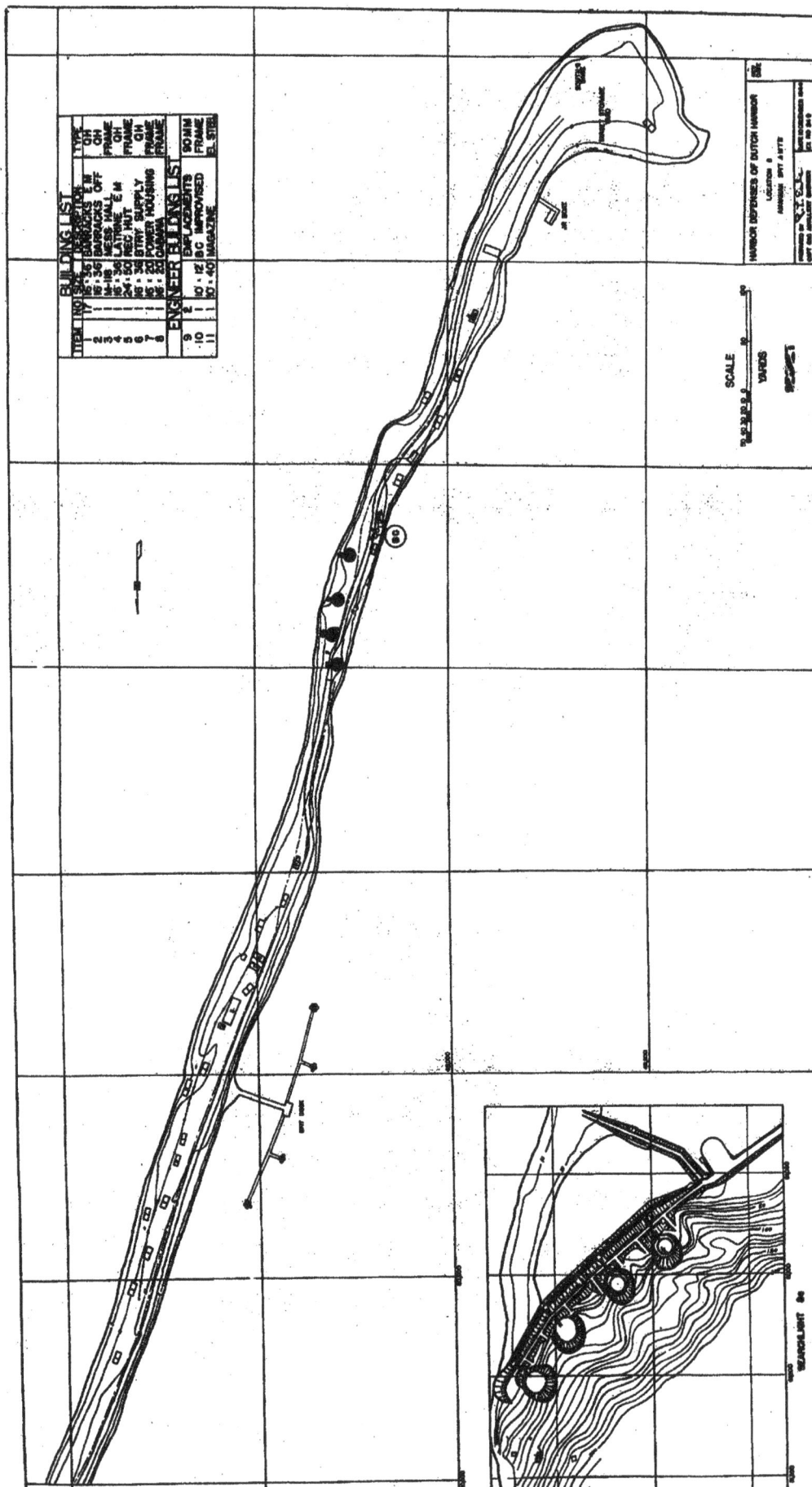

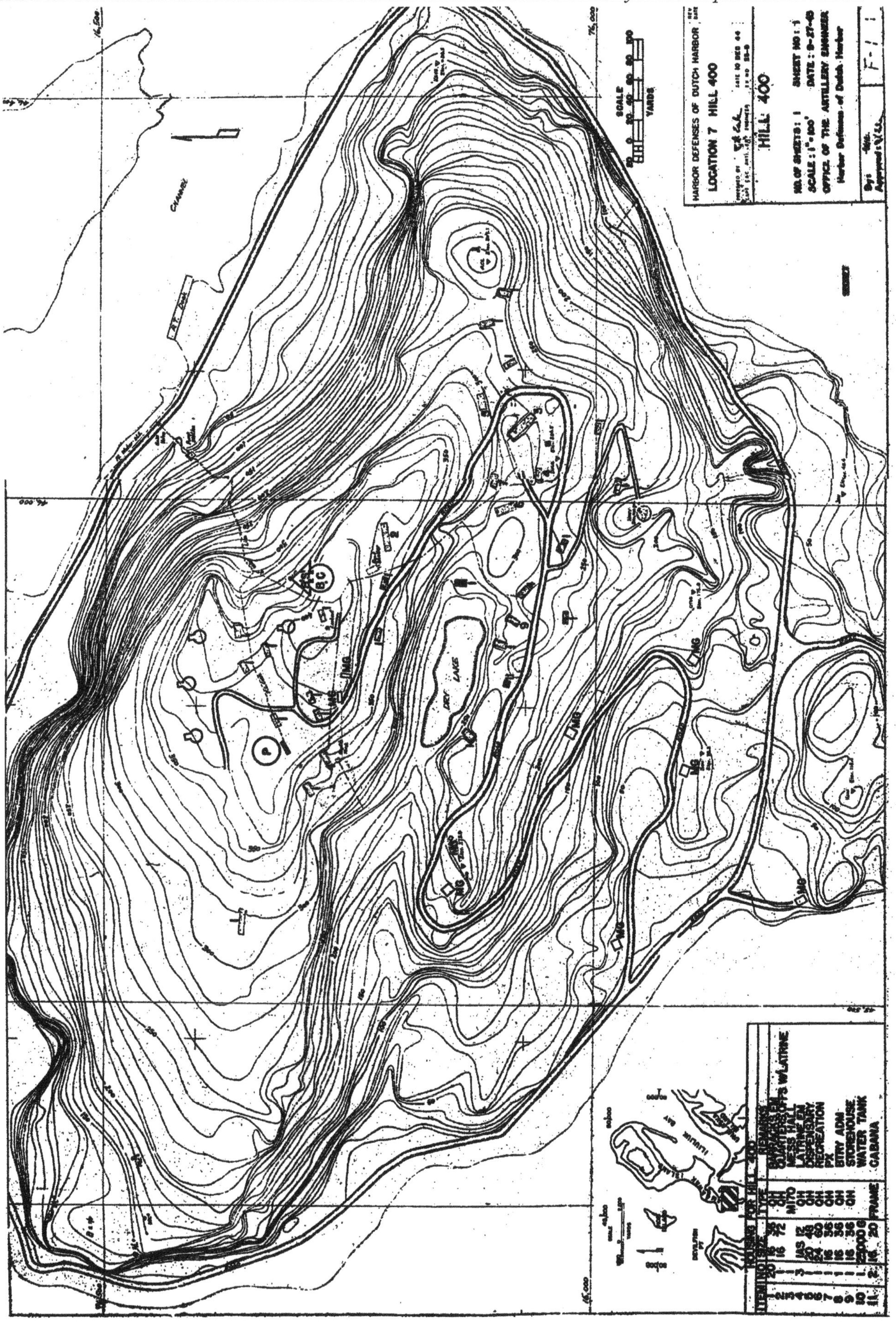
HARBOR DEFENSES OF DUTCH HARBOR
LOCATION 7 HILL 400
HILL 400
SCALE
YARDS
DRY LAKE
BC
MG

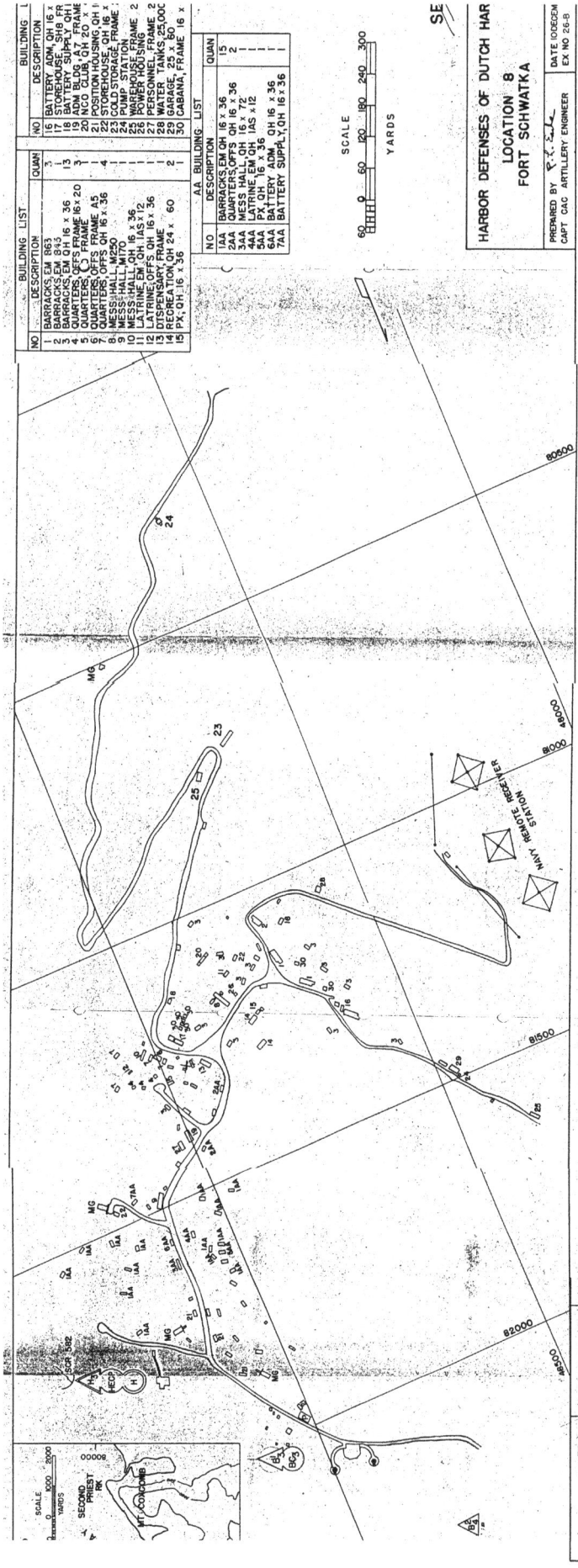
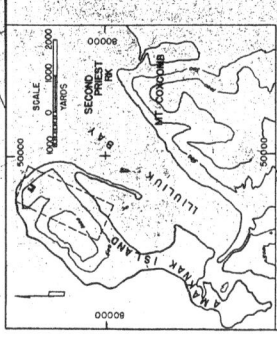
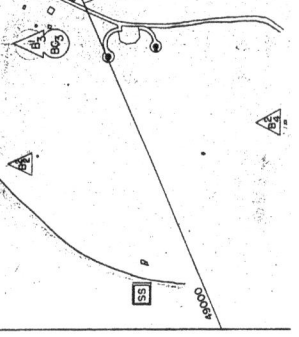

HARBOR DEFENSES OF DUTCH HARBOR
LOCATION 8 HECP, HDCP, HDOP 3, SCR 582
HARBOR DEFENSE RADIO STATION

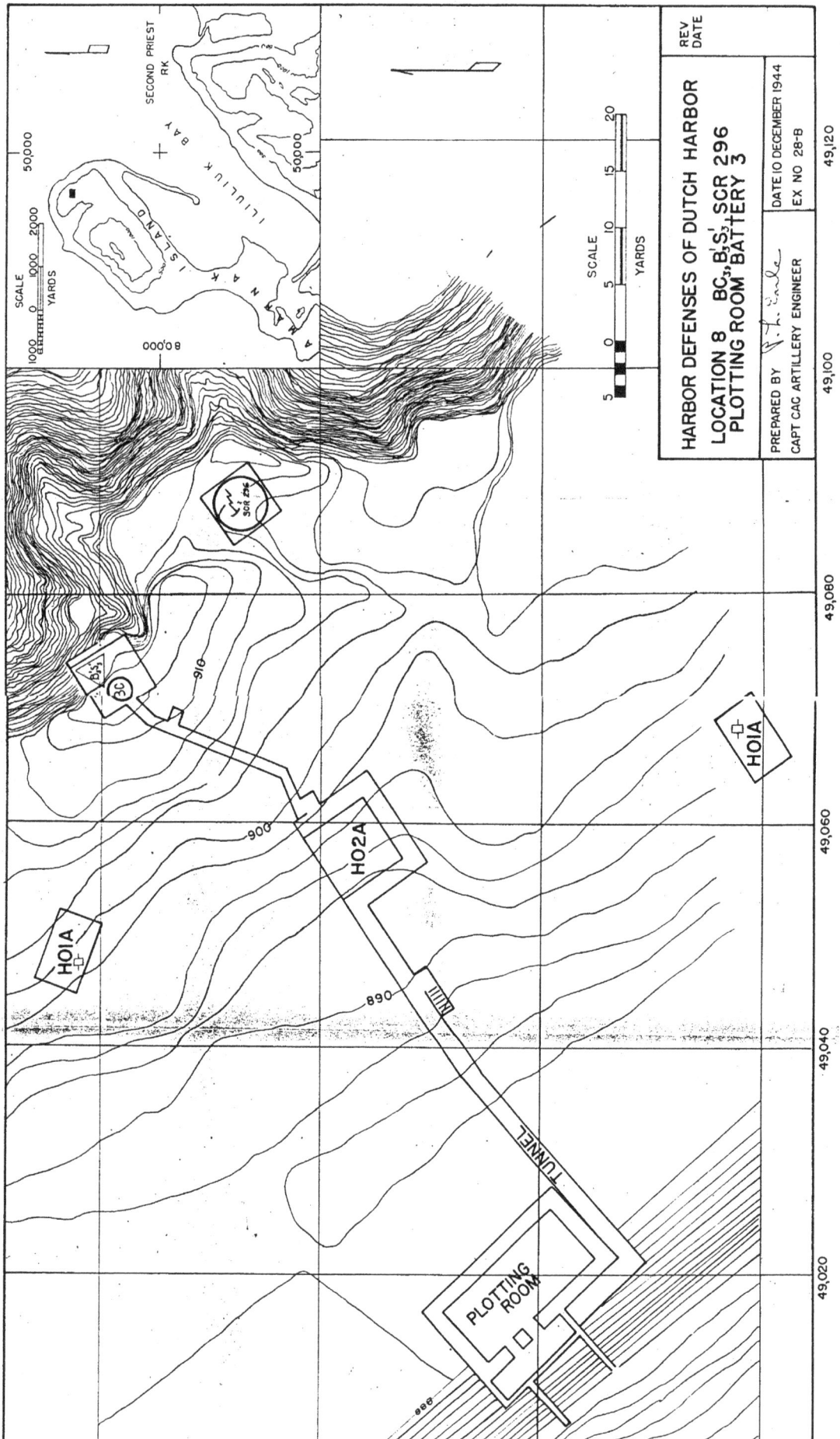

HARBOR DEFENSES OF DUTCH HARBOR

LOCATION 8   BC, B'₁S'₁ SCR 296
PLOTTING ROOM BATTERY 3

PREPARED BY
CAPT CAC ARTILLERY ENGINEER

DATE IO DECEMBER 1944
EX NO 28-B

REV
DATE

SCALE

YARDS

PLOTTING ROOM

TUNNEL

HO2A

HOIA

HOIA

SCR 296

B'₁S'₁

BC

SECOND PRIEST RK

IIULIUK BAY

N A K   I S L A N D

SCALE
YARDS

49,120

49,100

49,080

49,060

49,040

49,020

8,0,000

50,000

50,000

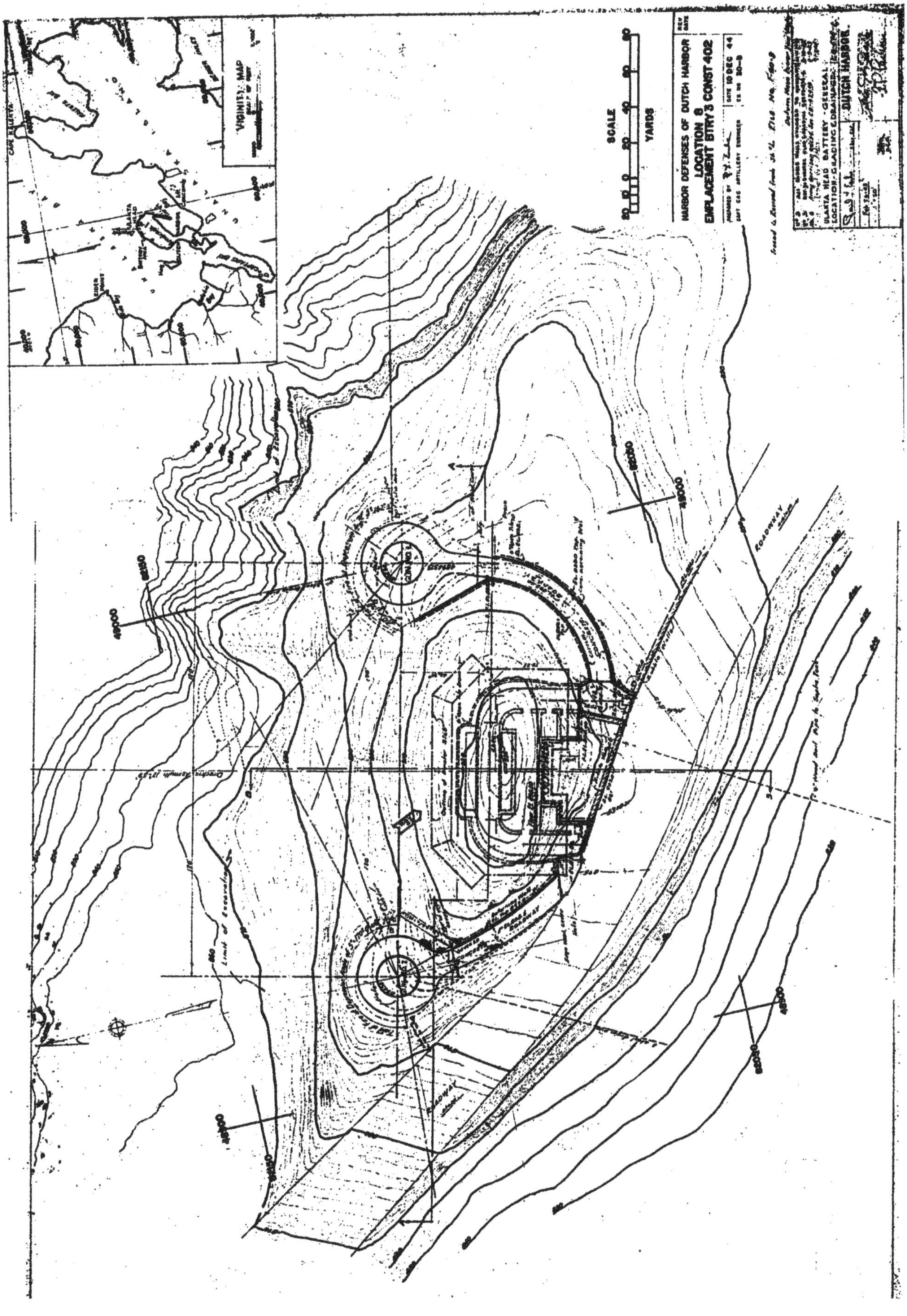
HARBOR DEFENSES OF DUTCH HARBOR
LOCATION 8
EMPLACEMENT BTRY 3 CONST 402
SCALE
YARDS
VICINITY MAP
DUTCH HARBOR
ROADWAY

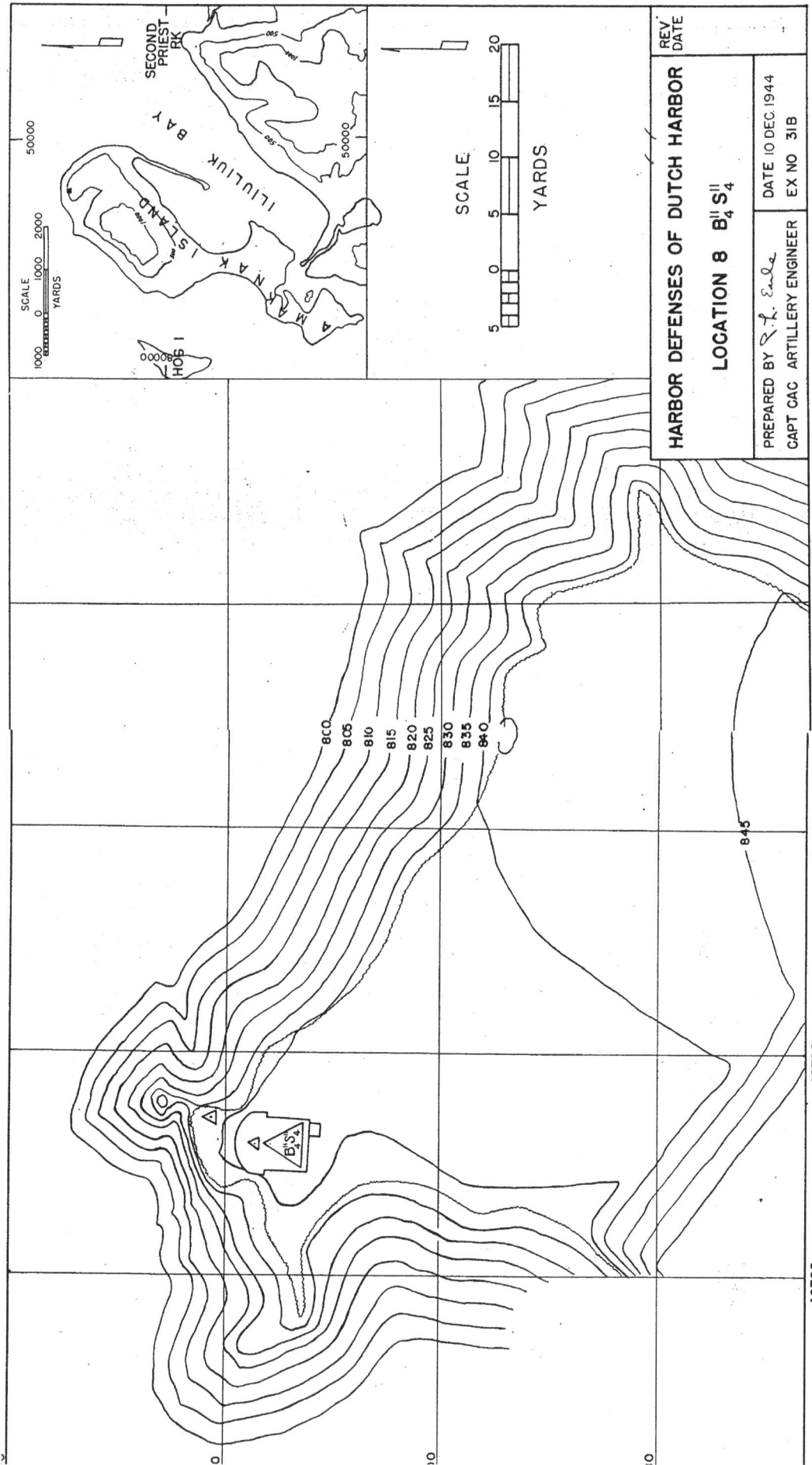

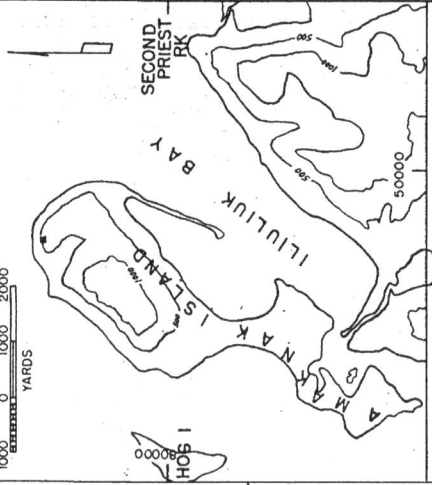

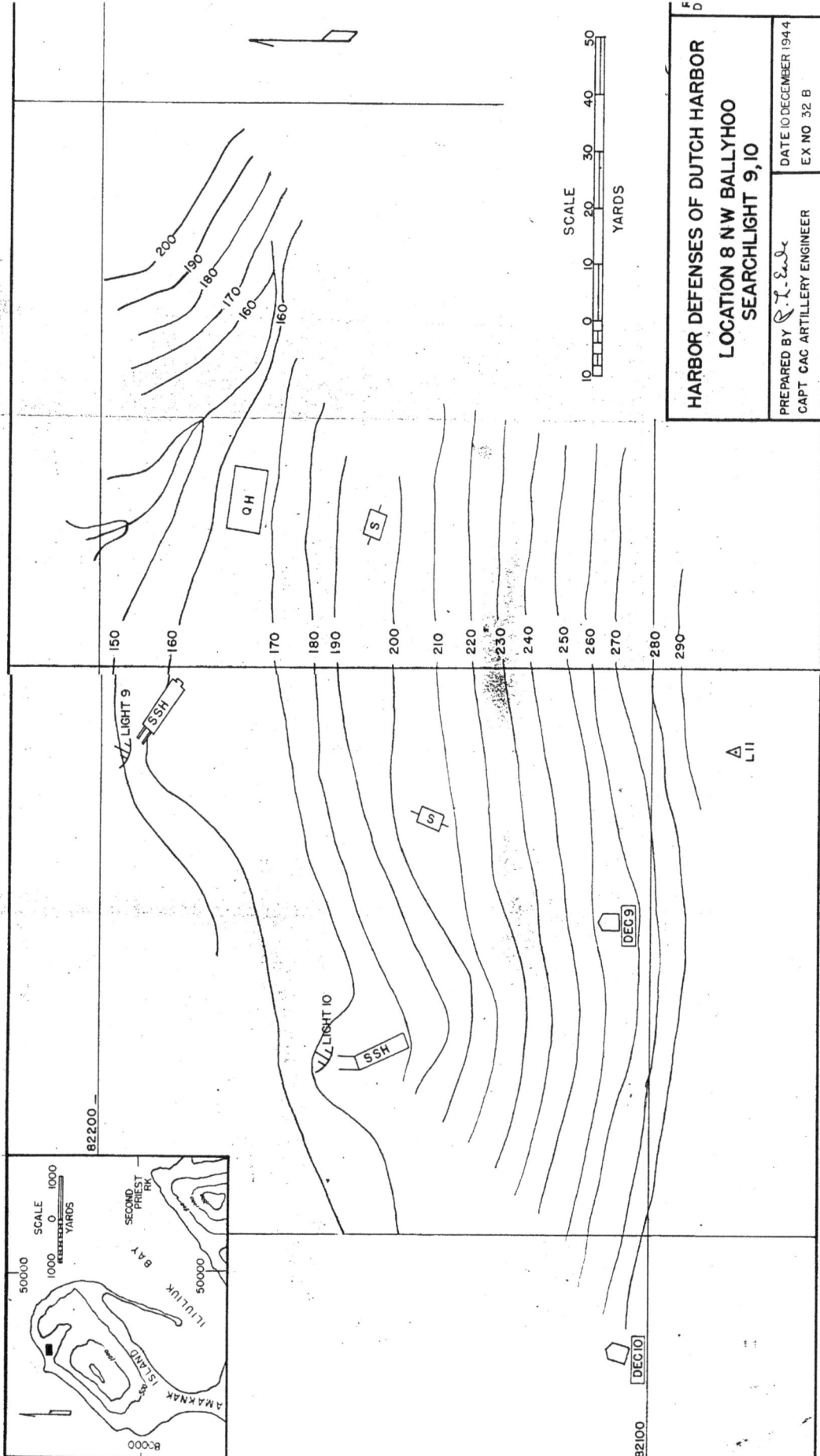

HARBOR DEFENSES OF DUTCH HARBOR
LOCATION 8 NW BALLYHOO
SEARCHLIGHT 9, 10

SCALE

YARDS

PREPARED BY
CAPT CAC ARTILLERY ENGINEER

DATE 10 DECEMBER 1944
EX NO 32 B

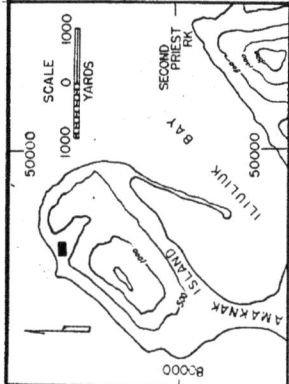

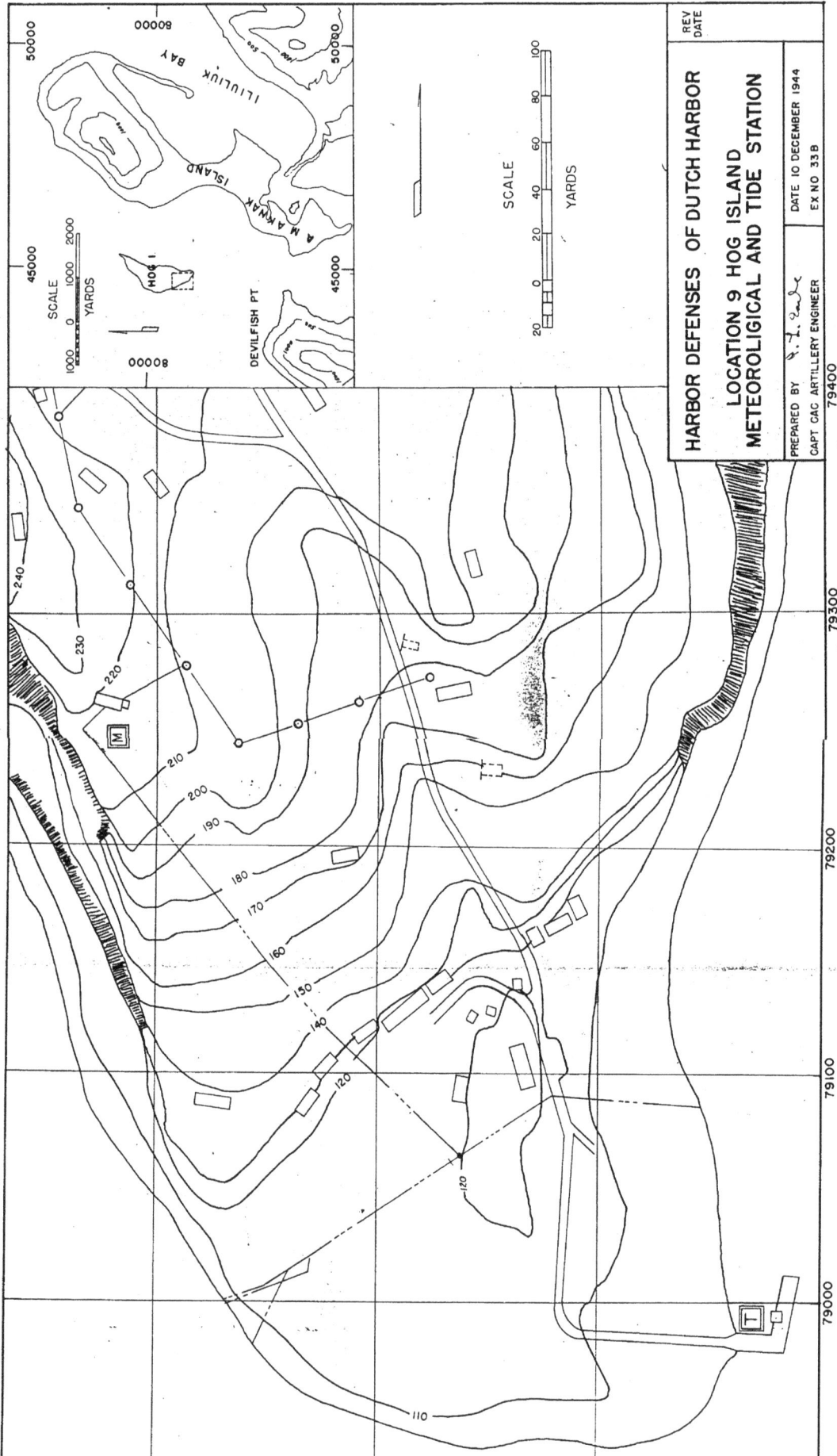

ILIULIUK BAY

AMAKNAK ISLAND

DEVILFISH PT

SCALE
1000   0   1000   2000
YARDS

80000

50000

45000

HOG I.

HARBOR DEFENSES OF DUTCH HARBOR

LOCATION 9 HOG ISLAND
METEOROLIGICAL AND TIDE STATION

REV
DATE

DATE 10 DECEMBER 1944

EX NO 33 B

PREPARED BY

CAPT CAC ARTILLERY ENGINEER

SCALE
20   0   20   40   60   80   100
YARDS

79400

79300

79200

79100

79000

240

230

220

210

200

190

180

170

160

150

140

120

120

110

M

T

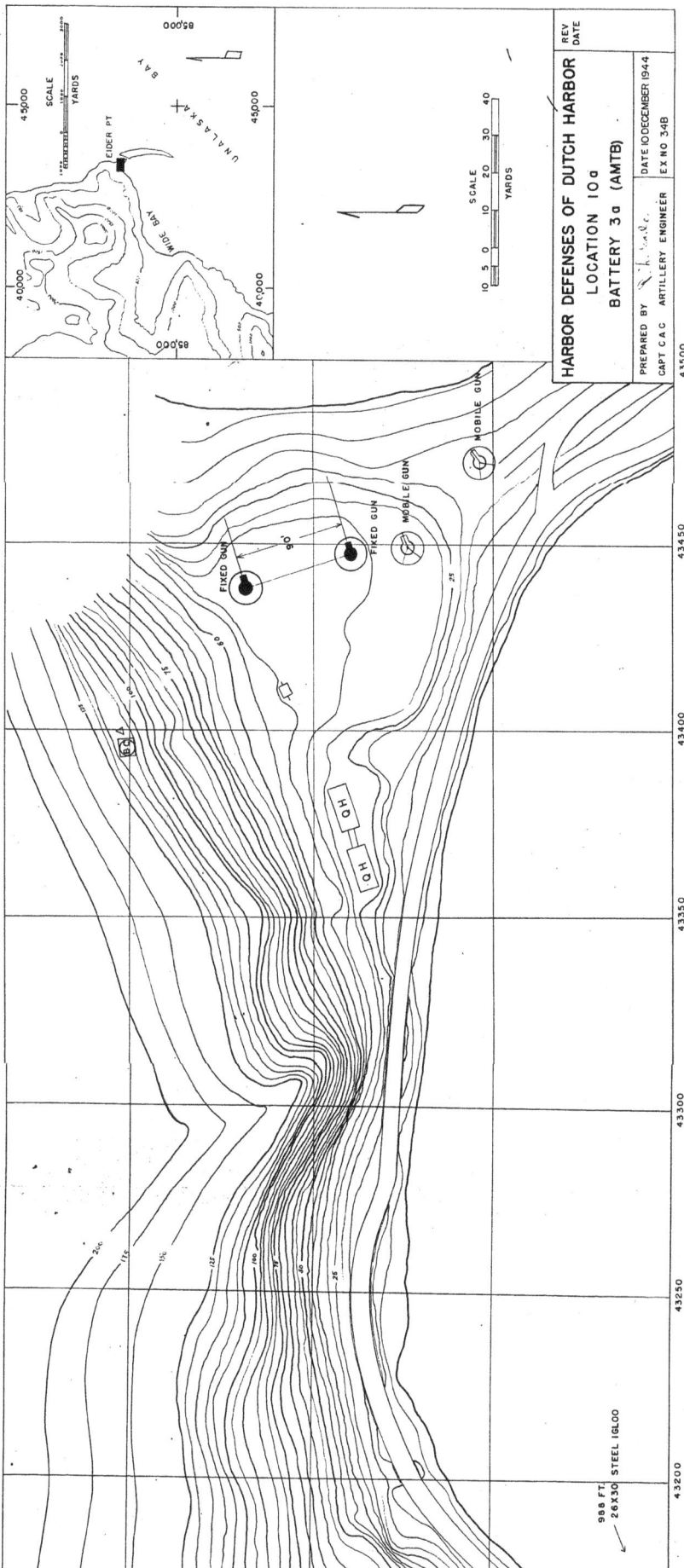

HARBOR DEFENSES OF DUTCH HARBOR
LOCATION 10a
BATTERY 3a (AMTB)

DATE 10 DECEMBER 1944
EX NO 34B

PREPARED BY
CAPT C A C   ARTILLERY ENGINEER

REV
DATE

SCALE
YARDS

10 5 0 10 20 30 40

UNALASKA BAY

EIDER PT

WIDE BAY

SCALE
YARDS

MOBILE GUN

MOBILE GUN

FIXED GUN

FIXED GUN

QH

QH

988 FT.
26X30 STEEL IGLOO

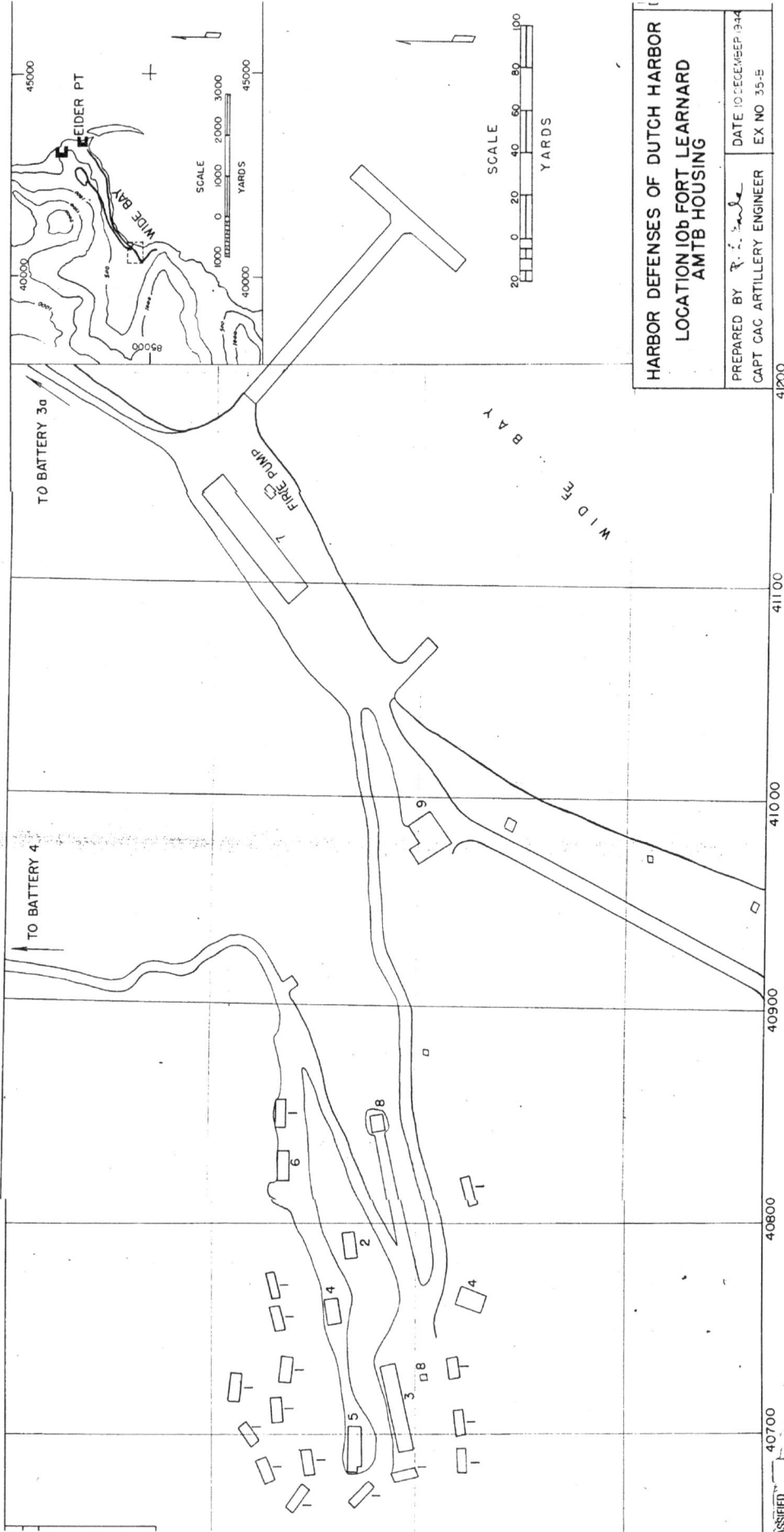

HARBOR DEFENSES OF DUTCH HARBOR
LOCATION 10b FORT LEARNARD
AMTB HOUSING

	BUILDING LIST		
ITEM NO	TYPE	SIZE	DESCRIPTION
1	16	QH	16×36	BARRACKS, EM
2	1	QH	16×36	QUARTERS, OFFS
3	2	M250		MESS HALL
4	1	L84		LATRINE
5	1	SH8	16×36	STOREHOUSE
6	1	QH		BTRY ADM
7	1		40×200	WAREHOUSE
8	2			POWER HOUSING
9	1	FRAME	56×43	MACHINE SHOP

PREPARED BY

CAPT CAC ARTILLERY ENGINEER

DATE 10 DECEMBER 1944

EX NO 35-B

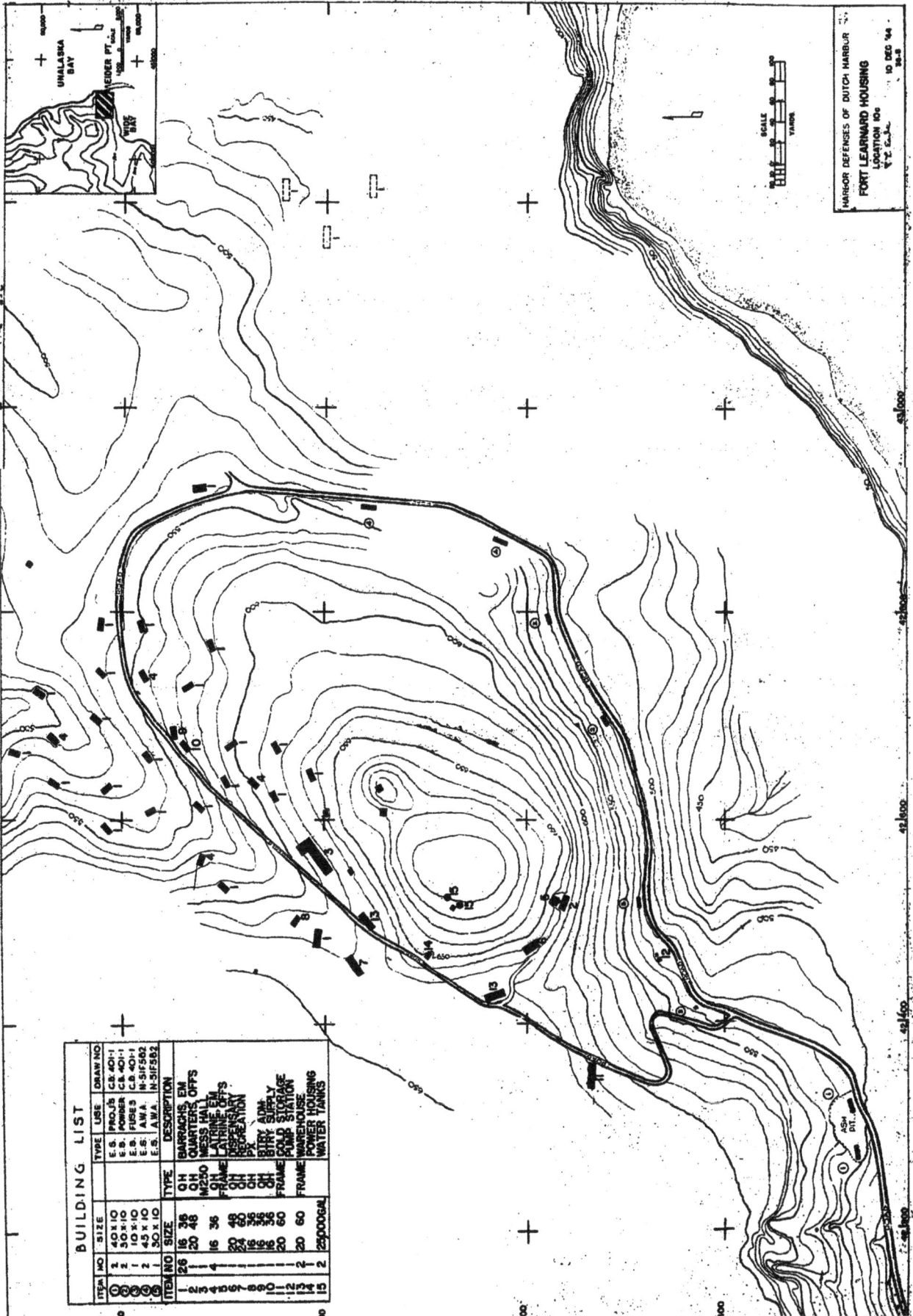

HARBOR DEFENSES OF DUTCH HARBOR
FORT LEARNARD HOUSING
LOCATION 10a
10 DEC '44

ALASKA SEACOAST DEFENSES.
EIDER POINT BATTERY
GRADING AND DRAINAGE PLAN
DUTCH HARBOR

HARBOR DEFENSES OF DUTCH HARBOR
LOCATION 10
EMPLACEMENT BTRY 4 CONST 298

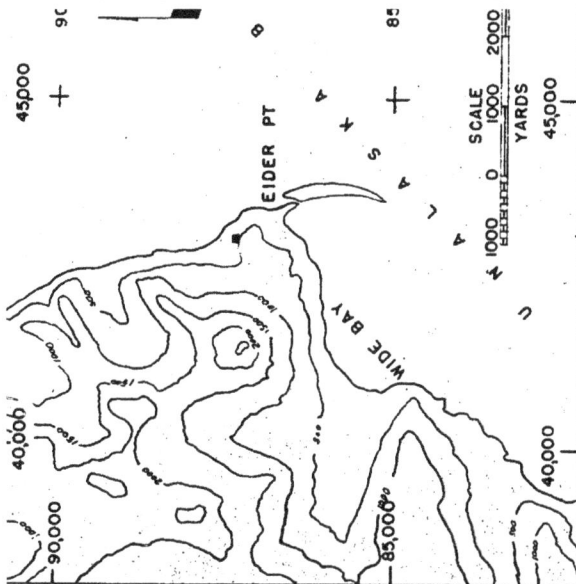

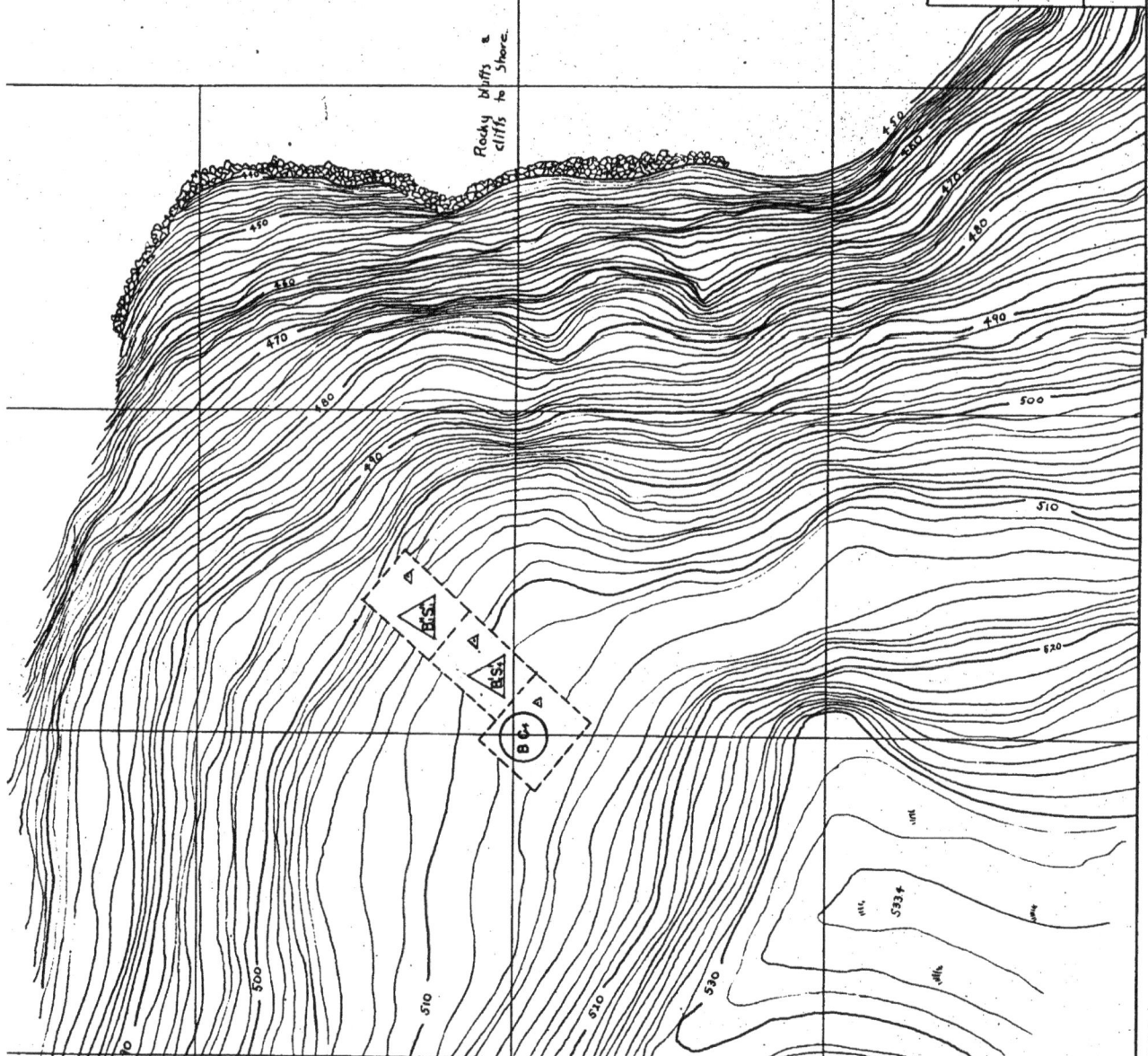

HARBOR DEFENSES OF DUTCH HARBOR

LOCATION 10   BC₄, B'S'₄, B"S"₄
FORT LEARNARD

PREPARED BY   DATE 10 DECEMBER 1944

CAPT. OAC ARTILLERY ENGINEER   EX NO 38-B

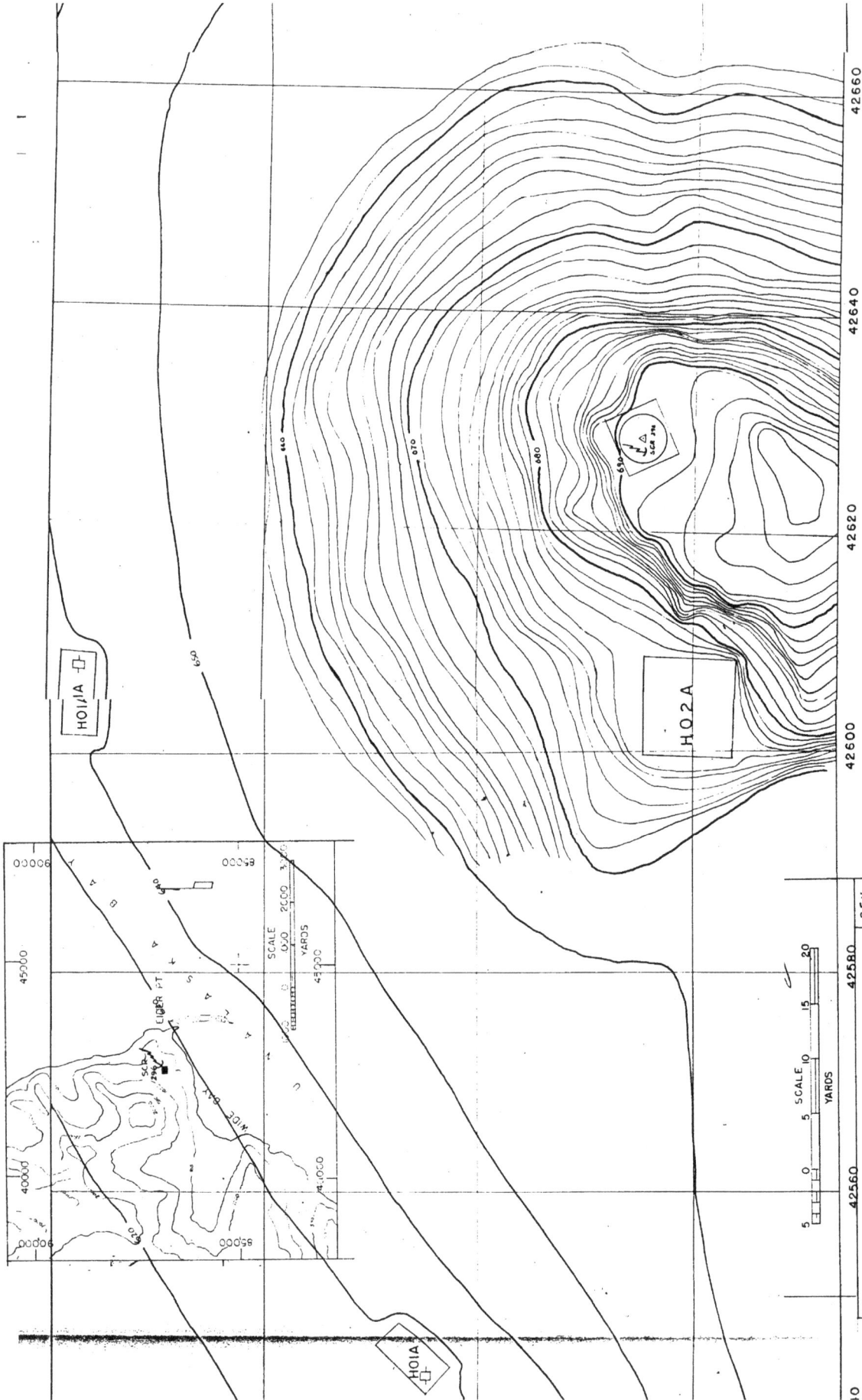

HARBOR DEFENSES OF DUTCH HARBOR

LOCATION 10   SCR 296
FORT LEARNARD

PREPARED BY   DATE 10 DECEMBER 1944
CAPT C A C ARTILLERY ENGINEER   EX NO 39-B

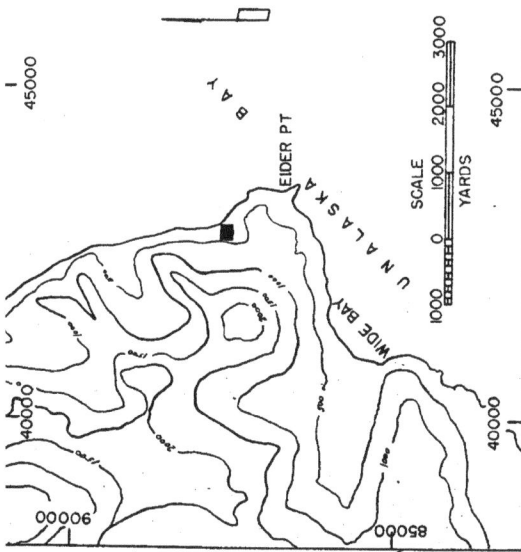

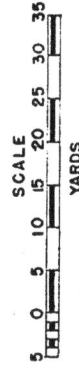

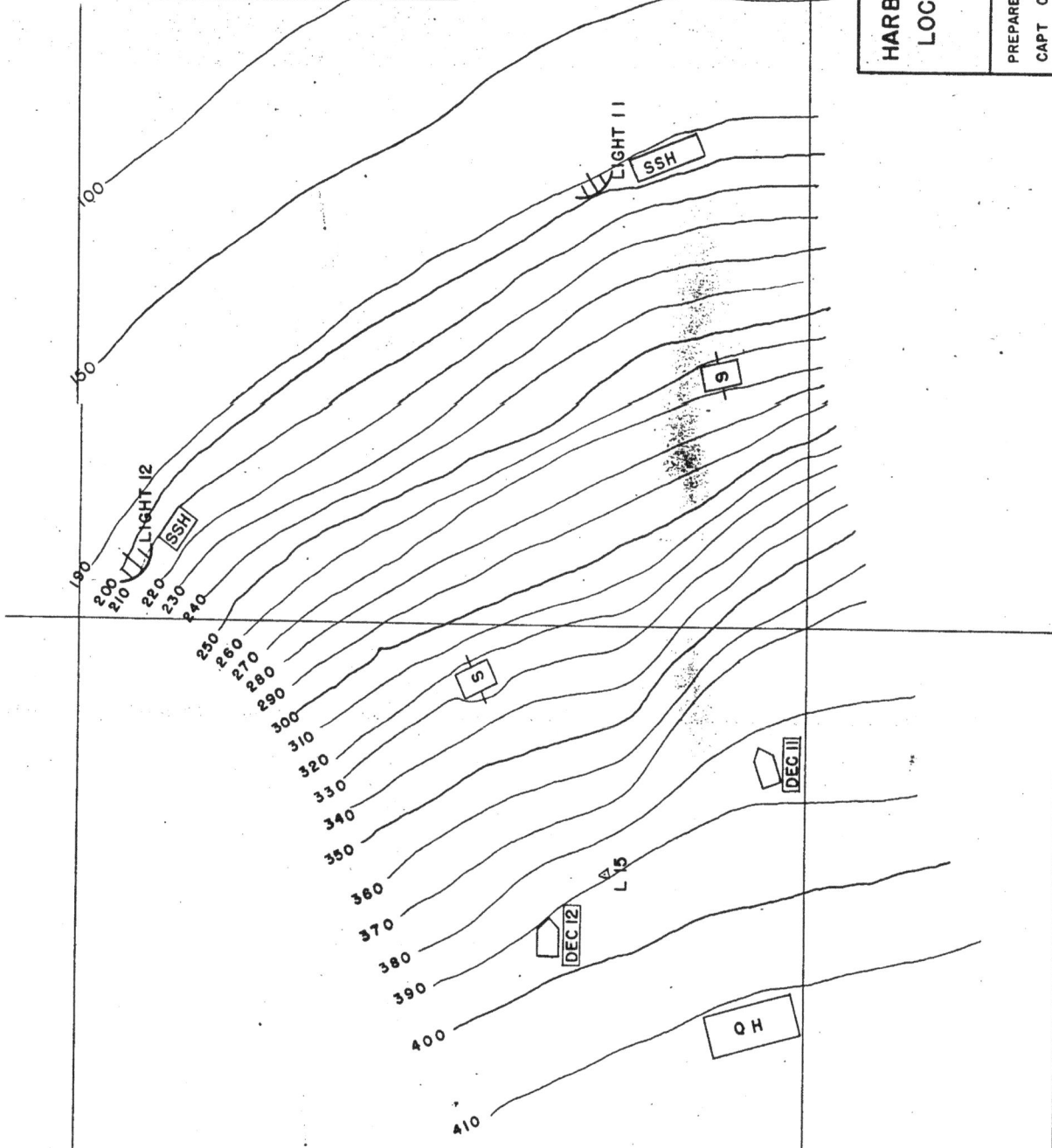

HARBOR DEFENSES OF DUTCH HARBOR
LOCATION 10   FORT LEARNARD
SEARCHLIGHT 11, 12

| PREPARED BY | DATE 10 DECEMBER 1944 |
| CAPT CAC ARTILLERY ENGINEER | EX NO 40-B |

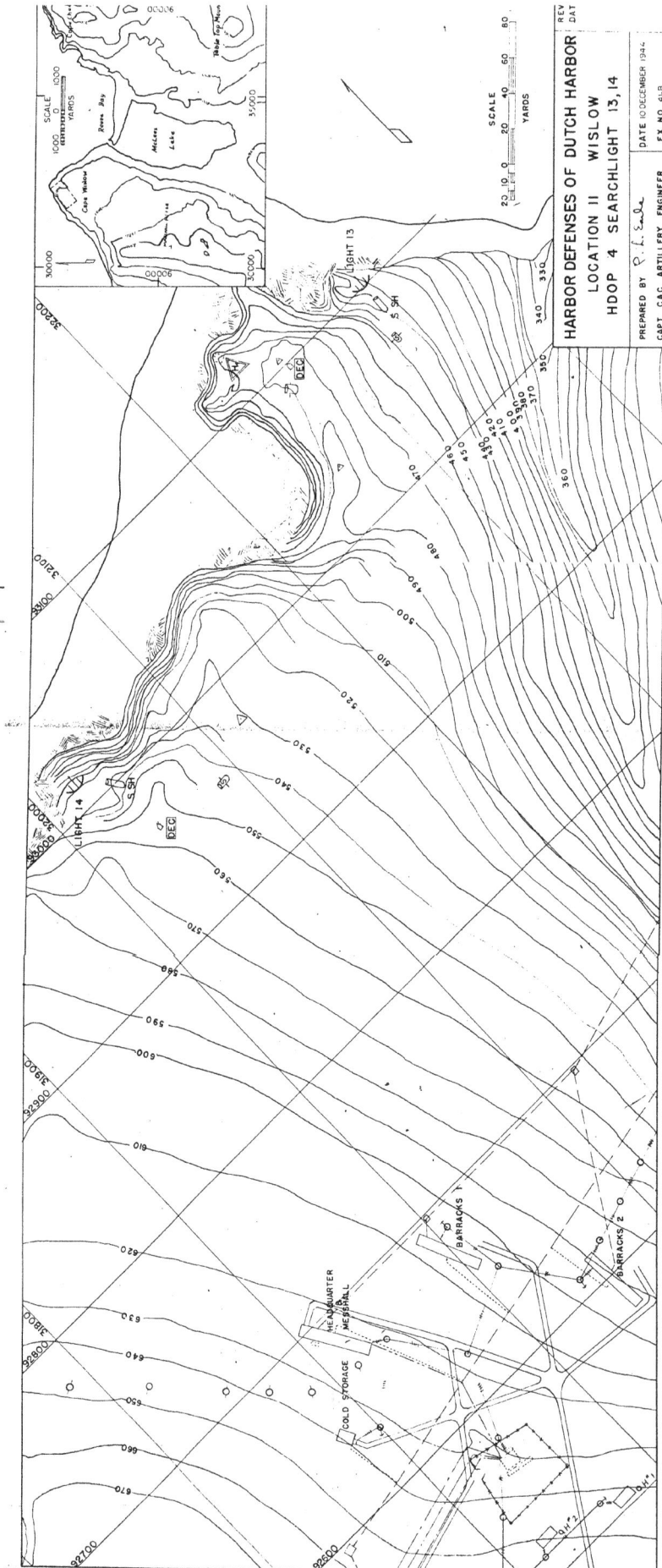

HARBOR DEFENSES OF DUTCH HARBOR
LOCATION II   WISLOW
HDOP 4 SEARCHLIGHT 13,14

PREPARED BY   CAPT CAC ARTILLERY ENGINEER
DATE 10 DECEMBER 1944   EX NO 41-B

OVERSEAS BASES – HAWAII

The Hawaiian Islands are thousands of miles west from the mainland United States. The defense of the islands was focused on the Pearl Harbor Naval Station on the island of Oahu. Oahu has a rugged coastline with Honolulu and Pearl Harbor its only major harbors. The climate is tropical and access to coast defense sites can be difficult. While construction of its harbor defenses started during the Endicott and Taft Programs, many of its defenses were the product of World War II and the response to the Japanese air attack on Pearl Harbor. As a result, the entire island became defended with beach defenses, artillery positions, mobile infantry and armor, and air assets. Important coast artillery forts were Fort Ruger, Fort DeRussy, Fort Kamehameha, Fort Barrette, Fort Weaver, Fort Hase, and other military reservations. The listing below only includes major sites with permanent coast defenses. An excellent description of these forts can be found in *Pacific Fortress* by Glen Williford or Osprey Fortress Series No. 8 *Defenses of Pearl Harbor and Oahu 1907-1950* by Terry McGovern and Glen Williford.

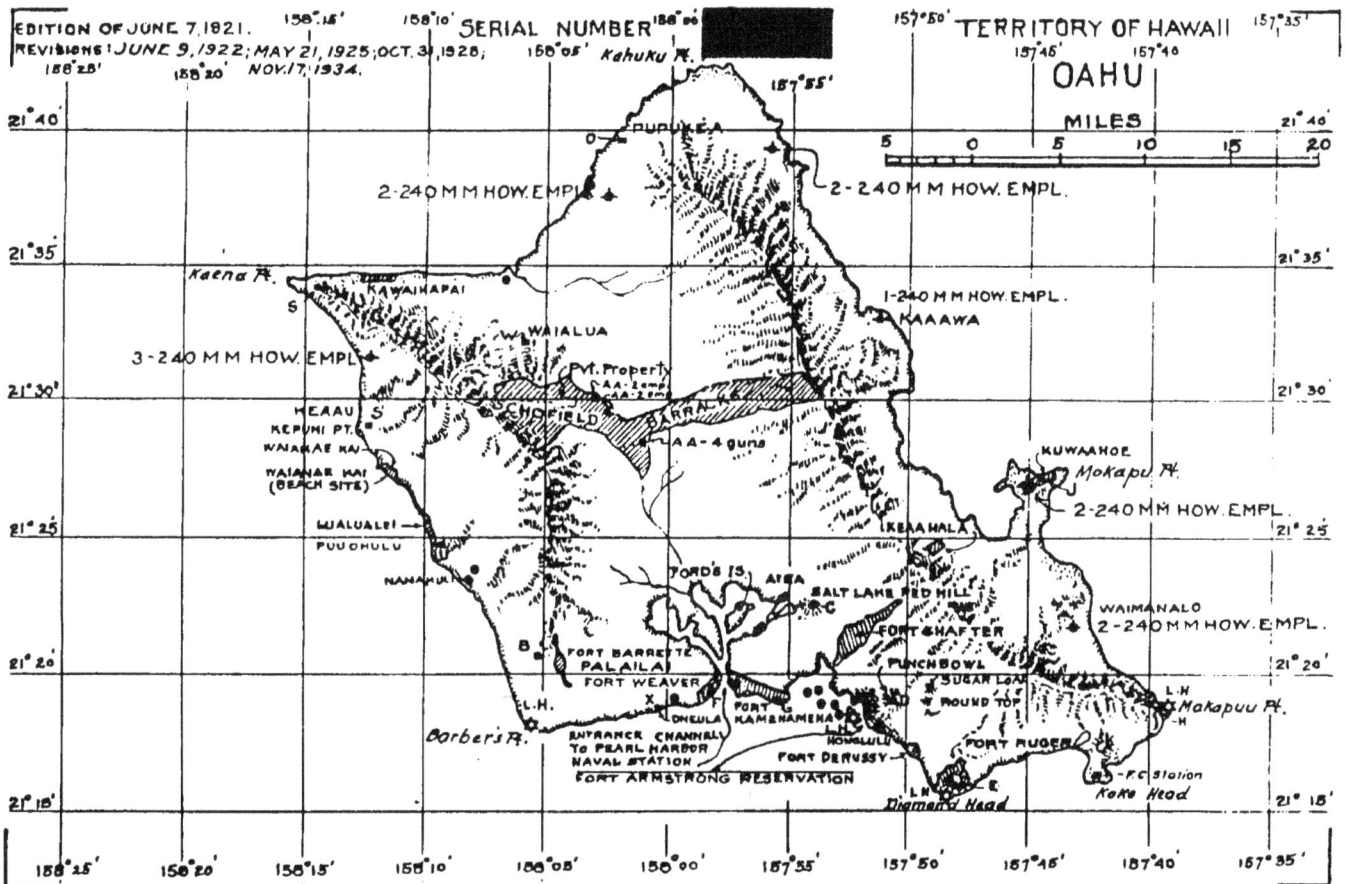

Williford, Glen, and Terrance McGovern. *Defenses of Pearl Harbor and Oahu 1907-50.* Fortress Series #8, Osprey Publishing Ltd., Oxford, GB, 2003.

Williford, Glen, *Pacific Fortress A History of the Seacoast Defenses of Hawaii,* CDSG Press, McLean, VA 2023

Dorrance, William H. *Fort Kamehameha, the Story of the Harbor Defenses of Pearl Harbor.* White Mane Publishing Co. Shippensburg, PA, 1993.

THE HARBOR DEFENSES OF HONOLULU – HAWAII
(INCLUDES OAHU'S EASTERN SHORE)

Fort Shafter (1899-present) is located to the north of the City of Honolulu on the ridgeline between Kalihi and Moanalua Valleys and down to the coastal plain at Mapunapuna. Originally named Kahauiki Military Reservation until 1905. When the post opened in 1907, it was named for Major General William Rufus Shafter (1835–1906), who led the United States expedition to Cuba in 1898. Four 3-inch AA gun batteries were built here in the 1920s. During World War II the post became the headquarters post for all U.S. Army forces in the Pacific region. The post's administration buildings, built in World War II, were known as the "Pineapple Pentagon". What began as a small cantonment for a single infantry battalion has now become the U.S. Army's base of operations for the entire Asia-Pacific region outside of Korea. Today, it's home to a garrison which includes the U.S. Army Support Command, Hawaii, and the U.S. Army Corps of Engineers, Pacific Ocean Division. Access is limited as active military base.

Sand Island Military Reservation (1916–1948), formerly known as Quarantine Island, is a small island within the city of Honolulu, Hawaii, United States. The island lies at the entrance to Honolulu Harbor. Two 3-inch AA guns were here in 1917, and two more were added in 1928. Also located here were four Panama mounts of Battery Sand Island (1937-1942), converted to AMTB Battery #4 (1943-1946), Battery Harbor (1942-1944) four 7-inch naval guns (two guns later moved to Fort DeRussy as display at Battery Randolph), and Naval Antiaircraft Shore Battery No. 9 (1942-1944) four 5-inch naval guns. Two searchlights were emplaced here. The 3-inch AA mounts may be buried in the southeast section of the park, the Panama mounts and AMTB mounts are destroyed. The location of the 5-inch AA mounts has not been determined. The 7-inch gun mounts and magazines are incorporated into a playground. The HECP tower still exists. The island is connected the mainland, and the state recreation area is open to the public.

Sand Island M.R. Gun Battery

- **AMTB-4**: An anti-motor torpedo boat battery of two fixed 90mm guns with M3 barbette carriage and two mobile 90mm M1 guns erected the Sand Island Military Reservation. It was built from late 1942 to June 1943. The fixed guns were on simple concrete gun blocks. It was also informally called Battery Sand. It was disarmed in mid or late 1945. Partial remains still exist at still exist at the Sand Island State Recreation Area. The battery is open to the public.

Fort Shafter 1938 (NARA)

HONOLULU and PEARL HARBORS

BATTERIES

Battery	
HASBROUCK	8 - 12" M
HARLOW	8 - 12" M
BIRKHIMER	4 - 12" M
RANDOLPH	2 - 14" D15.
SELFRIDGE	2 - 12"
DUDLEY	2 - 6"
JACKSON	2 - 6"
BOYD	
ADAIR	
S.C. MILLS	
HULINGS	
DODGE	
BARRI	
CHANDLER	2 - 3" P.
HAWKINS	2 - 3"
TIERNON	2 - 3"
HATCH	2 - 16" B.C.
GRANGER ADAMS	2 - 8 B.C.
CLOSSON	2 - 12 B.C.
WILLISTON	2 - 16" B.C.
A. Anti-aircraft guns	17 - 3"

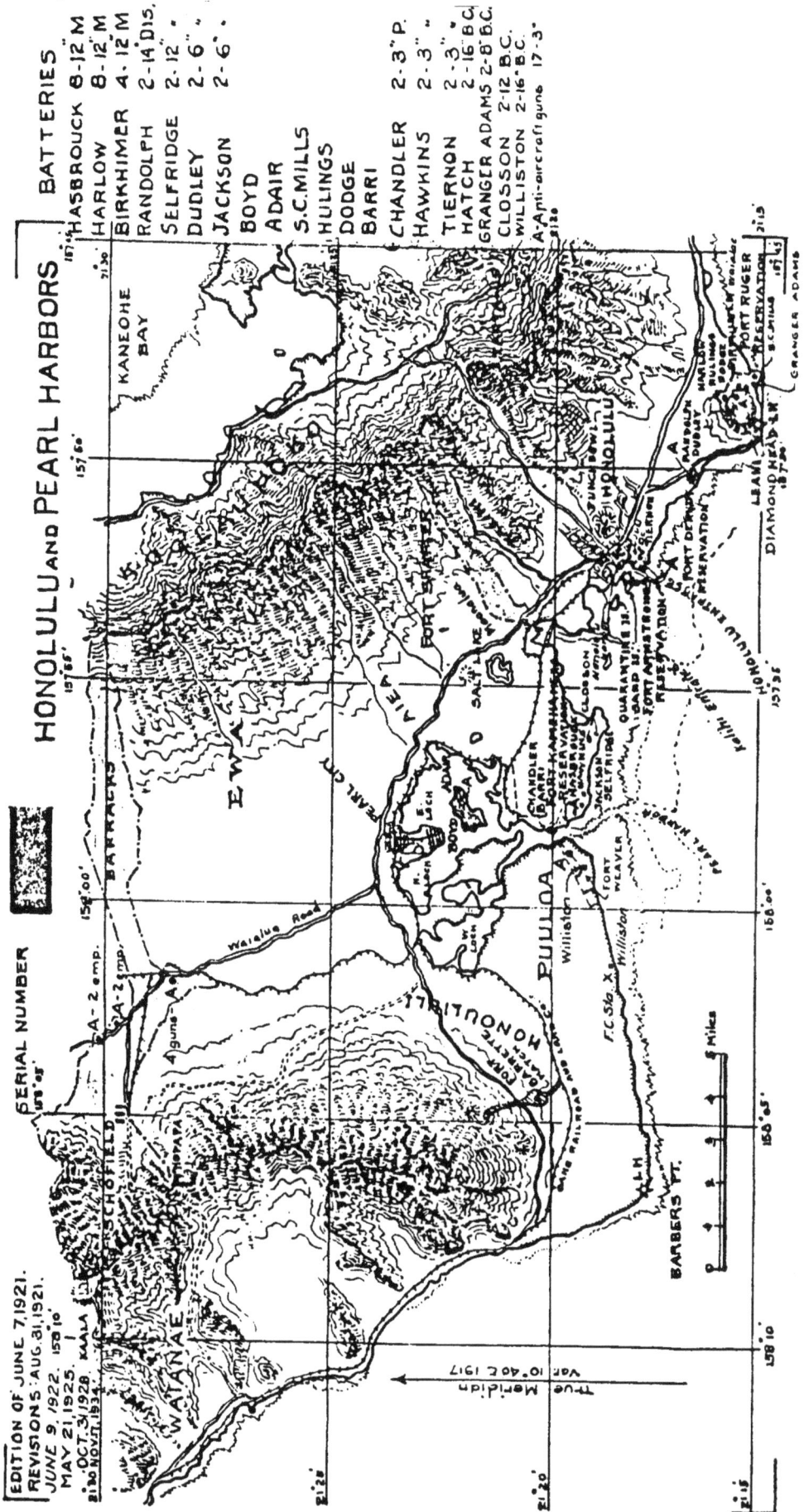

EDITION OF JUNE 7, 1921.
REVISIONS: AUG. 31, 1921.
JUNE 9, 1922.
MAY 21, 1925.
OCT. 31, 1928.
NOV. 11, 1931.

SERIAL NUMBER

HONOLULU HARBOR
SAND ISLAND

A ---- EMP. FOR 2 - 3"
ANTI AIR CRAFT GUNS

SERIAL NUMBER 124

EDITION OF JUNE 7, 1921.
REVISIONS

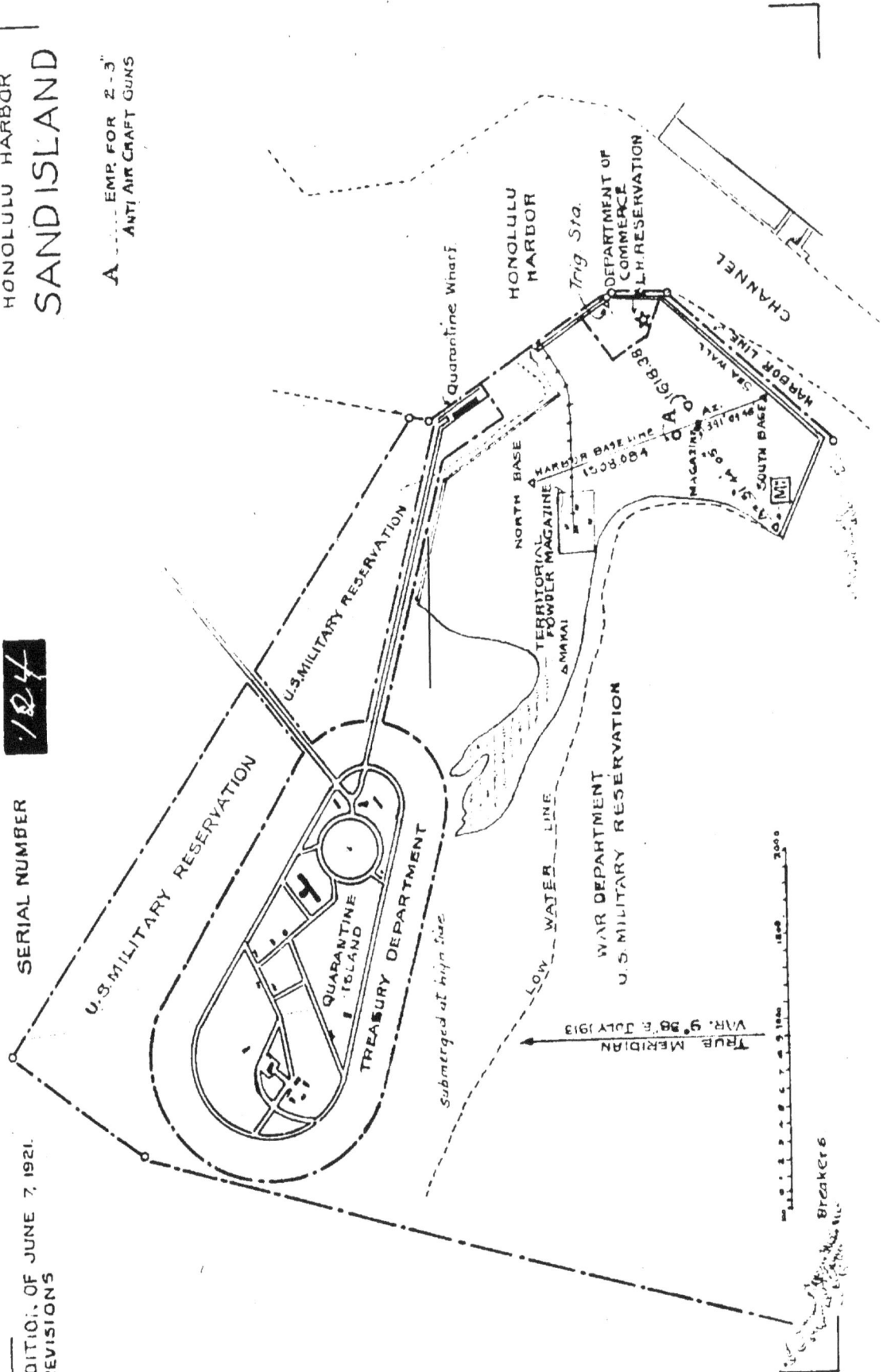

Quarantine Wharf

HONOLULU
HARBOR

Trg Sta.

DEPARTMENT OF
COMMERCE
L.H. RESERVATION

CHANNEL

NORTH BASE

HARBOR BASE LINE
1508.084

TERRITORIAL
POWDER MAGAZINE
△ MAKAI

MAGAZINE AZ

SEA WALL

HARBOR LINE

SOUTH BASE

U.S. MILITARY RESERVATION

U.S. MILITARY RESERVATION

QUARANTINE
ISLAND

TREASURY DEPARTMENT

Submerged at high tide

LOW WATER LINE

WAR DEPARTMENT
U.S. MILITARY RESERVATION

TRUE MERIDIAN
VAR. 9°.38' E. JULY 1913

Breakers

Battery Harbor and HECP on Sand Island (Terry McGovern)

The Punchbowl and Battery #304 (Terry McGovern)

Fort Armstrong (1899-1949/1974) is located at Honolulu, Oahu and was built on fill over Ku-akau-kukui reef in 1907 to protect Honolulu harbor. Named Fort Armstrong in General Orders 160, 1909, after Brigadier General Samuel C. Armstrong, who served during the U.S. Civil War. Located across the ship channel from Sand Island, it contained Battery Tiernon (1911-1943) which mounted 3-inch rapid fired guns with a mine casemate incorporated into the emplacement (destroyed). The present seawall was constructed 500 feet out from the original shoreline in 1948, and the area was backfilled, destroying in the process what was left of the old reef. The U.S. Army Corps of Engineers took over the post in 1949 until 1974. Kakaako Park was created over the landfill area, while the rest of area is a container port facility.

Fort Armstrong Gun Battery

- **TIERNON:** The only battery at the small Fort Armstrong post. This reservation at Kaakaukukui Reef had been set aside for submarine mining purposes by Presidential order on January 15, 1909. The emplacement was a fairly standard dual 3-inch battery but also had an extensive addition on its left flank for a four-room submarine mining casemate. It was built from April 1909 to late 1910, transfer being made on June 15, 1911 for $20,000. It was named in General Orders No. 59 on May 6, 1911 for Brigadier General John L. Tiernon, who just passed away in 1910. Throughout its service life it carried two 3-inch Model 1903 guns and pedestal mounts (#96/#96 and #95/#97). Though plans were made for reducing the status of the battery after 1935, it continued to serve until late in World War II. On December 21, 1943 it was listed as armed but not manned. The emplacement itself was destroyed in 1979 to allow expansion in the dock area. Nothing remains today.

Punchbowl Military Reservation (1899-1948) is an extinct volcanic tuff cone located north of the downtown of Honolulu, Hawaii. The location for Battery Punchbowl (1930s-1943) with two Panama mounts until 1943 when two more were added. One mount still exists. Two 3-inch AA gun batteries were built here in 1917. Fire-control station "D" (pre-World War II) was also here, three structures remain. Inside the Punchbowl was Camp Punchbowl, the cantonment area for the batteries. Toward the end of World War II, tunnels were dug through the rim of the crater for the placement of a modern 6-inch shielded barbette battery to guard Honolulu Harbor and the south edge of Pearl Harbor. Battery #304 was started in 1944 but never completed (tunnels still exist). The Punchbowl became the National Memorial Cemetery of the Pacific in 1949 and is open for day use but the harbor defenses sites are off limits.

Punchbowl M.R. Gun Battery

- **Battery #304:** This was one of the supplemental, permanent system emplacements authorized for Oahu relatively late, on February 2, 1944. Initially it was to be located at Sand Island, outside of Honolulu. However, a later site board of April 12, 1944 positioned it to a more favorable location at the Punchbowl on the western side of Honolulu. The plan adapted to the site featured gun platforms protected by overhead casemates with ammunition tunnels driven through the crater lip to access tunnels inside the Punchbowl crater. The casemates were on separate ridges on the crater exterior, making them unusually spaced at 320-feet between. Despite the lateness of the war and improving strategic situation, this work (and Battery 407 at Fort Ruger) made surprising progress. Much of the physical work of driving tunnels and building the casemates was done in 1944. Work, however, was suspended after the tunnels were driven and lined in mid-1944. The intended armament was on hand in 1944 and consisted of 6-inch guns M1(T2) and M4 barbettes (allocated #12/#27 and #13/#28). It is unlikely this was actually mounted in the battery. Work was abandoned after the war. The emplacement (tunnels) still exists and used as storage at the National Cemetery of the Pacific. The battery is not open to the public and special permission is required to visit.

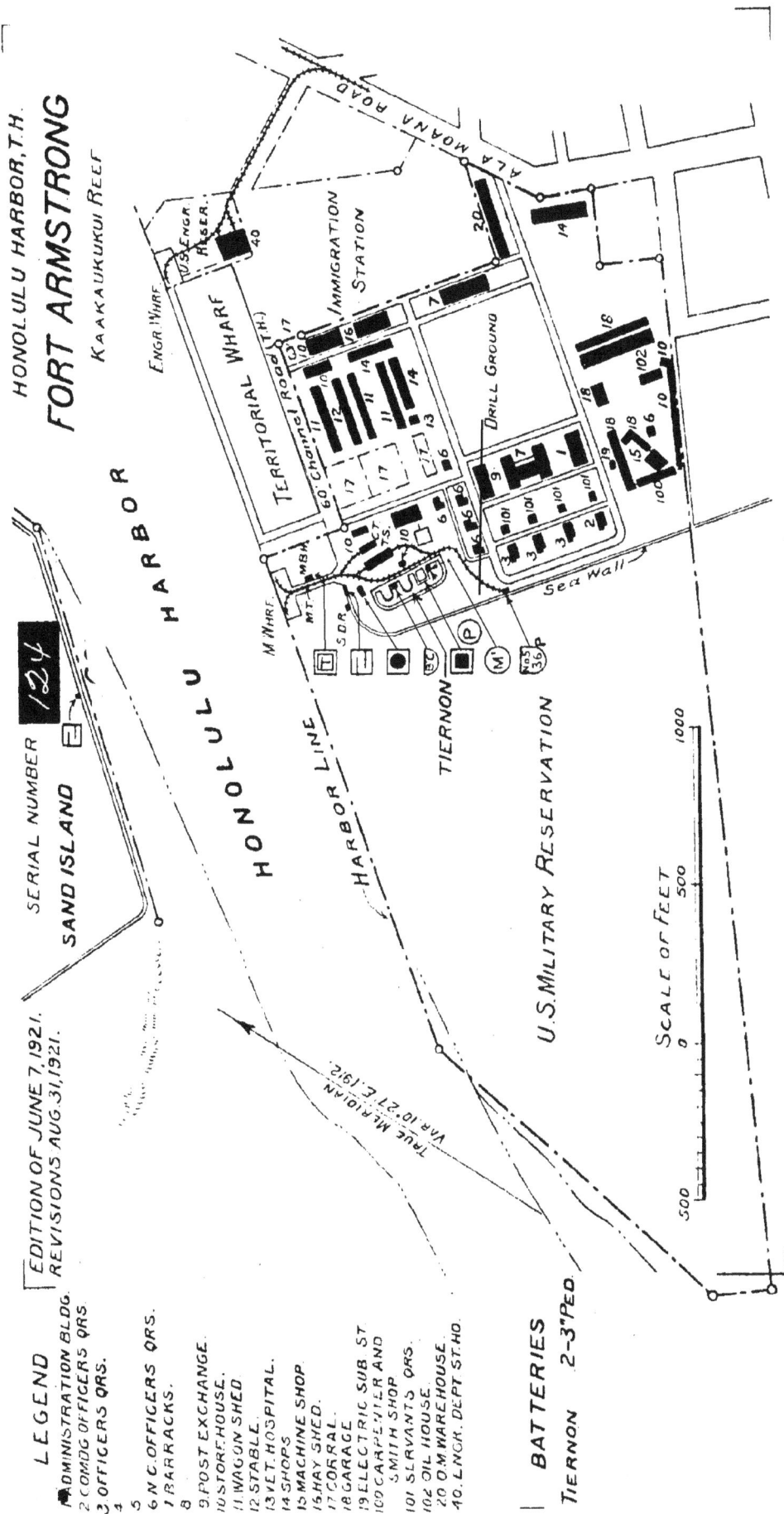

HONOLULU HARBOR,T.H.

FORT ARMSTRONG

KAAKAUKUKUI REEF

HONOLULU HARBOR

SERIAL NUMBER 124

SAND ISLAND

EDITION OF JUNE 7, 1921.
REVISIONS AUG. 31, 1921.

HARBOR LINE

TRUE MERIDIAN

VAR. 10°27' E. 1912.

U.S. MILITARY RESERVATION

TIERNON

DRILL GROUND

Sea Wall

TERRITORIAL WHARF

IMMIGRATION STATION

Territorial Road (T.H.)

60 Channel

ENGR WHARF

SCALE of FEET
500   0   500   1000

LEGEND

ADMINISTRATION BLDG.
2 COMDG OFFICERS QRS.
3 OFFICERS QRS.
4
5
6 N.C.OFFICERS QRS.
7 BARRACKS.
8
9 POST EXCHANGE.
10 STOREHOUSE.
11 WAGON SHED.
12 STABLE.
13 VET. HOSPITAL.
14 SHOPS
15 MACHINE SHOP
16 HAY SHED.
17 CORRAL
18 GARAGE
19 ELECTRIC SUB ST.
100 CARPENTER AND SMITH SHOP
101 SERVANTS QRS.
102 OIL HOUSE.
20 Q.M. WAREHOUSE.
40. ENGR. DEPT ST. HO.

BATTERIES

TIERNON 2-3" PED.

ISLAND OF OAHU, T.H.

PUNCHBOWL HILL

A - Platform for
Anti-aircraft gun - 2-3"

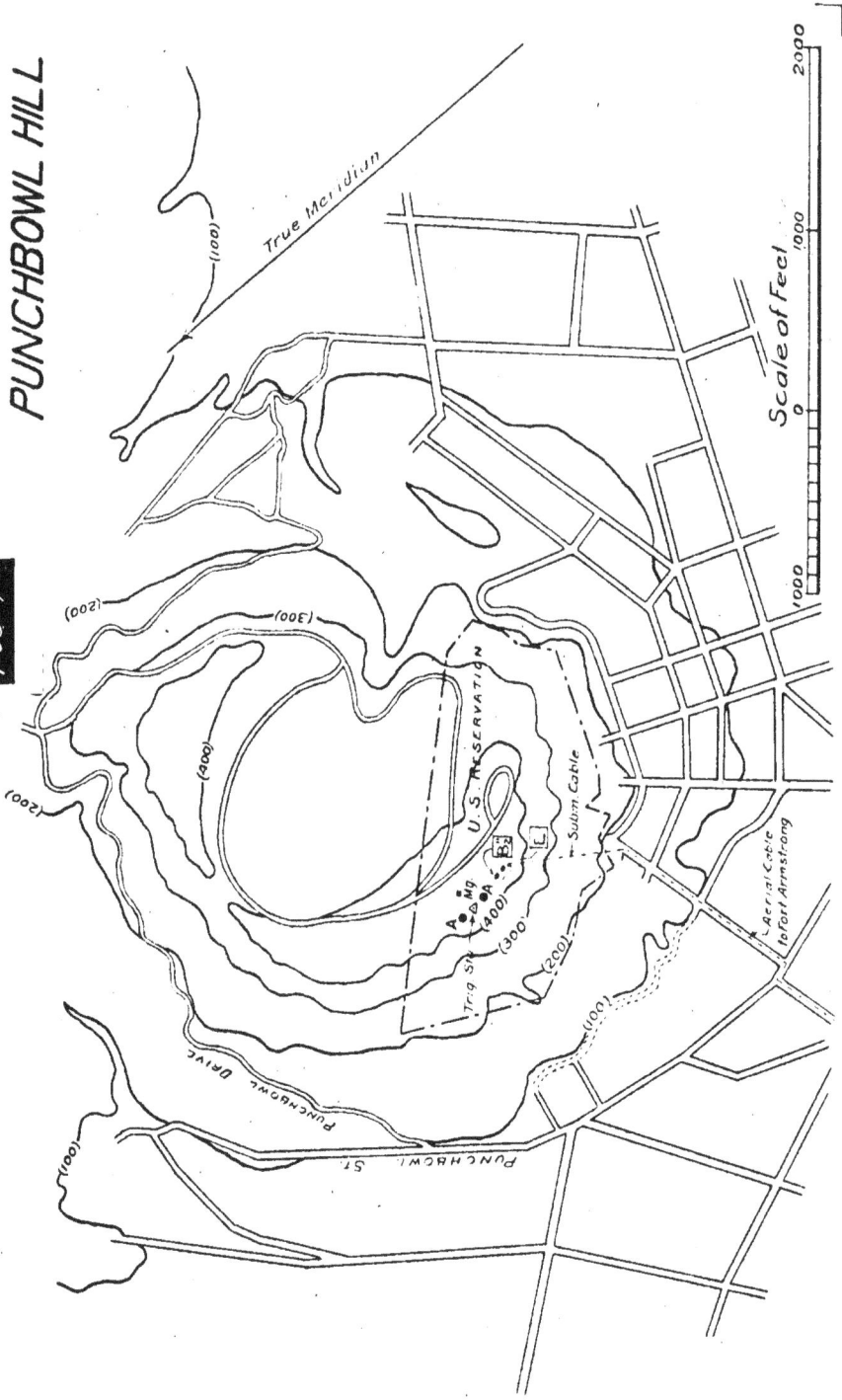

SERIAL NUMBER 124

EDITION OF AUG. 31, 1921.

Scale of Feet

True Meridian

U.S. RESERVATION

PUNCHBOWL DRIVE

PUNCHBOWL ST.

Subm Cable

Aerial Cable
to Fort Armstrong

Trig Sta

Fort DeRussy (1904-1949/present) is an active United States military reservation in the Waikiki area of Honolulu, Hawaii. Originally known as Kalia Military Reservation until 1909. Named Fort De Russy in General Orders 15, 1909, after Bvt. Brigadier General René Edward De Russy (Cullum 89), Superintendent of the United States Military Academy who served in the U.S. Civil War. The U.S. Army started construction of Battery Randolph (1913-1944) in 1911 as part of the Taft Program. The fort also contained Battery Dudley (1913-1946), which was destroyed in 1969, and AMTB Battery #5 (1943-1946), which has been also destroyed. AA gun emplacements were built in 1917. Became the U.S. Armed Forces Recreation Center in 1949. The Hale Koa Hotel (Army MWR resort property) was built in 1975. The U.S. Army Museum of Hawaii, established in 1976, is located inside Battery Randolph. Two 7-inch guns now on display in the battery's emplacements were originally from Sand Island. Also located on post since 1995 is the Asia-Pacific Center for Security Studies. Nearby at Ala Moana Beach Park was AMTB Battery Ala Moana (1943-1945) destroyed. The fort is open to public access, while the museum is open Tuesday – Saturday from 10am to 5pm.

Fort DeRussy Gun Batteries

- **RANDOLPH:** This battery for two 14-inch guns was the heaviest battery of the first generation of Oahu defenses. The construction plan was submitted on February 18, 1908. As one of the first new, 14-inch emplacements, it was of the 1908 type. This resembled closely just an enlarged 12-inch battery, with lower story magazines and ammunition hoists supplying the guns on upper-level platforms separated by 274.5-feet. Physical concrete construction was done between December 1908 and 1913. It was ready to transfer to service troops on October 13, 1913 for a cost of $428,893.89. The battery was named in General Orders No. 15 of January 13, 1909 for Major Benjamin H. Randolph, an artillery officer who died in 1907. The battery was armed with two 14-inch guns on Model 1907 disappearing carriages, initially of slightly different types. The first gun to arrive (September 1913) was the Model 1907 type gun (#1), emplaced on carriage Model 1907M1 #5. The second gun (not arriving and being mounted until 1915) was Model 1907M1 #3 emplaced on carriage Model 1907 #2. Shortly after installation (1917-19) the armament was altered to allow an increase in elevation from 15 to 20-degrees. Then in 1923 the type Model 1907 gun was dismounted and returned to the mainland and replaced with Model 1907M1 No. 1. The battery served between the wars, getting gas proofing installations in 1929, and splinter-proofing walls in 1937. These later additions added another $19,000 to the construction costs. It was still active at the start of World War II and received additional splinter and blast walls in early 1942. As newer batteries were commissioned, Randolph ceased being manned but was kept armed in reserve. The armament was scrapped on site in early 1946. Attempts were made to destroy the emplacement in the 1960s, but this was suspended after considerable damage had been done to some of the exterior elements. Modified for museum purposes, the battery survives as the central structure of the U.S. Army Museum of Hawaii. The battery is partly open to the public when visiting the museum when open.

- **DUDLEY:** A battery for two 6-inch guns built immediately adjacent to Randolph on its right (western) flank. It was of conventional Taft type designs, with two-gun platforms separated by 125-feet and a central traverse with magazines (at the same level as the loading platform) and support rooms. The plan was approved on March 27, 1909. Construction was done between July 1909 and the summer of 1913. Transfer was made on August 5, 1913 for a cost of $75,000. It was named in General Orders No. 59 of May 6, 1911 for Brigadier General Edgar S. Dudley, U.S. Army who died in 1911. It was armed with two 6-inch Model 1908 guns on Model 1905M1 disappearing

HONOLULU HARBOR, T.H.

FORT DE RUSSY

WAIKIKI BEACH

SERIAL NUMBER **124**

EDITION OF JUNE 7, 1921.
REVISIONS: JAN. 19, 1922.

LEGEND.
1. ADMINISTRATION BLDG.
2. COMDG. OFF. QUARTERS.
3. OFFICERS QUARTERS.
4. DISPENSARY.
5.
6. N.C.O. QUARTERS.
7. BARRACKS.
8. GUARD HOUSE.
9.
10. WIRELESS MAST.
11. CARP. & PLUMB. SHOP.
12. SIEGE GUNS SHELTER.
13. STABLE.
14. FIRE STATION.
15. LAUNDRY.
16. POST TEL. MANHOLE.
17. LAUNDRYMENS QRS.
18. ENLISTED MENS QRS.
19. GARAGE.
100. TOOL HOUSE.
101. MACHINE SHOP.
102. FUEL SHED.
103. TEAMSTERS HO.
104. FIRE APPARATUS.
     SHELTER.
105. BATH HOUSE.
106. SWIMMING
     PLATFORM.
107. CORRAL.

BATTERIES.
RANDOLPH   2-14" Dis.
DUDLEY   2-6" "
A-Antiaircraft gun-4-3".

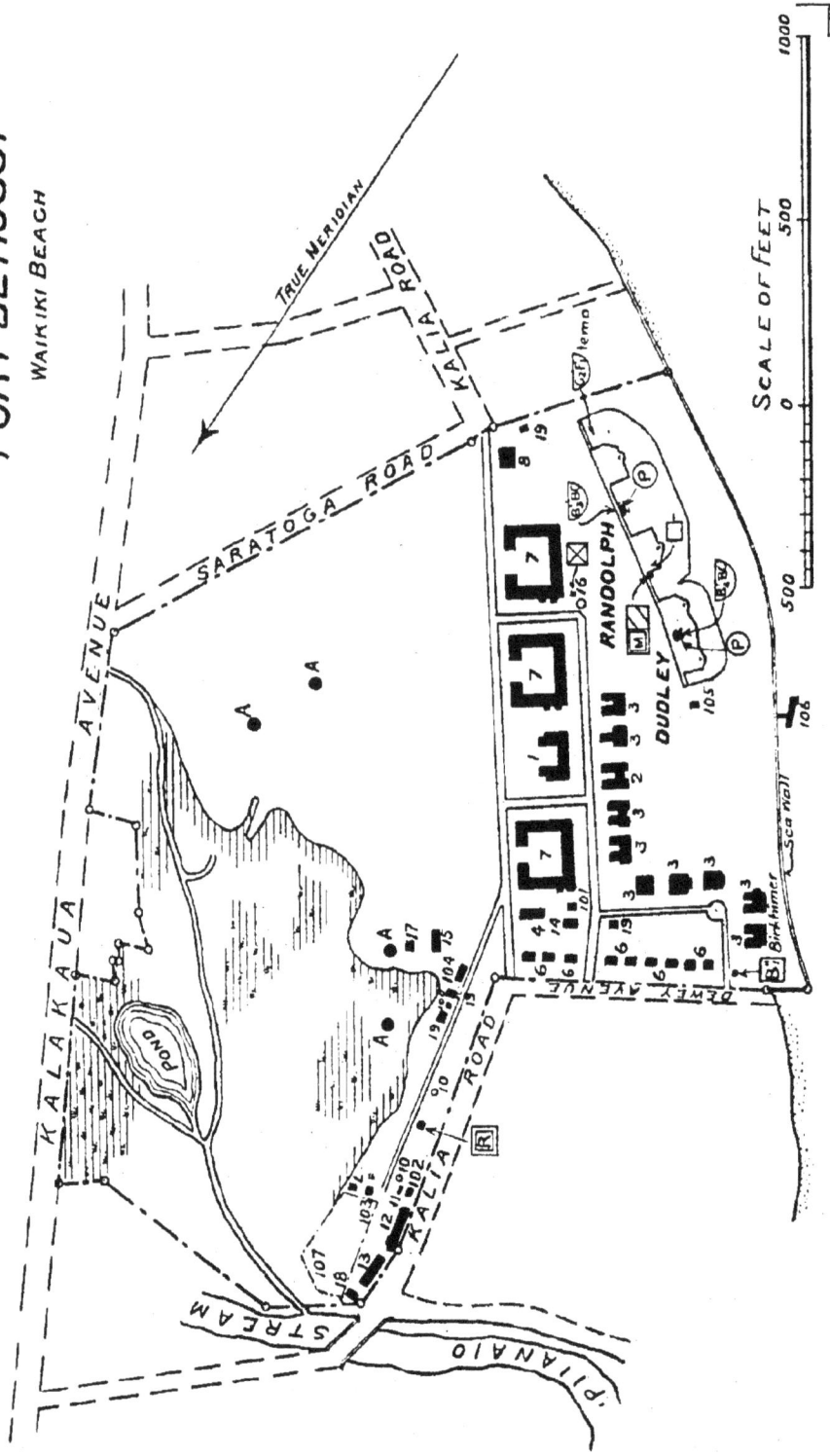

Fort DeRussy at Waikiki (Terry McGovern)

Diamond Head crater (Terry McGovern)

carriages (#1/#18 and #2/#19). It continued to serve with this armament until ordered disarmed. In the late 1930s the battery was given gas-proofing installations. The battery served up into the middle of World War II. On December 21, 1943 it was listed as armed but no longer manned. It was disarmed later in the war, probably in 1945. The battery structure was completely destroyed in later years.

• **AMTB-5**: An anti-motor torpedo boat battery of two fixed 90mm guns with M3 barbette carriage erected at Fort DeRussy. It was built beginning in late 1942 and completing in 1943. It consisted of simple concrete gun blocks, located near the beach southwest of Battery Dudley. It was disarmed in mid or late 1945. The battery no longer exist.

Fort Ruger (1906-1959) is located both inside and outside of Diamond Head (the remains of a tuff type volcano) which sits prominently near the eastern edge of Waikiki's coastline. Originally known as Diamond Head Military Reservation until 1909. Named Fort Ruger in General Orders 15, 1909, after Major General Thomas H. Ruger (Cullum 1633), a Superintendent of the United States Military Academy who served in the U.S. Civil War. Under the Taft Program, Battery Harlow (1910-1943) with eight 12-inch mortars was constructed outside the crater on its northern slope at the same time as the batteries at Fort DeRussey. The Kapahulu Tunnel (also referred to as the Mule Tunnel) was constructed in 1909. It runs 580 feet through the north wall of the crater, sloping fairly steeply uphill into the crater. Originally it was 5 feet wide and 7 feet high which is what was required for the passage of mules. Mules were used primarily to pull narrow gauge rail cars loaded with material in and out of the crater and to the various construction sites. In 1922 the tunnel was enlarged to 15 feet wide by 14 feet high. A Fire Control Switchboard that had been in a shed outside the tunnel was moved into a room carved into the wall about 100 feet from the outside end. In 1932 the tunnel was enlarged again, to 17 feet high to allow truck traffic. During the widening a larger cavern was cut into the wall of the tunnel at the downhill end, creating rooms for a Harbor Defense Command Post. The Kahala Tunnel was built through the east side of the crater wall during the 1940s and it is now the primary access to the crater. Several years later a Land Defense project was undertaken to protect both Pearl Harbor and Honolulu Harbor from infantry attack. This project resulted in the construction of Battery Birkhimer (1916-1943), armed with four 12-inch mortars, built within the crater to be able fire at both land and sea targets. Also, as part of this land defense plan was Battery Dodge (1915-1925) and Battery Hulings (1915-1925) both with two 4.7-inch guns, Battery Mills (1916-1925) with two 3-inch guns on Black Point, and an unnamed battery (1915-1919) with twelve 6-pounder field guns. The primary garrison area was outside the northern side of the crater (most buildings are now destroyed), including a balloon hanger that was located on the east-southeast corner of Fort Ruger. The fort received additional armament leading up to World War II. They included Battery Granger Adams (1935-1946) with two 8-inch guns built at Black (Kupikipikio) Point, and Battery 407 (1944) with two 8-inch guns located on the outside of the crater with tunnels bore from the inside of the crater. Also, added was Battery Ruger (1937-1943) with four 155mm GPF on Panama mounts outside of the crater. Four 3-inch AA gun batteries were built in the 1920s, including one battery at Black Point. Four searchlights were emplaced in various positions. The Diamond Head Hiking Trail to Point Le'ahi leads to fire-control station "E" (three pre-World War II and one World War II structures, including the Honolulu HD Fire Command Post). Station 1 at the top level was First Battle Command. Station 2 was below that and was the First Fire Command and served Battery Harlow. Station 3 was the Second Fire Command and served Battery Dudley at Fort DeRussy, Station 4 served Battery Randolph at Fort DeRussy. Seven storage tunnels were dug into the outside slope of Diamond Head along the northeast side. These tunnels have been used by a number of agencies over the years. The last military parcel of the reservation became a Hawaiian National Guard post in 1959. The fort's garrison area outside the crater is now used for a variety of public and private uses. Diamond Head State Monument encompasses over 475 acres, including the interior and outer slopes of the crater. This park is open to the public from 6am to 6pm, but an entrance and parking fee is required.

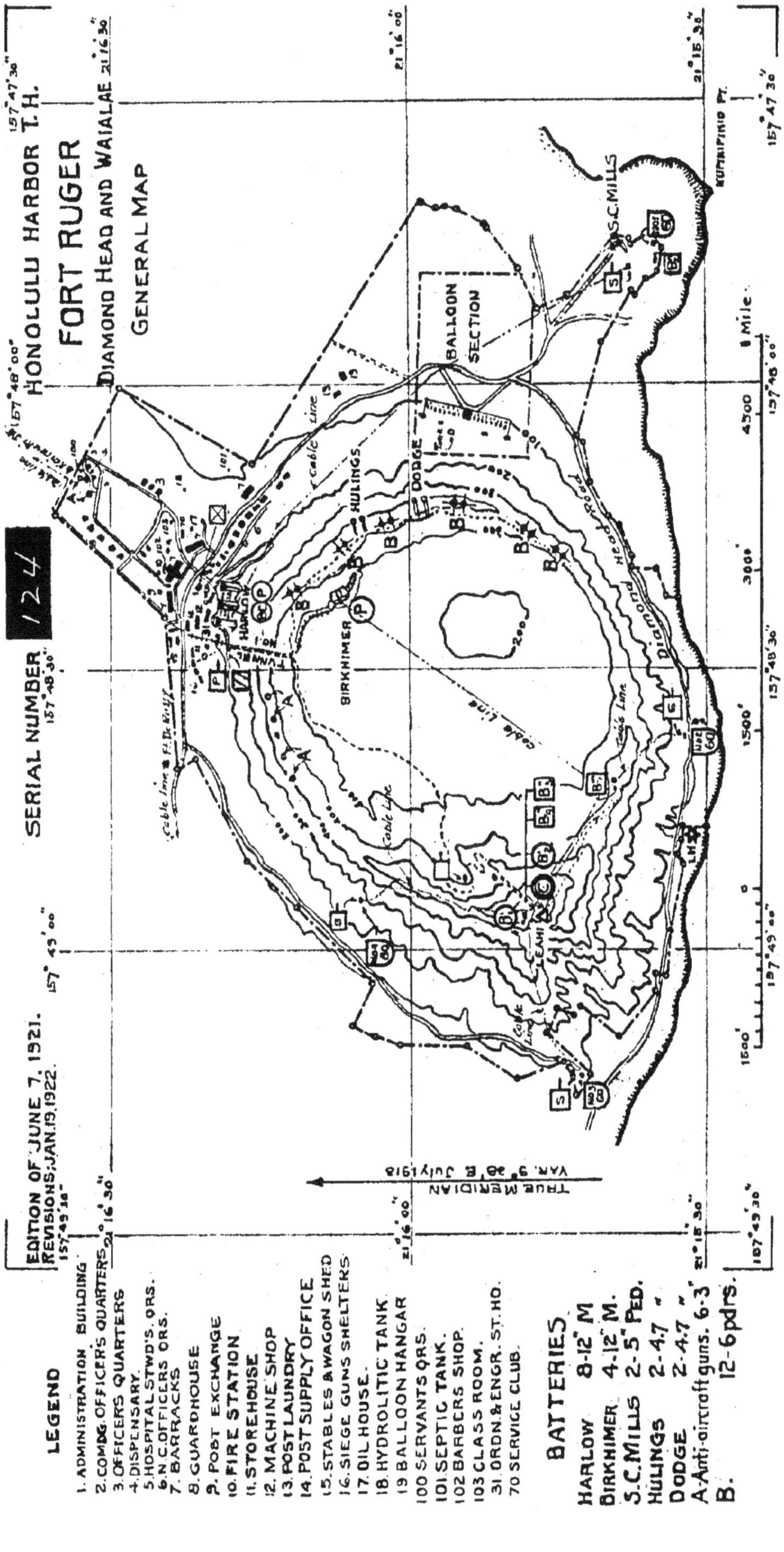

SERIAL NUMBER 124
HONOLULU HARBOR T.H.
FORT RUGER
DIAMOND HEAD AND WAIALAE
GENERAL MAP
EDITION OF JUNE 7, 1921.
REVISIONS: JAN. 19, 1922.
LEGEND
1. ADMINISTRATION BUILDING
2. COMDG. OFFICER'S QUARTERS
3. OFFICERS QUARTERS
4. DISPENSARY
5. HOSPITAL STWD'S. QRS.
6. N.C. OFFICERS QRS.
7. BARRACKS
8. GUARDHOUSE
9. POST EXCHANGE
10. FIRE STATION
11. STOREHOUSE
12. MACHINE SHOP
13. POST LAUNDRY
14. POST SUPPLY OFFICE
15. STABLES & WAGON SHED
16. SIEGE GUNS SHELTERS
17. OIL HOUSE
18. HYDROLITIC TANK
19. BALLOON HANGAR
100 SERVANTS QRS.
101 SEPTIC TANK
102 BARBERS SHOP
103 CLASS ROOM
31. ORDN. & ENGR. ST. HO.
70 SERVICE CLUB
BATTERIES
HARLOW 8-12" M
BIRKHIMER 4-12" M.
S.C. MILLS 2-5" PED.
HULINGS 2-4.7 "
DODGE 2-4.7 "
A. Anti-aircraft guns. 6-3"
B. 12-6pdrs.
TRUE MERIDIAN
VAR. 9° 30' E. July 1918
BALLOON SECTION
BIRKHIMER
HULINGS
DODGE
S.C. MILLS
LEAHI
TUNNEL
Diamond Head Road
Cable Line
KUPIKIPIKIO PT.
1 Mile
1500
0
1500
3000
4500

HONOLULU HARBOR, T.H.

FORT RUGER D-1

SCALE OF FEET

SERIAL NUMBER 124

EDITION OF JUNE 7, 1921.
REVISIONS: JAN. 19, 1922.

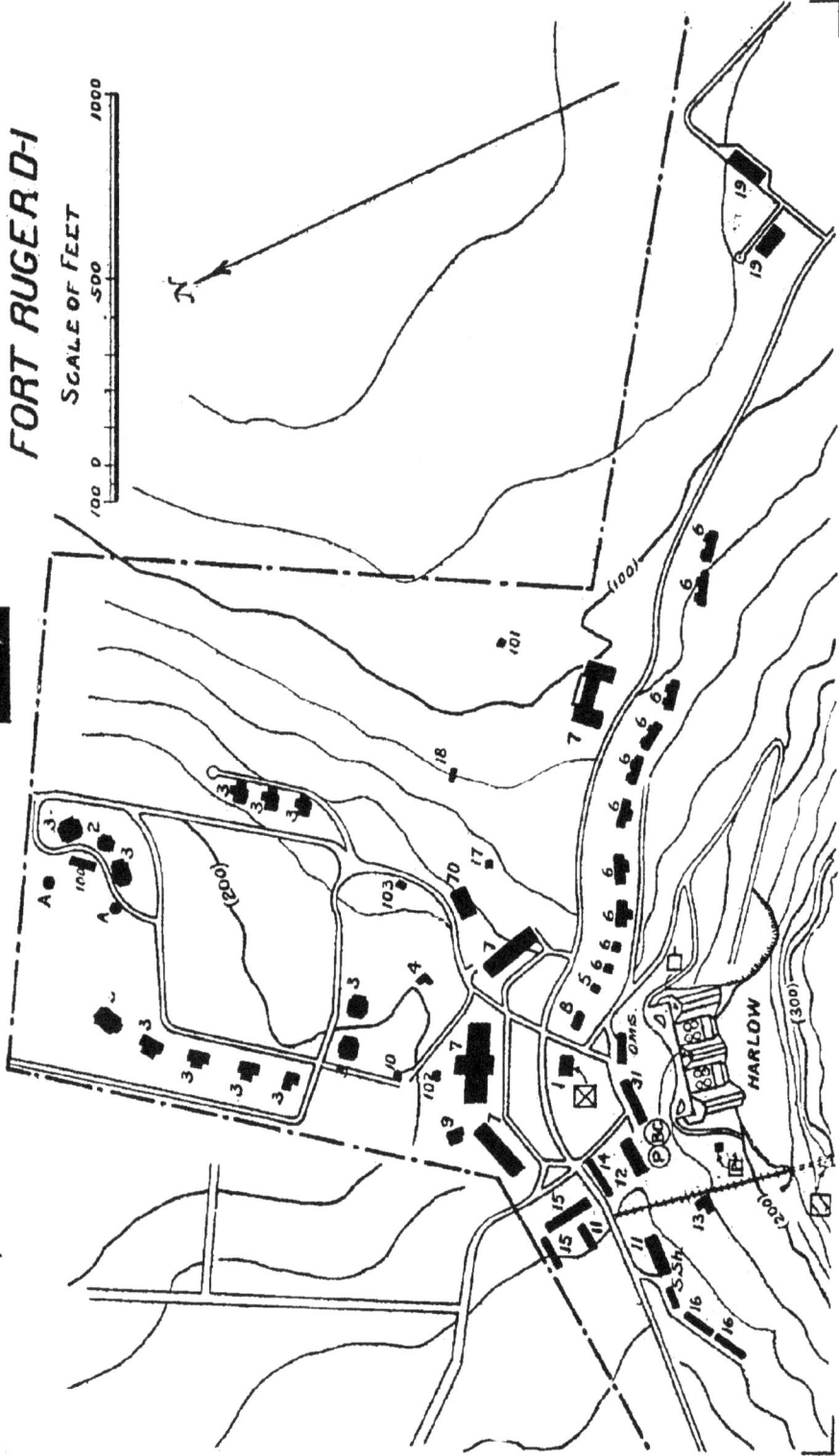

LEGEND

1. ADMINISTRATION BLDG.
2. COMDG. OFFICERS QRS.
3. OFFICERS QRS.
4. DISPENSARY.
5. HOSPITAL STWD'S QRS.
6. N.C. OFFICERS QRS.
7. BARRACKS.
8. GUARD HOUSE.
9. POST EXCHANGE.
10. FIRE STATION.
11. STORE HOUSE.
12. MACHINE SHOP.
13. POST LAUNDRY.
14. POST SUPPLY OFFICE.
15. STABLES & WAGON SHED.
16. SIEGE GUN SHELTER.
17. OIL HOUSE.
18. HYDROLITIC TANK.
19. BALLOON HANGAR
100. SERVANTS QRS.
101. SEPTIC TANK.
102. BARBER SHOP.
103. CLASS ROOM.
31. ORDN & ENGR. ST. HO.
70. SERVICE CLUB.

BATTERIES

HARLOW .... 8-12"M.
A–Anti-aircraft gun   2-3"

Fort Ruger Gun Batteries

- **HARLOW:** A battery for eight 12-inch mortars emplaced on the northern, outside edge of Diamond Head crater on the new Ruger reservation. The site and battery had been recommended for mortars frequently since 1901. The battery plan was submitted for approval on October 5, 1907. Work was done from 1908 to late 1909 for transfer on March 16, 1910 for a cost of $207,760. It was named on General Orders No. 15 of January 28, 1909 for Major Frank S. Harlow, Artillery Corps, who died in 1906. The emplacement was of conventional early Taft design with two pits for four mortars each and magazines in traverses adjacent and in between the pits. On the left flank was a lower-level power room. It was built with a complete mechanical data transmission system with displays on the traverse viewable by pit crews. One pit was armed with four 12-inch Model 1890M1 mortars on Model 1896M1 carriages (Pit B: #124/#261, #175/#243, #152/#262, and #181/#289 and Pit A had four Model 1890M1 mortars on Model 1896MII carriages (#180/#307, #178/#308, #179/#310, and #177/#309). Some of the mortars and carriages were relocated from other Endicott mortar batteries on the mainland. Upon proof firing several of the mortar racer rings were broken or damaged, necessitating extensive repairs. The battery served many years in essentially a primary seacoast defense role, but with some inevitable minor modifications to emplacement and armament. It was not designated for removal until authority given on March 11, 1946. After the war it served as both civil defense at National Guard storage, and it still intermittently used (even as a filming location). The emplacement still exists on national guard property. The battery is not open to the public.

- **BIRKHIMER:** A mortar battery conceived as an integral part of the land defense program for Oahu. It was located 500-feet south of Harlow, but inside the Diamond Head crater. It was named in General Orders No. 15 of April 25, 1916 for Brigadier General William E. Birkhimer, U.S. Army who died in 1914. Access to the battery was greatly improved by the building of a special "mule" tunnel excavated through the crater wall. Plans dated from the 1912 Macomb Land Defense Board. As submitted on December 22, 1914 they called for four mortars emplaced in two, two-gun pits with a traverse magazine between the pits and high exterior surrounding walls. It was designed for all-round fire, most importantly to fire on approaching enemy forces from the north and east, and also to have a secondary seacoast, anti-ship role. Work was done from 1915-16. It was armed with mortars and carriages re-located from the mortar battery at Fort DuPont, defenses of the Delaware. It received four 12-inch mortars Model 1890M1 on Model 1896M1 carriages (#119/#118, #98/#107, #62/#106, and #92/#103). Firing tests were conducted on June 29, 1916 and the results were not favorable. As feared the shock of firing in the enclosed, tight pits was highly destructive. Damage was done to windows, doors, and even structural parts. Dummies and tests to tethered animals also demonstrated hazard. The engineers supervising the tests could not understand the design concept was developed—especially the need for the protective enclosing walls when the entire battery was protected from direct fire by being in the deep volcanic crater in the first place. It was recommended that the battery be rebuilt. As a project for a reserve magazine in Fort Ruger was already pending, it was decided to move the mortars and carriages to an in-line series of open blocks just outside the magazine (which would still serve them via shell carts). The old battery pits would be roofed over and converted to provide the rquired magazine space. This was accomplished from February 1920 to June 1921. The new transfer date was June 17, 1921 for a cost of $189,534. IThe armement was not designated for removal until authority issued on March 11, 1946. Subsequently the magazines served numerous uses, most recently as a Hawaiian Civil Defense center, while the new mortar blocks are now a parking lot. As modified, it still exists. The battery is not open to the public.

- **DODGE**: A battery for two 4.7-inch Armstrong guns emplaced on the eastern side of the Diamond Head crater. It also dated from 1912 Macomb Land Defense Project. It and sister Battery Hulings were to be armed with 4.7-inch guns (while older, still capable of delivering a useful Shrapnel projectile) high on the eastern slope of Diamond Head crater covering land approaches from the east. It was built to the south of Battery Hulings at about a 300-feet elevation. It featured two casemates connected by a corridor with service magazines. Work was done from September 1914 to October 1915 for transfer on October 20, 1915 for a total cost of some $15,702.24. It was armed with two 4.72-inch Armstrong guns and pedestals (#11933/#11021 and #11009/#11019) originally mounted at Battery Ridgely, Fort St. Philip, New Orleans. It was named in General Orders No. 36 of June 9, 1915 for Major Theodore A. Dodge, U.S. Army. It did not serve long, the armament was still present in 1919, but gone by 1928. The emplacement was re-armed during World War II. In April 1942 two ex-navy MkIX 4-inch guns were installed at Battery Dodge. This required some cutting of the concrete overhead casemate to allow a 19-degree gun elevation, and of course a new block with foundation bolts. It is not clear when these in turn were removed, though most of this generation of emergency gun defense was taken out in 1944-45. The emplacement still exists on Hawaiian National Guard property. The battery is not open to the public.

- **HULINGS**: A battery for two 4.7-inch Armstrong guns emplaced on the eastern side of the Diamond Head crater. Hulings, Dodge, Birkhimer and the 6-pounder emplacements were all from the Oahu 1912 land Defense Project and designed to cover the eastern side and approaches to the Pearl Harbor-Honolulu-Diamond Head defensive "box." It was built between September 1914 to October 1915. Transfer was made on October 20, 1915 for a cost of $15,720.24. It was named in General Orders No. 36 of June 9, 1915 for Colonel Thomas M. Hulings of the 49th Pennsylvania Infantry serving during the Civil War. It was armed with two 4.72-inch Armstrong guns and pedestals, previously mounted at Battery Irons, Fort Armistead, Baltimore (#11005/#11016 and #11001/11015). The battery had two separate (by 76-feet) casemated embrasures connected with a tunnel and service magazines. Shortly after completion the armament was removed and was certainly gone by 1928. The structure was not thereafter used for armament but was a convenient storage space. Today it still exists and is controlled by the Hawaiian National Guard. The battery is not open to the public.

- A number of older Coast Artillery 6-pounder (2.24-inch, or 57mm) guns on wheeled field mounts were allocated for use with the 1912 Oahu Land Defense Project. Consequently, a series of small emplacement platforms were built along the 4.7-inch gun line defined by batteries Hulings and Dodge. A total of twelve prepared positions were built by 1921, likely constructed earlier in 1915-16. These were small, concrete parapets with wheeled carriage hold-down eyebolts and a few, spaced magazines. It is likely the guns were not mounted but kept stored nearby until an emergency warranted their placement. A transfer date and cost has not been located, nor was the battery named. They probably only served until the early 1920s when the 6-pdr gun was declared obsolete. The original concrete emplacements still exist, somewhat overgrown. This battery area is not open to the public.

- **Battery #407**: A World War II dual 8-inch battery intended for this location on the south face of Diamond Head crater, near but not actually a part of the Fort Ruger military reservation. It was one of the emplacements recommended early in the war to augment the coverage of remote sites of the island. It was designed for two covered casemates, as much for falling rocks from possible hits above the emplacement as anything else. Behind each casemate was a deep lined tunnel on separate, but adjacent ridges. Work proceeded rapidly during 1942-43 but was never brought to completion.

While the tunnels were driven and the gun houses built, the armament was never installed. It had been allocated two 8-inch ex-navy MkVIM3A2 guns without shields on M1 8-inch barbettes. One report states that guns #128 and #210 were sent to the fort in May 1944, but it is doubtful if they were ever installed. On May 31, 1945 it was listed as 75% completed, with work suspended. Now located on Hawaiian National Guard property, the battery is used a Joint Commend Center and storage, and as such still exists. The battery is not open to the public.

Battery Harlow Fort Ruger (Terry McGovern)

Battery Birkhiemer in Diamond Head crater (Terry McGovern)

Battery #407 Diamond Head crater (Terry McGovern)

Battery #407 Diamond Head crater (Terry McGovern)

Black Point Military Reservation (1916-1955) is located outside of Diamond Head to the east at Black Point which was originally called Point San Jose but was renamed by early settlers due to the dark oak woodland and bay. The U.S. Army built Battery Mills (2-3-inch) on Black Point in 1916 a part of the Oahu Land Defense Project. It was deactivated in 1925 and replaced by Battery Granger Adams (2-8-inch) in 1933-35. This battery remained active until 1946. In the 1920s, homes began to be built around the military reservation on Black Point, and it became a prestigious community of wealthy and influential people. The U.S. government returned the land to the State of Hawai'i in 1955. The land was later sold for private development, and reservation today is private homes.

Black Point M.R. Gun Batteries

- **S.C. MILLS:** A battery for two 5-inch pedestal guns erected south of the Fort Ruger reservation proper at Kupikipikio Point (also known as Black Point). The emplacement was recommended and part of the 1912 Land Defense Board project for the island. Battery S.C. Mills was built along with a new 60-inch searchlight position as the first use of this new reservation. Work was done from August 1914 to mid-1916, for transfer on December 12, 1916 at a construction cost of $30,560.58. The design featured two casemated gun positions and a long flank wall/parados along the southeast front. The battery fired to the northeast. It was named in General Orders No. 36 of June 9, 1915 for Colonel Stephen C. Mills, U.S. Army, who had died in 1914. It was armed with two 5-inch Model 1900 guns on Model 1903 pedestals (#20/#20 and #18/#21) previously mounted at Battery Gregg in the Delaware defenses. These guns were removed between 1921 and 1925, but the battery did continue to serve as the power station for the remaining searchlight and then in turn for Battery Granger Adams during the late 1930s and 1940s. Sometime postwar the battery was destroyed and residences were built over the site, nothing remains of the emplacement.

- **GRANGER ADAMS:** A battery for two 8-inch guns on railway top carriages placed near the Battery S.C. Mills location at Black Point. It consisted of a simple concrete gun blocks to take the dismounted Model 1918 railway top carriage with an adjacent concrete magazine. A separate plotting room some distance to the rear was also included and the disarmed 5-inch battery provided power and latrines. Work was done from October 5, 1933 to May 11, 1935 for transfer on March 2, 1935 for a construction cost of $116,832.70. It was named in General Orders No. 10 of November 23, 1934 for Brigadier General Granger Adams, Army artilleryman. It was armed with two 8-inch Model 1888MII guns on Model 1918 railway carriage (Watervliet tube #37/#24 and Watervliet tube #45/#25). It was active during the war, receiving additional splinter proofing in May 1942. Deletion of the emplacement and removal of armament came early postwar, probably in 1946 (in March 1946 the battery was listed as abandoned, but with armament still mounted in place). The emplacement was destroyed later, the site being used for residential home construction. No remains persist today.

Wiliwilinui Ridge Military Reservation (1942-1951) is located on the eastern side of Oahu along the spine of Ko'olau Mountain range, near the Waialae Iki neighborhood. Battery Wilridge / Kirkpatrick (1943-1948, renamed 1946) had two twin 8-inch naval gun in turret mount from the USS *Lexington*. The turret wells, magazines, plotting room, and battery commander's station still exist, located on the grounds of a private estate and on Bishop Estate land. There were also the four 155mm GPF on Panama mounts of Battery Wili (1942-1945) which were located 3,000 feet south of Battery Wilridge. A base-end station (1942) was once located nearby at Wailupe. Searchlights were emplaced at Wailupe Point. Battery Kirkpatrick's site is located near to the Wilwilinui Ridge Trail, but the battery site is on private property. To reach the trailhead (located here), you'll need to pass by the security guard at the Waialae Iki 5 community. There's only a handful of parking spots for hikers near the trailhead, so the guard passes out a certain number of laminated parking passes each day.

Wiliwilinui Ridge M.R. Gun Battery

- **KIRKPATRICK:** One of the four Oahu sites for two dual 8-inch naval turrets built during World War II. The U.S. Navy offered the 8-inch turrets being removed from aircraft carriers USS *Saratoga* and *Lexington* to the U.S. Army in Hawaii for coast defense on January 17, 1942. They were readily accepted; four sites being identified each to receive two dual-gun turrets. At each site would be two emplacements for the turrets (each with a turret roller ring, shell lift, and underground magazines) and separate underground plot, command, radar, and power plant rooms. This site was high on Wiliwiliui Ridge (over 1200-foot elevation). This was one of several north-south running ridges to the northeast of Diamond Head Crater. It was a very difficult site to work on, requiring new roads to access the site. Work started in May 1942 and was completed by December 26. 1942. It was armed with two turrets from USS *Lexington* carrying navy 8-inch, 55-caliber MkIXM2 guns (#519, 520, 521, and #522). It was named in General Orders No. 95 of August 27, 1946 for Lt. Colonel Lewis S. Kirkpatrick, Coast Artillery Corps. The battery served until put into caretaker status in 1946 and then deleted and armament scrapped in 1949. Part of underground structures were built over by a private residence in the early 2000s, but elements remain on private property. The battery is not open to the public.

Koko Head Military Reservation (1941-1945) is located on the headland that defines the eastern side of Maunalua Bay along the southeastern side of Oahu. On its western slope is the community of Portlock, a part of Hawai'i Kai, while Koko Head (at 642 ft) is the remains of an ancient tuff cone volcano. Located on Kuamookane Hill was Battery Koko Head (1941-1942) with four 155mm GPF guns on Panama mounts. It was under the command of Fort Ruger and was later replaced by Battery Wili. At Koko Saddle was to be location of Battery 305 with two 6-inch shielded barbette guns, but the battery was never built. Fire-control station "H'" (three pre-World War II structures, one World War II structure) was located on Koko Head, as well as an SCR-270 mobile radar (1941). An SCR-271A fixed radar (1942) was located on Koko Crater, with the operations and power rooms tunneled into the rim. Searchlights were emplaced at Kawaihoa Point. Visiting Koko Head requires climbing 1,048 steps to the top of the Koko Head Crater. The military used them previously as part of an incline tramway to transport supplies to the stations at the summit.

Koko Head M.R. Gun Battery

- **Battery #305:** This was another one of the supplemental, permanent system emplacements authorized for Oahu relatively late, on February 2, 1944. Originally it was proposed to place it at (either in front of, or in one plan actually physically replacing) Battery Dudley at Fort DeRussy. However, a later site board of April 12, 1944 changed that to a more favorable location known as Koko Saddle. This was a site between Koko Head and Koko Crater on the southeastern part of Oahu. While of somewhat difficult access, it had a much better field of fire and better opportunity for site security. Of somewhat lower priority, no physical work was ever started on this position or with building this battery. Apparently at some point the fixed armament intended for the battery was designated. A 1944 ordnance list has two 6-inch gun M1(T2) and M4 barbette carriages (#15/#29 and #16/#30). All thought of completing the battery was dropped late in the war, and the site eventually became part of the Koko Head District Park. The site is open to the public.

THE HARBOR DEFENSES OF PEARL HARBOR – HAWAII
(INCLUDES OAHU'S WESTERN SHORE)

Fort Kamehameha (1907-1949/1992) is located on Queen Emma Point, between Hickam Air Force Base and the Honolulu International Airport at the eastside of the entrance to Pearl Harbor. Originally named Fort Upton but was soon changed due to local public requests. Renamed Fort Kamehameha in General Orders 245, 1909, after Kamehameha, first king of unified Hawaii. Several Taft Program batteries were constructed to defend the naval station at Pearl Harbor. Batteries are Battery Selfridge (1913-1945), Battery Hasbrouck (1914-1943), Battery Jackson (1913-1943) and Battery Hawkins (1914-1943). The fort also had a mine casemate to control the mines placed in the Pearl Harbor channel. Starting in 1916, the garrison area was constructed that would include nine company-sized barracks, 28 officer quarters, and 18 married NCO quarters, along with supporting structures. As part of the Oahu's Land Defense Project, two additional batteries were added at Bishop Point, Battery Barri (1915-1924) and Battery Chandler (1915-1942), both have now been destroyed. During the Interwar Period, railway artillery was also used beginning in 1922 (used as a "dummy" battery in World War II). Two 3-inch AA gun batteries were built here in the 1920s. In 1920, two 12-inch guns on a new long range barbette carriage installed near Ahua Point. Battery Closson (1924-1948) this battery was casemated in 1942 as part of World War II defenses, while AMTB Battery #2 (1943-1946) and Battery Kam (1934-1943) with four mobile 155mm GPF guns were also added. Searchlights were emplaced on Queen Emma Point. The former U.S. Army reservation was transferred to Hickam Air Force Base in 1992. The eastern portion of the post became Mamala Bay Golf Course. The Air Force in 2009 recommended demolition of many of the former Army buildings, but this action has been put on hold though these buildings remain vacant. The former fort is part of the Joint Base Pearl Harbor-Hickam so public access is restricted, and the batteries are currently abandoned.

Fort Kamehameha Gun Batteries

- **SELFRIDGE**: This was the first new work for the Queen Emma Point fortification site. The approved plan called for a heavy, dual 12-inch gun battery here, and the engineering plan was submitted on May 10, 1907. It was located on the southern edge of the fort, firing almost due south. Construction was quickly started in 1907 and accomplished by 1909. Transfer to service troops was made on August 4, 1913 for a construction cost of $440,000. The battery was of conventional two-story Taft design with guns placed at 240-foot spacing, though with slightly increased height to help with water drainage. It was named in General Orders No. 15 of January 13, 1909 for 1st Lieutenant Thomas E, Selfridge, an early aviator who died in a crash in 1908. It was armed with two 12-inch Model 1895M1 Bethlehem guns on Model 1901 disappearing carriages (#12/#18 and #13/#19). At some point the tubes were exchanged, probably to manage bore wear. New gun tubes were Bethlehem #8 and #11. The gun carriages were modified to give an additional 5-degrees of elevation in 1915-17. The battery served actively up through World War II. In 1938 structural improvements were made to provide gas-proofing, and in common with numerous other batteries it obtained a rear blast wall right at the start of the war. The battery served throughout the war, even though it was obsolete its armament was retained as a sort of emergency reserve. It was authorized for elimination on March 11, 1946. Somewhat modified with construction to support nearby airfield activity, the emplacement is relatively intact on the Joint Base Pearl Harbor-Hickam. The battery is controlled by the Hawaiian Air National Guard and not open to the public.

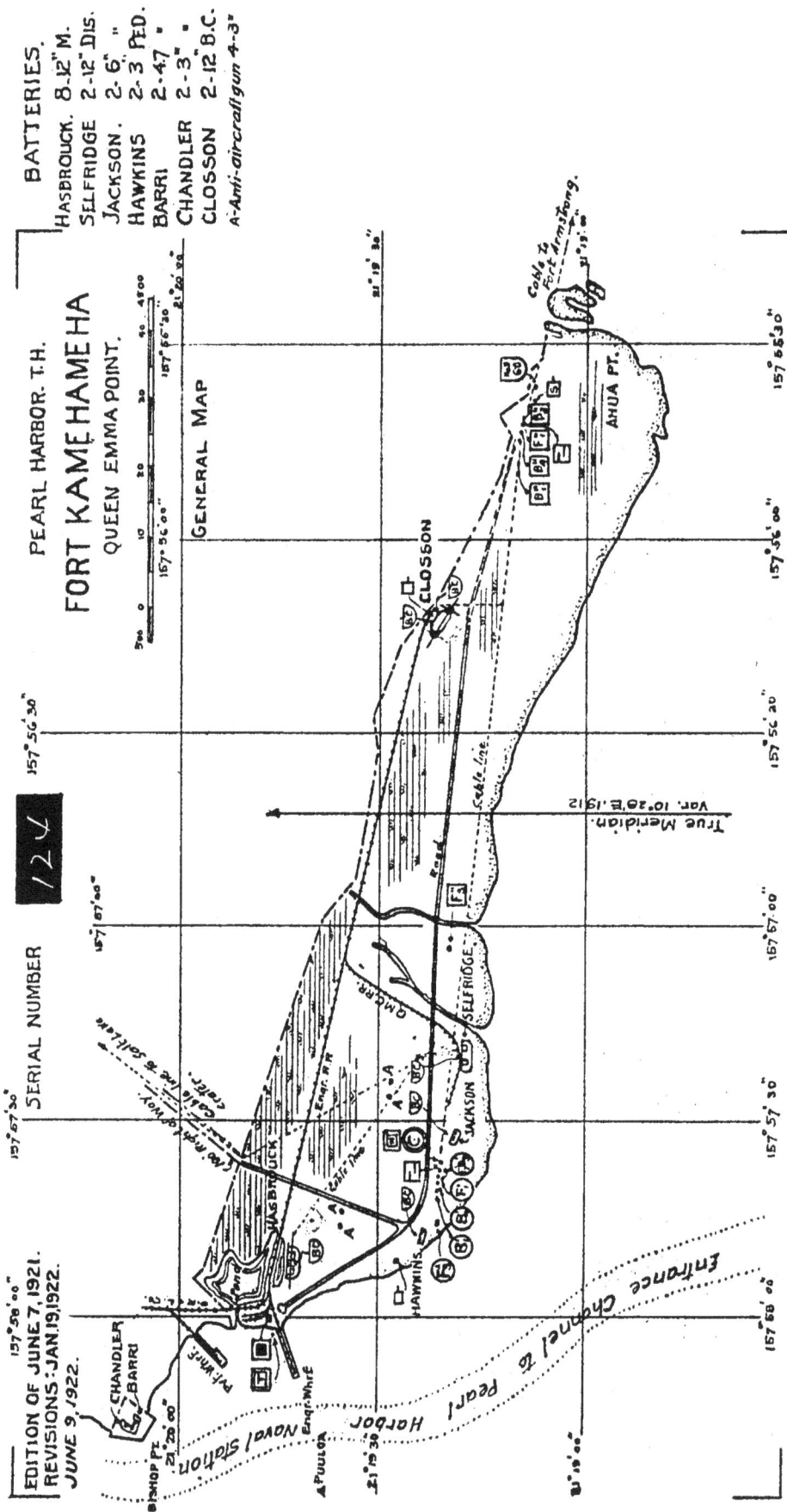

PEARL HARBOR. T.H.

FORT KAMEHAMEHA

QUEEN EMMA POINT.

GENERAL MAP

BATTERIES.

HASBROUCK. 8-12" M.
SELFRIDGE   2-12" Dis.
JACKSON.   2-6".
HAWKINS   2-3 Ped.
BARRI   2-47 "
CHANDLER   2-3 "
CLOSSON   2-12 B.C.
A-Anti-aircraft gun 4-3"

SERIAL NUMBER

EDITION OF JUNE 7, 1921.
REVISIONS: JAN. 19, 1922.
JUNE 9, 1922.

PEARL HARBOR, T.H.

FORT KAMEHAMEHA D-1.

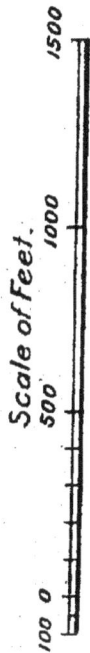

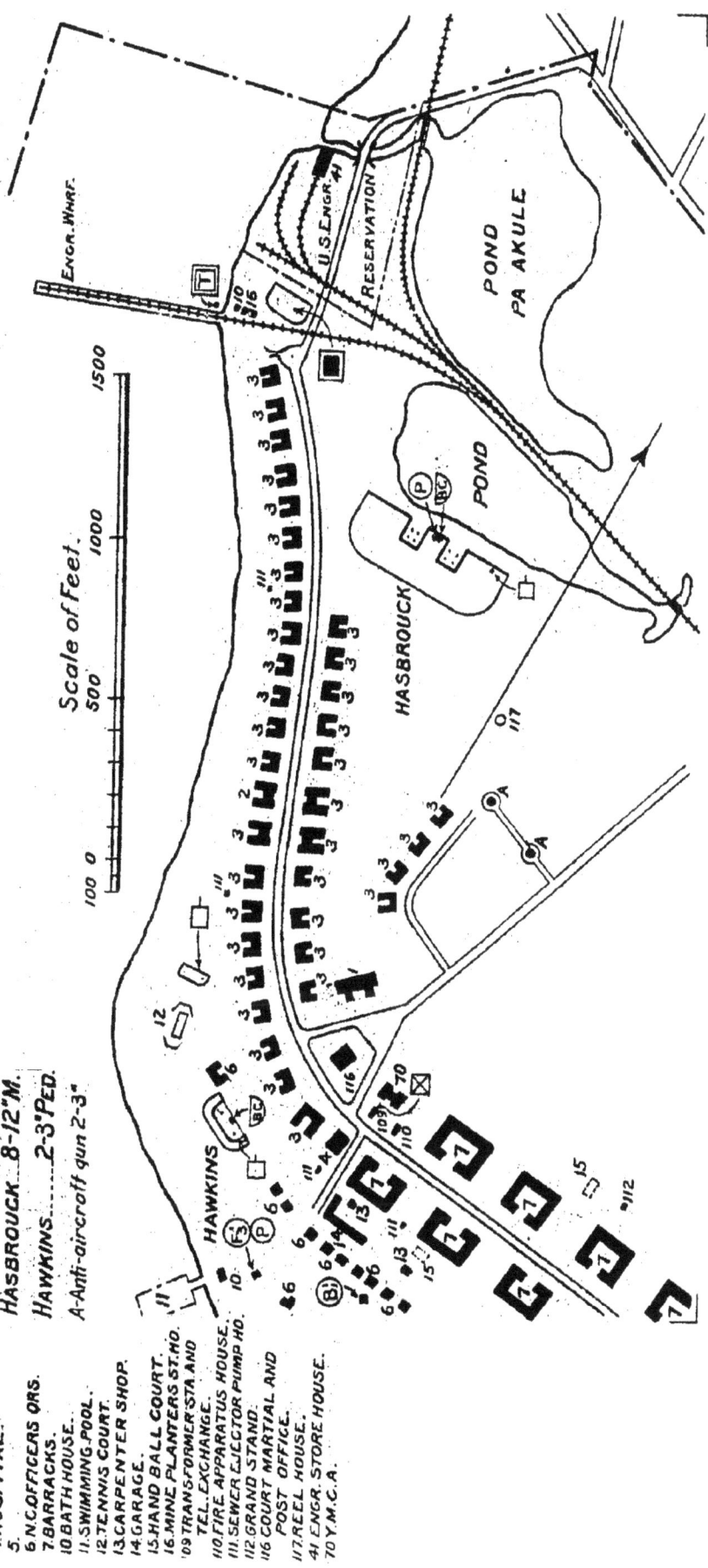

SERIAL NUMBER 124

EDITION OF JUNE 7, 1921.
REVISIONS: JAN. 19, 1922.

BATTERIES

HASBROUCK.... 8-12" M.
HAWKINS.......... 2-3" PED.
A-Anti-aircraft gun 2-3"

Scale of Feet.
1500
1000
500
100 0

POND PA AKULE

POND

HASBROUCK

HAWKINS

U.S. ENGR.

RESERVATION

Engr. Wharf.

LEGEND

1. ADMIN. BUILDING.
2. COMM. OFFICERS QRS.
3. OFFICERS QRS.
4. HOSPITAL.
5.
6. N.C. OFFICERS QRS.
7. BARRACKS.
10. BATH HOUSE.
11. SWIMMING POOL.
12. TENNIS COURT.
13. CARPENTER SHOP.
14. GARAGE.
15. HAND BALL COURT.
16. MINE PLANTERS ST. HO.
109. TRANSFORMER STA. AND TEL. EXCHANGE.
110. FIRE APPARATUS HOUSE.
111. SEWER EJECTOR PUMP HO.
112. GRAND STAND.
116. COURT MARTIAL AND POST OFFICE.
117. REEL HOUSE.
41. ENGR. STORE HOUSE.
70. Y.M.C.A.

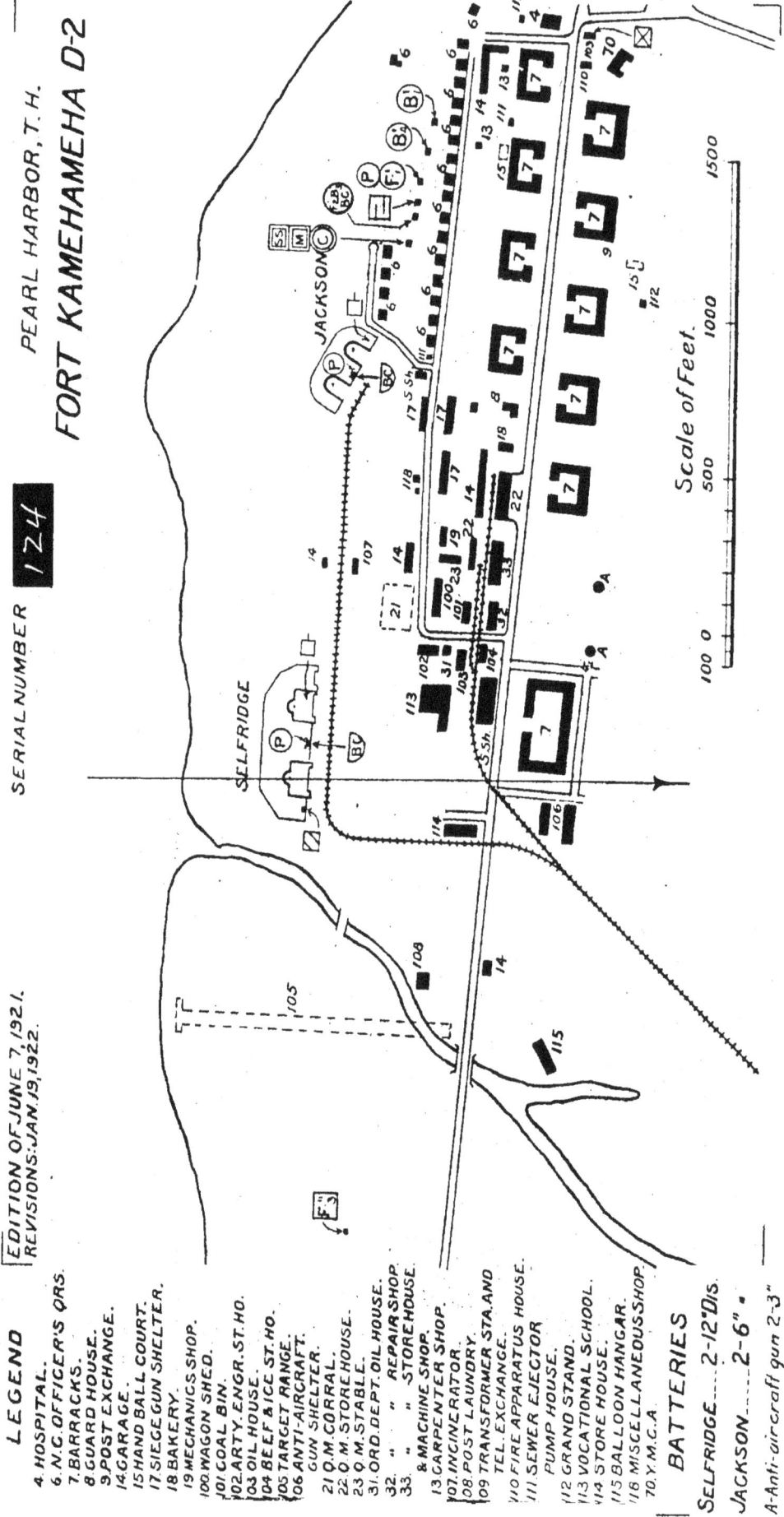

PEARL HARBOR, T.H.

FORT KAMEHAMEHA D-2

SERIAL NUMBER **124**

EDITION OF JUNE 7, 1921.
REVISIONS: JAN. 19, 1922.

JACKSON

SELFRIDGE

Scale of Feet.

100   0   500   1000   1500

LEGEND

4. HOSPITAL.
6. N.C. OFFICER'S QRS.
7. BARRACKS.
8. GUARD HOUSE.
9. POST EXCHANGE.
14. GARAGE.
15. HAND BALL COURT.
17. SIEGE GUN SHELTER.
18. BAKERY.
19. MECHANICS SHOP.
100. WAGON SHED.
101. COAL BIN.
102. ARTY. ENGR. ST. HO.
103. OIL HOUSE.
104. BEEF & ICE ST. HO.
105. TARGET RANGE.
106. ANTI-AIRCRAFT.
       GUN SHELTER.
21. Q.M. CORRAL.
22. Q.M. STORE HOUSE.
23. Q.M. STABLE.
31. ORD. DEPT. OIL HOUSE.
32. "   " REPAIR SHOP.
33. "   " STOREHOUSE.
        & MACHINE SHOP.
13. CARPENTER SHOP.
107. INCINERATOR.
108. POST LAUNDRY.
109. TRANSFORMER STA. AND
       TEL. EXCHANGE.
110. FIRE APPARATUS HOUSE.
111. SEWER EJECTOR
       PUMP HOUSE.
112. GRAND STAND.
113. VOCATIONAL SCHOOL.
114. STORE HOUSE.
115. BALLOON HANGAR.
116. MISCELLANEOUS SHOP.
70. Y.M.C.A

BATTERIES

SELFRIDGE ---- 2-12" Dis.
JACKSON ---- 2-6 " "

A-Anti-aircraft gun 2-3"

Battery Selfridge Joint Base Pearl Harbor Hickam Field (Terry McGovern)

Battery Closson Joint Base Pearl Harbor Hickam Field (Terry McGovern)

- **HASBROUCK**: The mortar battery recommended for the Queen Emma Point reservation. The site (just inland from the engineer wharf) was selected in August 1909. After approval of plan, physical construction took place between 1909 and 1911. It was of conventional design, with magazines in the traverses between and on either side of the pits. Transfer was not made until November 10, 1914 for a recorded construction cost of $274,160.65. It was named in General Orders No. 59 of May 6, 1911 for Brigadier General Henry C. Hasbrouck, U.S. Army who had died in 1910. It was armed with eight new-type Model 1908 12-inch mortars and carriages (#16/#16, #15/#15, #14/#14, and #7/#13 and #20/#5, #17/#6, #19/#7, and #18/#8). The battery served throughout the interwar years. In 1935 the new harbor defense command post in the right flank of the emplacement was gas proofed. New blast and splinter protection was added in 1942. It was disarmed in early 1946, though it had not been manned for several years. After the war it was used for a number of military purposes, including at one point parachute packing. The emplacement still exists at the Joint Base Pearl Harbor-Hickam. The battery is currently abandoned and not open to the public.

- **JACKSON**: A battery for two 6-inch disappearing guns located just to the west of Selfridge on the Fort Kamehameha property. It was of conventional Taft design for 6-inch batteries, with the magazines at the same level as the loading platforms. Construction was made between December 1912 and June 1914. It was transferred to service troops on June 17, 1914 for $86,067.25. It was named in General Orders No. 72 of November 1913 for Brigadier General Henry Jackson, U.S. Army. It was armed with two 6-inch guns Model 1908 on Model 1905MII disappearing carriages (#7/#32 and #13/#33). The battery served up into the middle of World War II. On December 21, 1943 it was listed as armed but no longer manned. Apparently, the guns were removed subsequent in 1944 or 1945. The emplacement still exists on the property of the military Joint Base Pearl Harbor—Hickam. The battery is currently abandoned and not open to the public.

- **HAWKINS**: A battery for two 3-inch guns intended to cover the mine field and entrance to Pearl Harbor from the Fort Kamehameha military reservation. Plans for the battery were submitted on September 28, 1912. It was of conventional 3-inch type and closely followed approved plans. Construction was accomplished between December 1912 and March 1914. Transfer was made to service troops of March 6, 1914 for a building cost of $22,200. It was named in General Orders No. 72 of November 1913 for Brigadier General Hamilton S. Hawkins of both Civil War and Spanish American War service and former commandant at West Point. The battery was armed with two 3-inch Model 1903 guns and pedestals (#97/#97 and #98/#98). The battery served throughout the interwar period, not being deleted until 1943. On December 21, 1943 it was listed as armed but not manned. The armament was removed and discarded in 1946. While the emplacement was not subsequently reused for armament, it still exists on the Joint Base Pearl Harbor–Hickam. The battery is currently abandoned and not open to the public.

- **CLOSSON**: Ambitious plans for modernizing the Oahu defenses were cast by the 1915 Board of Review Program. Initially three dual, long-range 12-inch gun batteries were planned for the island—a set of two at Ahua Point, Barber's Point, and on the north shore. Eventually just a single dual battery was authorized and funded, near Ahua Point here at the eastern extreme of Fort Kamehameha. It was built from September 1917 to early 1920, for transfer on May 4, 1920 at a cost of $300,727.42. The battery was of typical two-gun design of this generation, with an open back and completely exposed, open gun blocks. The battery was armed with two 12-inch Model 1895M1A4 Watervliet guns on Model 1917 long-range barbette carriages (#1/#26 and #30/#27). It was named in General Orders No. 13 of March 27, 1922 for Brigadier General Henry W. Closson who died in 1917. Subsequently a number of changes of gun tubes were made, and differing records offer different arrangements. Additional tubes #63 Watervliet, #20 Watervliet and #1 Bethlehem are mentioned (though always on carriages #26 and #27). Overhead casemating of the battery and its gun blocks was authorized on April 14, 1941, and accomplished by mid-1942. Serving as an important part of the defenses through the war, the emplacement was not declared excess until May 10, 1948 and soon disarmed. The battery was used for numerous purposes in the subsequent sixty years. While somewhat modified, particularly in the internal rooms, the battery still exists at the joint Base Pearl Harbor—Hickam. The battery is currently abandoned and not open to the public.

- **BARRI – CHANDLER:** The Land Defense Board for Oahu in 1912 called for a new light battery for Bishop Point on the western side of Fort Kamehameha. Property was transferred from the U.S. Navy to the U.S. Army in 1915. the batteries were to cover approaches from the west to the defensive redoubt or "box" envisioned to cover the forts and population centers stretching from Fort Kamehameha to Fort Ruger. Of course, it also had a secondary seacoast role in covering the entrance to Pearl Harbor. Plans were submitted for a complicated battery for four rapid-fire guns on October 29, 1914. The emplacement was a two-story structural bunker. On one side were two casemates for Armstrong 4.72-inch seacoast guns and on the other were similar casemates for two 3-inch Model 1903 pedestal guns. Work was done from September 1914 to September 1915. The 4.72-inch guns were transferred on September 26, 1915 for a cost of $29,043.38. It was armed with Armstrong guns (purchased in 1898 at the start of the Spanish American War) on pedestals (#10999/#9018 and #11000/#9083), These had been transferred here from Battery Griffin at Fort Hamilton, New York, The second half of the battery had Model 1903 guns and pedestals (#13/#56 and #93/#55). This section was transferred on June 9, 1915 for $26,923.37. The emplacements were named in General Orders No. 23 of April 27, 1915 for Captain Thomas O. Barri of the 11th Infantry killed at Gettysburg in the Civil War and for 2nd Lt. Rex Chandler, Coast Artillery Corps, who died in an aircraft accident in 1913. The batteries did not last as long as many of the other Oahu emplacements. Battery Barri with its 4.7-inch guns was disarmed early, being gone by 1924. Chandler lasted until the mid-1930s, the guns and pedestals being taken for spares and eventually re-emplaced in a new AMTB battery at Wailea in early 1942. The entire emplacement structure was destroyed in the mid-1950s, its place being taken by the new Air Force Officer's Club. Nothing now remains of the emplacement.

- **AMTB #2:** An anti-motor torpedo boat battery of two fixed 90mm guns with M3 barbette carriage erected at Fort Kamehameha. Apparently, it was emplaced on the southwest side of the reservation, directly covering the entrance to Pearl Harbor. It was begun in late 1942, completing in July 1943. It consisted of simple concrete gun blocks. It was disarmed in mid or late 1945. It is not believed that any remains still exist.

Pearl Harbor Military Reservation (1917-1939) is located on Ford Island in the center of Pearl Harbor. Under the command of Fort Kamehameha, Battery Adair and Battery Boyd (both 1917-1925), each mounting two 6-inch guns in casemates, were located on Ford Island as part of Oahu Land Defense Project for the naval base at Pearl Harbor, covering the East and Middle Lochs to the north. Battery Adair, located on Nob Hill, was sealed up and converted to the basement of the base admiral's quarters (Quarters K) in 1936. It was used as an air-raid shelter during the Japanese attack in 1941. Battery Boyd, located between San Jacinto Street and Ranger Loop, was converted for ammunition storage by the U.S. Navy. Also located here was Naval Antiaircraft Shore Battery No. 7 (1942-1944) with four 5-inch naval guns. Ford Island became the site of the Army Air Corps' Luke Field (1919-1939) until transferred to the U.S. Navy as Ford Island Naval Air Station (1939-1962). Today, Ford Island is used by multiple Navy tenants and has limited access except for the USS Missouri and Pearl Harbor Aviation Museum.

PEARL HARBOR, T.H.

BISHOP POINT

SERIAL NUMBER 124

EDITION OF DEC. 26, 1919.
REVISIONS: JUNE 7, 1921.
AUG. 31, 1921.

BATTERIES.
BARRI .... 2-4.7" PED
CHANDLER .. 2-3" "

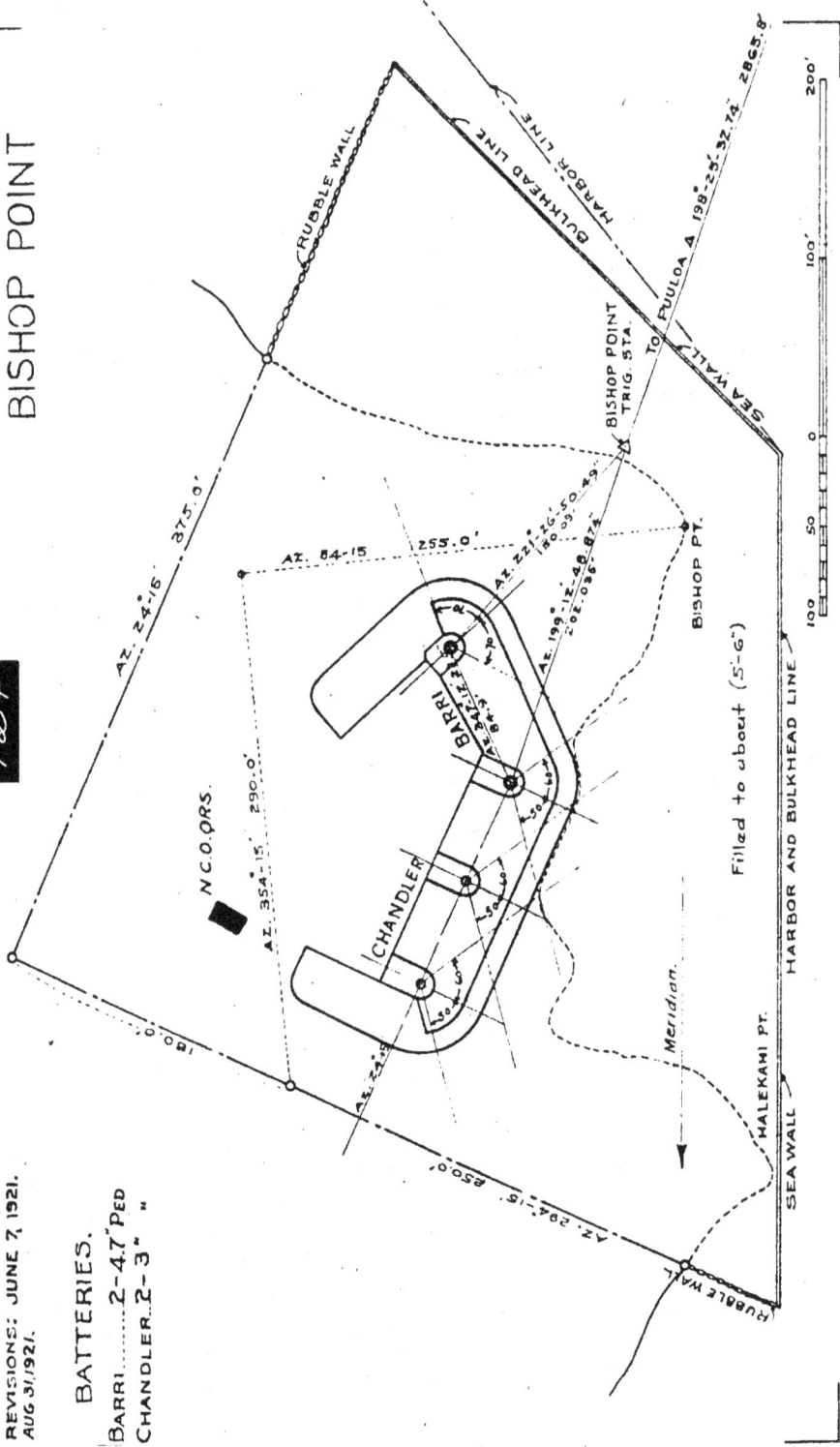

N.C.O. QRS.

RUBBLE WALL

BULKHEAD LINE

HARBOR LINE

To PUULOA △ 198°-25'-37.74' 2865.8'

BISHOP POINT TRIG. STA.

SEA WALL

BISHOP PT.

Filled to about (5'-6')

CHANDLER

BARRI

Meridian

HALEKAMI PT.

SEA WALL

HARBOR AND BULKHEAD LINE

RUBBLE WALL

AZ. 24-15   375.0'

AZ. 84-15   255.0'

AZ. 35A-15   290.0'

180.0'

AZ. 294-15   850.0'

Pearl Harbor M.R. Gun Batteries

- **ADAIR:** A new battery for two 6-inch guns emplaced on Ford Island as part of the recommendations of the Oahu Land Defense project. This project called for two batteries to be erected on the new Ford Island reservation, recently purchased property from the Oahu Sugar Company. The batteries were to fire landward, in a northerly direction to engage possible hostile forces approaching the defensive southern perimeter around Pearl Harbor, Honolulu, and Fort Ruger. The older 6-inch Vickers guns were selected, as their relatively low muzzle velocity was well suited to firing shrapnel or high explosive rounds. Also, the emplacements were intended to be casemated as they were expected to receive heavy return fire. This battery was built at the northeastern point of the island. Work was done from August 1916 to December 1917 for transfer on December 17, 1917 for a cost of $59,045.16. The armament consisted of two 6-inch Vickers guns, originally purchased in 1898 and originally emplaced in mainland forts (one came from Fort Williams and one from Fort Adams). These were guns #12123/#11159 and #12137/#11163. It was named in General Orders No. 13 of January 16, 1917 for 1st Lt. Henry R. Adair, killed on June 11, 1916 during the Mexican Punitive Expedition. While the emplacement was armed in 1919 it served only a short while, the guns were gone by 1925. In postwar years it was converted first to a personnel shelter with a new side entry and eventually was built over with senior officer quarters. Heavily modified and almost unrecognizable, parts of the battery still exist in the restricted navy housing area of Ford Island. The battery is not open to the public.

- **BOYD:** A new battery for two 6-inch guns emplaced on Ford Island as part of the recommendations of the Oahu Land Defense project. This project called for two batteries to be erected on the new Ford Island reservation, recently purchased property from the Oahu Sugar Company. The batteries were to fire landward, in a northerly direction to engage possible hostile forces approaching the defensive southern perimeter around Pearl Harbor, Honolulu, and Fort Ruger. The older 6-inch Vickers guns were selected, as their relatively low muzzle velocity was well suited to firing shrapnel or high explosive rounds. Also, the emplacements were intended to be casemated as they were expected to receive heavy return fire. This battery was built on the northwest shore, with a field of fire to the northwest. Work was done from April 1916 to February 1917 for transfer on February 17, 1917 for a total cost of $44,608.10. It was casemated as part of its original design. The armament consisted of two Vickers 6-inch guns originally purchased in 1898 and emplaced at Battery Bankhead, Fort Adams. It carried guns and carriages #12134/#11158 and #12138/#11164. It was named in General Orders No. 13 of January 16, 1917 for Captain Charles T. Boyd, killed on June 11, 1916 during the Mexican Punitive Expedition. The battery was armed only for a short period, probably being disarmed sometime between 1923 and 1925. The site was subsequently used for munitions storage. The earth cover of the magazines was at some point removed, but the basic emplacement structurally still exists but on restricted navy property at Fort Island. The battery is not open to the public.

Aliamanu Military Reservation (1934-1948/present) is located in and around the Āliamanu Crater, also known as Leilono Crater or North Crater, which is a volcanic tuff cone in the Salt Lake neighborhood of Honolulu that rises to an elevation of 760 feet. A major defense project of the mid-1930s was the construction of ammunition tunnels into the sides of Aliamanu Crater, called Aliamanu Ammunition Storage Depot at that time. Intended for centralized storage of army ammunition, eight tunnels were dug in 1934 and additional 35 magazines were completed in 1937. Fire-control station "C" (three structures, pre-World War II) was located on the rim of the crater. In 1939, the Army constructed the Hawaiian Department Com-

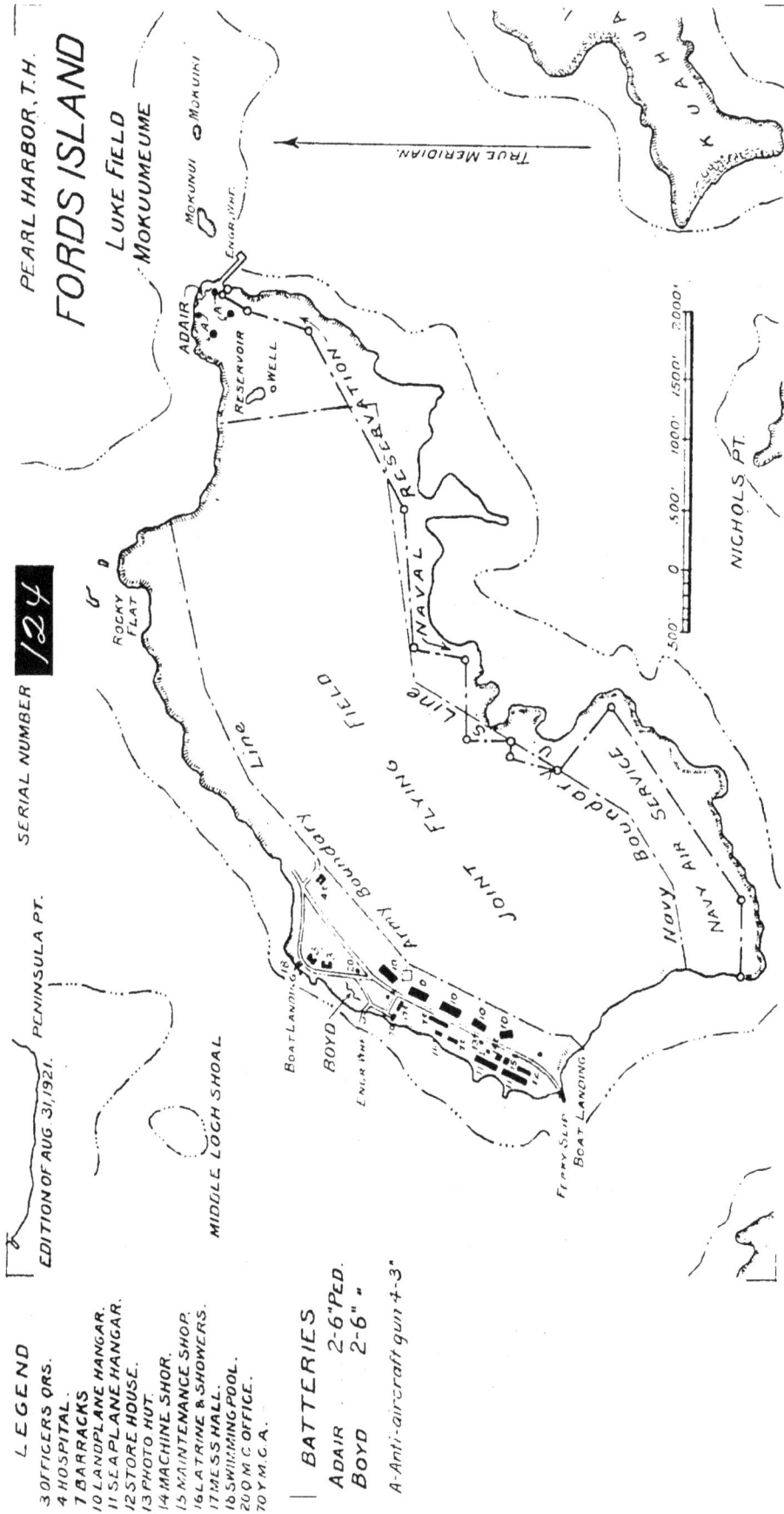

PEARL HARBOR, T.H.

FORDS ISLAND

LUKE FIELD
MOKUUMEUME

TRUE MERIDIAN

SERIAL NUMBER 124

EDITION OF AUG. 31, 1921.

PENINSULA PT.

LEGEND

3 OFFICERS QRS.
4 HOSPITAL.
7 BARRACKS
10 LANDPLANE HANGAR.
11 SEAPLANE HANGAR.
12 STORE HOUSE.
13 PHOTO HUT.
14 MACHINE SHOP.
15 MAINTENANCE SHOP.
16 LATRINE & SHOWERS.
17 MESS HALL.
18 SWIMMING POOL.
20 Q.M.C. OFFICE.
70 Y.M.C.A.

BATTERIES

ADAIR   2-6" PED.
BOYD   2-6"   "

A - Anti-aircraft gun 4-3"

MIDDLE LOCH SHOAL

ROCKY FLAT

ADAIR

RESERVOIR

WELL

NAVAL RESERVATION LINE

JOINT FLYING FIELD

Army Boundary Line

NAVY AIR SERVICE

Navy Boundary Line

BOAT LANDING

BOYD

Engr. Hut.

FERRY SLIP
BOAT LANDING

NICHOLS PT.

NAVY AIR

0   500   1000   1500   2000'
500'

LUKE

MOKUNUI   MOKUIKI
Engr. Hut.

mand Post (located in three tunnels with a portal through the northern rim from the inter rim). In October 1941, work was started next to the Army's Command Post to construct a very large underground Joint Army-Navy Command Post; although not completed at the time of the Japanese attack on Pearl Harbor, the post was shortly after putting into service by the island's command. Also here were three 4-inch naval guns (1941-1943), and Battery Aliamanu (1942-1944) with four 155mm GPF on Panama mounts. The Salt Lake Military Reservation's southern boundary abutted the Aliamanu Crater located here was Battery Salt Lake / Burgess (1942-1948) with two twin 8-inch naval guns in turret mounts. The gun mounts were later destroyed/buried, the magazine tunnels probably still exist. To alleviate continued family housing shortages in the early-1970s, the Army, Navy, and Marines developed a joint housing project at Āliamanu Military Reservation, the existing ammunition was moved to the Lualualei storage depot, and the crater was transformed into a 2,600-unit housing development. Most of the crater floor is now used for military housing. Over 150 World War II era ammunition bunkers and other storage bunkers are still located here. No public access to underground tunnels and the reservation access is limited to residents.

Aliamanu M.R. Gun Battery

- **BURGESS:** One of the four Oahu sites for two dual-gun 8-inch naval turrets built during World War II. The U.S. Navy offered the 8-inch turrets being removed from aircraft carriers USS *Saratoga* and USS *Lexington* to the U.S. Army in Hawaii for coast defense on January 17, 1942. They were readily accepted; four sites being identified each to receive two turrets. At each site would be two emplacements for the turrets (each with a turret roller ring, shell lift, and underground magazines) and separate underground plot, command, radar, and power plant rooms. This site was just east of Salt Lake, on the eastern side of Pearl Harbor. Work was started on February 10, 1942 and completed on September 20, 1943. Guns were emplaced on October 6, 1942. It was armed with two turrets from USS Saratoga carrying navy 8-inch, 55-caliber MkIXM2 guns (#516, #510, #518, and #517). It was named in General Orders No. 95 of August 27, 1946 for Colonel Louis R. Burgess, Coast Artillery Corps. The battery served until put into caretaker status in 1946 and then deleted and armament scrapped in 1949. Underground structures were built over by commercial residential housing in later years, no remains can still be seen.

Fort Weaver (1899-1948/present) is located on Iroquois Point on the left side of the Pearl Harbor Channel across from Fort Kamehameha. Originally named Iroquois Point Military Reservation until 1922. Named Fort Weaver in General Orders 13, 1922, after Major General Erasmus M. Weaver Jr. (Cullum 2563), Chief of the Coast Artillery Corps (1911-1918). During the Interwar Period, construction of Battery Williston (1924-1948) began in October 1921 and was transferred for service in September 1924. This was a two gun 16-inch all round fire (ARF) battery emplaced in the open on circular concrete pads. After World War II, the guns were scrapped, and the battery sites was used for the Iroquois Point Elementary School. The fort received four 155mm GPF guns on Panama mounts as Battery Weaver (1934-1944). Two four-gun 3-inch AA batteries were built in 1927. One set of four was relocated to Fort Barrette in the 1930s. AMTB Battery #1 (1943-1945) with 90mm guns was installed during World War II. This area was developed by U.S. Navy for housing in the 1950s, adjacent to the present-day USMC Puuloa Rifle Range. The Fort Weaver property is still owned by the U.S. Navy, and the homes are now leased to a property management company on a for-profit commercial basis. Access to site is restricted to residents.

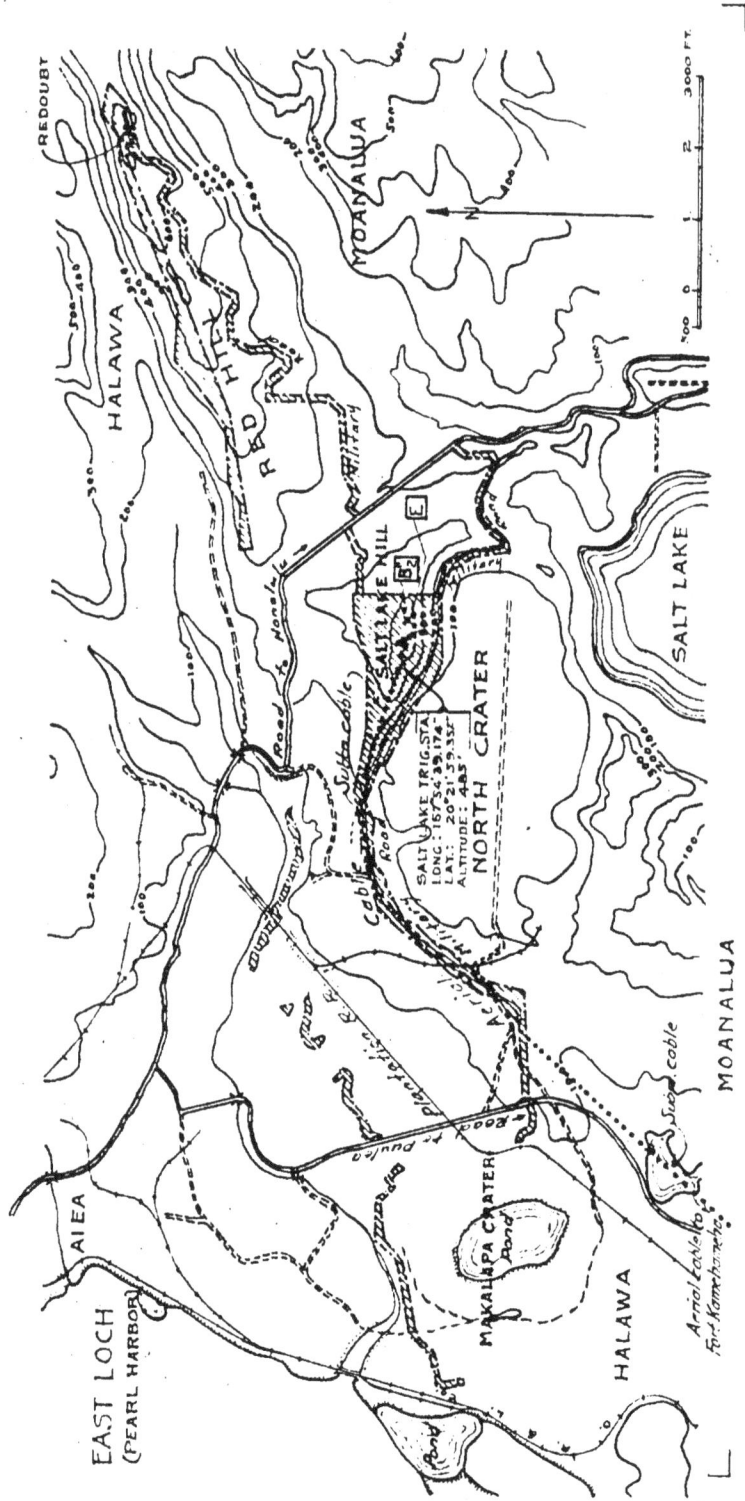

LEGEND

EDITION OF DEC. 26, 1919.
REVISIONS: JUNE 7, 1921.
REVISIONS. AUG. 31. 1921.

U.S. MILITARY RESERVATIONS

SERIAL NUMBER

124

ISLAND OF OAHU, T.H.

MAKALAPA CRATER, SALT LAKE,
RED HILL.

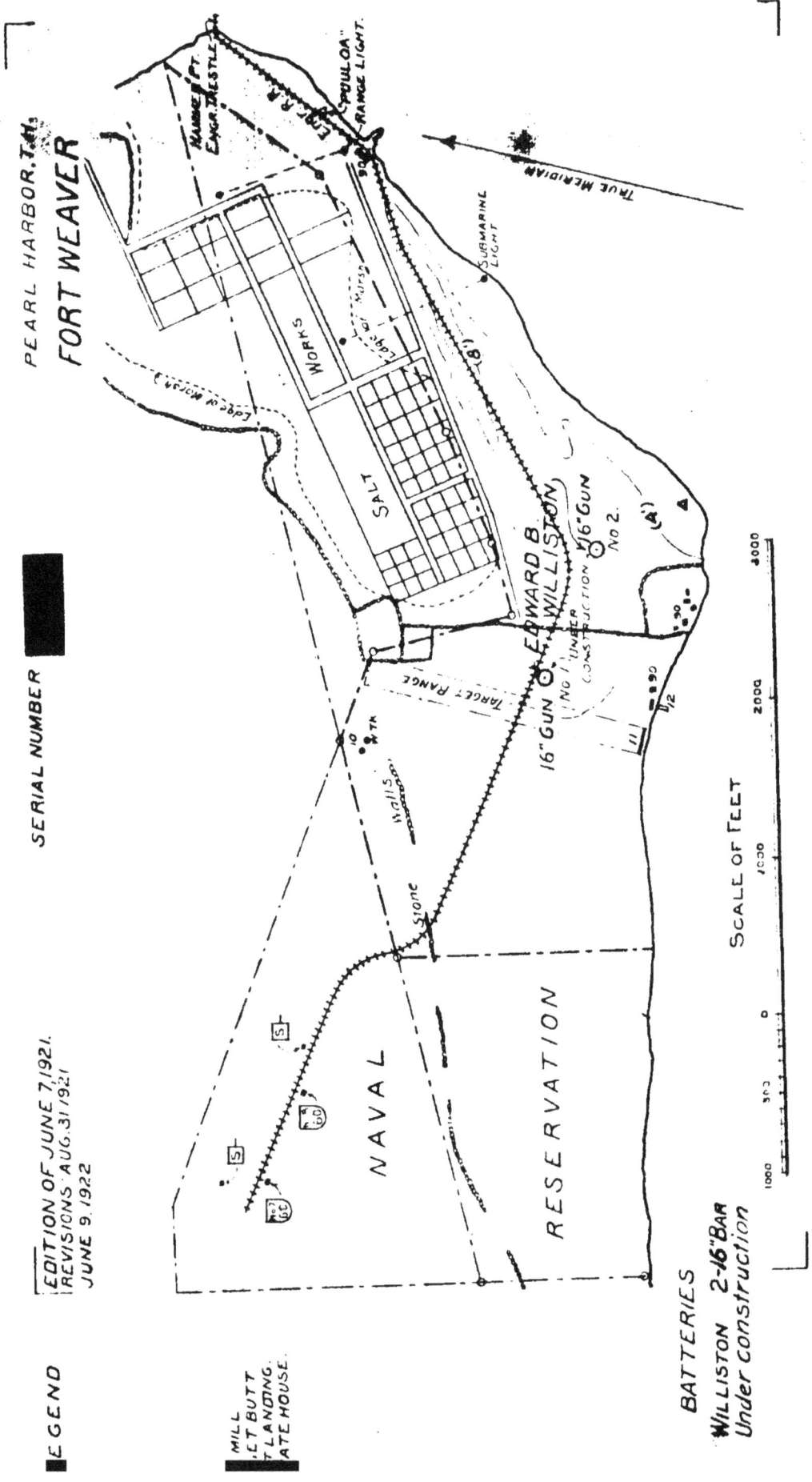

PEARL HARBOR.T. Mfrs.
FORT WEAVER

EDITION OF JUNE 7,1921.
REVISIONS·AUG.31,1921
JUNE 9,1922

SERIAL NUMBER

LEGEND

MILL
.LT BUTT
T LANDING.
ATE HOUSE.

BATTERIES
WILLISTON 2-16"BAR
Under construction

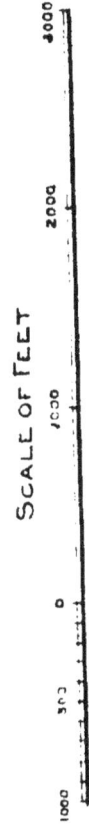

SCALE OF FEET

Fort Weaver Gun Batteries

- **WILLISTON**: An installation for a pair of Army 16-inch guns at the lower, western entrance to Pearl Harbor. The site was recommended for a very heavy battery by the 1915 Board of Review Program. Consistent with contemporary engineering designs, it was composed of two widely-spaced gun blocks to provide all-round fire for 16-inch barbettes. These blocks were connected by a dedicated rail line to the dispersed magazines. It solely depended on dispersal and camouflage for protection. Work was done from October 1921 to September 1924. Transfer was made on September 19, 1924 for a construction cost of $121,549.72. It was named in General Orders No. 13 of March 27, 1922 for Brigadier General Edward B. Williston. The battery was armed with two 16-inch Army Model 1919MII guns on Model 1919 barbette carriages (#6/#4 and #7/#3). It was one of only three dual 16-inch gun batteries of the 1920s with Army built gun tubes, and an extremely powerful addition to the defenses. A large, protected plotting room was added in September 1923 at a cost of $65,271.65. In late 1941 a project was undertaken to bombproof the magazines. Additionally, a program to develop and install armored "tunnel turrets" to the guns was begun. Unlike the other Hawaiian 16-inch Battery Hatch, the emplacement was never proposed for casemating—the army was unwilling to compromise the extensive field of fire of the battery, that could essentially cover all of Oahu Island. The tunnel turrets were eventually fabricated and shipped to Oahu for installation late in World War II, but due to some lack of parts this was never actually accomplished, the turrets were stored at Fort Weaver instead. The battery had the usual network of fire control stations, and during the war was equipped with fire control radar. A study was conducted in late 1944 about relocating the battery to new emplacements near Aliamanu Crater, but at this late date it was never implemented. It served through the war, not being deleted until the final abandonment of the harbor defense in 1948. Consequently, the guns and carriages were scrapped. Much of the site was either destroyed or covered over in 1958 to allow for naval housing and an elementary school, though the battery's protected plotting room and several magazine buildings still survive.

- **AMTB-1** An anti-motor torpedo boat battery of two fixed 90mm guns with M3 barbette carriages and two mobile 90mm M1 guns erected at Fort Weaver. It was begun in late 1942, completing in mid-1943. It consisted of simple concrete gun blocks. It was disarmed in mid or late 1945. No remains are believed to still exist.

Fort Barrette (1931-1948) is located in Kapolei, Hawaii, the island of Oahu. The fort is built on Puʻuokapolei, an ancient extinct cinder cone. Originally named Kapolei Military Reservation until 1934. Renamed in honor of John Davenport Barrette, who was a brigadier general in the coastal defenses of the United States Army during World War I. Battery Hatch (1934-1948) was completed in 1935 and mounted two 16-inch MK2 guns. During World War II, Battery Hatch's two guns were casemated when two concrete gun houses were constructed in 1942. Four 3-inch AA guns were transferred to Fort Barrette in the 1930s from Fort Weaver. From January 1961 to March 1970, the 298th Air Defense Artillery Group, HI ARNG, used Fort Barrette as a support base for the nearby Nike Hercules missile battery (double site, 24 missiles) at Pālehua (OA-63) on Makakilo. The fort was transferred to the City & County of Honolulu in 1994 as the Kapolei Regional Park. The battery area is leased to the Bushwhacker's Archery Club and access is restricted. A reserve magazine complex was located in a gulch a mile north of Fort Barrette. The magazine buildings are used today by a nursery.

PEARL HARBOR, T.H.
FT. BARRETTE
TRACT NO.1

SCALE IN FEET

300' 200' 100' 0 300' 600'

SERIAL NUMBER

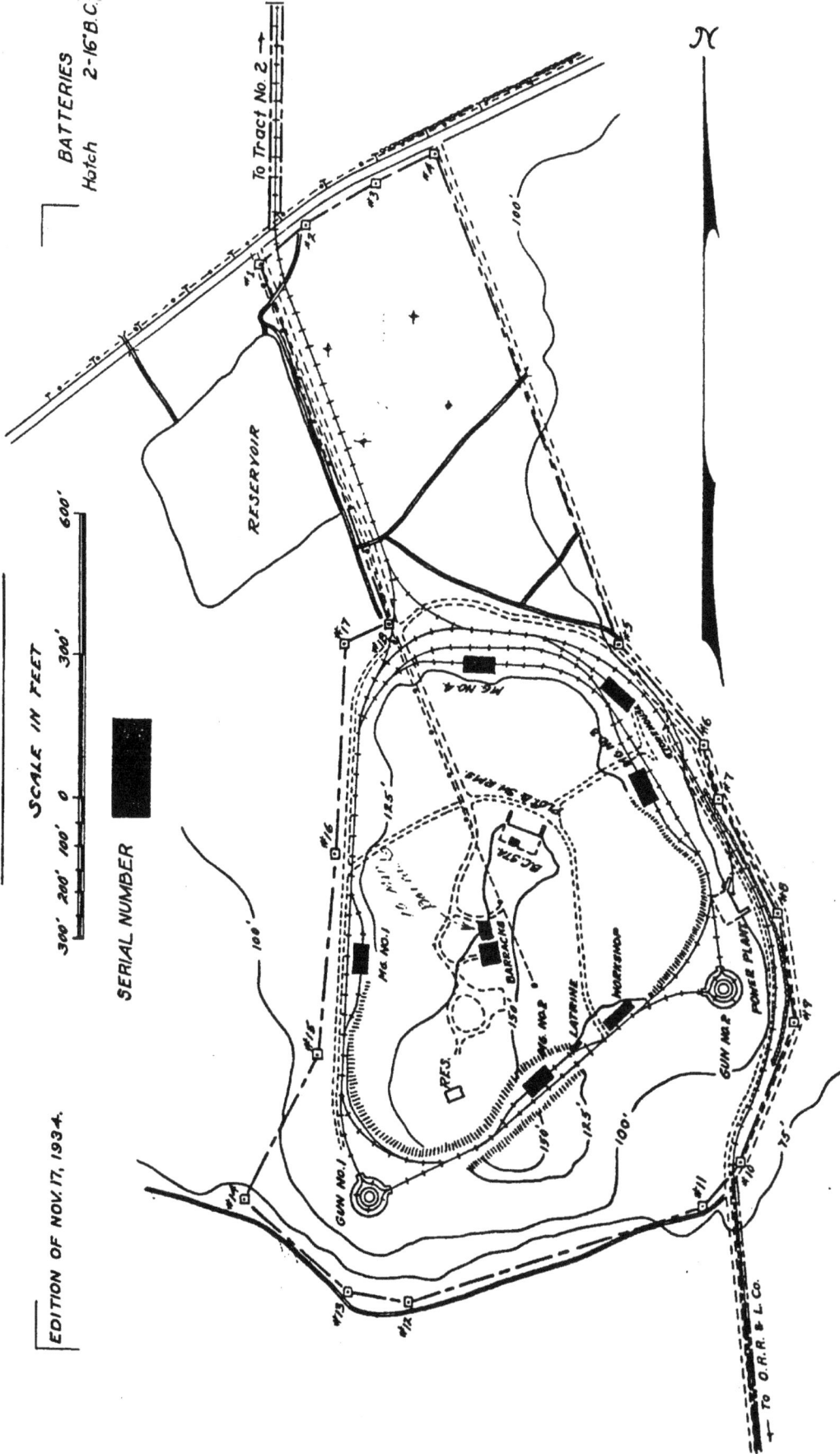

EDITION OF NOV. 17, 1934.

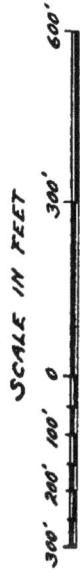

BATTERIES
Hatch 2-16"B.C.

To Tract No. 2 →

RESERVOIR

100'

100'

N

GUN No.1

MG. No.1

MG. No.4

MG. No.2

MG. No.3

BARRACKS

LATRINE

WORKSHOP

GUN No.2

POWER PLANT

B.C. STR.

TR. D & BN. RMS.

RES.

125'

150'

125'

100'

75'

#1

#2

#3

#4

#5

#6

#7

#8

#9

#10

#11

#12

#13

#14

#15

#16

#17

← To O.R.R. & L.Co.

Fort Barrette Gun Battery

- **HATCH**: A second pair of long-range 16-inch guns was proposed for Oahu during the interwar years—in fact one of the few new seacoast projects approved and funded for the U.S. Army in this decade. Plans were for a new, unprotected battery for two guns to be emplaced on a new reservation at the southwest of Oahu near Barber's Point. While under construction the site was known as Kapolei battery. By this time the army was using navy, 50-caliber 16-inch gun tubes on a slightly modified model 1919 barbette carriage for their standard heavy gun installations. Work was done from July 5, 1931 to June 12, 1935 for transfer on July 6, 1935. Construction cost was calculated at $730,558.64 for the emplacement, another $114,095.32 for the dispersed magazines, and $41,522.16 for the combined power and plotting rooms. It featured gun blocks separated by some 900-feet, connected by railroad to four separate magazines (located in the protected gulch nearby). It was named on General Orders No. 10 on November 23, 1934 for Brigadier General Henry J. Hatch, Coast Artilleryman. It was armed with two 16-inch MkIIM1 guns on Model 1919M1 carriages (#63/#8 and #62/#7). This armament was carried throughout the battery's service history, never being changed. However, two 16-inch barrels were stored near the emplacements as spare or reserve tubes if a change of barrel became required. In 1941 a project was approved to protect the gun blocks by providing them with overhead, casemated protection. It was recognized that the 360-degree, all-round fire capability would be sacrificed, but the tradeoff was accepted. Work was done in 1941-43. The battery served throughout World War II, not being deleted until 1948. The casemates and protected structures still exist at the Kopolei Archery Range section of the Kopolei Regional Park. The battery is not open to the public. Permission is required to visit the battery.

Kahe Point Military Reservation (1940-1948) is located on a ridge about 320 feet above and 900 feet behind Kahe Point Beach. Battery Arizona (1945) with a salvaged three-gun turret from the USS Arizona but was never fully completed. This battery was constructed using a Navy triple 14-inch gun turret removed from the sunken battleship USS *Arizona* (BB-39). Battery construction started in 1943 and was almost completed 1 Aug 1945 when construction was halted. The battery structures were mostly underground and included the central barbette that extended 70 feet underground and mounted the three-gun turret on the top. The battery was placed in caretaker status at the end of the war and abandoned by 1948. The guns and the turret were scrapped soon after. Also here were the four 155mm GPF guns on Panama mounts of Battery Kahe (1942-1944). Searchlights were emplaced on the shore. The reservation is now part of the City of Honolulu's landfill and Hawaiian Electric Kahe Point powerplant and public access is restricted.

Kahe Point M.R. Gun Battery

- **ARIZONA**: One of the two triple-gun 14-inch turrets removed from the sunken battleship USS *Arizona* and emplaced as coast artillery. The U.S. Navy had offered two turrets from the wrecked ship as early as January 16, 1942 to the U.S. Army. While much technical work was required to repair the turrets, barbettes, gun tubes, and ammo lifts by the navy salvage teams, the army selected sites and built the emplacements. Location selection was approved on October 13, 1942. This location was on a ridge on the island's west coast, in particular offering good coverage to a portion of the island only lightly defended. In May 1943 emplacement design was finalized and ground breaking started. It would feature a 28-foot diameter shaft to hold the barbette, racer and roller apparatus and reconditioned turret with guns. The ship's ammunition lifts would descend to a magazine level. Underground tunnels would connect to these magazines, a switchboard room, power plant (with two 125-kw gasoline engines), and a plotting room. Because of the site's isolation rudimentary habitability items were also included—a small galley, first aid station, latrines, and bunk racks in the long corridors. The primary gallery was of very long length; necessary in order to connect to a vertical shaft connecting to the above ground BC and fire control radar site. Construction work, carrying a very high priority, was accomplished through the last half of 1943 and early in 1944. On December 6, 1944 it was reported that most of the structural work was complete, though wiring, equipment installation, placing the foundation ring and final assembly of the turret still needed to be accomplished. The turret was armed with three 14-inch guns MkVIIIM4 guns on the MIIM4 mount. It was named Battery Arizona, for the source ship of the turret, USS *Arizona*. It was nearing completion at the very end of the war, though running about three months behind its sister Battery Pennsylvania. Final completion was then estimated for January 1, 1946. On August 1, 1945 further work was suspended, though the emplacement and equipment were carefully preserved in case of future need. It was never manned or turned over to service troops. Within a couple of years, the project was completely deleted, and the turret and guns scrapped. The emplacement still exists, though in an area restricted due to proximity to a commercial dump at the Kahe Point landfill. The battery is not open to the public.

Kapolei Regional Park (Fort Barrette) (Terry McGovern)

Battery Arizona Kahe Point (Terry McGovern)

Puu O Hulu Military Reservation (1923-1946) is located on Waianae Coast of Oahu at Puu-O-Hulu-Kai. The reservation on top of the ridge was created for fire control stations for several artillery batteries from Fort Kamehameha to up the Waianae Coast. On top, five fire control posts were built, and they were served with a wire hoisting system and several utility buildings. Fire-control station "U" (1924) was first, followed by stations more in 1929, 1934 and 1940. Also here was an AA warning station (1940). These stations served: Battery Williston, Battery Closson, Battery Burgess, Battery Hatch, Battery Arizona, and several other smaller stationary batteries, other 155mm gun batteries and several railway batteries on the leeward coast of Oahu. These stations have become a popular hiking trail, referred to as the Pink Pillbox, which is publicly assessable. Lower down the ridge near the ocean was Battery Hulu (1942-1945) with two 7-inch casemated naval guns, with plotting room, BC station, radar and searchlight stations; and Battery 303 (1944) for two 6-inch in a tunnel complex that was not completed. The site is owned by the Honolulu Civil Defense Agency and is currently abandoned and has restricted public access.

Puu O Hulu M.R. Gun Batteries

- **Battery #303**: A dual 6-inch battery of the special project for Oahu erected near Mailili Point on the lower rise of the Puu-O-Hulu promontory. The battery originated as a site for two army emplaced, ex-navy 7-inch pedestal mounts. This temporary battery was the starting point for new Battery Construction #303. A new protected magazine with support and power rooms would be tunneled out behind the already existing gun blocks. The blocks themselves would be changed to carry the standard 6-inch, shielded barbettes and guns. Most of this new work was accomplished starting in May of 1944. It was allocated two 6-inch M1 guns and M4 barbettes (guns M1(T2) #10 and #11 allocated in June 1944). However, the blocks were never changed, today they still hold the foundation bolts for navy 7-inch guns. The 7-inch guns may have been removed in April 1945 for transfer to Brazil as part of an armament sale to that country. In May of 1945 it was reported that the conversion work was 53% completed but suspended on that date. After the war the battery was abandoned. The emplacement still exists, and since it has remained in government hands, a certain amount of original equipment still survives inside the magazines. The battery is abandoned and closed to the public.

Fire control stations at site "U" above Pu"u O Hulu

Battery #303 Battery Hulu Pulu'Hulu (Terry McGovern)

Battery #303 Battery Hulu Mailili (Terry McGovern)

The Harbor Defenses of Kane'ohe Bay – Hawaii (includes Oahu's northern shore)

Fort Hase (1918-1949/present) is located on the eastern portion of the Mokapu Peninsula on the island of Oahu. Originally known as Kuwaahoe Military Reservation, and then Camp Ulupau in 1940 until formally named 1942. Named Fort Hase in 1942, after Major General William F. Hase, a Chief of Coast Artillery (1934-1935). The U.S. Navy established a small seaplane base there as Naval Air Station Kaneohe Bay in 1939. The U.S. Army moved onto the reservation in 1941, and the portion of the reservation occupied by the Coast Artillery became known as Fort Hase in 1942. Battery Kuwaahoe with two 240mm howitzers (1927-1941) was the first battery on Ulupa'u Head (at the foot of west rim, site destroyed in 1992). There was also East Beach Battery (1941-1942) with four mobile 155mm GPF guns in Ulupa'u Crater, replaced by Battery East (1942-1944) with 155mm GPF guns on Panama mounts. Fire-control station "J" (two pre-World War II structures, one World War II structure) was located on Ulupa'u Head. Temporary Battery Puka (1942-1944) (three 3-inch naval guns) was on the south side of Ulupa'u Head, and temporary Battery Kii (1942-1944) (two 3-inch naval guns) was on the north side of Ulupa'u Head. Battery Pennsylvania (1945-1948) was built on Ulupa'u Head and consisted of a salvaged Navy triple 14-inch gun turret removed from the sunken battleship USS *Arizona* (BB-39). Battery construction started in 1943 and was completed just after the war ended. The battery structures were mostly underground and included the central barbette that extended 60 feet underground and mounted the three-gun turret on the top. The steepness of the crater constrained the underground features of the emplacement. The battery was placed in caretaker status at the end of the war and abandoned by 1948. The guns and the turret were scrapped soon after. On the then Kaneohe Naval Air Station (1939-49) (the western portion of the Mokapu Peninsula) was Battery French / 301 (1944-1948) with two 6-inch guns were installed using a standard 1940 Program battery design. The battery commander station was on the roof of the magazine, the radar on top of that – though built to resemble a water tank as camouflage. Additional coast artillery was North Beach Battery (1941-1942) with four 155mm GPF mobile guns, replaced by Battery Pyramid (1942-1944) with 155mm GPF on Panama mounts, AMTB Battery Pyramid Rock (90mm) (1943-1945), and Battery Sylvester (1941-1944, dismounted 1942) with two 8-inch railway guns. The HDCP (1942) was located on Pu'u Hawai'i Loa (tunnels still exists). Public access is restricted on the base. Other nearby batteries include Battery Papaa (1942-1944) with four 155mm GPF on Panama mounts and Battery De-Merritt / 405 (1944-1948), both on Pu'u Papaa (private property). Battery 405 had two 8-inch Mark VI M3A2 guns mounted on M1 Barbette Carriages placed on a hillside with the magazines built into tunnels. Battery construction started in early 1943, was completed in early 1944. A fire-control station (1943) was located on Kalaheo Hill, also on Papaa Ridge.

Fort Hase Gun Batteries

- **PENNSYLVANIA**: One of the two triple-gun 14-inch turrets removed from sunken battleship USS *Arizona* and emplaced as coast artillery. Fort Hase was the U.S. Army name for its defensive post (replacing the name Camp Ulupau) overlooking the naval aviation base at Kaneohe Bay. The U.S. Navy had offered two turrets from the wrecked ship as early as January 16, 1942 to the U.S. Army. While much technical work was required to repair the turrets, barbettes, gun tubes, and ammo lifts by the Navy salvage teams, the U.S. Army was busy selecting sites and building emplacements. Location selection was approved on October 13, 1942. This location, though with a good field of fire, was constrained to a steep, narrow ridge with two small promontories—one would be used for the turret and the other for the BC and radar. Searchlight and an SCR-268 air warning radar previously on the site had to be removed. In April 1943 a joint Army-Navy project office was established to supervise the work for both turret sites. In May 1943 emplacement design was finalized and groundbreaking started. It would feature a 28-foot diameter shaft to hold the barbette, racer and roller apparatus and reconditioned turret with guns. The ship's ammunition lifts would descend to a magazine level. Underground tunnels would connect to these magazines, a switchboard room, power plant (with two 125-kw gasoline engines), and a plotting room. Because of the site's isolation rudimentary habitability items were also included—a small galley, first aid station, latrines, and bunk racks in the long corridors. Construction work, carrying a very high priority, was completed through the last half of 1943 and early 1944. On December 6, 1944 it was reported that most of the structural work was complete, though wiring, equipment installation, and final assembly of the turrets still needed to be accomplished. The turret was armed with three 14-inch guns MkVIIIM4 guns on the MIIM4 mount. It was named Battery Pennsylvania, for the sister ship of USS *Arizona*. It was nearing completion at the very end of the war, in fact proof firing of the guns was conducted in early September, 1945. At that point further work was suspended, though the emplacement and equipment were carefully preserved. It was never manned or turned over to service troops. Within a couple of years, the project was completely deleted, and the turret and guns scrapped. The emplacement still exists, though in an area restricted due to proximity to a firing range at the Marine Corps Base Hawaii. The battery has been sealed and is not open to the public.

- **FRENCH**: A dual 6-inch barbette battery at the Kaneohe Fort Hase, east of Pyramid rock on the beach berm. It was one of the initial defensive units recommended for the Bay in April 1941. Unlike the other Oahu 6-inch sites, it closely followed the standard 200-series design plan, although it did have a BC station atop the traverse magazine that was connected by a stairway with the interior corridor. It was constructed in 1941-1942. During the construction and early service, it was referred to as simply Battery Construction No. 301 It was named in General Orders No. 96 on August 27, 1946 for Forrest J. French, Coast Artillery Corps. It was armed with two 6-inch Model 1903A2 on Model M1 barbette carriages (#2/#90 and #12/#91). This armament had been shipped to the battery on March 12, 1943. Transfer date has not been located but was presumably in early 1943 for a cost approximating $360,000. One gun tube had to be replaced in 1943, gun M1903A2 #51 being supplied as the replacement in November 1943. The battery served out the war, though in reduced manning status starting in February 1945. It was disarmed postwar, probably in 1946-48. The emplacement still exists near military housing at the Marine Corps Base Hawaii. The battery is not open to the public.

- **SYLVESTER**: A battery for four 8-inch guns emplaced at Kaneohe Bay. These were model 1888 guns on the railroad Model 1918 top carriage emplaced as fixed barbette guns (in other words they were no longer on their railroad cars). In this regard this was similar to the guns emplaced at Black Point in the 1930s and Battery Kahuku in northern Oahu. Four 8-inch railway mounts had been assigned shortly pre-war to Kaneohe Bay and were known as Battery Ulupau. On August 21, 1941 it was authorized to change them to fixed firing positions at a new location on the Kaneohe Bay reservation. That work was well underway by mid-March 1942. It had 8-inch Model 1888 guns on Model 1918 top carriages #9/#9, #12/#15, #24/#20, and #57/#19. For most of the war it was manned by Company C of the 41st Coast Artillery Regiment. In early 1945, with heavy demands for manpower and the overlap with completing batteries #301, #302, and #405, it was recommended that this battery be abandoned and the material scrapped. Preparation for that was well underway in January 1945, but it was not reported completed until a report of March 11, 1946. Apparently, the site was built over with the subsequent heavy development at the post, no trace of the battery site remains today.

Battery Pennsylvania, Fort Hase (Terry McGovern)

Battery French Fort Hase (Terry McGovern)

- **AMTB-7**: An anti-motor torpedo boat battery of four fixed 90mm guns with M3 barbette carriage erected near Pyramid Rock at Fort Hase in Kaneohe Bay. It was built in late 1942 and was the first of this type of emplacement in Oahu. In addition to its torpedo boat role, it functioned as the Examination Battery for the Defenses of Kaneohe Bay. It consisted of simple concrete gun blocks. It was disarmed in mid or late 1945. One of the gun blocks remains today, but the battery site is not open to the public.

- **DEMERRITT**: An original battery for two 8-inch barbette guns for the defenses of Kaneohe Bay. It was authorized in April 1941 as Battery Construction #405 for a small, separate reservation known as Puu Papaa on a steep hillside southwest of the Fort Hase reservation. It had a field of fire with directrix of 230-degrees. This was the only non-casemated 8-inch battery planned for Oahu. It had a unique plan, once again utilizing the easily tunneled rock of Oahu. Open concrete pads held the (unshielded) gun barbettes. Behind each pad a tunnel was driven into the hillside, cross-laterals containing magazines and other support rooms connected the main shafts. Gun separation was 240-feet, site elevation was 125-feet. It was authorized for construction in October 1941 and completed in late 1943. The battery was armed with two 8-inch navy guns MkVIM3A2 on M1 barbette carriages (#212/#15 and #218/#16). Mounting of the guns was reported accomplished on March 25, 1943. These were open mount guns originally used on the battleship USS *Minnesota*. Some damage was done during the first test firing, but it was soon repaired. The battery served throughout the war. It was eventually named on General Orders No. 96 of August 27, 1946 for Colonel Robert E. DeMerritt, a Coast Artillery Officer. The battery was deleted after the war, probably in 1948-1949 and the armament scrapped. The reservation was eventually sold, and emplacement was used for several years as a commercial mushroom farm. It still exists today on private property but can be visited by special arrangement.

Kualoa Military Reservation (1942-1952) is located near the Kualoa Ranch at the north end of Kane'ohe Bay. Near the ruins of the Kualoa Sugar Mill on the north-side of the Kualoa airfield, on Lae O Ka Oio (Kaoio) Point, was Battery LOKO (1942-1944) with four 155mm GPF on Panama mounts. Anti-Motor Torpedo Boat Battery #8 (1943-1945, destroyed) with 90mm guns was located south on Kualoa Point. Searchlights were emplaced on Kaoio Point. Tours of the gun sites are given by guides at Kualoa Ranch. Battery Cooper (#302) was located on private property near the Kualoa Airfield.

Kualoa M.R. Gun Batteries

- **COOPER**: The original coast defense plan for the Defenses of Kaneohe Bay was approved in April 1941. It featured one fixed dual 8-inch battery (Battery #405) and two dual 6-inch batteries (#301 at Fort Hase and this #302 at Kualoa Ranch). Battery Cooper was located on private property at Lae-O-Kaoio near the Kualoa Airfield on the northern side of Kaneohe Bay. It was constructed with difficulty on the face of a cliff at several hundred feet of elevation. To protect it from falling rock it was given light concrete casemating. Behind the casemates tunnels were driven into the rock connecting with cross tunnels to magazine spaces. It had a battery commander's station above it on the cliff face connected by internal access shaft. Work on construction was authorized in October 1941, most of it was done in 1942. It was named on General Orders No. 96 of August 27, 1946 for Colonel Avery J. Cooper, Coast Artillery Corps. In 1944 it carried two 6-inch Model 1903A2 guns on Model M1 barbette carriages (#7/#92 and #36/#93). It served through the end of the war into the early postwar period. It was probably disarmed in 1948. The emplacement still exists on private property of the Kualoa Ranch. The battery is open to the public through guided tours at the Kualoa Ranch.

- **AMTB-8**: An anti-motor torpedo boat battery of two fixed 90mm guns with M3 barbette carriage erected at the Kualoa Point Military Reservation. It was built in late 1942, completing probably in November of 1942. It consisted of simple concrete gun blocks. It was disarmed in mid or late 1945. The battery has been removed.

Kahuku Ranch Military Reservation (1941-unknown), near Kahuku Ranch near Naalehu, Ka ʻū, on the northeastern park of Oahu. Battery Ranch (1941-1945) with four 155mm GPF guns on Panama mounts, located on Kolaeokahipa Ridge (still exists), New Battery Laie (1941-1942, converted to dummy position) railway firing position on the north-side of town, and Battery Kahuku (1942-1944 dismounted 8-inch railway guns, shields added to the guns in 1944) located east of town northwest of Makahoa Point. The BC station for Battery Kahuku was located at Monument 305 on Cross Mountain, west of town. Search-lights were emplaced near Battery Kahuku. Fire-control station "K" (two structures 1939, one structure 1942) was located on Punamano Heights south of Kahuku Point, adjacent to a later USAF radar station (still remains). Searchlights were emplaced at Kahuku Point. These structures are all closed to the public.

Kahuku Ranch Gun Battery

- **KAHUKU**: A battery for four fixed firing positions for the 8-inch guns and railway top carriages emplaced at Kahuku Mills. This was the site of a sugar mill, just south of the northern tip of Oahu. Authority to build fixed firing positions was proposed on August 21, 1941. Work was directed started on February 3, 1942. Four concrete foundations to hold the traversing top carriage were built with at least one of the positions having a surrounding parapet wall. Assigned were Model 1888 guns on Model 1918 top carriages #5 West Point Foundry/#42, #7 Watervliet/#34, #48 Watervliet/#41, and #54 Watervliet/#17. It was manned by a detachment of Company D of the 41st Coast Artillery Regiment and was assigned Tactical No. 19. With a shortage of manpower and overlap with authorized Battery #408, Kahuku was authorized for scrapping in early 1945. After the war most of the emplacement was destroyed, but one gun platform and parapet wall still exist. The battery site is open to the public.

Paumalu Military Reservation (1942-unknown) is located on the North Shore of Oahu, Hawaii, specifically in the Koʻolauloa District, near Sunset Beach. A salvaged two-gun turret from the USS *Oklahoma* was planned to be located here in 1944, but was replaced by Battery 408 (never built). A railway battery firing position was also to be built near here. Searchlights were emplaced here. Fire-control station "Lena" (WWII) was located nearby on Paumalu Heights, above Sunset Beach, west of the COMSAT site (between Paumalu and Pakulena Gulchs). A concrete machine-gun pillbox is also located here.

- *Battery #408* (planned): One of the additional permanent system, modern batteries authorized for Oahu on February 2, 1944. This battery was needed to provide long-range, fixed defense to the north shore. Its location site varied over time but was finally fixed at Paumalu near Wailalee. Site survey was done in May of 1944. It was to be a conventional cut and cover work, but the gun platforms were to be three feet higher than the roof, being served by ammunition hoists. However, as the war situation improved late in 1944, no work was ever started on the construction of this battery. According to one source 8-inch guns MkVI #126 and #234 were held in anticipation for issue to the battery, but they were not likely ever supplied to the work site. Work was indefinitely deferred on May 31, 1945—when it was reported as just 1% completed. No evidence of the battery still exists.

Opaeula Military Reservation (1942-1948) is located about four miles inland from Haleiwa at Opaeula Camp was Battery Opaeula / Carroll G. Riggs (1942-1948, renamed 1946) with two twin 8-inch guns with turret mounts. Built as a World War II concrete coastal gun battery with four 8-inch Mark IX M2 Navy guns mounted in two turrets removed from the aircraft carrier USS *Lexington* (CV-2). The battery was taken out of service 26 Dec 1944 and held in caretaker status until it was deactivated 2 Apr 1945. The guns and carriages were processed for salvage in 1948. One mount location has been destroyed. The magazines, plotting room, and battery commander's station still exist on private property. Nearby was Battery Pine (1942-1945) with four 155mm GPF on Panama mounts. No remains.

Opaeula M.R. Gun Battery

- **RIGGS**: One of the four Oahu sites for two dual 8-inch naval turrets built during World War II. The U.S. Navy offered the 8-inch turrets being removed from aircraft carriers USS *Saratoga* and USS *Lexington* to the U.S. Army in Hawaii for coast defense on January 17, 1942. They were readily accepted; four sites being identified each to receive two dual turrets. At each site would be two emplacements for the turrets (each with a turret roller ring, shell lift, and underground magazines) and separate underground plot, command, radar, and power plant rooms. Both guns in a turret traversed and elevated together. All four guns of the battery were proof fired 6 Aug 1942. The proof firing included three reduced charge rounds and three full-service rounds. Structures at the battery included the two turrets, each supported by a connected underground powder and projectile magazine. Two underground plotting rooms were built, one for the battery and one planned as a Command post and a fire control switchboard for the Lexington Gun Group (Battery Brodie and Battery Opaeula). A separate building was built to house the SCR-296A fire control radar equipment. A fire control station was built on a steel tower behind the turrets. Further back from the rear of the turrets was the underground generator building and the troop barracks and support buildings. This was part of the second pair of emplacements, using the turrets from USS *Lexington*. This site was known as Opaeula, on the eastern side of Oahu's central plain. Its major field of fire was to the north. The battery was reported completed on August 6, 1942. It was armed with two turrets carrying navy 8-inch, 55-caliber MkIXM2 guns (#496, #497, #498, and #503). It was named in General Orders No. 95 of August 27, 1946 for Colonel Carroll G. Riggs, CAC, who was killed in an aircraft crash in Australia on December 18, 1943. It served through the rest of the war but was abandoned in late 1945. The armament was scrapped in 1949. Most of the emplacement elements still exist on private property. The battery is not open to the public.

Brodie Camp Military Reservation (1942-1948) is located near Brodie Camp Four in Wahiawa, a central valley that lies between the two mountain ranges on either side of Oahu. The guns and carriages were processed for salvage in 1948. Site buried and located on private property surrounded by pineapple fields.

Brodie Camp M.R. Gun Battery

- **RICKER**: One of the four Oahu sites for two dual 8-inch naval turrets built during World War II. The U.S. Navy offered the 8-inch turrets being removed from aircraft carriers USS *Saratoga* and USS *Lexington* to the U.S. Army in Hawaii for coast defense on January 17, 1942. They were readily accepted; four sites being identified each to receive two dual turrets. At each site would be two emplacements for the turrets (each with a turret roller ring, shell lift, and underground magazines) and separate underground plot, command, radar, and power plant rooms. This was part of the first pair of emplacements, using the turrets from USS *Saratoga*. Work was started on March 10,

Brodie Camp (Terry McGovern)

Battery Cooper Kualoa (Terry McGovern)

Battery Cooper Kualoa (Terry McGovern)

Site of Battery #409 Kaena Point (Terry McGovern)

1942. It was located on a site known as Brodie Camp in the northern part of Oahu's central plain. The primary directrix was to the north, but of course the turrets could fire with 360-degree fields. With a high priority, it was structurally complete by late 1942. It was armed with two turrets from USS *Saratoga* carrying navy 8-inch MkIXM2 guns (#495, #507, #508, and #509). It was named in General Orders No. 96 of August 27, 1946, for Lt. Colonel George W. Ricker, who was killed in an airplane crash in December 1941. It served throughout the war but was abandoned at war's end in 1945 and the turrets were scrapped in 1949. Some of the emplacement elements still exist on private land of Dole Pineapple. The battery is not open to the public.

Kaena Point Military Reservation (1923-1948/present) is located at the westernmost point of Oahu at Ka'ena Point. Also known as Camp Kaena. Located on Pu'u Pueo was Battery Kaena (1942-1944) with two 4-inch naval guns until 1943 when two 155mm GPF guns were installed on Panama mounts. No remains. A salvaged three-gun turret from the USS *Oklahoma* was planned for this location in 1944, but was replaced by Battery 409 (1945, tunnels nearly completed, guns never mounted). Begun in mid-1943, the facility was designed to support two 8-inch naval guns and army M1 barbette carriages. Fire-control station "S" (1923, one structure) was also located here (now ruins). A second structure (1934) built below still exists. Also here was a SCR-271A radar (1942), with the operations and power rooms tunneled into the mountain. The trail to Kaena Point follows an old railroad bed and former dirt road that ran along the westernmost point of Oahu. The trail leads to Kaena Point Natural Area Reserve, a remote and scenic protected area harboring some of the last vestiges of coastal sand dune habitat on the island, and home to native plants and seabirds. Special permits are available to allow 4-wheeled drive vehicles to drive to the point. The battery is abandoned, and tunnel entrances are blocked with earth.

Kaena Point Gun Battery

- **Battery #409** A dual 8-inch battery approved relatively late in World War II as one of the intermediate batteries required to enhance Oahu's defenses, particularly for outlying areas. The project was authorized for construction on January 8, 1944. The site selected was at the northwestern extreme of the island, at Kaena Point. It was intended to provide direct coverage of the western approaches to the island, especially to provide additional protection for the Lualualei Ammunition Depot and Mokule'ia Airfield. It would have been of a unique design layout. Two tunnels driven into the base of the cliff at Kaena at an elevation of 125 feet that would have terminated at the surface in open platforms for the guns. Inside, at a lower level would have been placed the magazines. The tunnel complex, designed to house all support operations, powder magazines, and electrical generators and compressors, was composed of two access tunnels connected internally by two traverse tunnels. All chambers were 15 feet high and 15 feet wide. The northern access tunnel was the longest at 200 feet; the southern access tunnel extended underground for 40-50 feet; and the two traverse tunnels were 75-85 and 100 feet long. The tunnel entrances were spaced 300 feet apart and were accessed by an 18 foot-wide, 2,483-foot-long gravel road that approached the tunnels from the northwest. Remains of mining machine is still located in one of the tunnels. A BC station was located on the cliff above. Work was delayed and not started until late in the war. The tunnels were driven, but never lined, the blocks were poured but no base rings set. On June 6, 1945 it was only 60% complete when the project was abandoned in 1945. It was to have been armed with two of the navy 8-inch MkVIM3A2 guns on M1 open barbettes. The tunnels are all that survive on the remote site and earth has been placed over the tunnel entrances to close access. Reportedly, some of this earth has been removed for access.

Battery DeMerritt (#405) Fort Hase Oahu (Terry McGovern)

Diamond Head fire control stations (Terry McGovern)

OVERSEAS BASES – PHILIPPINES

The defenses of the Manila and Subic Bays focused on the island forts that stretched across the mouth of each bay. These rugged, volcanic islands were good sites for coast artillery and their support facilities. The climate is tropical and access to coast defense sites can be difficult. The construction of these harbor defenses started during the Endicott and Taft Programs and continue until the 1922 Washington Naval Treaty that froze any further development of those fortifications. Some preparations were made in 1940 and 1941 to improve the defense of the islands. The Harbor Defenses of Manila Bay were the last bastion of the American defenders against the Japanese invasion of the Philippines, finally surrendering on May 6, 1942. The island forts were bombarded and fought over in 1945 by the returning American troops, the only U.S. harbor defenses to see action since the War of 1812. Important coast artillery forts were Fort Mills, Fort Hughes, Fort Drum, Fort Frank, and Fort Wint. The listing below only includes major sites with permanent coast defenses. An excellent description of these forts can be found in *Pacific Rampart* by Glen Williford or Osprey Fortress Series No. 4 *American Defenses of Corregidor and Manila Bay 1898-1945.* by Terry McGovern and Mark Berhow.

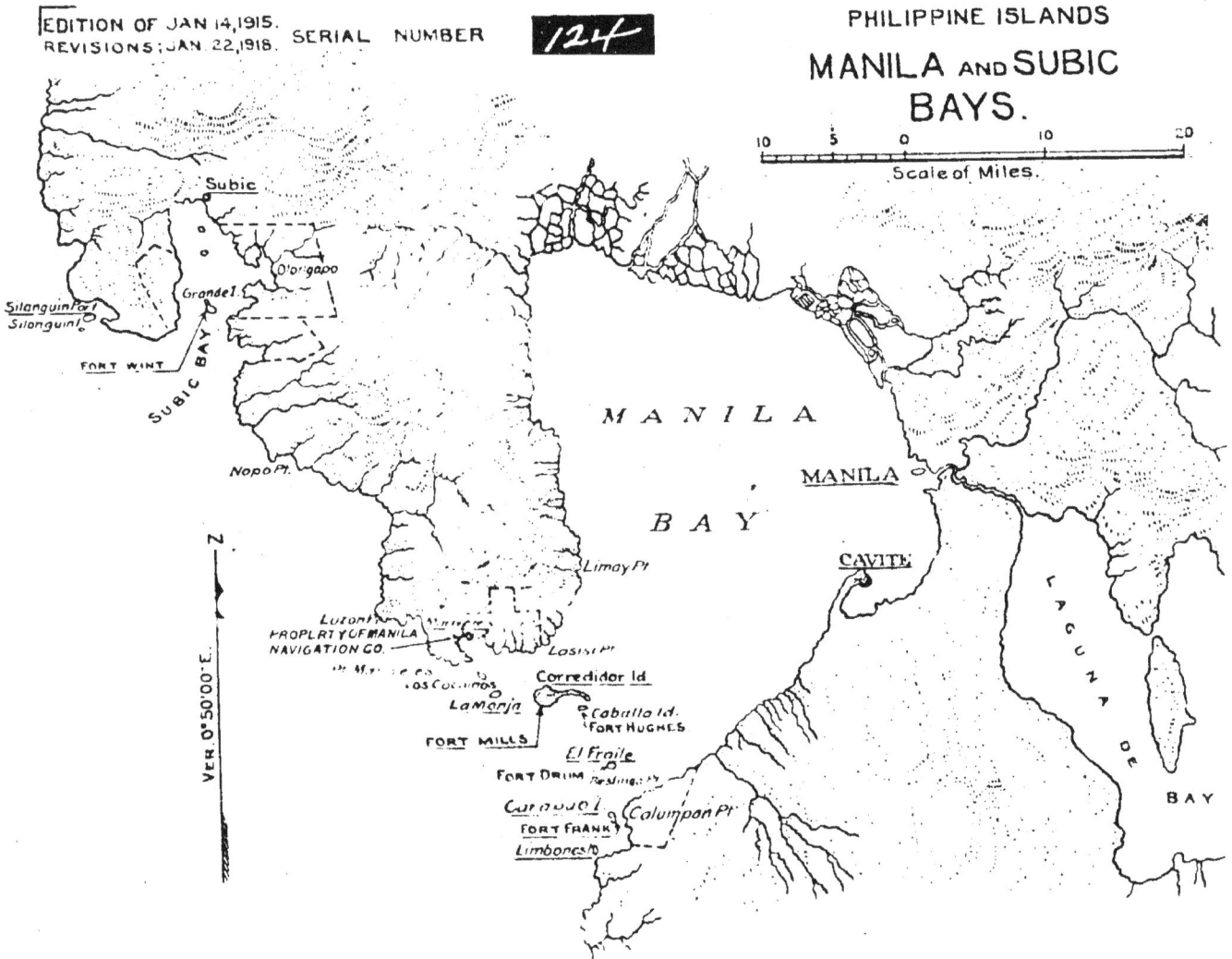

Williford, Glen. *Pacific Rampart A History of Corregidor and the Harbor Defenses of Manila and Subic Bays,* CDSG Press, McLean, VA 2020

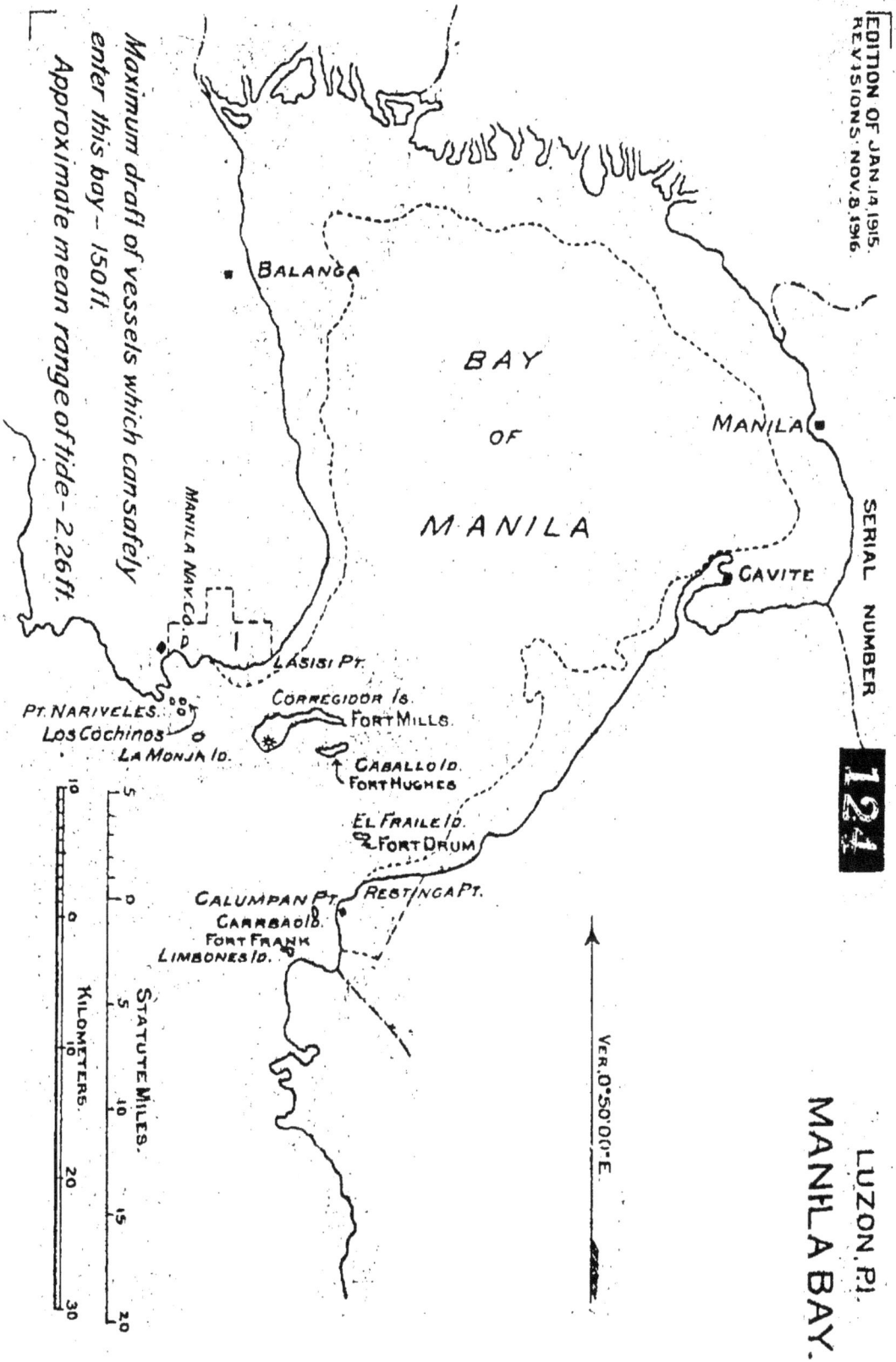

EDITION OF JAN.14,1915.
REVISIONS NOV.8.1916.

BALANGA

BAY

OF

MANILA

MANILA

CAVITE

MANILA NAV. CO.

LASISI PT.

CORREGIDOR IS.
FORT MILLS

CABALLO ID.
FORT HUGHES

EL FRAILE ID.
FORT DRUM

PT. NARIVELES.
LOS COCHINOS
LA MONJA ID.

CALUMPAN PT.
CARRBAO ID.
FORT FRANK
LIMBONES ID.

RESTINGA PT.

SERIAL NUMBER

124

Maximum draft of vessels which can safely
enter this bay – 150 ft.
Approximate mean range of tide – 2.26 ft.

10

5

0

STATUTE MILES.

5

10

15

20

0

0

5

10

KILOMETERS.

20

30

VER. 0°50'01" E.

LUZON, P.I.
MANILA BAY.

THE HARBOR DEFENSES OF MANILA BAY – PHILIPPINES

Fort Frank (1902-1949) is located on Carabao Island which is near the southern end of the entrance to Manila Bay about 0.5 miles off the coast of Maragondon in the province of Cavite. Built between 1908 and 1919. Named in General Orders 15 of 1909 for Brigadier General Royal T. Frank, Batteries here are Battery Greer (one 14-inch DC gun), Battery Crofton (one 14-inch DC gun), Battery Koehler (eight 12-inch mortars), Battery Hoyle (two 3-inch guns), Battery Ermita (1941-1942, four 3-inch AA guns), and Battery Frank (North) (1937-1942, four 155mm GPF guns on Panama mounts). There were also three 75mm field guns emplaced for beach defense in 1941. There were three or four searchlight stations, and twelve fire-control stations located on the island. Fort Frank was heavily engaged in the Japanese invasion of the Philippines. On 31 January 1942 the fort's mortar battery bombarded mainland positions in the Pico de Loro Hills where the Japanese had emplacing artillery. The 75 mm guns were also able to engage mainland targets. The Japanese began bombarding Fort Drum and Fort Frank on 6 February 1942. Fort Frank was surrendered, along with all other US forces in the Philippines, on 6 May 1942, after destruction procedures were executed on its guns to prevent their use by the enemy. The Japanese may have put Battery Greer back in action. In April 1945, during the American liberation of the Philippines, Fort Frank was heavily bombarded with 1,000 lb. bombs and napalm (among other ordnance) in preparation for re-capture. On 16 April 1945 the US Army landed on Fort Frank to find that the Japanese had successfully evacuated the island. The fort still lies abandoned as it was since 1945. Most of the guns survived the war intact, but all were salvaged for scrap steel in the 1970s. The Philippine government took control of the island in 1946 and occupied it for a short time. Scrappers have removed much of metal from the fort, even the rebar imbedded in concrete. Philippines Marines in Cavite have control of island so public access to the island fort is not officially allowed.

Fort Frank Gun Batteries

- GREER: Two 14-inch guns on disappearing carriages had been assigned to Carabao Island since the earliest of the Taft plans for Manila Bay. The problem was that the narrow, mostly-plateau island could not easily accommodate a structure with such a large footprint. In July 1908 a compromise was found in splitting the battery and building two separate, single-gun batteries. Greer was on the northern section of the island. It was situated to straddle the top of the island with a trunnion height of 127-feet. It fired to the northwest with a full command of the main, south channel. The plan was adapted from the left-flank single emplacement, with magazine at a lower level on the right. Several different features made it a unique emplacement. There was a second, basement-level magazine served by chain hoists for reserve ammunition. On the right flank, but still under cover of parapet, was the fort power plant. Work was done from December 1908 to late 1909 on the concrete. It was transferred on October 24, 1913 for $393,500.95. Late in the construction a lengthy cut-and-cover entrance tunnel was added to protect the way from Koehler to Greer, incorporating racks for bunks and access tunnels to fire control stations off the flanks. The battery was armed with one 14-inch gun Model 1907M1 #2 on Model 1907 disappearing carriage #3. The battery was named in General Orders No. 15 of January 28, 1909 for Colonel John F. Greer, an ordnance officer who died on September 19, 1907. The battery was modified in about 1917 to allow the carriage to elevate to 20-degrees verses the previous 15. In addition to modifying the carriage, concrete had to be removed in the gun pit in front of the carriage to accommodate this change. Because it could not fire to the south, the battery was little used during the 1942 siege, though of course it received incoming fire and was periodically damaged. At times parts were removed to keep Battery Crofton in service. Except for salvaging small parts, the Japanese did nothing with the emplacement

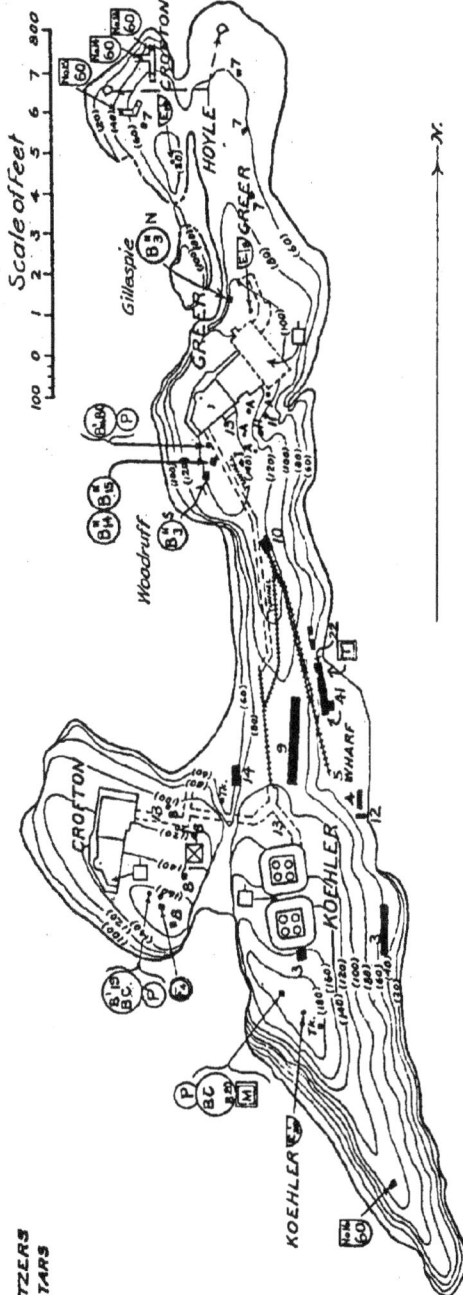

MANILA BAY, P.I.

FORT FRANK

CARABAO ISLAND

Scale of Feet

BATTERIES
KOEHLER.....8-12"M.
GREER......1-14"D is.
CROFTON...1-14" "
HOYLE=...... 2-3" Ped
A-Anti-aircraft gun 4-3"
Guns not mounted

ON ACTIVE STATUS

LEGEND   EDITION OF AUG. 12, 1921.
1. FLAG POLE.   REVISIONS: MAY 8, 1925.
2.   MAR. 21, 1929, APRIL 11, 1935
3. OFFICERS QRS.
4. N.C. OFFICERS-ENLIS. & CIVIL QRS.
5. DERRICK ENGINE.
6.
7. MAG. & EMPL. 7" SIEGE HOWITZERS
8. "   "   -   " MORTARS
9. POST EXCHANGE.
BARRACKS +- OFFICE
0. CABLE ENGINE HOUSE.
1. AMMUNITION SHELTER
    FOR A.A.C. GUNS.
2. BATHS AND TOILET.
3. BOMBPROOF.
4. PUMPING & DISTILLING PLANT.
1. ENGR. STORE HOUSE.
2. Q.M. STORE HOUSE.

SERIAL NUMBER

during the occupation. In the 1970s the gun and carriage were removed by illegal scrappers. The emplacement still exists on restricted Philippine military property. The battery is not open to the public and accessing the island fort is difficult.

- **CROFTON**: The second of the pair of single 14-inch gun emplacements for Carabao Island. This battery was located on the separate "knob" extending off the western side of the island, but still of sufficient height to allow placement with a trunnion height of 129-feet. It fired more to the west than Greer, covering all of Limbones Bay around to El Fraile Island. The emplacement itself was a fairly conventional left flank emplacement, with the adjacent, lower-level magazine and ammunition service via chain hoists. As part of the bomb-proofing project for Fort Frank, the battery also had an extensive entry tunnel with bunk racks and crew shelter rooms. Work was done from January 1909 to mid-1910. It was transferred on October 24, 1913 for a cost of $393,788.66. The access tunnel was transferred separately on the same date for an additional $62,507.91. The battery was armed with one 14-inch gun Model 1907M1 #4 on Model 1907 disappearing carriage #4. The battery was named in General Orders No. 15 of January 28, 1909 for Captain William C. Crofton, 1st U.S. Infantry who died in 1907. The battery was modified in about 1917 to allow the carriage to elevate to 20-degrees verses the previous 15. In addition to modifying the carriage, concrete had to be removed in the gun pit in front of the carriage to accommodate this change. Because its field of fire covered at least part of the Cavite mainland, the battery was more heavily used in the 1942 campaign than Battery Greer. It fired on and received fire from Japanese artillery batteries on the southern shore. At times the battery was kept in action by borrowing parts from Battery Greer. It was still serviceable at the time of surrender and was disabled by its crew prior to turnover. Except for salvaging small parts, the Japanese did nothing with the emplacement during the occupation. In the 1970s the gun and carriage were removed by illegal scrappers. The emplacement still exists on restricted Philippine military property. The battery is not open to the public and accessing the island fort is difficult.

- **KOEHLER**: The mortar battery emplaced at Fort Frank. The only location on the island which could accommodate a standard dual-pit mortar battery was on the topside to the south of the island, with a height of 114-feet. The plan adopted was for two four-mortar pits, surrounded on three sides with magazines and enclosed on the fourth side with support room in a closed parados. It had the spacings and room assignments of standard types but was modified to adapt to the needs of a fully-protected site on an exposed island. Work was done in January 1909 through 1910. Transfer was made on October 24, 1913 for a cost of $381,601.70. The battery had bomb-proof access galleries extending to the west towards Battery Crofton and to the north towards Battery Greer (but not fully connected to either). It was armed with eight late-model M1908 mortars and carriages (#8/#1, #21/#9, #9/#2, #23/#10, #10/#3, #24/#11, #11/#4, and #25/#12). It was named in General Orders No. 15 of January 28, 1909 for 1st Lieutenant Edgar F. Koehler, 9th U.S. Infantry who was killed in the Philippines in 1899. The battery was very active during the 1941-1942 campaign, especially with its ability to fire all-around towards the enemy artillery batteries on the south shore. On February 6, 1942 a muzzle burst killed one scout and disabled that mortar. Then on March 15th the battery was disabled for several days. Still, it was in operation at the time of surrender, though sabotaged prior to the Japanese arrival. Except for salvaging small parts, the Japanese did nothing with the emplacement during the occupation. In the 1970s the mortars and carriages were removed by illegal scrappers. The emplacement still exists on restricted Philippine military property. The battery is not open to the public and accessing the island fort is difficult.

Fort Frank 1930s (NARA)

Caraboa Island (Fort Frank) 1990s (Terry McGovern)

Battery Crofton Fort Frank (Terry McGovern)

Battery Greer Fort Frank (Terry McGovern)

Battery Koehler Fort Frank (Terry McGovern)

- **HOYLE**: A 1919 searchlight-protection battery for Fort Frank. Starting in April of 1916 the local command sought authority to relocate the 3-inch guns from Fort Wint to new positions. The U.S. Navy had withdrawn most of its workers and equipment from the Olongapo yard, and while the heavier guns were to remain, the U.S. Army sought to use the 3-inch guns more productively than in an abandoned post. The searchlights at Fort Mills and Fort Frank were vulnerable to quick-moving light attack craft and it was thought essential to provide protective batteries to ensure their continued wartime use. The searchlights at the northern point of Fort Frank (No. 13, 14, 15) were considered most vulnerable. The War Department was reluctant at first to authorize the removal of the Fort Wint guns, instead wishing to use them to arm local transports. However, this option fell through, and on October 3, 1918 authority was granted to use the six guns from Fort Wint (two were to stay behind) to arm new searchlight-protection batteries. At Frank the plan was to build two separate single gun platforms; one on each of the two low, northern lobes of the island. No. 1 was to be at 52-foot height and had no parapet, while No. 2 was a little lower at 35-foot trunnion height and had a short parapet wall around the loading platform. There were no magazines constructed, the nearby searchlight emplacements providing that space. Work was done in early 1919. Transfer was made on December 19, 1919 for a cost of $2,269. It was armed with two 3-inch Model 1903 guns and pedestals (#33/#13 and #29/#12). It was named in General Orders No. 13 of March 27, 1922 for Brigadier General Eli D. Hoyle, U.S. Army who died in 1921. During the 1930s one gun was transferred to Fort Mills and likely placed in storage. During the 1941-1942 campaign the Mills gun was shifted to Fort Drum where it was erected for a while on the fantail on January 12, 1942. The other gun disappeared during the Japanese occupation. The gun blocks still exist on restricted Philippine military property. The battery is not open to the public and accessing the island fort is difficult.

Fort Drum (1902-1945) is located on El Fraile Island which is situated at the mouth of Manila Bay in the Philippines, due south of Corregidor Island. A Spanish two-gun battery (4.7-inch Armstrong QF BLR) was located here in 1898. It fired the first shots of the Battle of Manila Bay. Fortifications on Manila Bay's El Fraile Island was an integral part of the American defensive plans. Early in the American occupation a mining casemate was built on the island. Starting in 1908 grander schemes were advanced. In June 1908 a general proposal was approved for two 14-inch, two 6-inch, and four 3-inch guns. In February 1909 a concept emerged for the use of heavily armored turrets on the island—as a way of both providing protection in an exposed location and in getting a maximum field of fire. Twelve and 14-inch inch turrets were alternately proposed. By mid-year plans had settled of what had to be done. The island would be leveled and then rebuilt of reinforced concrete through use of a cofferdam. A ship-shaped structure would result. On top of the "deck" facing the broad western channel would be two dual 14-inch army-designed and built turrets. They would be step mounted—the upper would fire over the lower. Two casemated 6-inch guns would be built into the structure on each of the north and south side to protect the mine fields and defend against small boats trying to get in close. Searchlights, a mining casemate, and accommodation for the crew would be included. It was anticipated that the garrison would be present only during war conditions, but at those times a total of 20 officers and 300 men would need quarters, mess, laundry, and medical facilities. Work was done reducing the rock foundation in 1909-1911, construction of the main structure accomplished in 1912-1914. It took several more years to install the armament and other required equipment. Transfer of the entire work was made on June 17, 1918 for an engineering construction cost (exclusive of armament and ammunition) of $2,391,694.67. Named in General Orders 245 in 1909 after Brigadier General Richard C. Drum, who served with distinction during the Mexican American War and the American Civil War, and died on October 15, 1909. The result was a 350-by-144-foot concrete structure, with concrete thickness varying from 25 to 36-feet thick. The fort had a caged tower with three searchlight and several fire control stations. Tower was removed during World War II. During the period of 1918 to 1941 the fort was not regularly garrisoned but normally occupied with a small maintenance detachment and hosting periodic firing exercises by crew. Additional armament was added through Battery Exeter (1934-1942, two 3-inch AA guns), and New Battery Hoyle (1941-1942). Of course, it was fully occupied and used during the 1941-1942 siege. The heavy guns fired on Bataan and received return fire and damage, but nothing serious enough to disable the armament. This nearly impregnable stronghold resembles a battleship, hence the popular name the "Concrete Battleship". Alone, the fort remained effective until the moment of Corregidor's surrender in May 1942. After sabotaging the armament, it was surrendered with the rest of the defenses on May 6, 1942. The Japanese only occupied the fort again in late 1944, assigning about 65 sailors from the sunken battleship Musashi to the island. None of the armament was usable. American forces retook the fort. The fort still lies abandoned as it was since 1945. No effort was made to restore the fort after the war. Shell damage and magazine explosions from the American recapture in 1945 are still very evident. Scrappers have removed much of interior of the fort. Philippines Marines in Cavite have control of island so public access to the island fort is not officially allowed.

Fort Drum Gun Batteries

- **MARSHALL**: The lower of the two 14-inch turrets emplaced on the western side of the deck of Fort Drum. Its traverse was limited by the rise in the deck to the rear, it could fire in a 230-degree arc. It was named in General Orders No. 13 of March 27, 1922 for Brigadier General William L. Marshall, who died in 1920. It was armed with a turret for two 14-inch Model 1909 army guns on M1909 turret carriages (#1/#1 and #2/#2). This turret was received at the fort on May 28, 1916 and the entire mechanism assembled by mid-1917. During World War II siege it was hit by a 240mm shell that opened a seam in the protective plate. The guns were manned until surrender on May 6, 1942, at which time they were disabled by loading with sandbags and being fired. The turret has been gutted by scrappers so the floors in the turret are missing. The battery is accessible, but the fort is closed to the public.

- **WILSON**: The upper of the two 14-inch turrets emplaced on the western side of Fort Drum. The turret could revolve a full 360-degrees but would obviously be obstructed by the mast or other structures on the deck. It was named in General Orders No. 13 of March 27, 1922 for Brigadier General John M. Wilson who died in 1919. It was armed with a turret for two 14-inch Model 1909 army guns on M1909 turret carriages (#3/#3 and #4/#4). This turret was received at the fort on May 28, 1916 and the entire mechanism assembled by mid-1917. Little damage was done during the first siege, but the guns were sabotaged by loading with filled sandbags and being fired just before being surrendered on May 6, 1942. The turret has been gutted by scrappers so the floors in the turret are missing. In recent years the 14-inch barrels have fallen down the turret well, so they are concealed with the turret. The battery is accessible, but the fort is closed to the public.

- **McCREA**: The northern pair of casemated 6-inch guns, arranged one above the other. The magazines for this battery were on the fort barracks level, just behind the casemates. It was armed with two 6-inch Model 1908M1 guns on Model 1910 barbette mounts behind shields (#1/#2 and #3/#1). These were received in 1914 and mounted soon after. It was named in General Orders No. 13 of March 27, 1922 for Brigadier General Tully B. McCrea who died in 1918. Being on the northern side of the fort, these were relatively unexposed during the campaign but sabotaged in May 1942 before the surrender. During the retaking in April 1945 the battery was seriously damaged when the explosion took place inside the fort, the armored deck cover being blown completely away. The

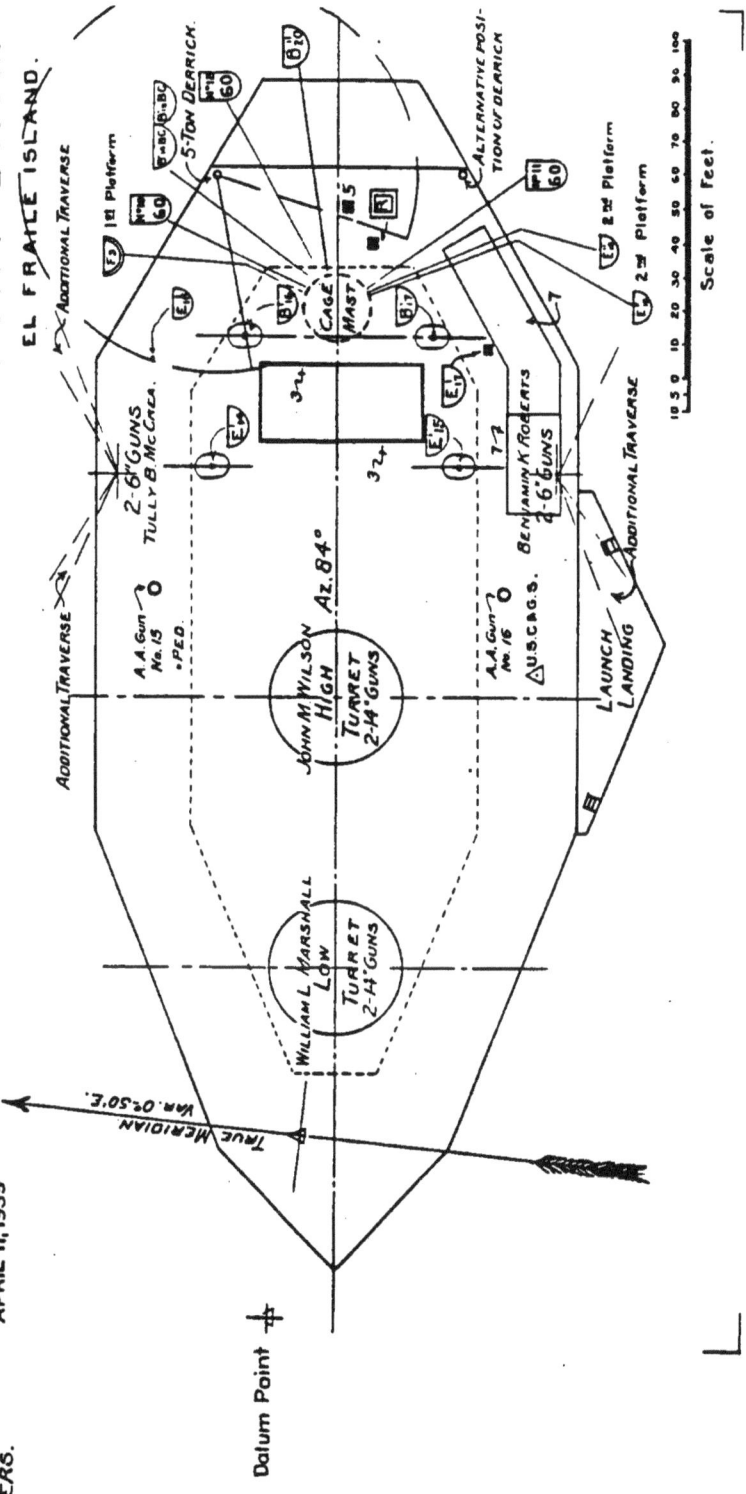

MANILA, LUZON, P.I.

FORT DRUM

EL FRAILE ISLAND.

BATTERIES
Wilson......2-14".
Marshall...2-14".
McCrea......2-6"P.
Roberts.....2-6"Pₜ
A.A.Guns   2-3"

Active Status

Scale of Feet.
10 5 0 10 20 30 40 50 60 70 80 90 100

SERIAL NUMBER

LEGEND
3.QRS. & OFFICE.
5.DERRICK HOIST.BLDG.
7.QUARTERS.

EDITION OF OCT. 24,1919.
REVISIONS: AUG.12,1921.
MAY 8,1925; MAR. 21,1929.
APRIL 11, 1935

TRUE MERIDIAN
VAR. 0° 50'E.

Datum Point

WILLIAM L. MARSHALL
LOW
TURRET
2-14" GUNS

JOHN M. WILSON
HIGH
Az. 84°
TURRET
2-14" GUNS

A.A.Gun
No. 15
PED.

2-6" GUNS
TULLY B. McCREA.

A.A.Gun
No. 16
U.S.C.& G.S.

BENJAMIN K. ROBERTS
2-6" GUNS

LAUNCH
LANDING

ADDITIONAL TRAVERSE

ADDITIONAL TRAVERSE

ADDITIONAL TRAVERSE

CAGE
MAST

5-TON DERRICK

ALTERNATIVE POSI-
TION OF DERRICK

1ˢᵗ Platform
2ⁿᵈ Platform
2ⁿᵈ Platform

Fort Drum 1930s (NARA)

Batteries Mashall and Wilson Fort Drum 2020 (Terry McGovern)

Fort Drum 2020 (Terry McGovern)

casemate has been gutted by both the magazine explosion and scrappers so the guns and floors in the casemate are missing. The battery is accessible, but the fort is closed to the public.

- **ROBERTS**: The southern pair of casemated 6-inch guns, arranged one above the other. The magazines for this battery were on the fort barracks level, just behind the casemates. It was armed with two 6-inch Model 1908M1 guns on Model 1910 barbette mounts behind shields (#4/#3 and #2/#4). These were received in 1914 and 1915 and were fully mounted in 1916. It was named in General Orders No. 13 of March 27, 1922 for Brigadier General Benjamin K. Roberts, U.S. Army who died in 1921. The battery was seriously damaged by 240mm howitzer shell hits during the March and April 1942 siege and further damaged on purpose right before the surrender on May 6, 1942. In April 1945 the exterior of the casemates and their armor were heavily damaged again by the fire from USS *Phoenix*. The casemate has been heavily damaged by shelling and by scrappers with one gun missing. The battery is accessible, but the fort is closed to the public.

- **HOYLE (NEW)**: Not an original battery for Fort Drum, this unit of a single 3-inch Model 1903 pedestal gun was relocated to the fort during the 1941 campaign. Concern about the dead space being the cage mast to the rear of the fort arose in January 1942. On January 12 a concrete bolt foundation was poured and mounted on the fort's "stern" for a gun to be relocated here from Battery Hoyle at Fort Frank. On the next day it went into action against small Japanese boats—reportedly one of which was hit. While it was not used extensively, it was damaged 35 minutes before the surrender by shell fragments hitting the breech. The gun was removed by the Japanese occupation forces. The base of pedestal still remains today. The battery is accessible but the fort is closed to the public.

Fort Hughes (1902-1946) is located on Caballo Island which is about a mile south of Corregidor Island near the entrance to Manila Bay in the Philippines. A Spanish three-gun battery (6-inch Armstrong BLR) was located here in 1898, behind a stone parapet on the eastern end of the island. It was not in action during the American Navy's attack of Manila in 1898. American fortifications were built between 1904 and 1919. Named in General Orders 254 of 1909 for Major General Robert Patterson Hughes, a veteran of the American Civil War, Spanish American War, and the Philippine–American War. Batteries are Battery Woodruff (one 14-inch DC gun), Battery Gillespie (one 14-inch DC gun), Battery Craighill (four 12-inch mortars), Battery Leach (two 6-inch DC guns), Battery Fuger (two 3-inch guns), Battery Williams (1942, two 155mm GPF guns on Panama mounts), Battery Idaho (1941-1942, four 3-inch AA guns), and Battery Hooker (1942, one 155mm GPF gun on Panama mount). The guns and mortars for Batteries Woodruff, Gillespie, and Craighill remain on site. There were two searchlight stations, and 15 fire-control stations located on the island. Battery Fuger's two guns were removed by the Japanese and emplaced in the Malinta Tunnels on Corregidor. Battery Idaho's four AA guns were removed by the Japanese and emplaced at Clark Field on the mainland. Battery Leach was destroyed by American bombardment in 1945. One gun tube still remains. One Japanese 120mm gun remains on the island. The island was handed over to the Philippine government in 1946 and is currently a Philippine Naval Ammunition Depot and access is restricted.

Fort Hughes Gun Batteries

- **GILLESPIE**: One of two single 14-inch disappearing guns emplaced as part of the original defense plans for Caballo Island. Plans were approved on June 14, 1911. The island was too small to take a dual emplacement, so Hughes' two 14-inch guns were split into single gun batteries. This unit was emplaced on the steep ridge top near the western extreme of the island, entailing considerable excavation and the construction of an access road and cable line along the inside of the island facing Corregidor. Despite these constraints, the right flank emplacement still was able to closely follow

MANILA BAY, P.I.

FORT HUGHES

CABALLO ISLAND

BATTERIES

CRAIGHILL......4-12"M.
GILLESPIE....1-14"Dis.
WOODRUFF....1-14" "
LEACH.........2-6" "
FUGER........2-3"Ped.

A-Anti-aircraft gun 4-3"
Guns not mounted.

SERIAL NUMBER

LEGEND   EDITION OF AUG. 12, 1921.
          REVISIONS: MAY 8, 1925.
          MAR. 21, 1929, APRIL 11, 1935

1. FLAG POLE.
2. SIGNAL LIGHT.
3. TRENCH AND MAGAZINE.
4.   "   MACH.GUN EMPL.
5. SIEGE ST. NOS. 1, 2, & 3.
6. CONCRETE TANKS.
   (FRESH AND SALT WATER)
7. OFFICERS QUARTERS.
8. ANTENNA POLES
9. BARRACKS (ON TOP OF Q.M.STRUCTURES-27)
10. CABLE ENGINE.
11. AMMUNITION SHELTER.
    FOR A.C. GUNS.

2.
20. OLD LIGHT HOUSE BLDGS.
27. Q.M. STRUCTURES, PAINT & REPAIR SHOPS.
3. SITES FOR T' HOWITZERS

Scale of Feet

ACTIVE STATUS

type plans. It was two-story in layout, with lower magazines passing shells to the loading level using hoists. The battery was also designed to use the latest, Model 1910 14-inch guns of 40-caliber length. It fired mostly to the south. Trunnion elevation was over 250-feet. Work was generally complete by mid-1913, transfer was made on December 31, 1914 for a cost of $297,281.90. It was armed with one 14-inch Model 1910M1 gun #15 on Model 1907M1 disappearing carriage #20. The gun tube remained throughout the battery's service life, but the carriage was altered for increased elevation and range soon after installation in 1919-1920. It was named in General Orders No. 63 of September 2, 1914 for Major General George L. Gillespie, U.S. Army who died in 1913. During the 1941-42 campaign the battery was little used. Its field of fire to the south severely limited its use for counterbattery work against Bataan, and the Japanese Navy did not try to penetrate the bay during the campaign. It was relatively undamaged at the surrender, though rendered inoperable by its crew. Surprisingly intact, with gun and carriage are in poor condition and partly encased in concrete, the emplacement still exists at the Philippine Navy base at Caballo Island. The battery and fort are closed to the public. Special permission from the Philippine Navy is required to visit the island.

- **WOODRUFF**: The second single 14-inch disappearing gun emplaced at Fort Hughes on Caballo Island. Plans were approved simultaneously with Battery Gillespie on June 14, 1911. It was emplaced mid-island, tucked into the base of the massive hogback that occupied the western portion of the small island. The battery was at just a crest elevation of 34-feet and fired to the south across the expanse of Manila Bay. There was enough room at this location to adopt a modified plan using adjacent-level magazines, avoiding the use of ammunition hoists. There was, however, a lower storage level, with ramps leading to the loading platform to aid the supply of ammunition to the magazines in peace time. Work was done in 1911-1912. Transfer came on December 31, 1914 for a cost of $213,737.20. It was armed with a single 14-inch gun Model 1910 #8 on Model 1907M1 disappearing carriage #17. The carriage was altered for increased elevation and range soon after installation, in 1919-1920. The battery was named in General Orders No. 63 of September 2, 1914 for Brigadier General Carl Augustus Woodruff, who died in 1913. For a considerable period in the 1920s and 1930s the battery (and fort) was only irregularly garrisoned by caretaker detachments. Woodruff saw relatively little action during the 1941-1942 campaign, it was only sparingly used in counterbattery fire. It was lightly damaged by Japanese action but suffered more heavily during the ground fighting in 1945 when American forces retook the island. The emplacement still exists, with its heavily rusted gun and carriage at the battery at the Philippine Navy base on Caballo Island. The battery and fort are closed to the public. Special permission from the Philippine Navy is required to visit the island.

- **CRAIGHILL**: The four-gun mortar battery emplaced at Fort Hughes. It was located on shelf part way up the inside passage along the island's hogback. Engineers were directed to prepare plans on September 27, 1913, who complied with a site plan on December 11, 1913. Considerable changes were necessary to fit a standard plan of two pits with two mortars each into the available space. One pit was located in front of the other, the rear one being ten-foot lower in floor elevation. The pits were entirely surrounded by parapets. A gallery along the right side continued a covered road up the hogback slope and also contained the cable-pull for heavily loaded carts. Extra rooms were fit into the flanks to act as bombproofs for crew shelter. Magazines were built with a capacity of 500 rounds per mortar. Work was approved and conducted in 1914-1915. Transfer was not made until June 30, 1919 for a cost of $298,028.49. It was armed with Model 1912 mortars on Model 1896MIII carriages (#40/#41, #39/#40, #41/#38 and #38/#39). The battery was named in General Orders No. 15 of April 25, 1916 for Brigadier General William P. Craighill, former Chief of Engineers who died in 1909. The battery was mostly in caretaker status in the 1920s and 1930s. During the 1941-1942 campaign the mortars were extensively used by the Americans for counterbattery work. While it received considerable attention from Japanese fire (including dismounting of one mortar by a 240mm hit on May 6, 1942) the battery suffered relatively little damage. However, during the retaking of the island in early April 1945 fuel oil was pumped into the structure by the Americans and detonated. The battery was heavily damaged by the ensuing fire. In damaged condition, but with the armament still in place, the emplacement still exists at the Philippine Navy base on Caballo Island. The battery and fort are closed to the public. Special permission from the Philippine Navy is required to visit the island.

- **LEACH**: A Taft Program battery for two 6-inch disappearing guns emplaced at Fort Hughes. Orders to proceed with the construction were given on April 26, 1912. The location finally settled on was a position to the north of a shallow lake on the southern side of the island, the battery firing to the south. It followed conventional mimeograph plans for late disappearing batteries, with same-level ammunition service. Work was done from May 1912 to late 1913, for transfer made on December 31, 1914 at a construction cost of $70,137.33. It was named in General Orders No. 63 of September 2, 1914 for Colonel Smith Stallard Leach, Corp of Engineers who died in 1909. It was armed with two 6-inch guns Model 1908 on Model 1905MII disappearing carriages (#5/#20 and #6/#21). This armament was never changed. The only action the battery was involved in during the war was the firing on Japanese motor launches on March 8th and 14th, 1942. Not damaged during the first campaign, the magazine was blown up by a U.S. aerial bomb in early 1945, wrecking the battery. Only partial remains of the emplacement and some wrecked gun and carriage parts remain at the Philippine Navy base on Caballo Island. The battery and fort are closed to the public. Special permission from the Philippine Navy is required to visit the island.

- **FUGER**: A dual 3-inch pedestal gun battery emplaced at Hughes as part of the Land Defense project for the harbor defenses. Four new emplacements for standard Model 1903 coast defense pedestal guns were recommended and approved in the 1910 Land Defense Program. After some discussion two new dual batteries were approved, two at Battery Maxwell Keys at Fort Mills, and the two here. They were emplaced on the northeast side of the island, on a slope overlooking the sandy eastern flatlands of the island, firing to the east. Their primary mission was to protect against landings on the eastern end of either Caballo or Corregidor Island. The emplacement was a greatly simplified battery to keep cost down. It was just two simple platforms and a lower-level traverse magazine between them. The plan was submitted in mid-1911. Work was done during 1912. Transfer was made on December 31, 1914 for a cost of $8,262.47. The battery was named on General Orders No. 63 of September 2, 1914 for Lieutenant Colonel Frederick Fuger, U.S. Army who died in 1913. It was armed with two 3-inch Model 1903 guns and pedestals (#99/#98 and #100/#99). Because of its position, the battery did not fire much during the campaign in 1942. There was only one major action, firing on motor launches during the night of March 14, 1942. Basically, intact after the 1942 campaign, the Japanese removed the armament during their occupation and re-located it elsewhere. There is no record of fighting here in the 1945 retaking. At some point the emplacement was filled in, but the outline of the platforms can still be identified at the Philippine Navy base on Caballo Island. The battery and fort are closed to the public. Special permission from the Philippine Navy is required to visit the island.

Fort Hughes 1930s (NARA)

Caballo Island (Fort Hughes 2020 (Terry McGovern)

Battery Woodruff Fort Hughes (Terry McGovern)

Battery Gillespie Fort Hughes (Terry McGovern)

Saysain Point Gun Battery

- **Saysain Point Battery:** General Douglas MacArthur had requested armament to emplace in the Philippines at various key straits as part of an "Inland Seas Project." His proposals were approved in early 1941, and a quantity of 155mm GPF guns and seven 8-inch Model 1888 guns with their Model 1918 top carriages were approved for shipment in mid-April 1941. This armament was without units or troops—those would be organized by the Philippine Army in the islands. The permanent emplacements were just starting construction when the war began. Five of the 8-inch guns were intercepted in transit from Fort McKinley, but two were received in the Bataan peninsula in January 1942. A battery for two guns was planned on the western shore of Bataan in order to interdict any enemy sea traffic in the local bays and passage down the coast. Two concrete platforms were constructed by a section of the 803rd Engineer Battalion (Aviation). These were entirely open emplacements, without even a berm of other protection. A small magazine and wooden CRF station were included. In the second week of January the guns arrived. However, one emplacement was found to have a faulty arrangement of foundation bolts, its gun carriage would not fit and eventually it was transferred to Corregidor Island to go to an emergency battery there. The other gun was put into service. Installed by February 15th, with a crew of mostly Philippine Army coast artillery students it fired on enemy ship targets a number of times, including February 15th, 25th, March 2nd and 3rd, and April 2nd and 3rd. While this area was not occupied by the enemy prior to the Bataan surrender, the crew was given orders to destroy the gun (which they did) and surrender to the Japanese at the end of the campaign. There are no recognizable remains at the site, which is now private property.

Fort Mills (1902-1946) is located on Corregidor Island at the entrance to Manila Bay in the Philippines. Corregidor was nicknamed "The Rock" and served as the headquarters for the harbor defenses of Manlia Bay. The island is divided geographically into "Topside", "Middleside", "Bottonside", and the "Tail". A Spanish bastioned three-gun battery (aka Spanish Fort) (8-inch Armstrong MLR) was located at Punta Talisay (Battery Point) in 1898. A second battery, Bateria Bocana (one 6-inch gun), was located just west of Malinta Hill on the southern shore. Several field guns were located near the hospital and the wharf. The Spanish garrison was quartered at San Jose on "Bottomside". The Lighthouse was used by the Spanish as a signal station and command post. Most of the island's American fortifications were built between 1904 and 1919. It was named in General Orders 77 in 1908 for Brigadier General Samuel Meyers Mills Jr., Chief of Artillery 1905–1906, but was not fully garrisoned until 1918. Camp Avery (1908) was located on the plateau overlooking "Bottomside" (the "Stockade Level"), near the old Spanish "fort". The camp was used by the Philippine Scouts to guard the convict labor used to build the military structures. The camp's buildings survived until destroyed by the Japanese in 1942. Kindley Field was located on the "Tail". The "Mile-Long Barracks" and the Parade Ground were on "Topside". Additional barracks were on "Middleside". The wharves and docks, and the mining facilities were located on "Bottomside". The civilian community lived at San Jose on "Bottomside". A massive tunnel complex was built into Malinta Hill from 1931-1934. The mine casemate was located in James Ravine. Manila Bay and Subic Bay had U.S. Army-operated minefields as well as naval mines. In Manila Bay, two controlled minefields were placed, one extending west from Corregidor to La Monja Island, and the other extending north from Corregidor to the Bataan Peninsula east of Mariveles Bay. Both of these were operated from Corregidor. A buried Army Radio Station (1933) was located near Battery Geary. A buried Navy Radio Intercept Station (1936) was located on Monkey Point. Batteries on the island include Battery Hearn (one 12-inch LRBC gun), Battery Smith (one 12-inch LRBC gun), Battery Way (four 12-inch mortars), Battery Geary (eight 12-inch mortars), Battery Cheney (two 12-inch DC guns,), Battery Wheeler (two 12-inch DC guns), Battery Crockett (two 12-inch DC guns),

Battery Grubbs (two 10-inch DC guns), Battery RJ-43 (one 8-inch railway gun), Battery Morrison (two 6-inch DC guns), Battery Ramsay (three 6-inch gun), Battery James (four 3-inch guns), Battery Keyes (two 3-inch guns), Battery Cushing (two 3-inch guns), Battery Hanna (two 3-inch guns), and 155mm batteries Battery Martin (1937, two guns on blocks), Battery Hamilton (South) (1941-1942, three guns PM), Battery Kysor (North) (1938-1942, two guns PM), Battery Rock Point (1937-1942, two guns PM), Battery Sunset (1937-1942, four guns PM), Battery Stockade (1941-1942, two guns field), Battery Monja (1936-1942, two guns, one casemated), Battery Concepcion (1937-1942, three guns PM), Battery Levagood (1941-1942, three guns PM), and Battery Ordnance Point (1941-1942, four guns PM). Three 75mm field guns were casemated on Malinta Hill. Seventeen more were located along the shoreline for beach defense, as well as a number of 37mm guns. At Fort Mills from the 1930s until early 1942 several 155mm fixed (Panama Mount) and field emplacements were prepared for 155mm GPF guns. These included locally named batteries of Concepcion, Levagood, Lubang, Martin, Monja, North (also known as Kysor), Ordnance Point, Rock Point, Stockade, Sunset, South (also known as Hamilton), and West. During the campaign this type of gun was organized into roving batteries (usually of just one or two guns) and most often referred to with the last name of their commanding officer: Byrne, Dawes, Farris, Fulmer, Gulick, Lehr, Rose, and Wright. There were eight 60-inch searchlight positions on the island, plus several 36-inch locations. There were numerous fire control stations (about 40) located throughout the island. An additional location was established across the channel on the Bataan coast at Los Cochinos. In the 1920s nine battery emplacements were constructed for 3-inch Model 1917 AA guns, but only two such guns were ever received, and mounted for a while pre-war at Battery Morrison Hill. Fort Mills did, however, receive many mobile 3-inch AA guns, some of which were emplaced in prepared field positions with protected batteries. The batteries carried their city name renditions from the alphabetical designations of the regiment's batteries: Boston, Chicago, Denver, Flint, Globe, Hartford. During the war the Japanese attempted, apparently with some success to restore some of the American guns to service. Several 3-inch seacoast guns were relocated into tunnel positions on Malinta Hill. Also they emplaced some of their own guns on the island, including at least one battery of modern 120mm dual-purpose AA guns with its own protected magazine. During World War II, Fort Mills was the primary location of the Battle of Corregidor in the Japanese invasion of the Philippines in 1941–42, and of the recapture of Corregidor in February 1945 by the Americans. The US government turned over Corregidor and Fort Mills to Philippines government in 1948 and island become a memorial to the battles in 1942 and 1945. Permanent memorials for the Japanese, American, and Filipino soldiers have been constructed over the years. For many years the Corregidor Foundation operated the island as tourist attraction for the Department of Tourism, while the Ministry of Defense owned the island. Daily ferry service was available from Manila (about 35 miles away) allowing for one day tours as well as several overnight lodgings. Recently, the ferry service has been discontinued and the overnight lodging closed.

Fort Mills Gun Batteries

- **WHEELER:** One of the early 12-inch disappearing batteries emplaced as part of the heavy armament on the top of the western bluff on Corregidor Island. Funds were made available in mid-1904, though survey work and considerable discussion for sites followed. The battery was sited southwest of the future parade ground, bearing on the main channel to the southwest. It was on a high ridge point, trunnion heights being at 545-feet. The battery is of typical early Taft design, with two spacious platforms for the guns, each a flank emplacement to give maximum coverage around the curvature of the island. Lower-level magazines in the central traverse were served by modern chain hoists. Work was begun in September 1904; the majority of the concrete construction being done within a year. It was finally transferred on February 2, 1911 for a cost of $290,046.75. While not initially being part of the plan, during construction a large battery commander's station and plot-

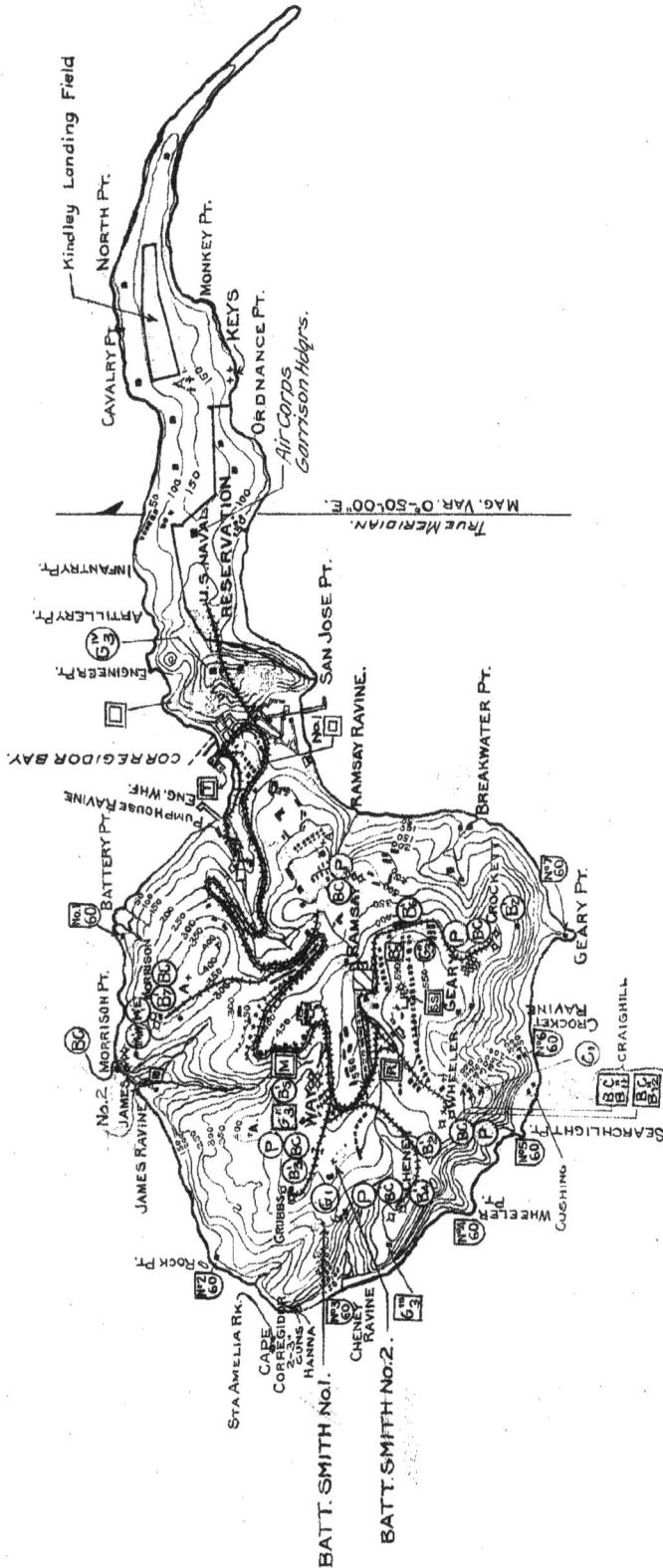

DEFENSES OF MANILA BAY

FORT MILLS

CORREGIDOR ISLAND, P.I.

GENERAL MAP.

BATTERIES

GEARY	8-12" M.
WAY	4-12" M.
WHEELER	2-12" DIS.
CROCKETT	2-12" "
CHENEY	2-12" "
GRUBBS	2-10" "
RAMSAY	3-6" "
MORRISON	2-6" "
JAMES	4-3" "
SMITH #1	1-12" BAR.
SMITH #2	1-12" BAR.
KEYS	2-3"
HANNA	2-3"
CUSHING	2-3"

A. Antiaircraft Gun Blocks

Razor Is.

EAST PT.

HOOKER PT.

EDITION OF JAN. 14, 1915.
REVISIONS: DEC. 7, 1915;
NOV. 8, 1916; JAN. 22, 1918.
OCT. 24, 1919; MAY 8, 1925.
MAR. 21, 1929. APRIL 11, 1935

SERIAL NUMBER

Active Status.

5000 FT.

2500

0

2500

Kindley Landing Field

North Pt.

Cavalry Pt.

Monkey Pt.

KEYS

Ordnance Pt.

Air Corps
Garrison Hdqrs.

True Meridian,
MAG. VAR. 0°-50'-00" E.

U.S. NAVAL RESERVATION

INFANTRY PT.

ARTILLERY PT.

ENGINEER PT.

San Jose Pt.

CORREGIDOR BAY.

ENG. WHF.

PUMPHOUSE RAVINE

Battery Pt.

RAMSAY RAVINE

Breakwater Pt.

Geary Pt.

Morrison Pt.

James Ravine

Rock Pt.

Cheney Ravine

Wheeler Pt.

SEARCHLIGHT PT.

Cushing

Crockett Ravine

CRAIGHILL

CAPE CORREGIDOR
2-3" HANNA 2 GUNS

STA. AMELIA RK.

BATT. SMITH No. 1.

BATT. SMITH No. 2.

BATTERIES

WAY	4-12"M.
GEARY	8-12"M.
WHEELER	2-12"Dis.
CHENEY	2-12" "
CROCKETT	2-12" "
FRANK G. SMITH	2-12"Bar.
GRUBBS	2-10"Dis.
MORRISON	2-6" "
RAMSAY	3-6" "
KEYS	2-3"Ped.
JAMES	4-3" "
CUSHING	2-3 "
HANNA	2-3 "

A-A. Gun block ID-3"
on which 2 have
A.A. guns mounted

RAZOR Is.
HOOKER PT.

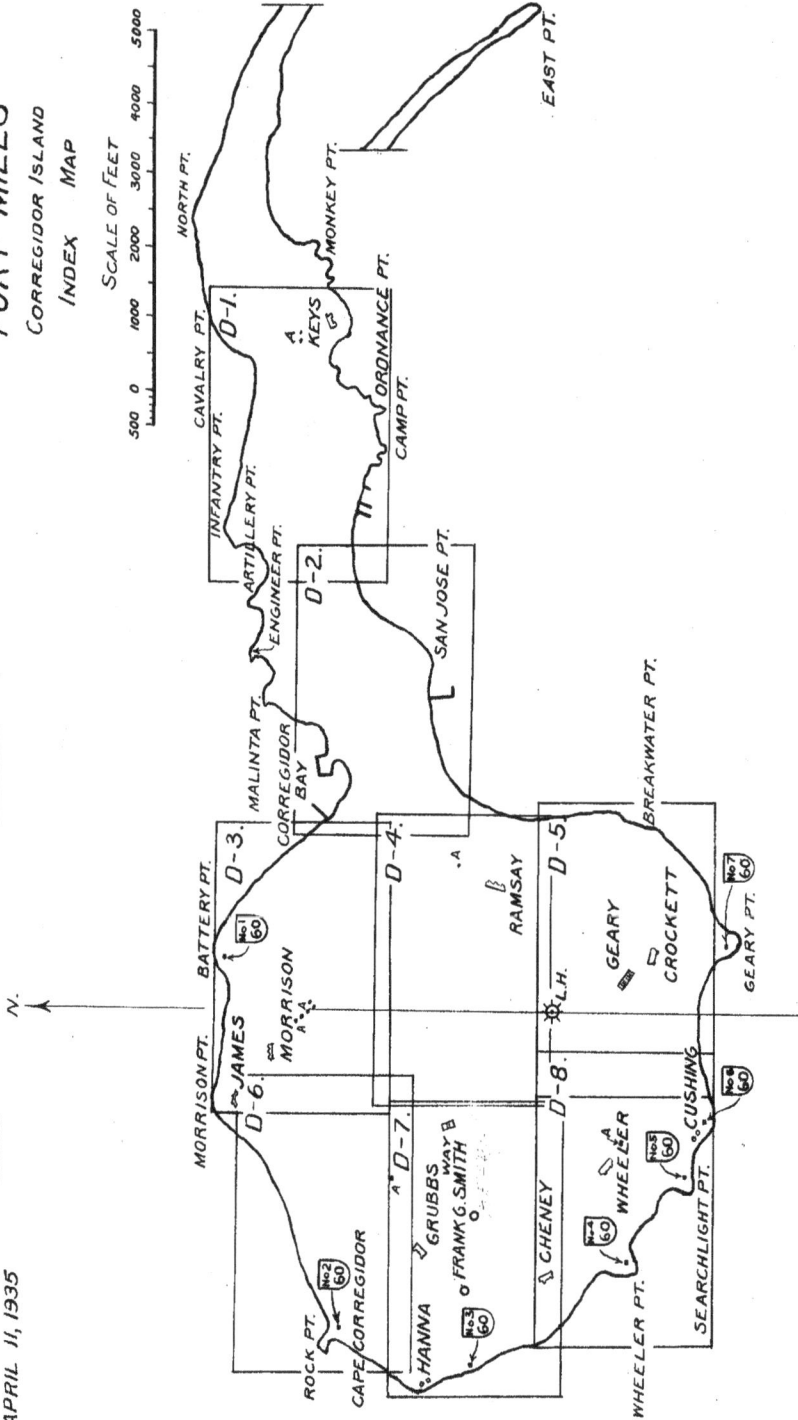

ACTIVE STATUS.

MANILA BAY, P.I.
FORT MILLS
CORREGIDOR ISLAND
INDEX MAP
SCALE OF FEET
500 0 1000 2000 3000 4000 5000

SERIAL NUMBER

N.

EDITION OF AUG. 12, 1921.
REVISIONS; MAR. 21, 1929.
APRIL 11, 1935.

NORTH PT.
CAVALRY PT.
INFANTRY PT.
ARTILLERY PT.
ENGINEER PT.
MALINTA PT.
CORREGIDOR BAY
BATTERY PT.
MORRISON PT.
MORRISON A.A.
JAMES
ROCK PT.
CAPE CORREGIDOR
HANNA
GRUBBS
WAY
FRANK G. SMITH
CHENEY
WHEELER
WHEELER PT.
SEARCHLIGHT PT.
CUSHING
CROCKETT
GEARY
GEARY PT.
BREAKWATER PT.
RAMSAY
L.H.
SAN JOSE PT.
CAMP PT.
ORDNANCE PT.
KEYS
MONKEY PT.
EAST PT.

D-1.
D-2.
D-3.
D-4.
D-5.
D-6.
D-7.
D-8.

No.1 60
No.2 60
No.3 60
No.4 60
No.5 60
No.6 60
No.7 60

MANILA BAY P.I.

FORT MILLS D-I.

CORREGIDOR ISLAND

Scale of Feet

100 0 1 2 3 4 5 6 700

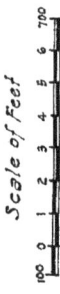

EDITION OF AUG. 12, 1921
REVISIONS: MAY 8, 1925;
MARCH 21, 1929;
APRIL 1935

SERIAL NUMBER

INFANTRY POINT

2-75 mm. Gun Empl.

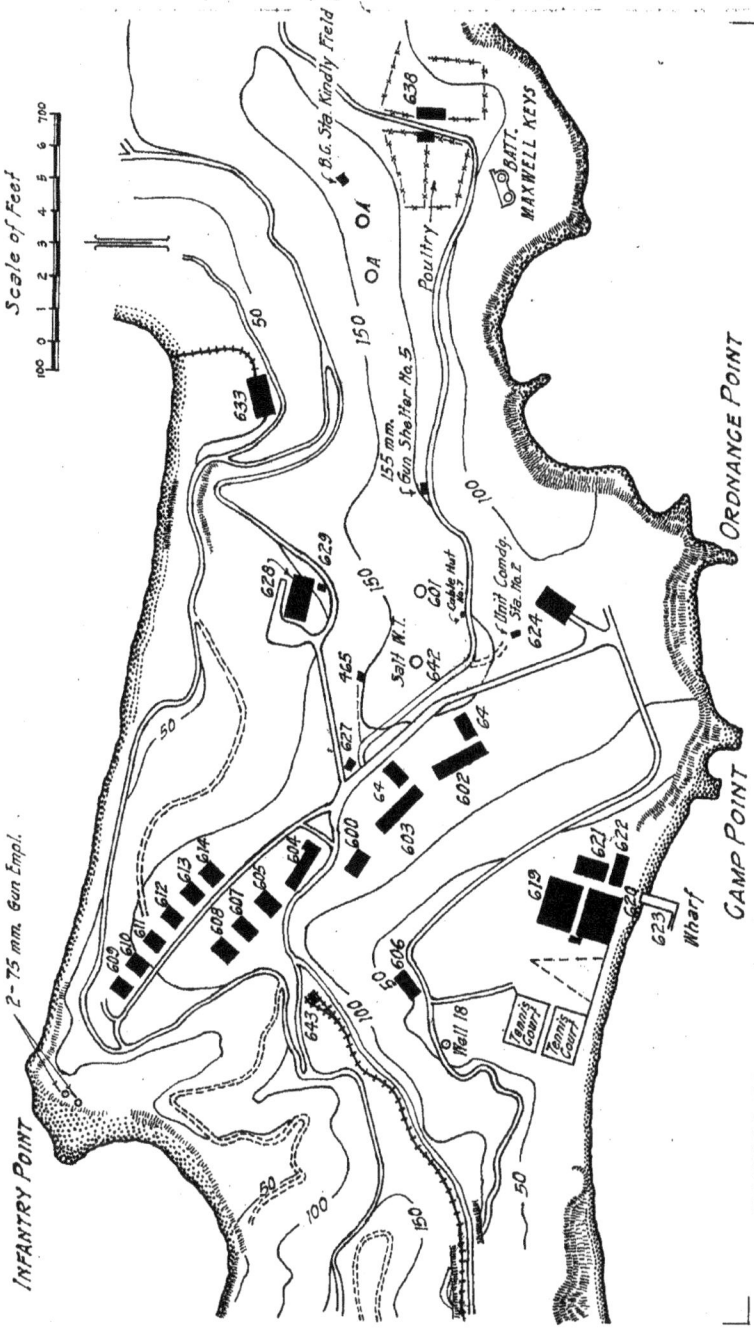

B.G. Sta. Kindly Field

Poultry

155 mm.
Gun Shelter No. 5

Salt W.T.

Cable Hut

Unit Comdg.
Sta. No. 2

Unit Comdg.
Sta. No. 2

Tennis Court

Tennis Court

Well 18

Wharf

ORDNANCE POINT

CAMP POINT

MAXWELL KEYS

B. ATT.
MAXWELL KEYS

BATTERIES

MAXWELL KEYS 2-3" PED.

A-A. a. Gun Block 2-3"

75 mm Gun Empl.-2

Active

LEGEND

64	KITCHEN.
465	OFFICERS QRS.
600	ADMINISTRATION BLDG., 92 od C.A. (P.S.).
601	FRESH WATER TANK No. 6.
602	BARRACKS.
603	"
604	N.G. OFFICERS QRS.
605	OFFICERS QRS.
606	DOPE SHED
607	OFFICERS QRS.
608	"
609	"
610	"
611	"
612	"
613	"
614	"
619	GARAGE & REPAIR SHOP, 92 od C.A. (P.S.)
620	
621	TRUCK GARAGE, 92 od C.A. (P.S.)
622	PUMPHOUSE
623	WHARF
624	GARAGE, 92 od G.A. (P.S.)
627	FIRE APPARATUS.
628	WAREHOUSE
629	FRESH WATER PUMPING STA.
633	WAREHOUSE.
638	HOG FARM BLDG.
642	SALT WATER TANK.
643	CHECKER & GAR. STA.

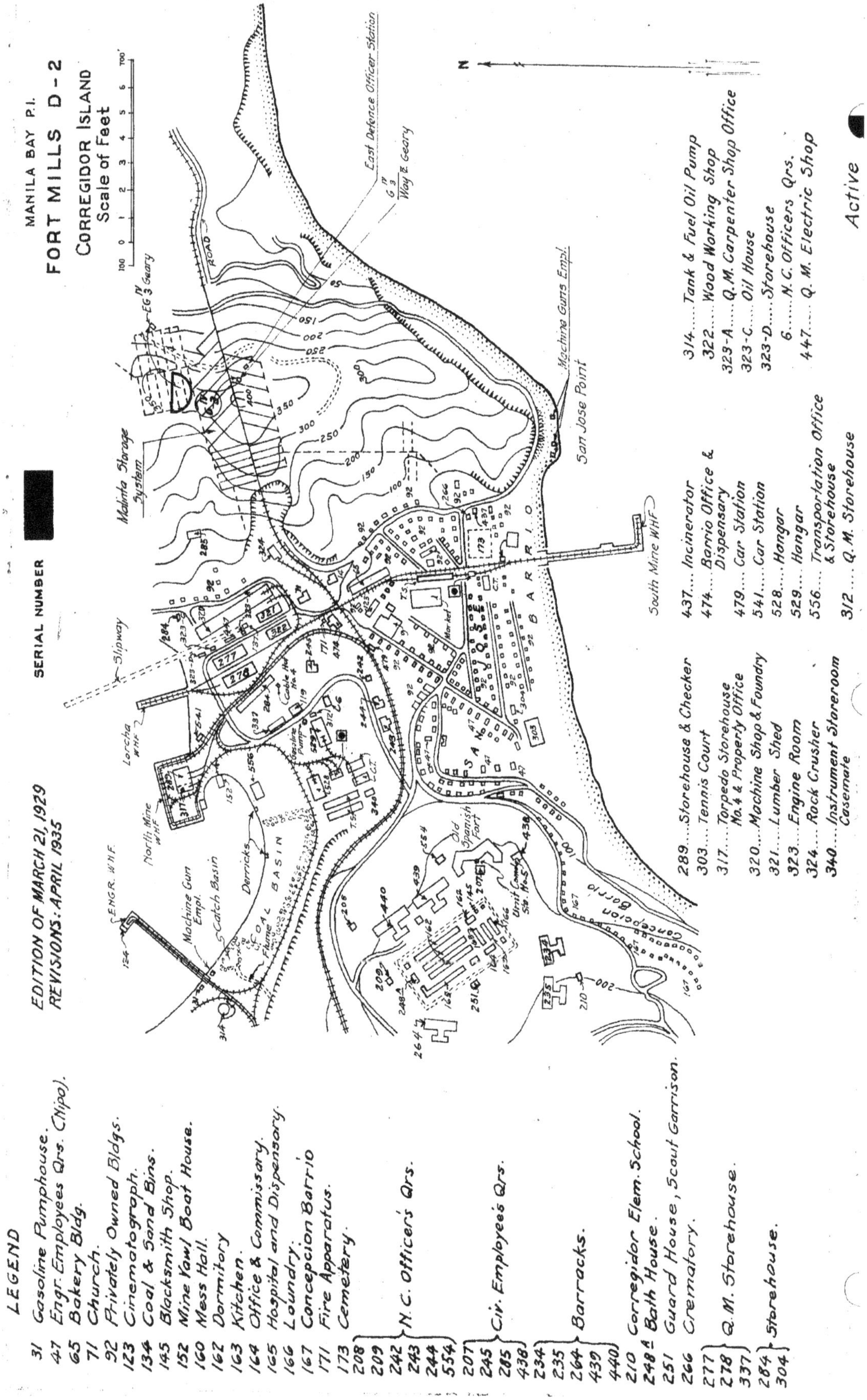

MANILA BAY P.I.

FORT MILLS D-2

CORREGIDOR ISLAND
Scale of Feet

EDITION OF MARCH 21, 1929
REVISIONS: APRIL 1935

SERIAL NUMBER

LEGEND

31   Gasoline Pumphouse.
47   Engr. Employees Qrs. (Nipa).
65   Bakery Bldg.
71   Church.
92   Privately Owned Bldgs.
123   Cinematograph.
134   Coal & Sand Bins.
145   Blacksmith Shop.
152   Mine Yawl Boat House.
160   Mess Hall.
162   Dormitory
163   Kitchen
164   Office & Commissary.
165   Hospital and Dispensary.
166   Laundry.
167   Concepcion Barrio
171   Fire Apparatus.
173   Cemetery.
208
209
242
243 } N. C. Officer's Qrs.
244
554
207
245
285 } Civ. Employees Qrs.
438
234
235
264 } Barracks.
439
440
210   Corregidor Elem. School.
248   Bath House.
251   Guard House, Scout Garrison.
266   Crematory.
277
278 } Q.M. Storehouse.
337
264 } Storehouse.
304

289   Storehouse & Checker
303   Tennis Court
317   Torpedo Storehouse
       No.4 & Property Office
320   Machine Shop & Foundry
321   Lumber Shed
323   Engine Room
324   Rock Crusher
340   Instrument Storeroom
       Casemate

314   Tank & Fuel Oil Pump
322   Wood Working Shop
323-A   Q.M. Carpenter Shop Office
323-C   Oil House
323-D   Storehouse
6   N.C. Officers Qrs.
447   Q.M. Electric Shop

437   Incinerator
474   Barrio Office &
       Dispensary
479   Car Station
541   Car Station
528   Hangar
529   Hangar
556   Transportation Office
       & Storehouse
312   Q.M. Storehouse

Active

East Defence Officer Station
Way E Geary
Malinta Storage System
Machine Guns Empl.
San Jose Point
South Mine WHF
Slipway
Locha WHF
North Mine WHF
Machine Gun Empl.
Scatch Basin
COAL BASIN
Derricks
Old Spanish Fort
Concepcion Barrio
BARRIO
N

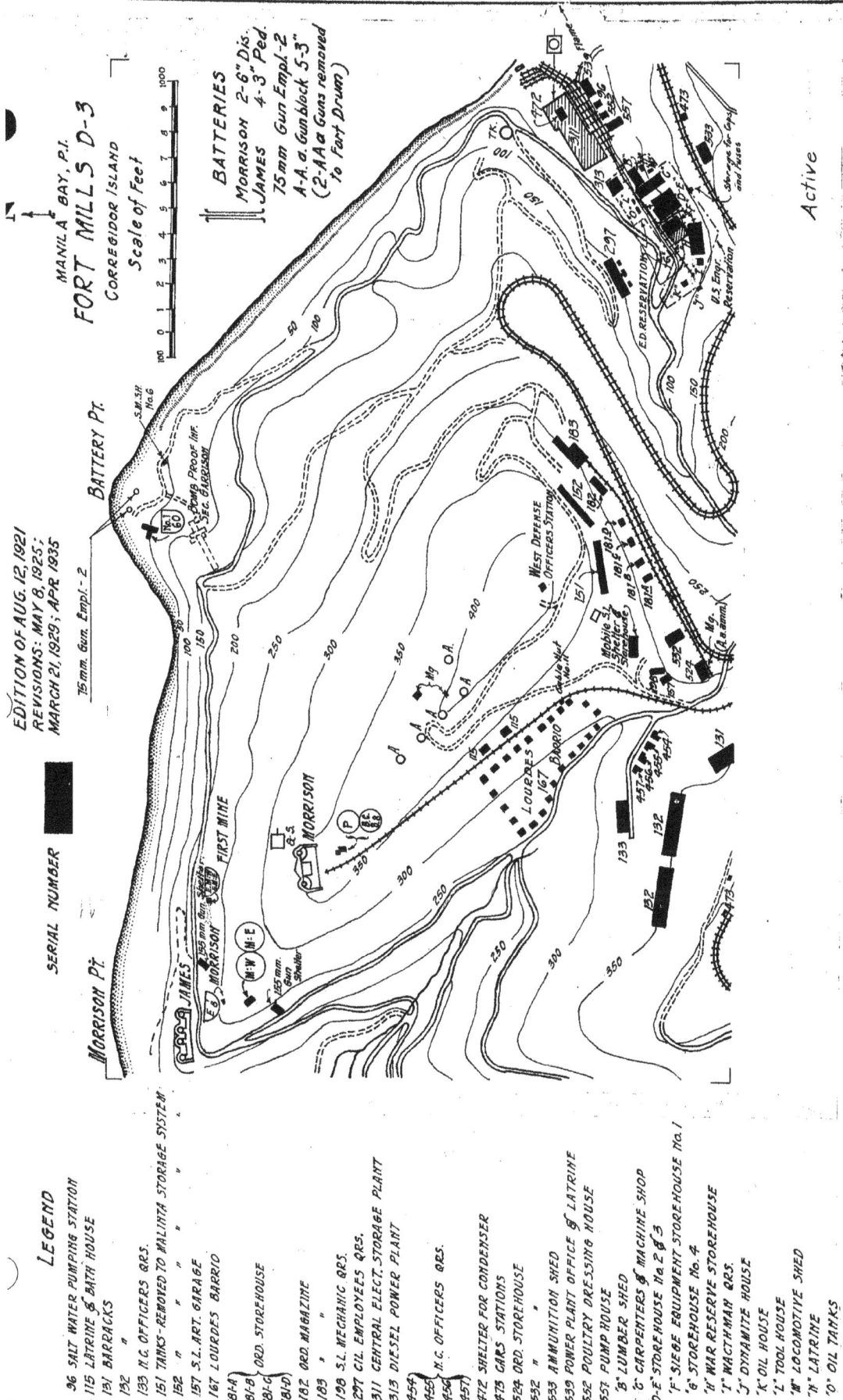

MANILA BAY, P.I.
FORT MILLS D-3
CORREGIDOR ISLAND
Scale of Feet

EDITION OF AUG. 12, 1921
REVISIONS: MAY 8, 1925;
MARCH 21, 1929; APR. 1935

SERIAL NUMBER

BATTERIES
Morrison 2-6" Dis.
James 4-3" Ped.
75mm Gun Empl-2
A-A. a Gunblock 5-3"
(2-AAa Guns removed
to Fort Drum)

Active

LEGEND

36 SALT WATER PUMPING STATION
115 LATRINE & BATH HOUSE
131 BARRACKS
132
133 N.C. OFFICERS QRS.
151 TANKS-REMOVED TO MALINTA STORAGE SYSTEM
152
157 S.L. ART. GARAGE
167 LOURDES BARRIO
181-A
181-B } ORD. STOREHOUSE
181-C
181-D
182 ORD. MAGAZINE
185
198 S.L. MECHANIC QRS.
297 OIL EMPLOYEES QRS.
311 CENTRAL ELECT. STORAGE PLANT
313 DIESEL POWER PLANT
454
455 } N.C. OFFICERS QRS.
456
457
472 SHELTER FOR CONDENSER
475 GAS STATIONS
524 ORD. STOREHOUSE
552
553 AMMUNITION SHED
539 POWER PLANT OFFICE & LATRINE
552 POULTRY DRESSING HOUSE
557 PUMP HOUSE
"B" LUMBER SHED
"D-E" STOREHOUSE No.2 & 3
"F" SIEGE EQUIPMENT STOREHOUSE No.1
"G" STOREHOUSE No. 4
"H" WAR RESERVE STOREHOUSE
"I" WACHMAN QRS.
"J" DYNAMITE HOUSE
"K" OIL HOUSE
"L" TOOL HOUSE
"M" LOCOMOTIVE SHED
"N" LATRINE
"O" OIL TANKS
"C" CARPENTERS & MACHINE SHOP

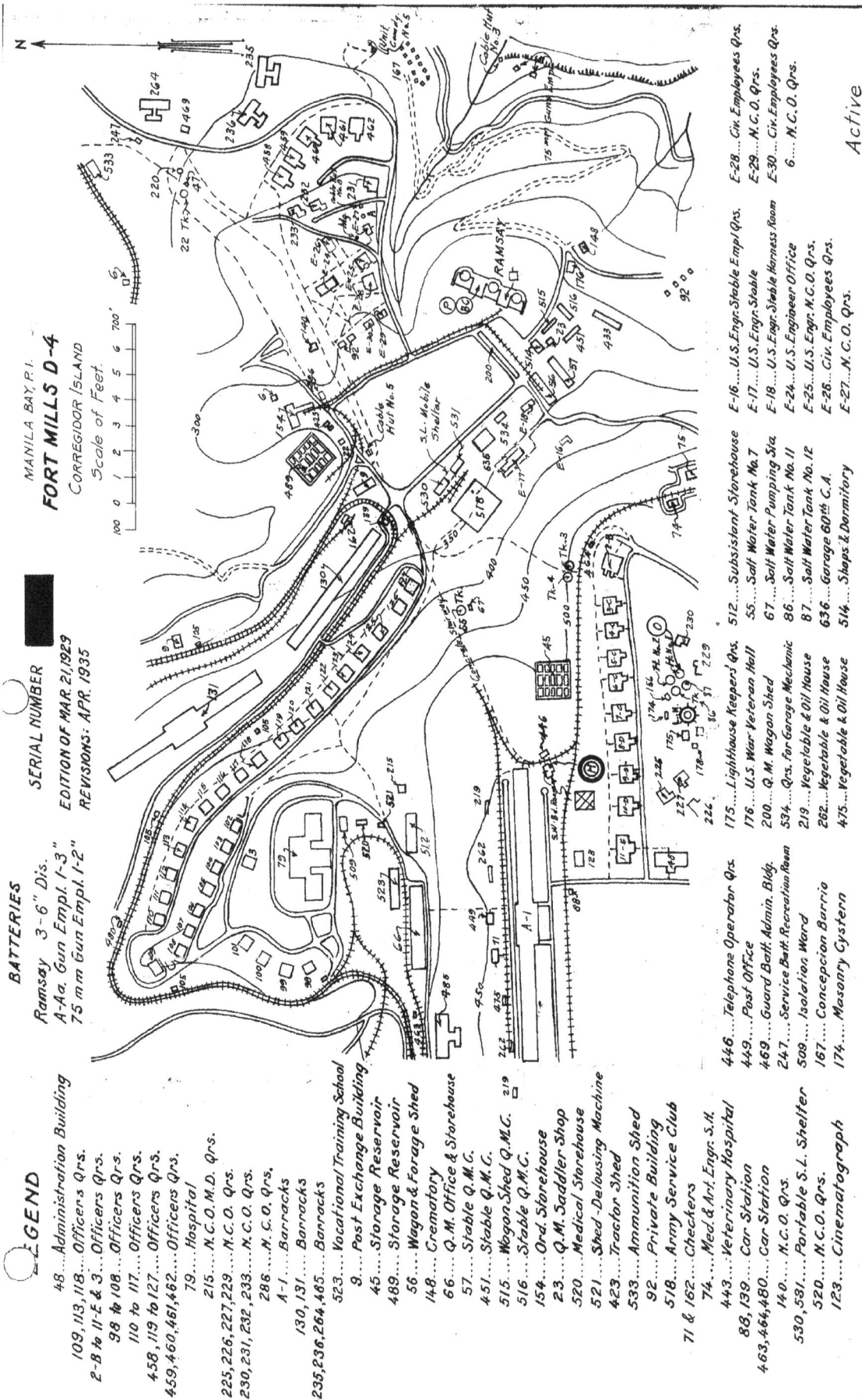

MANILA BAY P.I.
FORT MILLS D-4
CORREGIDOR ISLAND

Scale of Feet.

SERIAL NUMBER

EDITION OF MAR. 21, 1929
REVISIONS: APR. 1935

BATTERIES
Ramsay 3-6" Dis.
A-Aa. Gun Empl. 1-3"
75 mm Gun Empl. 1-2"

LEGEND

48...Administration Building
109,113,118...Officers Qrs.
2-B to II-E & 3...Officers Qrs.
98 to 108...Officers Qrs.
110 to 117...Officers Qrs.
458,119 to 127...Officers Qrs.
459,460,461,462...Officers Qrs.
79...Hospital
215...N.C.O.M.D. Qrs.
225,226,227,229...N.C.O. Qrs.
230,231,232,233...N.C.O. Qrs.
286...N.C.O. Qrs.
A-1...Barracks
130,131...Barracks
235,236,264,485...Barracks
523...Vocational Training School
9...Post Exchange Building
45...Storage Reservoir
489...Storage Reservoir
56...Wagon & Forage Shed
148...Crematory
66...Q.M. Office & Storehouse
57...Stable Q.M.C.
451...Stable Q.M.C.
515...Wagon Shed Q.M.C.
516...Stable Q.M.C.
154...Ord. Storehouse
23...Q.M. Saddler Shop
520...Medical Storehouse
521...Shed-Delousing Machine
423...Tractor Shed
533...Ammunition Shed
92...Private Building
518...Army Service Club
71 & 162...Checkers
74...Med & Ant. Engr. S.H.
443...Veterinary Hospital
88,139...Car Station
463,464,480...Car Station
140...N.C.O. Qrs.
530,531...Portable S.L. Shelter
520...N.C.O. Qrs.
123...Cinematograph

446...Telephone Operator Qrs.
449...Post Office
469...Guard Batt. Admin. Bldg.
247...Service Batt. Recreation Room
509...Isolation Ward
167...Concepcion Barrio
174...Masonry Cystern

175...Lighthouse Keepers' Qrs.
176...U.S. War Veteran Hall
200...Q.M. Wagon Shed
534...Qrs. for Garage Mechanic
219...Vegetable & Oil House
262...Vegetable & Oil House
475...Vegetable & Oil House

512...Subsistant Storehouse
55...Salt Water Tank No.7
67...Salt Water Pumping Sta.
86...Salt Water Tank No.11
87...Salt Water Tank No.12
636...Garage 60th C.A.
514...Shops & Dormitory

E-16...U.S.Engr.Stable Empl.Qrs.
E-17...U.S.Engr.Stable
E-18...U.S.Engr.Stable Harness Room
E-24...U.S.Engineer Office
E-25...U.S.Engr. N.C.O. Qrs.
E-26...Civ. Employees Qrs.
E-27...N.C.O. Qrs.

E-28...Civ. Employees Qrs.
E-29...N.C.O. Qrs.
E-30...Civ. Employees Qrs.
6...N.C.O. Qrs.

Active

MANILA BAY, P.I.

FORT MILLS D-5

CORREGIDOR ISLAND

Scale of Feet

100 0 1 2 3 4 5 6 700'

EDITION OF AUG. 12, 1921
REVISIONS: MAY 8, 1925;
MAR. 21, 1929; APR. 1935

SERIAL NUMBER

N

BREAKWATER POINT

GEARY POINT

Active

BATTERIES

GEARY 8-12"M
CROCKETT 2-12" Dis.

RAMSAY

CROCKETT

GEARY

Caddy House

Swimming Pool

Crofton

Flag Pole

WAY

550 WAY

LEGEND

12-B	Officer's Qrs.
13-C	" " "
14-C	" " "
15-D	" " "
16-D	" " "
17-D	" " "
18-D	" " "
19-D	" " "
20-D	" " "
21-C	" " "
22-C	" " "
23-D	" " "
24-D	" " "
25-D	" " "
2G-D	" " "
48	Administration Bldg.
75	Q. M. Storehouse
76	" " "
77	" " "
78	" " "
8C	Salt Water Tank No.11.
87	" " " No. 12.
90	Treasury P.I. Vault.
163	Car Station.
174	Masonry Cistern.
175	Lighthouse Keeper Qrs.
177	Tennis Court.
178	Signal Station.
179	Phil.Treas. Guard Qrs.
224	N.C.O. Qrs.
225	" " "
22C	" " "
227	" " "
544	Pumphouse Civ. Gov't.
551	Corregidor Club.
552	Servant Qrs.
553	Store Room.

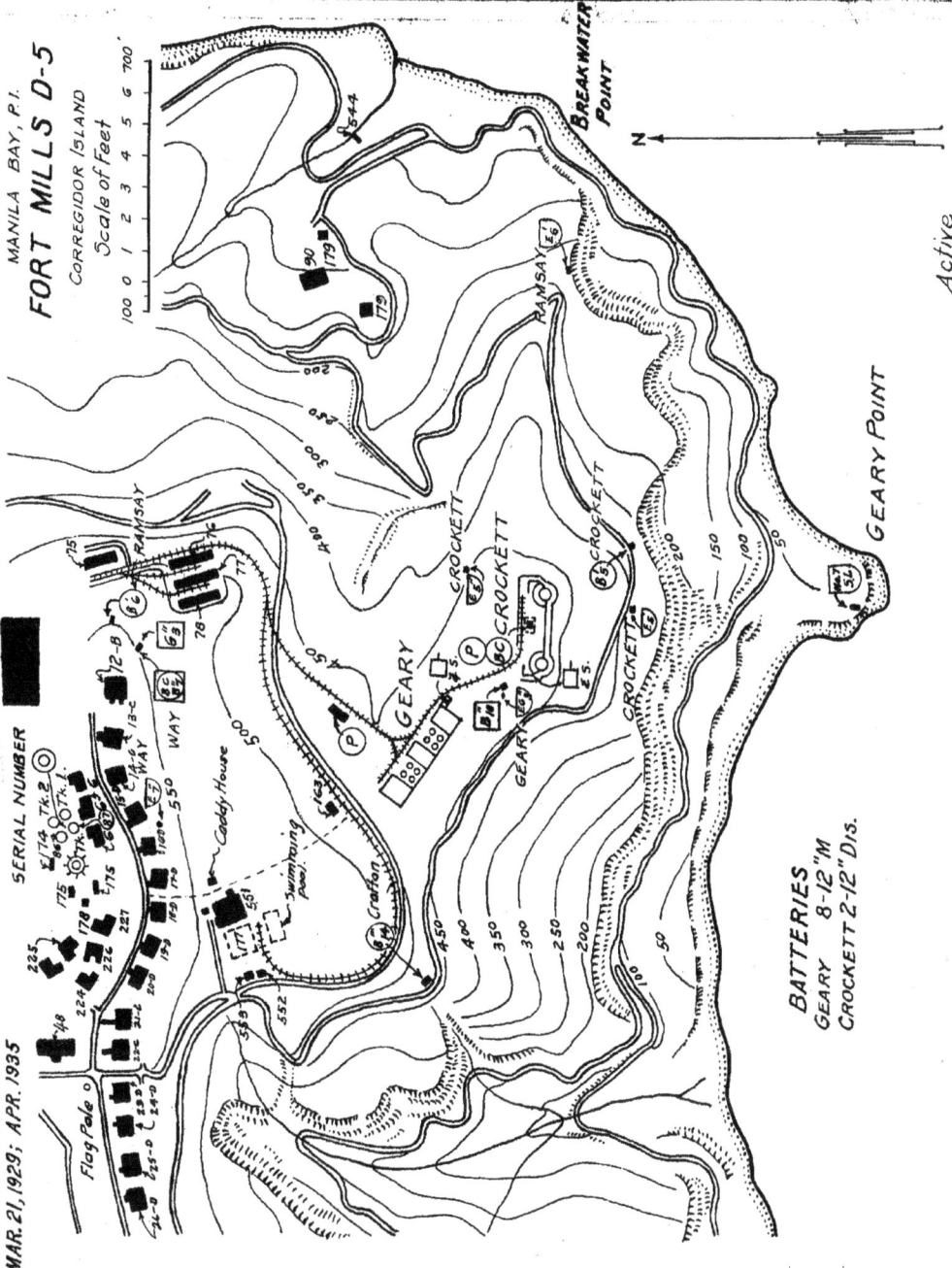

MANILA BAY, P.I.

FORT MILLS D-6

CORREGIDOR ISLAND
Scale of Feet

EDITION OF AUG. 12, 1921
REVISIONS: MAY 8, 1925;
MAR. 21, 1929; APR. 1935

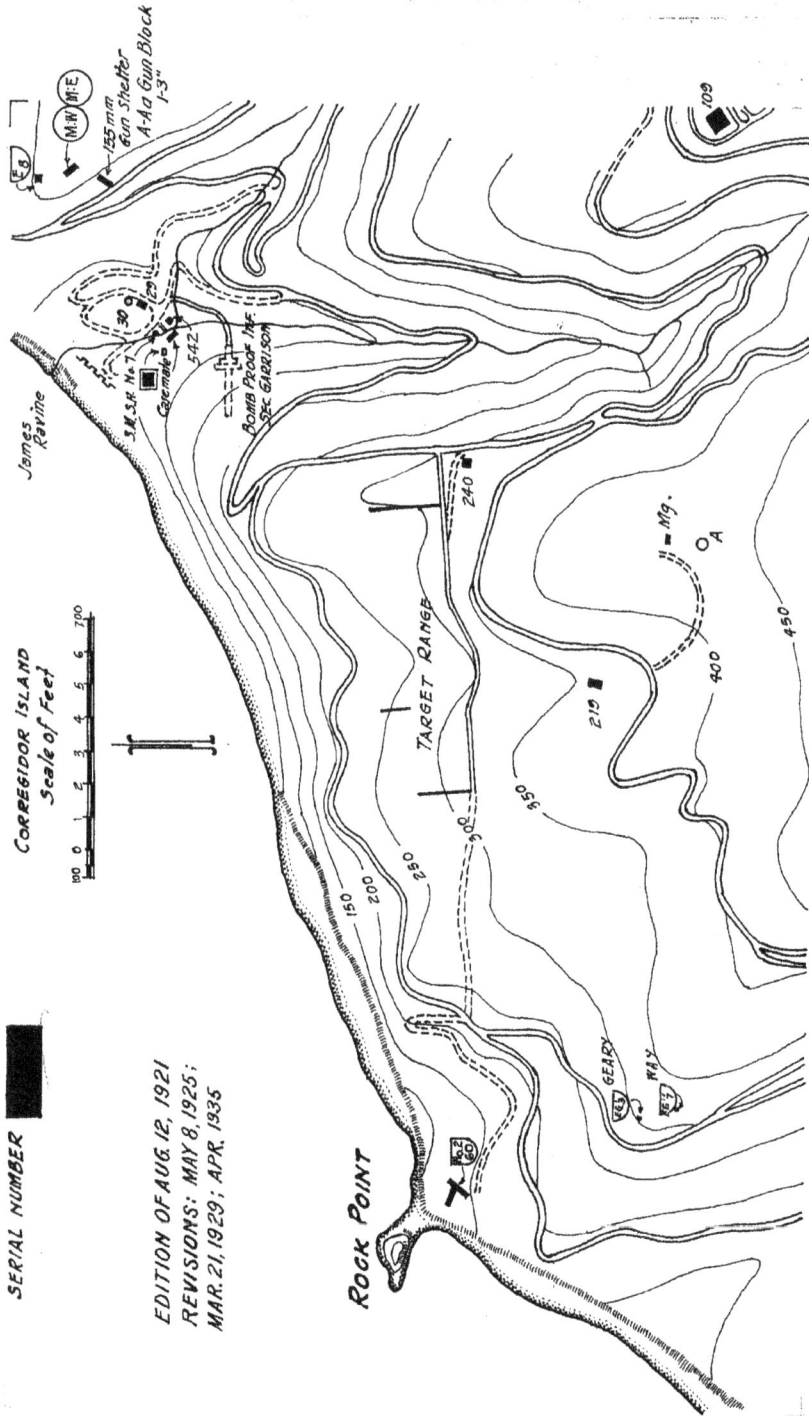

SERIAL NUMBER

LEGEND

29 JAMES RAVINE PUMPHOUSE
30 "   "   WATER TANK
109 OFFICERS QRS.
219 VEGETABLE OIL HOUSE
240 RANGE HOUSE
542 PUMP HOUSE, JAMES RAVINE

Active

Rock Point

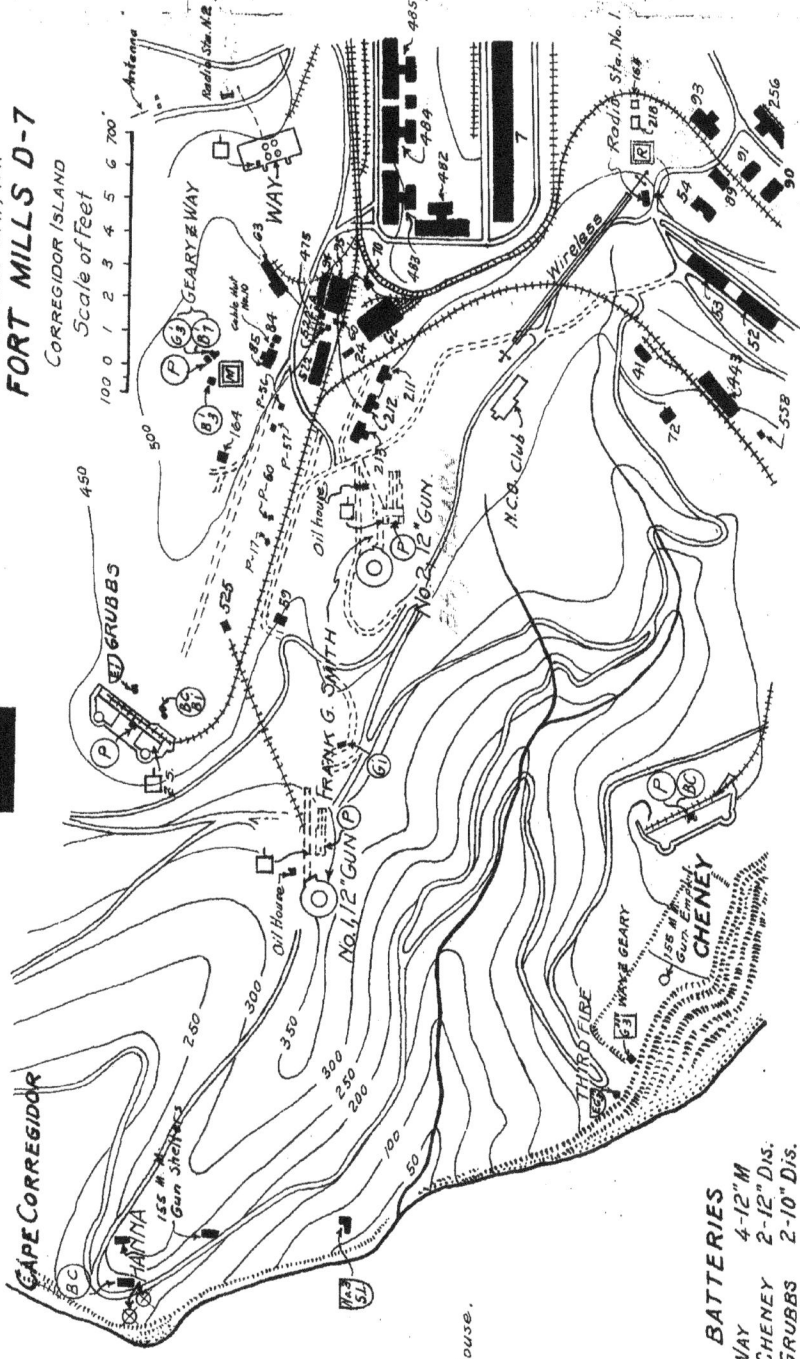

MANILA BAY, P.I.
FORT MILLS D-7

CORREGIDOR ISLAND

Scale of Feet

EDITION OF AUG. 12, 1921
REVISIONS: MAY 8, 1923;
MAR. 21, 1929; APR. 1935

SERIAL NUMBER

BATTERIES
WAY   4-12" M.
CHENEY   2-12" Dis.
GRUBBS   2-10" Dis.
FRANK G. { No.1, 12" Bar
            { No.2, 12" Bar
SMITH   2-12" Bar
HANNA   2-3" Ped.
155 mm Gun Empl. I.

Active

LEGEND

164	N.C. Officer's Qrs.
52	
53	
54	
89	
90	
91	
93	
211	
212	
213	
256	
482	
483	Barracks.
484	
485	
7	
63	Ord. Mach. Shop.
84	Civ. Employees Qrs.
85	"
59	Crematory.
62	Car Barn.
94	Garage Q.M.C.
95	Tool House Q.M.C. & Art. Cable Storehouse.
24	Q.M. Car Barn Tool House.
41	Ordnance Magazine.
70	Dispatcher's Office.
522	Auto Shed.
522-A	Ord. Tractor Garage.
443	Storehouse, Art. Engr.
72	Art. Engineer Office.
475	Vegetable, Oil House.
525	Cable Engine Shed.
558	Oil House, Art. Engr.
218	Post Improvement Bldg.
164	Wood Saw Shed.

MANILA BAY, P.I.
FORT MILLS D-8
CORREGIDOR ISLAND
Scale of Feet

EDITION OF AUG. 12, 1921
REVISIONS: MAR. 21, 1929;
APR. 1935

SERIAL NUMBER

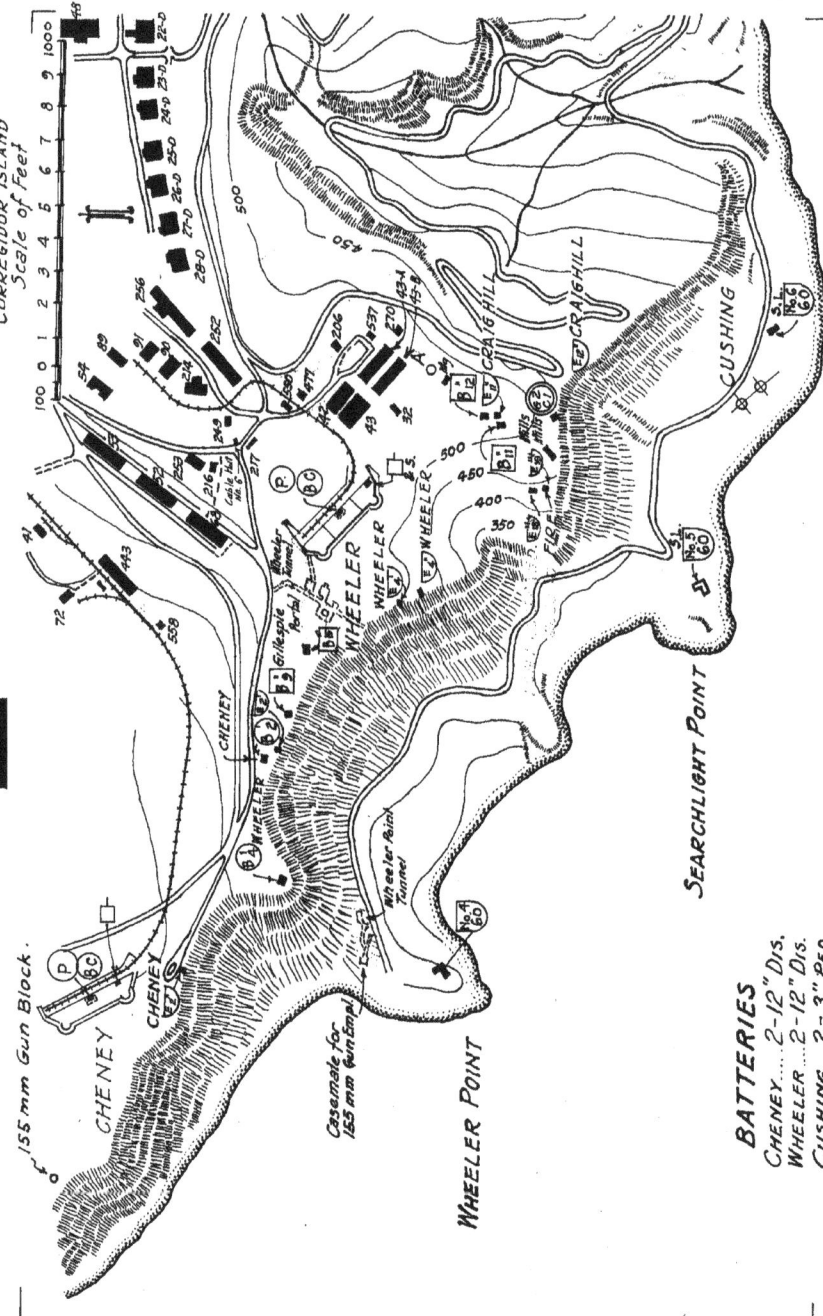

155mm Gun Block.

CHENEY

CHENEY

WHEELER POINT

SEARCHLIGHT POINT

WHEELER

WHEELER

CRAIGHILL

CRAIGHILL

CUSHING

Active

Gillespie Portal

Wheeler Point Tunnel

Casemate for 155mm Gun Empl.

LEGEND

22-D OFFICERS QRS.
23-D   "   "
24-D   "   "
25-D   "   "
26-D   "   "
27-D   "   "
28-D   "   "
32   OIL HOUSE
41   ORD. MAGAZINE
42   GUN-SHED 60th C.A. (A.A.)
43-A ORD ARM REPAIR SHOP
43-B   "   "   "   "
48   ADMINISTRATION BLDG.
52   N.C. OFFICERS QRS.
53   "   "   "   "
54   "   "   "   "
68   "   "   "   "
72   ART. ENGINEER OFFICE
89   N.C. OFFICERS QRS.
90   "   "   "   "
91   "   "   "   "
206   "   "   "   "
214   "   "   "   "
216   "   "   "   "
217   "   "   "   "
249 TEL. REPAIR SHOP
282 ORD. OFFICE and STOREHOUSE
283 N.C. OFFICERS QRS.
256   "
270 POWDER STOREHOUSE
443 STOREHOUSE ART. ENGINEER
477 VEGETABLE and OILHOUSE
537 N.C. OFFICERS QRS.
550 ORDNANCE MAGAZINE
558 OIL HOUSE ART. ENGINEER

BATTERIES

CHENEY..... 2-12" DIS.
WHEELER.... 2-12" DIS.
CUSHING.... 2-3" PED.
155mm Gun Empl.-1
A-Aa. Gun Block 1-3"

ting room was built immediately behind the center traverse. It was proof fired on January 27, 1909 and was the first major gun battery completed in the Philippines by the Americans. It was armed with two 12-inch Model 1895M1 guns on Model 1901 disappearing carriages (Bethlehem tubes #8/#13 and #7/#12). It was named in General Orders No. 77 of May 12, 1908 for Captain David P. Wheeler, 22nd U.S. Infantry, who was killed in 1904 in the Philippines. The carriages were modified in about 1916 to give an extra five degrees of elevation and improve their range capabilities. In the 1930s the tubes were changed; apparently newly relined tubes replaced the worn ones. By 1936 the battery had Bethlehem #7/#13 (the tube was relined in 1935) and #36/#12. Battery Wheeler saw extensive combat service in the 1941-1942 campaign. On March 24, 1942 an aerial bomb hit one gun on its racer, breaking several rollers. The gun was repaired, but it was subsequently found difficult to traverse. Gun No. 2 was in action until the surrender and was rendered inoperable by its crew on May 6th. With POW help the Japanese were able to rebuild one gun and carriage into serviceable condition by cannibalizing parts from the other—including dragging the gun tube of No. 1 across the traverse parapet to a position near No. 2. However, it was not fired by the Japanese in the 1945 campaign; though heavy infantry fighting occurred around the emplacement during the fighting in February of that year. The emplacement still exists, with remains of the armament at the Corregidor Memorial Park. The battery is open to the public, but in an overgrown condition.

- **CHENEY**: Another early dual 12-inch disappearing battery emplaced at the bluff top on the western extreme of Corregidor Island. It fired to the west, and as the most northern of the three dual 12-inch emplacements, it had the best field of fire on the northern channel and parts of southwestern Bataan. The siting challenge was difficult. Originally it was thought that the battery would have to be emplaced as two separate, single emplacements—each one on a separate ridge point at the edge of the bluff. Eventually with a little extra excavation, a site for a type emplacement for two guns was prepared. Work was begun in November 1904 clearing and excavating, concrete work began on March 27, 1905. It was a standard emplacement with two flank positions to give a wide field of fire. During construction a large BC station was added immediately to the rear of the central traverse. Trunnion heights were calculated at 419-feet. On the right flank a lengthy retaining wall extended to the rear of the emplacement where the battery access road ended. Work was done from 1905 to 1906. Transfer was made on July 23, 1910 for a total cost of $267,200. It was named in General Orders No. 77 of May 12, 1908 for 1st Lieutenant Ward Cheney, 4th U.S. Infantry who was killed in service in the Philippines in 1900. The battery was armed with two 12-inch Model 1895M1 Bethlehem guns on Model 1901 disappearing carriages (#3/#16 and #11/#17). In about 1916 the Model 1901 carriages were modified to give an extra five degrees of elevation. Sometime in the late 1920s gun tube #3 was replaced with a Model 1895 Watervliet tube (#37). Then in July 1934 tube #11 was replaced with Watervliet tube #12. With its ability to fire on Bataan, the battery saw considerable combat action in 1942. While suffering light damage on January 8th from a bomb hit and on April 10th from a shell hit, the battery was never put totally out of action. Personnel were removed from it as intended beach defense troops on May 5, 1942, but disabling sabotage to the carriages and removal of breech blocks was accomplished before the surrender on May 6th. The Japanese salvaged some small parts from the battery in 1942-45 and reportedly restored gun No.1 to some sort of serviceable status. The heavily damaged emplacement and armament still exist at the Corregidor Memorial Park. The battery is open to the public, but in an overgrown condition.

- **CROCKETT**: Another early 12-inch disappearing battery emplaced on Corregidor, the southernmost of the three heavy batteries initially erected. Crockett was the last of the three big batteries, and the most southern. It fired more to the south, southwest primarily to cover the main channel entrance. It was located quite close to mortar Battery Geary at the terminus of a natural cut. Trunnion elevation was 481-feet. The battery was of standard design, with two flank emplacement platforms to maximize the field of fire. Begun a little later than Wheeler and Cheney (in early 1907 vs. late 1904) an integral battery commander station was included in a position on the roof of the parapet between the two guns. Central traverse magazines used chain hoists to lift shells and shot to the loading level. Work was done in 1907-1909, Transfer was made on February 2, 1911 for a cost of $290,046.75. The battery was named in General Orders No. 77 of May 12, 1908 for 2nd Lieutenant Allen T. Crockett, 21st U.S. Infantry, who was killed in 1901 during service in the Philippine Insurrection. The battery was armed with two 12-inch Model 1895M1 Bethlehem guns on Model 1901 disappearing carriages (#15/#14 and #9/#15). In about 1916 the carriages were modified to give an additional five degrees of elevation. The armament served through the late 1920s, though earmarked for replacement by newly lined gun tubes. This was accomplished probably in the late 1930s, with new tubes Bethlehem #13 (relined in 1939) and Watervliet #27 (relined in 1938). Due to its southern field of fire, it was little used during the 1942 siege. However major damage was done to the battery by a Japanese 240mm howitzer barrage on April 24, 2942. The No. 1 gun was permanently disabled by a direct hit on its recoil cylinder, and fires were started in the hoist galleries. More damage was done by an aerial bomb on May 3rd, which pierced the loading platform floor. Personnel was taken from the battery as reinforcing infantry during the invasion of May 5th. With the No. 1 gun wrecked, only the No. 2 gun was disabled by the Americans before the surrender. The Japanese managed to put one gun back into service in late 1942, but there is no evidence that it was ever fired by them. Further damage was done to the emplacement during the retaking in February 1945. With damaged armament, the emplacement still exists at the Corregidor Memorial Park. The battery is open to the public.

• **GRUBBS:** This was the only 10-inch gun battery emplaced at the Manila Bay harbor defenses. It was located well inland in the west central part of the reservation, firing to the northwest to cover the Boca Chica (Little Channel) entrance between Corregidor and Bataan. It was begun after the start of the three 12-inch batteries, in 1907. The initial history of this battery is a little different. When a project for three 10-inch guns in a battery at Macmany Point at Subic Bay was cancelled in early 1907, the three guns and carriages were released for other employment. Two went on to arm Battery Warwick at Fort Wint, but the third was available for Corregidor. As the 1907 political crisis with Japan was at its height, it was decided to emplace this armament at the future site for a 10-inch battery to enhance the defense during the emergency. A block was poured in November 1907 and the gun and carriage mounted in May 1908. Even a combined concrete battery commander's and primary station was built at where it was thought it would later be needed for a full battery. Then another full appropriation was granted for Fort Mills construction, and the project for a standard dual 10-inch battery was begun at the site. The approved plan was generally of type plan, with two platforms and shared lower-level magazines. Ammunition service was with chain hoists. Work was done in the summer of 1909, incorporating the block and mounted gun and carriage as the No. 1 emplacement. However, the BC and primary station was found too close to the battery, was destroyed and a new one built slightly to the rear with a new plot room in late 1909. This station was transferred on February 11, 1911. On that same date the battery itself was transferred at a cost of $212,397.86. It was armed with two 10-inch Model 1895M1 Watervliet guns on Model 1901 disappearing carriages (#25/#14 and #22/#16). It was named in General Orders No. 77 of May 12, 1908 for 1st Lieutenant Hayden Y. Grubbs, 6th U.S. Infantry, who was killed during the Philippine Insurrection in 1899. For most of its service life it was manned by a coast defense battery of Philippine Scouts. At the start of World War II, the battery was in reduced status and not manned; but it was put into service again with the transfer here of Battery C of the 91st Coast Artillery from Battery Morrison in early April 1942. Once it began firing on Bataan it was very quickly put out of action by Japanese counterbattery fire. A bomb hit on April 11th damaged the power plant, and on April 16th a direct hit took out the No. 2 gun (No. 1 never went into service due to mechanical problems). The battery commander's station was also knocked out, the emplacement was abandoned. During the occupation one gun was partially restored by the Japanese, but never actually used. Additional damage occurred during the retaking in February 1945. The emplacement and its armament still exist at the Corregidor Memorial Park. The battery is open to the public.

- **WAY**: A battery for four 12-inch mortars built on protected high ground at Fort Mills. It was one of the first batteries built in the defenses. Plans were approved in 1904 and the site clearing began that October. Using the concrete plant that had just finished at Battery Wheeler, concrete construction began in early 1905, completing late that year. The site was 700-yards from Wheeler, with an elevation of 450-feet. Initially it was considered so far inland and so high that it was safe from any return fire from warships and built only with a minimum of protection (just 30-inches of protection to the magazine roofs). It consisted of just one open pit with four mortars, with an earthen parapet to the north and adjacent concrete magazines on either flank. There was no BC station or plotting room at the battery, those were to be located elsewhere where a clear observation of the channel and field of fire could be made. Transfer was made on July 26, 1910 for a cost of just $72,000. It was armed with four 12-inch mortars Model 1890M1 on Model 1896 carriages (#173/#158, #174/#151, #170/#241, and #172/#150). The mortars were mounted by the end of March 1909. These were newly produced tubes, but the carriages came from a variety of arsenals or repaired pieces taken from continental batteries. It was named in General Orders No. 77 of May 12, 1908 for 2nd Lieutenant Henry N. Way, 4th U.S. Infantry who was killed during the Philippine Insurrection in 1900. Almost immediately it was recognized that the relatively unprotected magazine might be safe from low-trajectory ship fire, but not from high angle fire from howitzers mounted on the mainland. In 1911 suggestions were made to add concrete and earth protection to the magazine roofs. That was accomplished, along with sealing up the sides of the pit to provide a more fully protected pit. Also, a long data booth was built immediately to the rear of the pit. These modifications were completed in 1914 for an additional cost, bringing the total expense to $112,969.79. After World War I the battery was seldom used; in fact, at times the earthen slopes became heavily overgrown. The emplacement was not manned at the start of the 1941-1942 campaign. Finally on April 29, 1942 Battery "E" of the 59th Coast Artillery arrived from Bataan and put the battery back into service. At that point only three mortars were serviceable, but these fired intensely until all but one was put out of action during the late April and early May bombardment. It was ultimately the final fixed battery still capable of firing at the time of the surrender. Apparently, the Japanese managed to put one mortar back into service, but it was not manned nor fired during the retaking in February 1945. The emplacement, in relatively good condition with its four mortars, still exists at the Corregidor Memorial Park. The battery is open to the public.

- **GEARY**: A battery of two pits for eight 12-inch mortars built on the southwestern side of the Corregidor Topside reservation. Detailed planning for this work began when the project for a mortar battery in Subic Bay was cancelled in 1907 and the funds made available for another mortar battery for Manila Bay. In November 1907 a site was selected close to Battery Crockett on the southwest side of Topside in a shallow, well-protected ravine. Using the nearby concrete plant providingmaterial for the 12-inch battery, work was able to get underway rather quickly in the summer of 1908. The plan called for a more conventional two-pit design (at least as compared to Battery Way). The magazines were located to the sides and in the central traverse between the two wide pits. A new cut was made to allow a new access road to go past Geary and on to Crockett. Work was done for transfer on June 6, 1911 for $145,198.27. The battery was named in General Orders No. 77 of May 12, 1908 for Captain Woodbridge Geary, 13th U.S. Infantry who was killed in 1899 in the Philippines. It was intended to mount the modern Model 1908 mortars and carriages in all eight emplacements. There was a delay in manufacturing and shipping these guns. As there was a new crisis with Japan in 1910, the Secretary of War intervened on August 12, 1910 and ordered that four older Model 1890M1 mortars on Model 1896 carriages be sent immediately to Fort Mills to be emplaced at Battery Geary. Even with shipping delays they were reported mounted and ready for service in November 1910. The mortars came from Battery Whitman at Fort Andrews, which would get the displaced four Model 1908s in return. At Battery Geary the base rings already emplaced in concrete of Pit A had to be dug out and replaced with the rings for the Model 1896 carriage. Pit A received these mortars and carriages, all tubes Bethlehem manufacture, #31/#104, #32/#116, #33/#210, and #40/#294. Pit B received a little later its intended four Model 1908 mortars and carriages (#2/#17, #12/#18, #13/#19, and #22/#20). As it was considered a well-protected site, like Battery Way it had less overhead protection and the entire battery was not constructed with rebar reinforcement. In 1911 plans were started to add additional overhead concrete and earth cover. That was finished in mid-1914, bringing the total cost of the battery to $177,026.28. Battery Geary was one of the most important gun installations for the Americans in the 1942 siege. These were powerful mortars capable of firing in any direction. They were heavily used throughout the battle and attracted plenty of return artillery fire. On January 6th a bomb hit an incomplete personnel shelter killing 31, including most of the battery's operating crew. In late January the mortars of Battery Geary delivered very effective fire to help liquidate the Japanese landing at Longoskawayan Point pocket, being the first major firing of a coast artillery piece in combat since the Civil War. During the intense artillery bombardment of April and May 1942 the emplacement suffered badly. A direct hit by a 240mm howitzer shell detonated the battery's magazine on May 2, 1942. A terrific explosion occurred, destroying the magazine, killing 27 men, and tossing the guns and parts of the battery large distances. The devestated emplacement still exists at the Corregidor Memorial Park. The battery is open to the public.

- **MORRISON**: An important dual 6-inch disappearing battery built on a hill on the northern face of topside, facing due north across the channel and mine fields of the Boca Chica. This hill, with several other important tactical elements, was known as "Morrison Hill." Trunnion height of the battery was 362-feet. The battery plan strictly adhered to the late 6-inch designs, with a single traverse magazine and two flank-configured platforms. A small battery commander station was placed atop and at the rear of the traverse. Work was done from December 1907 to late 1908. It was transferred on November 10, 1910 for a construction cost of $79,845.99. The battery was named in General Orders No. 77 of May 12, 1908 for 1st Lieutenant John Morrison Jr., 4th U.S. Infantry who was killed in the Philippines in 1901. The battery was armed with two 6-inch guns Model 1905 on Model 1905M1 disappearing carriages (#31/#12 and #32/#13). In 1921 the battery received a new plotting room built into the side of the road directly to the rear of the battery, and a new enlarged battery commander station was placed on top. These works were transferred on November 19, 1921 for $10,458.27. Due to continual manpower shortages at the post the battery was not regularly manned, though it was periodically exercised. It was not in service at the start of the war, but with the occupation of Bataan by enemy forces that changed. It was manned by Battery "C" of the 91st Coast Artillery in early April 1945. Exchanging fire with Japanese pieces on Bataan, it was soon overwhelmed and by April 12th was knocked out of action. The emplacement was abandoned by the time of the surrender on May 6th to the Japanese. The emplacement, with its damaged ordnance, still exists at the Corregidor Memorial Park. The battery is open to the public, but in an overgrown condition.

- **RAMSAY**: An early battery for three 6-inch disappearing guns emplaced on the eastern side of the topside plateau. It had a bearing to the southeast, specifically covering the bay and minefield between Fort Mills and Caballo Island. The battery trunnion height was 343-feet. It was of conventional design; the wider traverse between emplacements No. 2 and 3 holding a shared magazine for two guns vs. the smaller one between No. 1 and 2. The two end positions were built as flank emplacements for a wider field of fire. Work was done from May 1907 to mid-1908. It was transferred on January 20, 1911 for a cost of $99,536.16. It was named in General Orders No. 77 of May 12, 1908 for 1st Lieutenant Charles Ramsay, 21st U.S. Infantry who was killed in the Philippines in 1901. The battery was armed with three 6-inch Model 1905 guns on Model 1905M1 disappearing carriages (#2/#10, #4/#9, and #33/#11). This armament was never changed. The battery was not manned during the siege of 1942 and was probably surrendered more or less intact, with minor sabotage to the armament. Severe damage did occur later in the war. Two American bomb hits (one in January and one in February 1945) detonated the traverse magazines. Huge craters were created where these stood, concrete blocks were thrown considerable distances, and the guns and carriages completely upset and wrecked. The destroyed battery is usually heavily overgrown, at the Corregidor Memorial Park. The battery is open to the public, but in an overgrown condition.

- **JAMES**: A battery for four 3-inch pedestal guns placed on the north shore of Corregidor for protection of the north channel mine field. It was placed on a bluff point just to the east of James Ravine (named for the battery) which had water facilities, a mining casemate, and possible enemy landing beach to protect. It followed in general type plans, but the sequence of four emplacements was bent in the center allowing for two each on slightly different bearings. A rather large BC station was also built to the rear of the central traverse, behind the rear parado (road) accessible by a bridge from the battery. Work was done from April 1908 to early 1909, for transfer on November 10, 1910 for a cost of $50,537.95. It was named in General Orders No. 77 of May 12, 1908 for 1st Lieutenant John F. James, 8th U.S. Infantry, who was killed on Leyte in 1903. It was armed with four 3-inch Model 1903 guns and pedestals (#2/#2, #3/#3, #4/#4, and #9/#5). During proof firing at the battery on July 10, 1909 gun #4 developed a longitudinal crack. It had to be returned to Watervliet Arsenal but was replaced with gun #8 which was fortunately on hand. In 1910 the guns were briefly dismounted and the emplacement fired upon during maneuvers to see if an opposing army could occupy Bataan and lay siege to the defenses. During World War I two guns and pedestals were removed and sent to arm a new searchlight protection battery at Fort Mills (Battery Hanna with #2/#2 and #3/#3). The two emplacements at Battery James were left unoccupied until rearmed in July 1936. It received tubes transferred from Battery Campbell in Long Island Sound and pedestals already spare in the Philippines (#28/#23 and #58/#103). The two guns in emplacements No. 3 and No.4 had their tubes replaced in the 1930s, resulting in thre gun and carriage combination of #41/#3 and #39/#4. At the start of the island's siege from Bataan in April 1942 this exposed battery on the north shore was soon put out of action. By April 20th it was no longer serviceable. A large personnel shelter tunnel behind the battery collapsed under fire on April 15th, causing the deaths of 42 soldiers. During the occupation the Japanese dismounted the guns for salvage or perhaps relocation, the emplacement is empty of all armament and partially demolished today at the Corregidor Memorial Park. The battery is open to the public, but in an overgrown condition.

- **MAXWELL KEYS:** One of the two emplacements for seacoast 3-inch Model 1903 guns emplaced as part of the 1910 Land Defense Plan for Manila Bay. This project called for emplacing four such guns primarily to cover waters and beaches susceptible to boat landings that were otherwise not covered with the fort's primary armament. After some discussion, two went to arm Battery Fuger at Fort Hughes, and two others were used to arm this emplacement. A site was decided upon on August 22, 1911—with a note that the armament was already in transit. It was sited on the island eastern tail of Corregidor. At an elevation of about 90-feet it was still on the tail's flat ridgeline or plateau and pointed southeast to cover the waters and eastern end of Fort Hughes. The plan was a simplified 3-inch type plan, in fact just two standard-sized platforms with low parapet wall and a single, protected room for magazine and storage in the central traverse. Work was done between February 1912 and September 1913. Transfer was made on September 10, 1913 for a cost of only $4,140.64. It was armed with two 3-inch Model 1903 guns and pedestals (#101/#100 and #100/#101). These were never changed. A little later the battery had a simple, open battery commander's station built above and behind it on the spine of the tail. Lighting was installed in 1931 for a cost of $479.78. It was named in General Orders No. 15 of April 24, 1916 for 2nd Lieutenant Maxwell Keys, killed in the Philippines in 1899. It is frequently misspelled even in army documents as Maxwell "Keyes." There is no evidence that it was used during World War II combat. It still exists, though often heavily overgrown, at the Corregidor Memorial Park. The battery is open to the public, but in an overgrown condition.

- **HANNA:** One of the two searchlight-protection batteries built during World War I at Fort Mills. Starting in April of 1916 the local command sought authority to relocate the 3-inch guns from Fort Wint to new positions. The U.S. Navy had withdrawn most of its workers and equipment from the Olongapo yard in early 1917, and while the heavier guns were to remain, the U.S. Army sought to use the 3-inch guns more productively than in an abandoned post. The searchlights at Fort Mills and Frank were vulnerable to quick-moving light attack craft and it was thought essential to provide protective batteries to ensure their continued wartime use. Batteries to protect lights No. 3 and 4, then the pair No. 5 and 6, and then the guns at the northern point of Fort Frank (No. 13, 14, 15) were considered most vulnerable. The War Department was reluctant at first to authorize the removal of the Fort Wint guns, but at Fort Mills it was decided to relocate two of Battery Janes' guns to this new position to protect lights No. 3 and 4. Work would only require new gun blocks and bolts, magazines were to be light temporary structures, and labor would be provided by prisoners. A position was selected high on the cliff top above Cape Corregidor, with a trunnion height of 219-feet. It consisted of two staggered cylindrical gun blocks and a buried magazine between. Work was done from November 1917 to early 1918, for transfer on March 29, 1919 for $1,716.73. It was named in General Orders No. 13 of March 27, 1922 for Captain Guy B. Hanna, a Coast Artillery Corp officer who died in a service accident on May 23, 1913. The battery was armed with two 3-inch guns Model 1903 and pedestals (#2/#2 and #3/#3 from Battery James). Later, probably in the mid-1930s, this was changed, and the battery was armed with #58/#3 and #3/#2. While not seeing extensive service during the siege, Hanna was knocked out by Japanese artillery fire from Bataan in April 1942. The Japanese dismounted the guns. In postwar years one gun block has fallen off the cliff face due to erosion, but the other still exists at the Corregidor Memorial Park. The battery is open to the public, but in overgrown condition and on the cliff's face.

- **CUSHING:** The second of the two searchlight-protection batteries built at Fort Mills. Starting in April of 1916 the local command sought authority to relocate the 3-inch guns from Fort Wint to new positions. The U.S. Navy had withdrawn most of its workers and equipment from the Olongapo yard in early 1917, and while the heavier guns were to remain, the U.S. Army sought to use the 3-inch guns more productively than in an abandoned post. The searchlights at Fort Mills and Fort Frank were vulnerable to quick-moving light attack craft and it was thought essential to provide protective batteries to ensure their continued wartime use. Batteries to protect lights No. 3 and 4, then the pair No. 5 and 6, and then the guns at the northern point of Fort Frank (No. 13, 14, 15) were considered most vulnerable. The War Department was reluctant at first to authorize the removal of the Fort Wint guns, instead wishing to use them to arm local transports. However, this option fell through, and on October 3, 1918 authority was granted to use the six guns from Fort Wint (two were to stay behind) to arm new searchlight-protection batteries. This site was designed to protect searchlights No. 5 and No. 6 on the southwest coast of Corregidor. The battery site was between these lights, right off the south coast road that ran parallel to the shore at about an 80-foot elevation. The battery consisted solely of two cylindrical gun blocks separated by 80-feet. The magazine was just a temporary wooden building. Work was done from January to March 1919. Transfer was made on December 19, 1919 for a cost of $2,309.59. It was named in General Orders No. 13 of March 27, 1922 for Brevet Lt. Colonel Alonzo H. Cushing, who died at Gettysburg in 1863. It was armed with two 3-inch Model 1903 guns and pedestals (#15/#9 and #14/#8, tubes moved here from Battery Flake at Fort Wint and carriages from Battery James at Fort Mills). Cushing was not damaged in the 1942 campaign. The guns were disabled at surrender by draining the cylinders of recoil fluid and firing. The Japanese removed the guns either for scrap or to salvage them. The emplacement still exists at the Corregidor Memorial Park. The battery is open to the public, but in an overgrown condition and with difficult access due to the erosion in front of the emplacement.

- **SMITH**: The 1915 Board of Review Program battery for two separate emplacements of 12-inch guns on long-range carriages added to the Fort Mills defenses in the early 1920s. They utilized the new long-range, Model 1917 barbette carriage recently adopted by the Ordnance Department. While other American batteries for this gun type were all dual emplacements (open gun platforms allowing all-round fire and heavily protected traverse magazines between), the lack of sufficient space topside on Corregidor forced the adoption of two single emplacements, each with its own individual protected magazine. Both emplacements were on the western side of the island, firing generally to the west. While initially both emplacements were named Battery Smith, in 1937 the most easterly of the pair was separated tactically and named Battery Hearn. The western emplacement was built in 1917-1919. It was transferred on June 27, 1921 for a cost of $145,833. The gun was on a totally open platform with apron that allowed 360-degrees of traverse. Adjacent was a protected, cut-and-cover structure for magazine, power room, plotting room, and latrine. It was named in General Orders No. 13 of March 27, 1922 for Brigadier General Frank G. Smith, U.S. Army who died in 1912. It was armed with one 12-inch Model 1895M1A4 Watervliet gun on a Model 1917 barbette carriage (#47/#30). In the mid-1930s the tube was exchanged for gun Model 1895 Watervliet #1. The battery was not heavily used during the 1941-1942 campaign, at least partially due to the desire to keep it hidden and protected from counterbattery fire and to reserve it for any potential incursion into Manila Bay by enemy ships. Nonetheless it received considerable damage from artillery fire during the siege. It was rendered disabled prior to surrender by draining the recoil fluid and firing; breechblock was also removed. The emplacement and carriage were further damaged by bombs in 1945 prior to the American landings in February. The emplacement is now heavily damaged, often overgrown, and of difficult access at the Corregidor Memorial Park. The battery is open to the public, but in an overgrown condition.

- **HEARN**: The 1915 Program battery for two separate emplacements of 12-inch guns on long-range carriages added to the Fort Mills defenses in the early 1920s. They utilized the new long-range, Model 1917 barbette carriage recently adopted by the Ordnance Department. While other American batteries for this gun type were all dual emplacements (open gun platforms allowing all-round fire and heavily protected traverse magazines between), the lack of sufficient space topside on Corregidor forced the adoption of two single emplacements, each with its own individual protected magazine. Both emplacements were on the western side of the island, firing generally to the west. While initially both emplacements were named Battery Smith, in 1937 the most easterly of the pair was separated tactically and named Battery Hearn. This eastern emplacement was built in 1917-1919. It was transferred on June 27, 1921 for a cost of $148,105. The gun was on a totally open platform with apron that allowed 360-degrees of traverse. Adjacent was a protected, cut-and-cover structure for magazine, power room, plotting room, and latrine. Originally part of Battery Smith, for a while it was known as Smith No.2, and then in General Orders No. 11 of October 9, 1937 it was named for Brigadier General Clint C. Hearn. It was armed with one 12-inch Model 1895M1A4 Watervliet gun on a Model 1917 barbette carriage (#44/#31). This tube was exchanged in the mid-1930s for gun tube Watervliet #8, and then probably again for tube Watervliet #6 in 1937. The battery was not extensively used during the 1941-1942 campaign, due to its vulnerability and the desire to husband it for any attempted naval penetration into the bay. It did sustain some damage during the campaign but was still operable at the end and was sabotaged by the American troops prior to surrender. It was the site for the famous photograph of Japanese conquerors celebrating with a banzai salute. The Japanese used POW labor to rebuild the gun using spare parts and items from Battery Smith, but it did not see firing action in 1945. Somewhat damaged, the gun and emplacement still exist at the Corregidor Memorial Park. The battery is open to the public.

- **RJ-43**: An emplacement for a single 8-inch gun emplaced during World War II campaign. The Philippines had been sent seven 8-inch Model 1888 guns on Model 1918 railway top carriages as part of the armament intended for the Inland Seas Project. This was a network of 8-inch and 155mm guns to be emplaced to cover the passages between various islands of the archipelago. The armament had been sent in early 1941 but work on the emplacements was only about to begin when war broke out. The 8-inch guns were being moved to Bataan during the retreat of December 1941 when five of them were disabled during an attack on their rail transport. Two guns reached the defenders; one was sent to a new emplacement on the west coast of Bataan, the final unit was sent to Fort Mills. In an attempt to use it, fort engineers built a simple concrete pad with foundation bolts to hold the carriage on the east side of Malinta Hill at a location close to Road Junction No. 43. Hence the informal name of RJ-43. Work was done in February 1942; the gun mounted a proof fired in March. Ammunition and crew shelter were in adjoining tunnels. No report confirms any service firing. One report lists it as having been subsequently destroyed in an air attack. Apparently, the carriage was scrapped by the Japanese. The emplacement block still exists, and the 8-inch gun tube lies on the north dock at the Corregidor Memorial Park. The battery is open to the public, but in an overgrown condition.

Middleside and Topside, Fort Mills, Corregidor, PI 1932 (NARA)

Middleside and Topside, Fort Mills 1932 (NARA)

Battery Hearn Fort Mills (Terry McGovern)

Battery Smith Fort Mills (Terry McGovern)

Battery Way Fort Mills (Terry McGovern)

Battery Geary Fort Mills (Terry McGovern)

Battery Crockett Fort Mills (Terry McGovern)

Battery Cheney Fort Mills (Terry McGovern)

Topside parade ground, Fort Mills (Terry McGovern)

Topside Barracks Fort Mills (Terry McGovern)

THE HARBOR DEFENSES OF SUBIC BAY – PHILIPPINES

Fort Wint (1902-1946/1992) is located on Grande Island at the entrance of Subic Bay, approximately 35 miles north of Manila Bay. Spanish batteries (four 15cm Ordonez Hontoria BLR guns) were originally located around Subic Bay. However, they were not yet completed in 1898, so the Spanish decided to abandon them. El Grande Island was the site of numerous Marine Corps temporary batteries in 1907. American fortifications were built between 1907 and 1910. Named in G.O, of 1908 for Brigadier General Theodore J. Wint. Batteries here are Battery Warwick (two 10-inch DC guns), Battery Woodruff (two 6-inch DC guns), Battery Hall (two 6-inch DC guns), Battery Flake (four 3-inch guns, two removed in 1930s), Battery Jewell (four 3-inch guns, two removed in 1930s), Battery Subic (1941, two 155mm guns on field mounts, transferred to Bataan in 1941), and Battery Cebu (1920-1941, four 3-inch AA guns, transferred to Bataan in 1941). There were two searchlight stations, and five fire-control stations located on the island. Fort Wint also had a mine casemate and mine loading facilities for controlled submarine mines for the Subic Bay channel. The island fort was abandoned in December 1941 and suffered little damage from the Japanese. Fort Wint was recaptured by U.S. forces in March 1945. Battery Hall was destroyed during American bombing in 1945. Its two guns still remain. The magazines of Battery Woodruff were destroyed after the war by an accidental explosion. Battery Warwick's two guns, as well as the four remaining 3-inch guns were transferred to the State of Washington (to Forts Casey and Flagler) in the 1960s. The post, as part of the Subic Bay U.S. Naval Base, was closed and transferred to the Philippines in 1992. Grande Island was operated for many decades after World War II by Subic Bay U.S. Navy Special Services as an on base resort for US military and civilians stationed throughout the Philippines and Southeast Asia. Today, the island is the Grande Island Resort which is accessed via a ferry from former naval station. Access to the former fort is through visiting the resort.

Fort Wint Gun Batteries

- **WARWICK:** The largest Taft battery emplaced at Grande Island in Subic Bay. Site selection for this unit was not easy, and various plans alternated between a location on the mainland or the island. Eventually it was decided to concentrate all the new gun batteries at the Fort Wint reservation on Grande Island. Authorization to start planning was initiated on February 20, 1905 from funds of July 12, 1904 authorizing three 10-inch and six 6-inch for Subic Bay. It was located on the island's highest land on the southwest corner and fired into the bay to the south. Submitted plans were approved for a battery of just two of the 10-inch guns on disappearing carriages. Physical work was done from 1907-1908. Transfer was made on December 30, 1910 for a cost of $254,896.22. It was of conventional type design. The battery was named in General Orders No. 77 of May 12, 1908 for Captain Oliver B. Warwick, 18th U.S. Infantry who was killed in the Philippines in 1899. It was armed with two 10-inch guns Model 1895M1 on Model 1901 disappearing carriages (#26/#13 and #28/#15). While proof firing the guns in their new emplacement, one gun (#26) suffered a catastrophic failure, splitting from the muzzle down. It had to be removed and returned to Watervliet for scrapping, being replaced with tube #28. While fully equipped and manned initially, the defenses of Subic Bay were soon put into a sort of long-term caretaker status. For many years in the 1920s and 1930s there was no active garrison. Just before World War II, Fort Wint was selected to be the training station for the new Philippine Coast Artillery. It was mostly Philippine trainees on station with American instructors when war started. It did not see combat but was abandoned when the U.S. forces retreated from the area in late 1941. The gun breech blocks were removed on December 24, 1941 just prior to evacuation. The Japanese did not make use of the emplacement, at least for armament, during their occupation. After reoccupation it remained in American army

SUBIC BAY, P.I.
FORT WINT
GRANDE ISLAND.

BATTERIES
WARWICK .... 2-10"DIS.
WOODRUFF .... 2-6" "
HALL .... 2-6" "
JEWELL .... 2-3"PED.
FLAKE .... 2-3" "
A-Anti-aircraft gun 2-3"
Guns not mounted

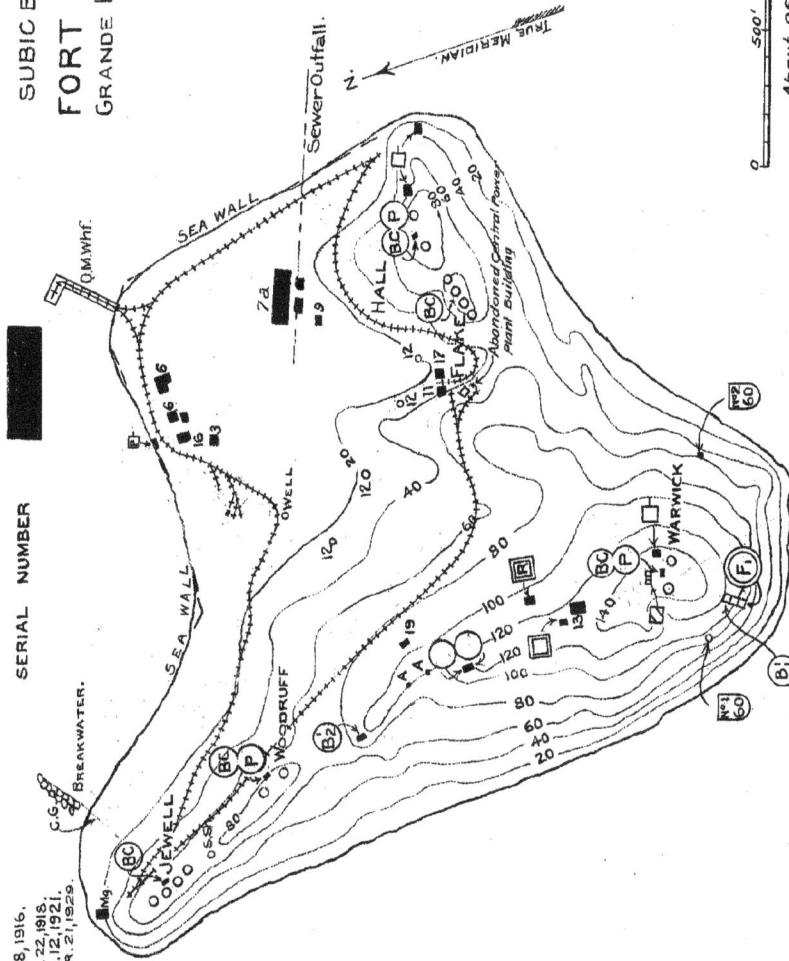

About 95 A.   Active Status.

0   500'   1000'

N.
TRUE MERIDIAN

SERIAL NUMBER

LEGEND.
1
2
3 OFFICER'S QUARTERS.
4
4a
5
6 N.C.OFFICERS' QRS.
6a
7
7a 1 COMPANY BARRACKS.
8
9 POST EXCHANGE.
10
11 ICE PLANT.
12 ARTESIAN WELL.
13 RESERVOIR.
14
15
15t.
16 OFFICE & STOREROOM
17 LAUNDRY.
18
21 Q.M. STOREHOUSE.
22
23
24
41
42
9C
19. AMMUNITION SHELTER
    FOR A.A.C.GUNS.

EDITION OF NOV. 8, 1916.
REVISIONS; JAN. 22, 1918.
OCT. 24,1918; AUG.12, 1921.
MAY 8, 1925; MAR. 21, 1929.
APRIL 11, 1935.

and navy hands for many years. In the late 1960s the two guns and carriages were removed from Battery Warwick and sent to Fort Casey in Washington State for use as display items at Fort Casey State Park. In Philippine hands since 1991 the emplacement was used as an air traffic radar site. It still exists, though somewhat damaged by settling on Grande Island. The battery is not open to the public.

- **WOODRUFF:** One of two dual 6-inch disappearing batteries built at Grande Island and Fort Wint. It was located on the northwest point of the island at about a 90-foot elevation, firing to the southwest. The submitted design followed closely the mimeograph for the Model 1903 type of disappearing emplacement, with two flank emplacements and a magazine in the common traverse. Work was done in 1906-1907. Transfer was made to service troops on November 9, 1910 for $88,062.11. It was armed with two 6-inch Model 1905 guns on Model 1905 disappearing carriages (#5/#1 and #8/#8). The battery was named in General Orders No. 77 of May 12, 1908 for 1st Lieutenant Harry A. Woodruff, U.S. Infantry who was killed on Mindanao in 1904. Like the other Fort Wint batteries, it served most of the interwar years in caretaker status. It was abandoned with the withdrawal of Fil-Am troops on December 24, 1941 and taken over by the Japanese. They did not reuse it as coast defense armament. It was retaken in January 1945. In August 1945 it suffered a devastating magazine explosion due to the carelessness of some visiting navy sailors. In later years the site was built over for local housing and the armament remains were scrapped. Some fragments and retaining walls can be seen on the property of the Grande Island Resort.

- **HALL:** The second dual 6-inch disappearing batteries built at Grande Island and Fort Wint. It was located on the eastern point of the island at about an 80-foot elevation, firing to the southwest. The submitted design followed closely the mimeograph for the Model 1903 type of disappearing emplacement, with two flank emplacements and a magazine in the common traverse. Work was done in 1906-1907. Transfer was made to service troops on December 7, 1910 for $83,168.10. It was armed with two 6-inch Model 1905 guns on Model 1905 disappearing carriages (#7/#7 and #6/#6). The battery was named on General Orders No. 77 of May 12, 1908 for 2nd Lieutenant Joseph Hines Hall, 17th Infantry who was killed on Mindanao in 1904. Like the other Fort Wint batteries, it served most of the interwar years in caretaker status. It was abandoned with the withdrawal of Fil-Am troops on December 24, 1941 and taken over by the Japanese. They did not reuse it as coast defense armament. It was retaken in January 1945. Sometime prior to July 1945 (most likely during the January 1945 intensive bombing campaign) the magazine exploded, heavily damaging the battery. In this condition it still exists, with the two guns and carriages still on the premises and at times heavily overgrown on the property of the Grande Island Resort. The battery is open to the public.

- **JEWELL:** One of two 4-gun, 3-inch batteries emplaced at Fort Wint for minefield coverage. This battery was emplaced on the northwestern end of the island, on the same ridge and on the right flank of Battery Woodruff. In most respects it was built to the design of standard types, with two interior and two flank emplacement platforms, though its battery commander station is more elaborate and set back from the traverse more than in most such emplacements. Work was done in 1908-1909. Transfer was made on November 9, 1910 for a cost of $43,201.25. Naming was done in General Orders No. 77 of May 12, 1908 in honor of 2nd Lieutenant James M. Jewell, 14th Cavalry, who was killed on Jolo in 1905. The original armament was four 3-inch Model 1903 guns and pedestals (#17/#10, #18/#11, #29/#12 and #33/#13). In 1917 the last two mentioned guns and carriages were removed for use to arm transports locally and eventually wound up arming searchlight-protecting Battery Hoyle in 1919 on Fort Frank. Then in October 1941, with the post's new mission as a training facility, the two empty emplacements received new guns to aid in training. These were an ex-navy 3-inch deck gun and a 6-pounder (57mm). With this mixed armament the battery was abandoned on December 24, 1941 when the U.S. forces evacuated Fort Wint. This armament remained when the island was retaken in January 1945. In the 1960s one of the Model 1903 3-inch guns and the navy 3-inch gun were sent to Washington State and put on display there at Fort Flagler. The original emplacement, with the single ex-navy 6-pounder still exists at the Grande Island Resort. The battery is open to the public.

• **FLAKE:** The second 4-gun, 3-inch battery at Fort Wint. It was emplaced on the eastern point of the island, inland and to the west of 6-inch Battery Hall. It was of standard design, with two flank and two interior emplacements and intervening traverse magazines. Work was done in 1908-1909. Transfer to service troops occurred on December 7, 1910 for a construction cost of $37,351.62. Naming was in General Orders No. 77 of May 12, 1908 for 2nd Lieutenant Campbell W. Flake, 22nd US Infantry who was killed on Mindanao in 1904. It was armed with four 3-inch guns Model 1903 on Model 1903 pedestal mounts (#11/#6, #12/#7, #14/#8, and #15/#9). Soon after construction it was kept in caretaker reserve status and rarely manned. In 1917 guns from emplacement No. 3 and No. 4 were removed for potential use on ship transports but eventually wound up being used as the armament for searchlight-protection Battery Cushing at Corregidor's Fort Mills. The two remaining guns and carriages served at the start of World War II at the training post of Fort Wint. During the reoccupation in 1945, gun No. 2 was damaged by a large fragment on the end of the gun tube and rendered unserviceable. Both of the remaining guns and carriages were removed in 1960 for transfer to Washington State to be displayed at a battery at Fort Casey. The empty emplacement still exists at the Grande Island Resort. The battery is open to the public.

Fort Wint prior to 1932 (NARA)

OVERSEAS BASES – PANAMA CANAL ZONE

The Overseas Bases – Panama Canal Zone is characterized by low, tropic seashore along the Caribbean Sea, a narrow canal passing through a large lake and then a mountain range, until you reach the semi-tropical Pacific Ocean with volcanic hills and islands. The coast defenses of the Panama Canal were focused around the canal entrances at Balboa and Cristobal. The climate is tropical and access to coast defense sites can be difficult. Construction of these harbor defenses started during the Taft Program, while defenses against air attack were developed during Interwar Period and during World War II. As a result, the entire Canal Zone became defended with coast artillery batteries, mobile infantry units, and air assets (both anti-aircraft batteries and pursuit aircraft). Important coast artillery forts were Fort Amador, Fort Grant, Fort Kobbe, Fort Randolph, Fort Sherman, and other military reservations. An excellent description of these forts can be found in Redoubt Press's *The American Defenses of the Panama Canal*. This book is available from Amazon.com. The American Canal Zone civilian government ceased to exist as such on October 1, 1979, however, the U.S. military did not turn over control of the territory to Panama until 1990 (not including canal operations or some of the bases). All remaining bases were closed as of August 1, 1999, and the official transfer of the canal occurred on December 31, 1999. The former bases have been converted to civilian usage or abandoned since that time.

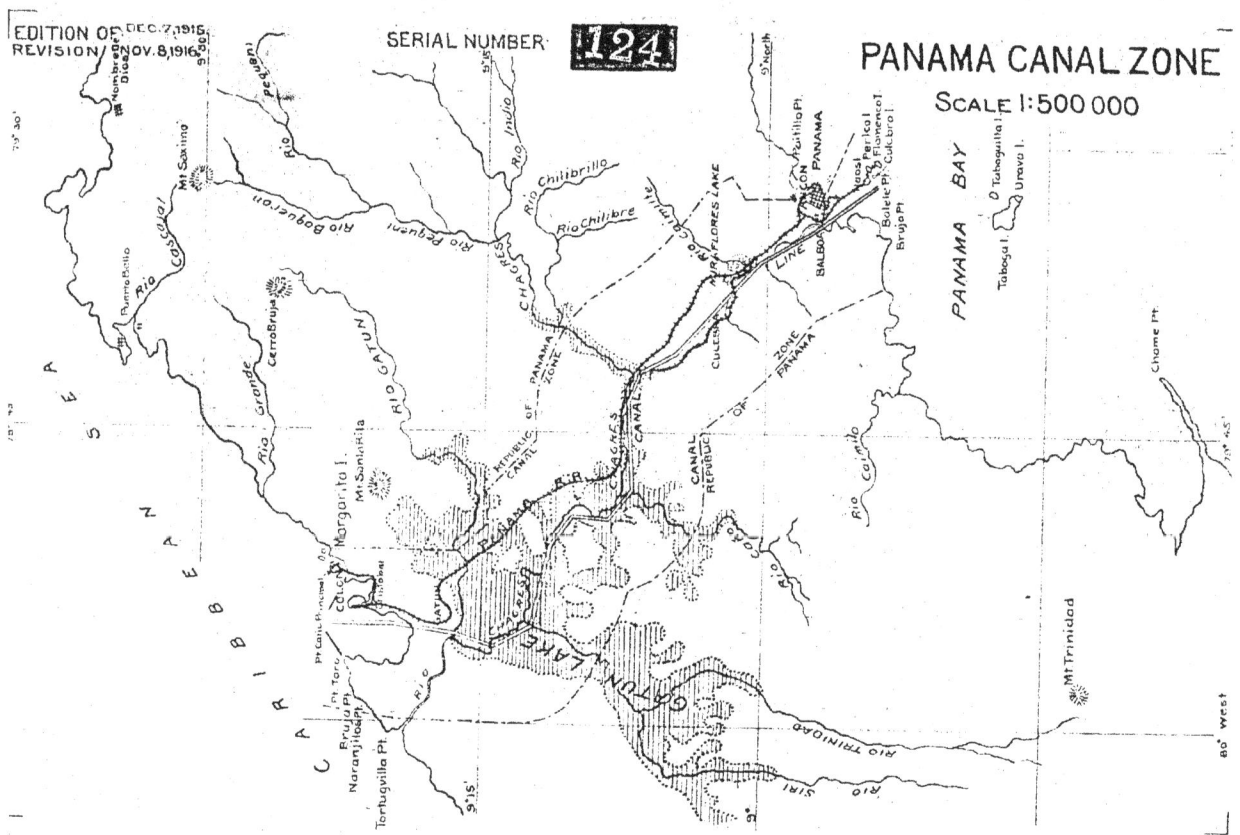

McGovern, Terrance. *The American Defenses of the Panama Canal.* Nearhos Publications and Redoubt Press. McLean, VA, 1999.

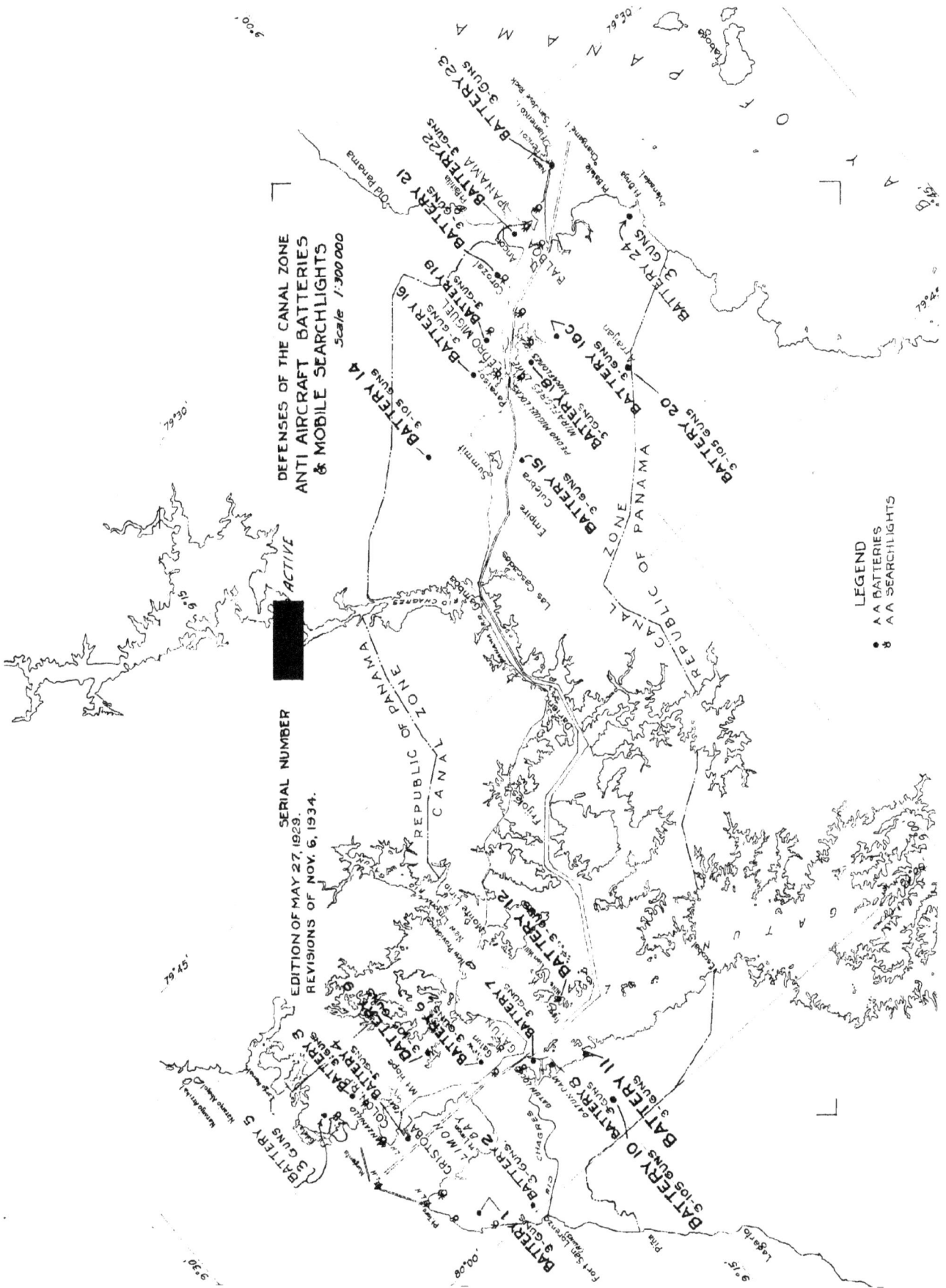

DEFENSES OF THE CANAL ZONE
ANTI AIRCRAFT BATTERIES
& MOBILE SEARCHLIGHTS
Scale 1:300 000

EDITION OF MAY 27, 1929.
SERIAL NUMBER
REVISIONS OF NOV. 6, 1934.

ACTIVE

REPUBLIC OF PANAMA

CANAL ZONE

REPUBLIC OF PANAMA

BATTERY 1
BATTERY 2
BATTERY 3
BATTERY 4
BATTERY 5
BATTERY 6
BATTERY 7
BATTERY 10
BATTERY 11
BATTERY 12

BATTERY 14 3.105 GUNS
BATTERY 15 3.GUNS
BATTERY 16
BATTERY 18
BATTERY 19
BATTERY 20 3.105 GUNS
BATTERY 21 3.GUNS
BATTERY 22 PANAMA 3.GUNS
BATTERY 23 3.GUNS
BATTERY 24 3 GUNS

LEGEND
• AA BATTERIES
✕ AA SEARCHLIGHTS

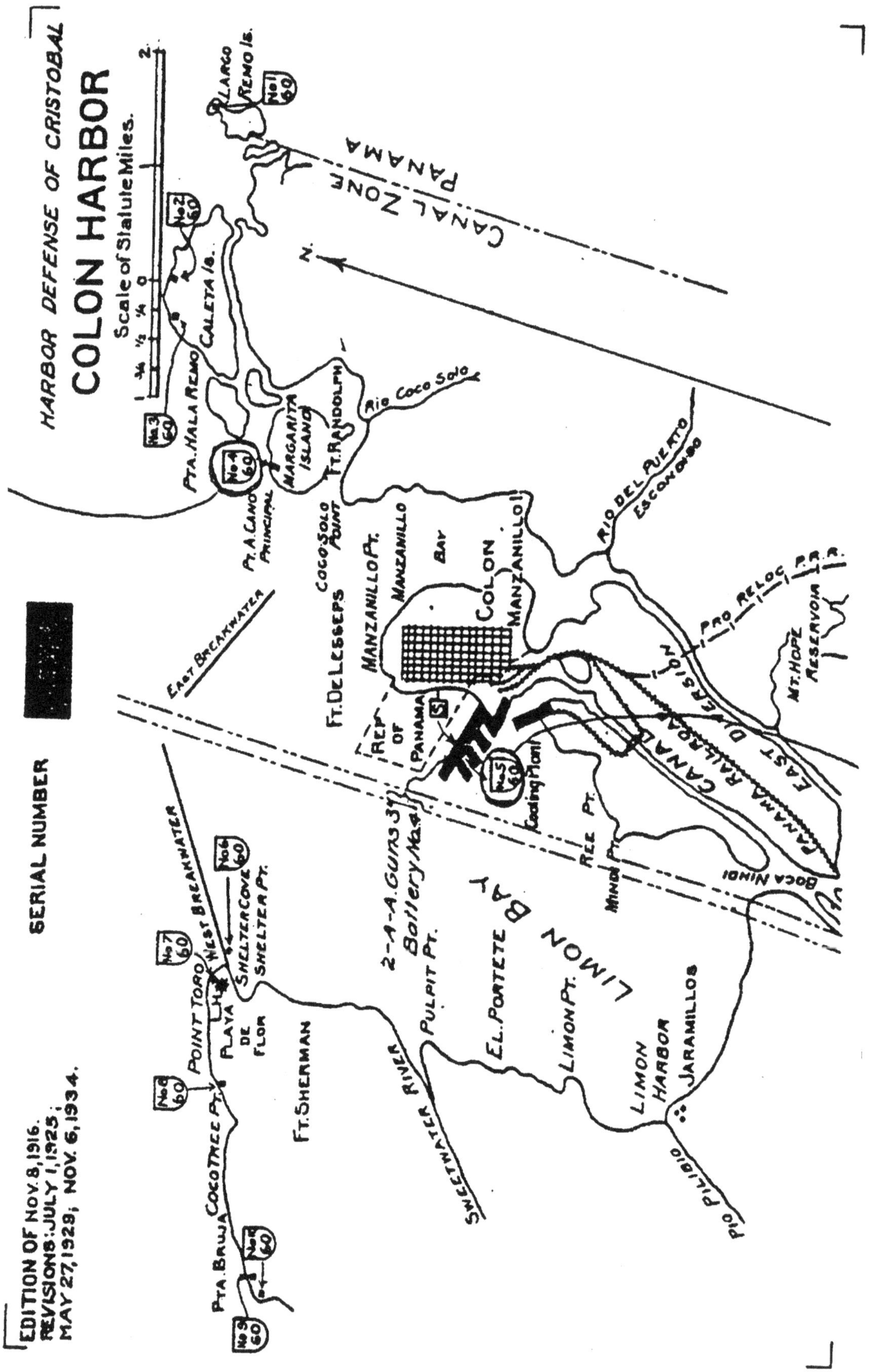

HARBOR DEFENSE OF CRISTOBAL

COLON HARBOR

Scale of Statute Miles.

¾ ½ ¼ 0   1   2

SERIAL NUMBER

EDITION OF NOV. 8, 1916.
REVISIONS: JULY 1, 1925;
MAY 27, 1929; NOV. 6, 1934.

CANAL ZONE
PANAMA

N

LARGO REMO Is.
No.1 60

No.2 60

PTA. HALA REMO
CALETA Is.

No.3 60

PTA. A. CANO
PRINCIPAL

No.4 60

MARGARITA ISLAND
Ft. RANDOLPH
Rio Coco Solo

CocoSolo Point
Ft. DE LESSEPS
MANZANILLO Pt.
MANZANILLO

BAY

COLON
MANZANILLO I.

RIO DEL PUERTO
ESCONDIDO

PRO RELOC. P.R.R.
EAST DIVERSION

MT. HOPE
RESERVOIR

EAST BREAKWATER

REP. OF PANAMA

No.5 60

Coaling Plant

Ree. Pt.

CANAL

PANAMA RAILROAD

Mindi Pt.

BOCA MINDI

No.6 60

WEST BREAKWATER

SHELTER COVE
SHELTER Pt.

POINT TORO
PLAYA DE FLOR

No.7 60

PTA. BRUJA
COCOTREE PT.

No.8 60

Ft. SHERMAN

SWEETWATER RIVER

2-A-A GUNS 3"
Battery/Mag.
PULPIT Pt.

EL PORTETE

LIMON Pt.

LIMON BAY

LIMON HARBOR

JARAMILLOS

RIO PIBIRIO

No.9 60

THE HARBOR DEFENSES OF CRISTOBAL – PANAMA CANAL ZONE (ATLANTIC/CARIBBEAN ENTRANCE)

Fort Randolph (1911-1979) is located on Margarita Island, actually a narrow-necked peninsula lying northeast of Colon. The Eastern part of the breakwater that protects Limon Bay extends out from the western shore of Margarita Island. Named in General Orders 153 in 1911 for Major General Wallace F. Randolph, who served as the first Chief of Artillery for the US Army. The largest guns of Fort Randolph were installed at Battery Webb. Situated on the northwest corner of the island, Battery Webb was equipped with two 14-inch rifles on disappearing carriages. The design of the battery has the standard magazines on same level as the loading platform, so ammunition service did not require lifts, but a large reserve magazine was located on a lower level under the battery. Battery Weed was erected a short distance to the southwest of Battery Webb. The battery consisted of two 6-inch rifles on disappearing carriages. To the east and slightly north of Battery Webb are Batteries Tidball and Zalinsky. These two batteries were each armed with four 12-inch mortars set in heavy concrete emplacements. On the northeastern tip of Margarita Island were two firing positions for Battery No. 1, the 14-inch railway guns. Since they could travel on the five-foot gauge tracks of the Panama Railroad, both guns could be used on either side, if required, although it was planned to keep one gun on each side. The circular firing positions themselves were to the east of the mortar batteries, while the magazines were located inside Battery Webb with overhead trolleys to bring the ammunition to rail carts to move shells and powder to the 14-inch railway guns. The fort also had four 155mm GPF guns on Panama mounts. The fort's garrison area was destroyed not long after the US Army turned over the reservation to Panama. Several development efforts have been undertaken but never completed. The most recently has been a large container port involving a large amount of landfill and leveling the island. The concrete batteries still survive as this project is hold for many years, but are still at risk of removal. There is no public access as the site is privately controlled.

Fort Randolph Gun Batteries

- **WEBB:** A battery for two 14-inch disappearing guns emplaced at the Margarita Island location on the northwest side firing to the northeast. E. E. Winslow of the U.S. Army Corps of Engineers submitted preliminary plans on September 9, 1911, and detailed final plans on June 26, 1912. Because of the low site and deep wells of 14-inch guns, it had to be built as a two-story structure. Adjacent to the guns were ammunition rooms calculated for a single engagement (roughly 90-rounds) with the lower-level rooms holding additional reserve ammunition and other service rooms. The battery had its own gasoline power generators in the battery structure. It was constructed from 1912 to 1914. Transfer was made on February 15, 1917 for a cost of $371,834.09. The battery was armed in late 1915 and carried two 14-inch Model 1910 guns on Model 1907M1 carriages (#9/#13 and #10/#14). Trunnion height was just 28.8 feet, gun centers were the standard distance for this gun size of 320-foot distance. These weapons were carried throughout the unit's service life. It was named on December 12, 1911 on General Orders No. 153 for Brevet Major General Alexander S. Webb of Civil War service who had just passed in 1911. During the 1920s unequal settlement of the structure pitched the guns forward, but by 1928 reverse settlement and repairs had mostly corrected this situation. In 1938 modifications were made in the magazine to store 14-inch railway ammunition and to support the two guns of this type also at Randolph. These modifications were transferred on March 14, 1928 for $9793. In 1946 the site was abandoned militarily and the armament removed, subsequently the emplacement was used for a variety of storage purposes. It still exists on commercial property in abandoned condition. The battery is closed to the public.

- **TIDBALL – ZALINSKI**: An emplacement for eight 12-inch mortars built as four adjacent tw mortar pits. E.E. Winslow submitted the preliminary plans along with those for Battery Webb on September 9, 1911. The design was for two adjacent batteries each of two pits. These mortars were on the northern side of the reservation, firing to the north. It was of conventional late Taft mortar emplacement design with a large forward parapet but no reverse parados. Battery Zalinski had the power plant for both batteries. Each of the eight mortars had an adjacent shell and powder room. On this low elevation, trunnion height was just 12.17 feet. The two western pits were named on December 12, 1911 in General Orders No. 153 Battery Zalinsky (named for Major General Edward Lewis Zalinski) and the two eastern Battery Tidball (named for Major General John C. Tidball of Civil War artillery service). Work was done from 1912 to January 1914 for transfer on October 6, 1916 for a cost of $171,364.15 (for each of the two similar batteries). The batteries were each armed with four Model 1912 mortars on Model 1896MIII carriages (Zalinsky #23/#23, #26/#24, #21/#25, and #1/#22, and Tidball #20/#18, #19/#19, #22/#20 and #24/#21). This armament was carried until directed salvaged on June 15, 1943. Postwar the area was used for a while as a low security containment area. The emplacement still exists and is used by several businesses and residences, though in a commercial area intended for development. The batteries are closed to the public.

- **WEED**: A battery for two 6-inch disappearing guns emplaced on the western side of the Margarita Island reservation covering the Limon Bay breakwater. Plans were submitted on September 16, 1912 by army engineer E.E. Winslow. The plan was a modification of recent mimeograph No. 59 but had raised magazines to allow adjacent-level ammunition service and concrete side walls replaced with gentle earthen slopes. To conceal the battery from small boats, the parapet was extended rearward on the left flank. A BC station was on the top of the traverse between the guns with plotting room immediately underneath. No power room was installed, emergency power coming from the one in adjacent Battery Webb instead. Gun centers were at 180-feet. Work was done from 1912 to 1914, and transfer to service troops made on October 6, 1916 for a total cost of $126,230.69. It was named in advance on December 12, 1911 in General Orders No. 153, in honor of Brig. Gen. Stephen H. Weed, U.S. Volunteers, Captain, 5th U.S. Artillery, who was killed in action, July 2, 1863, at Gettysburg. The battery was armed throughout its service life with two 6-inch Model 1908 gun tubes on Model 1905M1 carriages (#12/#16 and #15/#27). In later years it was little used, being last fired on October 8, 1929. By 1944 it was deleted and unmanned, and in 1946 authority was given to scrap the armament. The emplacement was not subsequently used, except perhaps as a local storage. It still exists today in abandoned condition on commercial land slated for development. The battery is closed to the public.

- **Railway Battery**: Two alternate sites were constructed for the pair of 14-inch railway guns employed in the Panama defenses, one in Fort Grant at Culebra Island and one here at Fort Randolph. Two specially reinforced gun blocks were built at the northeastern extreme of the island off of a spur track. Work was done from September 1928 to April of 1930, transfer being made on January 19, 1931 for a cost of $77,612.62. The blocks were on 42-foot diameter pads emplaced on piles. In 1937 additional sidings for plotting cars were also constructed. The guns (Model 1920 14-inch/50 #9/#2 and #8/#3) were mainly kept at Fort Grant but made at least one photo-documented trip to Randolph in 1932. The battery was apparently deleted shortly postwar and the guns were removed. The firing blocks remained in a heavily overgrown state until recently, but may have been destroyed by commercial development. The battery site is not open to the public.

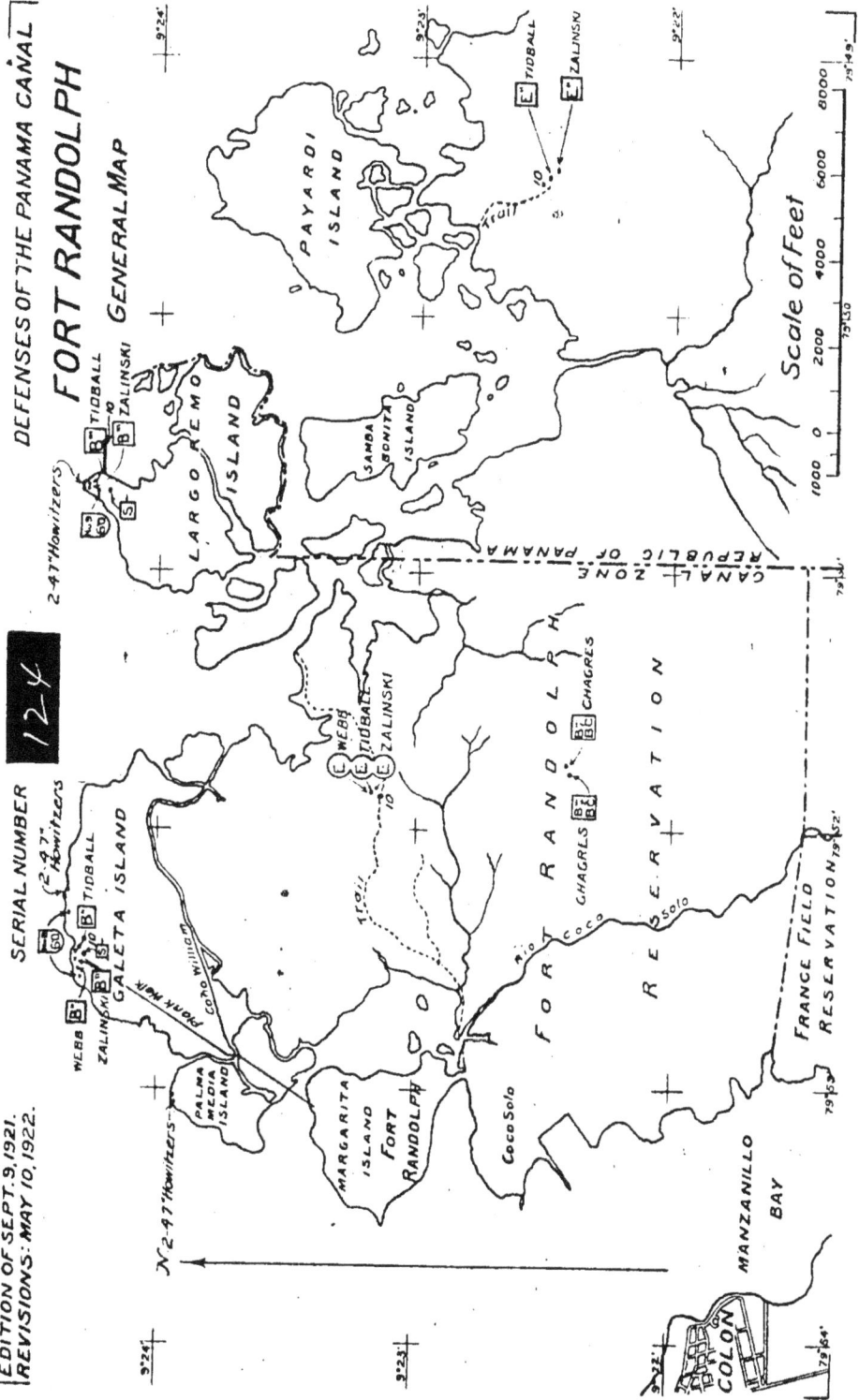

DEFENSES OF THE PANAMA CANAL

FORT RANDOLPH
GENERAL MAP

BATTERIES 6-4.7" H.

SERIAL NUMBER

EDITION OF SEPT. 9, 1921.
REVISIONS: MAY 10, 1922.

Scale of Feet

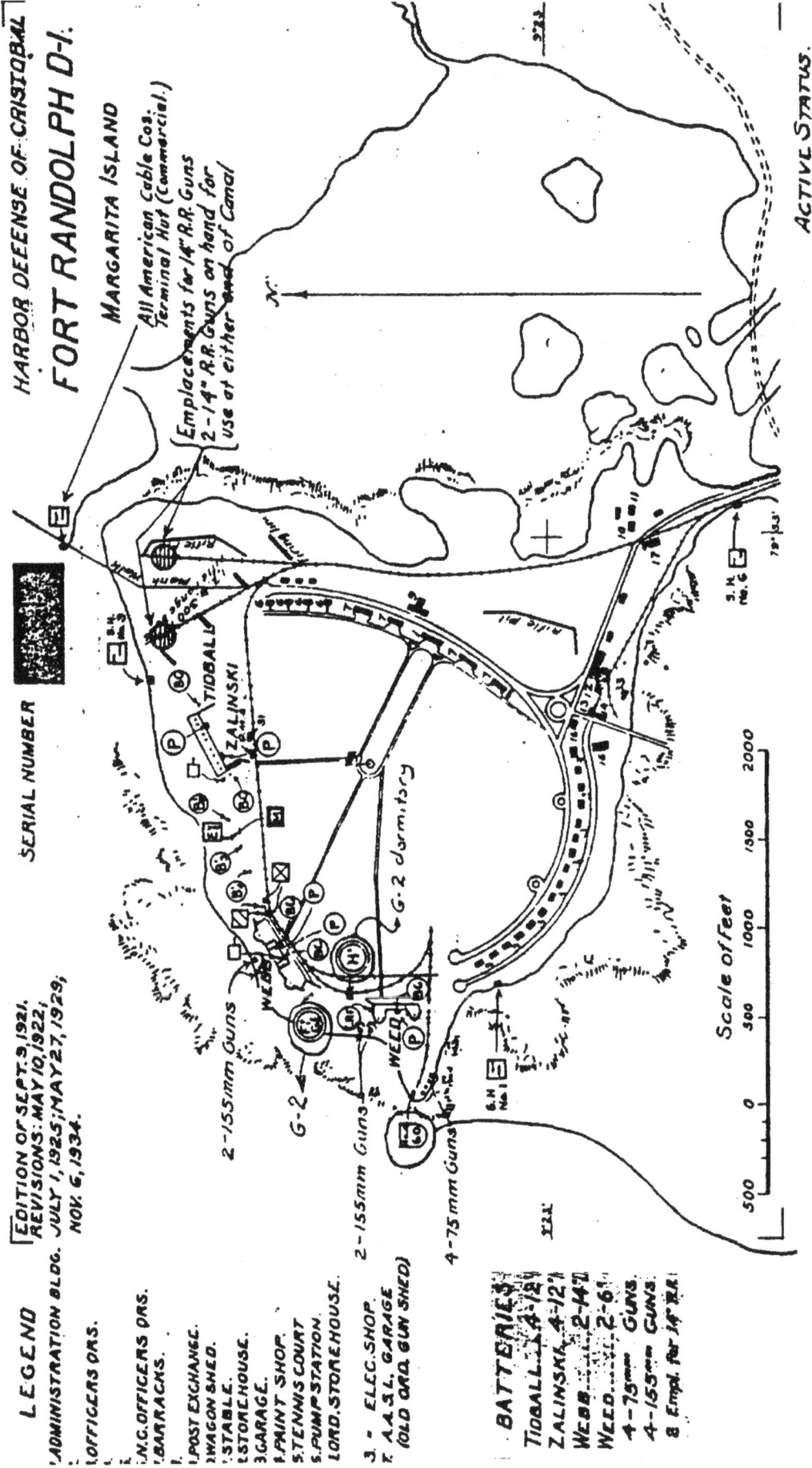

HARBOR DEFENSE OF CRISTOBAL

FORT RANDOLPH D-1.

MARGARITA ISLAND

All American Cable Cos.
Terminal Hut (Commercial.)

Emplacements for 14" R.R. Guns
2 - 14" R.R. Guns on hand for
use at either end of Canal

ACTIVE STATUS.

N

SERIAL NUMBER

LEGEND

EDITION OF SEPT. 9, 1921.
REVISIONS: JULY 1, 1925; MAY 10, 1922;
NOV. 6, 1934.

ADMINISTRATION BLDG. JULY 1, 1925; MAY 27, 1929;

OFFICERS QRS.

N.C. OFFICERS QRS.
BARRACKS.

POST EXCHANGE.
WAGON SHED.
STABLE.
STOREHOUSE.
GARAGE.
PAINT SHOP.
TENNIS COURT
PUMP STATION.
ORD. STOREHOUSE.

3. - ELEC. SHOP.
7. AA.S.L. GARAGE
(OLD ORD. GUN SHED)

BATTERIES
Tidball ... 4 - 12"
Zalinski .. 4 - 12"
Webb ..... 2 - 14"
Weed ..... 2 - 6"
4 - 75mm GUNS
4 - 155mm GUNS
8 Empl. for Inf. T.R.

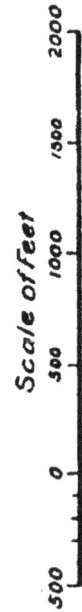

TIDBALL

ZALINSKI

G-2 Dormitory

2 - 155mm Guns

G-2

2 - 155mm Guns

4 - 75mm Guns

Scale of Feet

500   0   500   1000   1500   2000

Fort Randolph 1932 (NARA)

Fort DeLessups 1932 (NARA)

Fort DeLesseps (1911-1955) is located at Manzanillo Point adjacent to the Hotel Washington in the city of Colón. It was named in General Orders 153 of 1911after Ferdinand de Lesseps a French diplomat and later developer of the Suez Canal and the Panama Canal. Battery Morgan had two 6-inch rifles on barbette mounts. These mounts were only used at one other location for the two 6-inch batteries at Fort Drum in the Philippines. This battery supported guns on either flank (Battery Kilpatrick, Fort Sherman, and Battery Weed, Fort Randolph) in protecting the Atlantic approaches from attacks by light naval vessels, including minesweepers and submarines. The battery area is presently deserted, and the fort's buildings have been demolished to make way for commercial spaces. Nearby was located the Anti Motor Torpedo Boat Battery 3B (1942-1948) with four 90mm fixed mounts) on the Cristobal Mole along with two 3-inch AA guns, which has been built over many years ago.

Fort DeLesseps Gun Batteries

- **MORGAN**: The only battery for the small Fort DeLesseps at Manzanilla Point, Colon on the Atlantic side of the Panama Canal. First plans were submitted by engineer E.E. Winslow on September 18, 1912. The plans called for a battery of two 6-inch disappearing guns to be placed on the coral reef just west of the lighthouse. There were issues with the plan and location. On a cramped reservation (which was Panamanian, not part of the Zone). it was too close to the private hotel and swimming pool of the Panama Railroad Washington Hotel. Its closeness to the city might invite crowds or mobs attacking it, requiring Winslow to design an elaborate moat, two redoubts, and a protective defensive wall around the battery. The plan was refused, recommendations made that 6-inch pedestal guns be substituted which would need a much smaller footprint and that the dangers of attack could be accepted. Winslow submitted new plans on August 30, 1913. The plan was based generally on Battery Montgomery at Fort Monroe for a two-level battery with magazines below the loading platform. It was built just two hundred feet west of the Hotel Washington and fired to the northwest. Magazines could accommodate 900 rounds, and there were ready shelves near the guns, that were centered just 96.5-feet apart. It was constructed between 1914 and 1915, being transferred on February 5, 1917 for a cost of $111,416.25. It was named for Brigadier General Charles H, Morgan, of Civil War service. The battery was armed with a unique armament combination of two 6-inch Model 1908MII guns on Model 1910 pedestals (#6/#5 and #5/#6). Between the wars it was rarely manned and consistently had a problem with sand blowing into the emplacement. While the armament was mounted in 1928, it was rarely fired. Apparently, it served until about 1942 and then placed in reserved status after March 1944. New shields were developed for the carriage during the war, produced in mid-1943 and actually installed shortly thereafter. Eventually the fort site returned to Panamanian possession, however the original emplacement still exists. The battery is abandoned and open to the public but is considered a high security risk as one needs to travel through Colon to visit.

- **AMTB-3B**: An anti-motor torpedo boat battery of four fixed 90mm guns with M3 barbette carriage erected at Fort De Lesseps. It was located directly on the Cristobol mole extending as a breakwater into the anchorage. It was built from late 1942 to mid-1943. It consisted of simple concrete gun blocks. It was disarmed in 1948. Postwar the blocks were destroyed or built over; no trace remains today.

FORT DE LESSEPS

HARBOR DEFENSE OF CRISTOBAL

BATTERIES

MORGAN 2-6"M

Scale of Feet

100 50 0   200   400

LEGEND

EDITION OF SEPT. 9, 1921.
REVISIONS: MAY 10, 1922;
JULY 1, 1925; DEC. 11, 1926,
MAY 27, 1929; NOV. 6, 1934.

SERIAL NUMBER

ADMINISTRATION BLDG.
OFFICERS QRS.

N.C. OFFICERS QRS.
BARRACKS.
GUARD HOUSE.

TENNIS COURT.
BATHING POOL.
PAINT HOUSE.
OIL HOUSE.
SALVAGE HOUSE.
THEATER.
GARAGE.

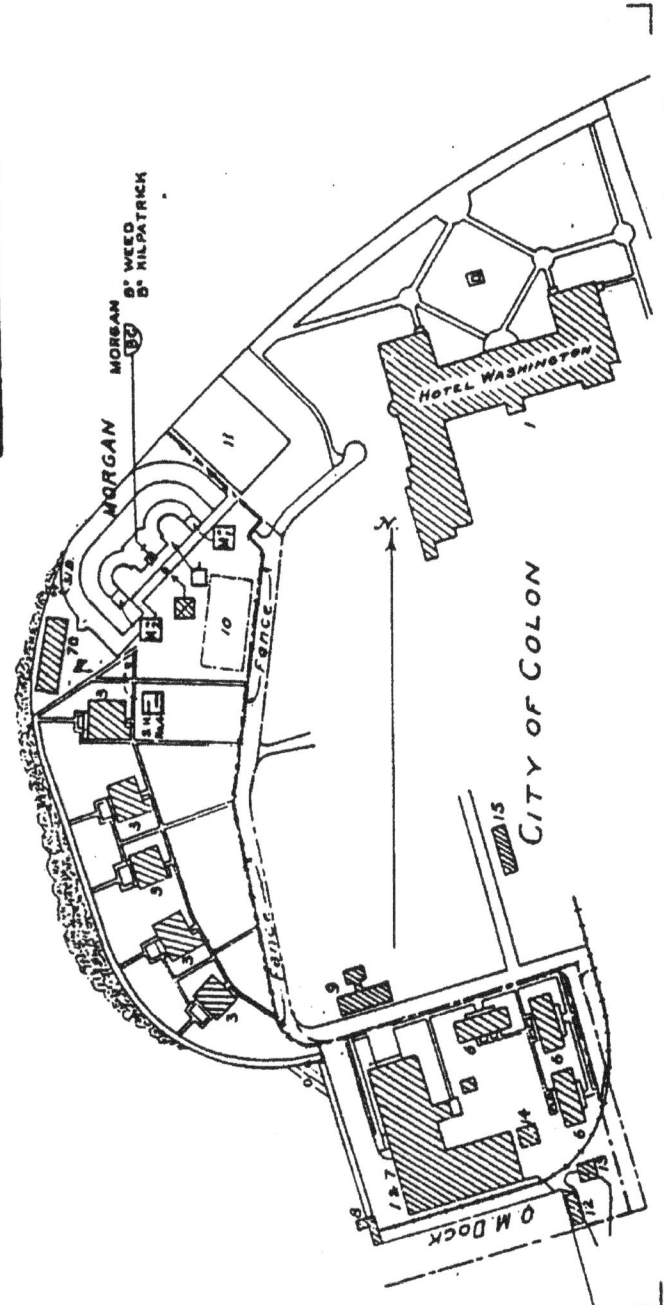

MORGAN

MORGAN
6" WEED
6" KILPATRICK

HOTEL WASHINGTON

CITY OF COLON

O. M. DOCK

ACTIVE STATUS.

Battery Morgan (Terry McGovern)

Fort Sherman (1911-1953/1999) is located across the harbor from Cristóbal at the entrance to Limon Bay. Fort Sherman was developed in the area lying generally between Limon Bay and the Chagres River, with the majority of its defense fortifications being constructed on the northern tip of land known as Toro Point. The Fort included 23,100 acres of land, about half of which was covered by jungle. Fort Sherman was named by War Department General Order No. 153 dated November 24, 1911, in honor of General William Tecumseh Sherman. Battery Mower contained a single 14-inch rifle mounted on a disappearing carriage. Battery Stanley directly supported Battery Mower, being situated about 200 yards to the west. Also equipped with a single 14-inch rifle, the emplacement and armament of Battery Stanley were nearly identical to that of Mower. Batteries Baird and Howard were constructed a short distance to the rear of Batteries Mower and Stanley. In each of the batteries were four 12-inch mortars A branch of the rail line that connected Batteries Mower and Stanley with the Shelter Cove docks also ran to Batteries Baird and Howard. Battery Kilpatrick was erected to the east of Batteries Baird and Howard, on the northern tip of Shelter Point. The battery was equipped with two 6-inch rifles on disappearing carriages The two guns covered the Toro Point Breakwater and supported the field of fires of Batteries Morgan and Weed. Also, located on Shelter Point was AMTB Battery 3C (90mm) and Battery W with 4-155mm guns on Panama mounts (1940) near the lighthouse. Located at Shelter Cove was the controlled submarine mine complex that included wharfs for the Army mine planter, yawls, DB boat, mine storehouses, loading room, cable tanks, and mine casemate (which was integrated into the Battery Kilpatrick emplacement). The harbor minefield consisted of 15 groups of 19 buoyant mines each in World War I; and 26 groups of 13 ground mines each (with hydrophones) in World War II, controlled by the mine casemates at Shelter Cove-Fort Sherman, and at Fort Randolph. Battery Pratt was erected approximately two and half miles southwest of the Toro Point battery complex. When construction first started in October 1916, the battery was designated Chagres Battery No. 1, since the area in which it was located is near Chagres River. There was a concern that modern battleships could bombard the Gatun Dam and Locks from the sea near these battery sites. These long-range guns would force the attacking warships out of range of these key Panama Canal structures.

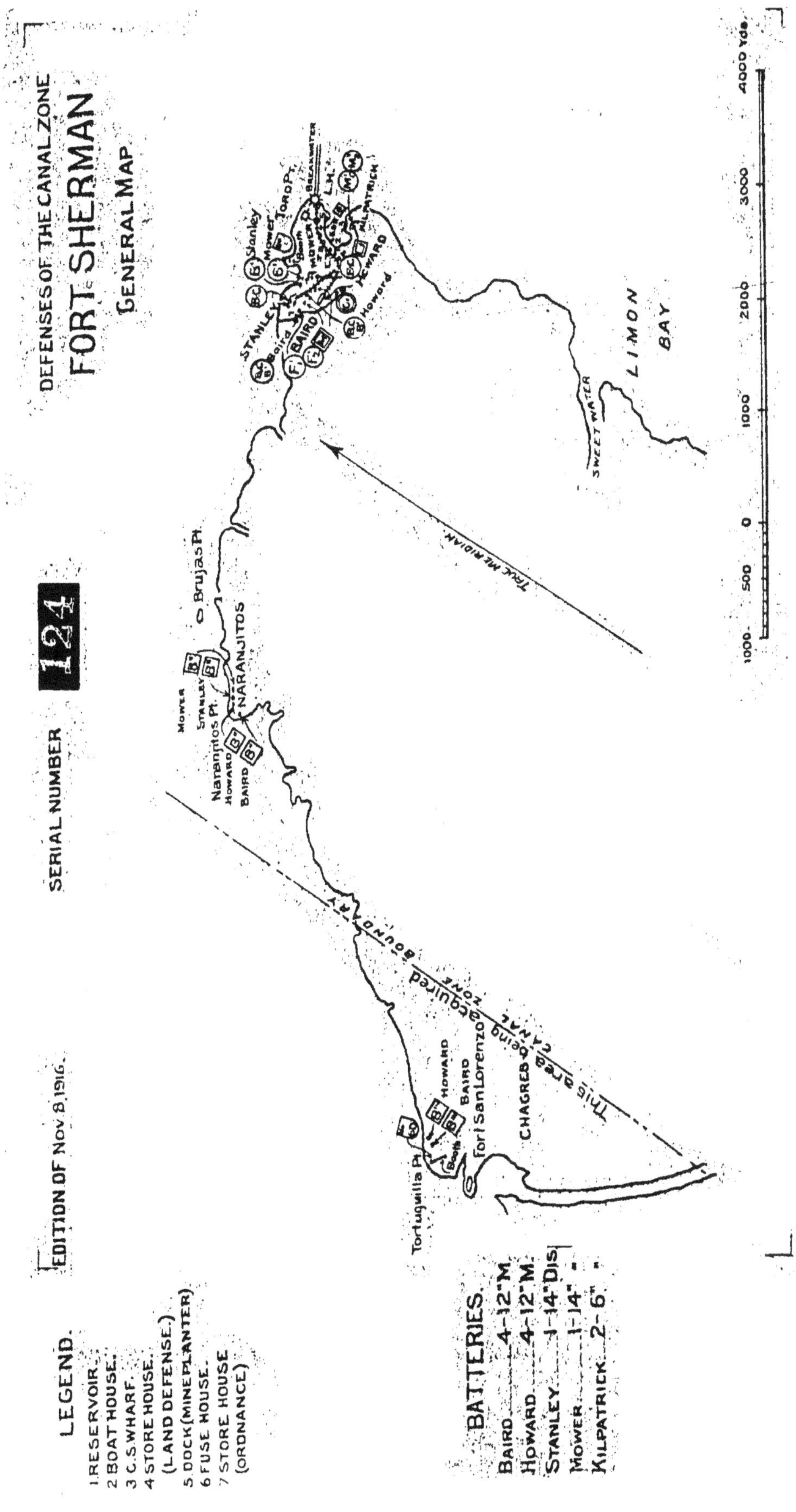

DEFENSES OF THE CANAL ZONE
FORT SHERMAN
GENERAL MAP

SERIAL NUMBER 124

EDITION OF Nov 8. 1916.

LEGEND.

1 RESERVOIR.
2 BOAT HOUSE.
3 C.S. WHARF.
4 STORE HOUSE.
   (LAND DEFENSE.)
5 DOCK (MINE PLANTER)
6 FUSE HOUSE.
7 STORE HOUSE
   (ORDNANCE)

BATTERIES.

BAIRD ———— 4-12"M.
HOWARD —— 4-12"M.
STANLEY —— 1-14" Dis.
MOWER ——— 1-14" "
KILPATRICK 2- 6" "

LIMON BAY

TRUE MERIDIAN

1000 500 0 1000 2000 3000 4000 YDS.

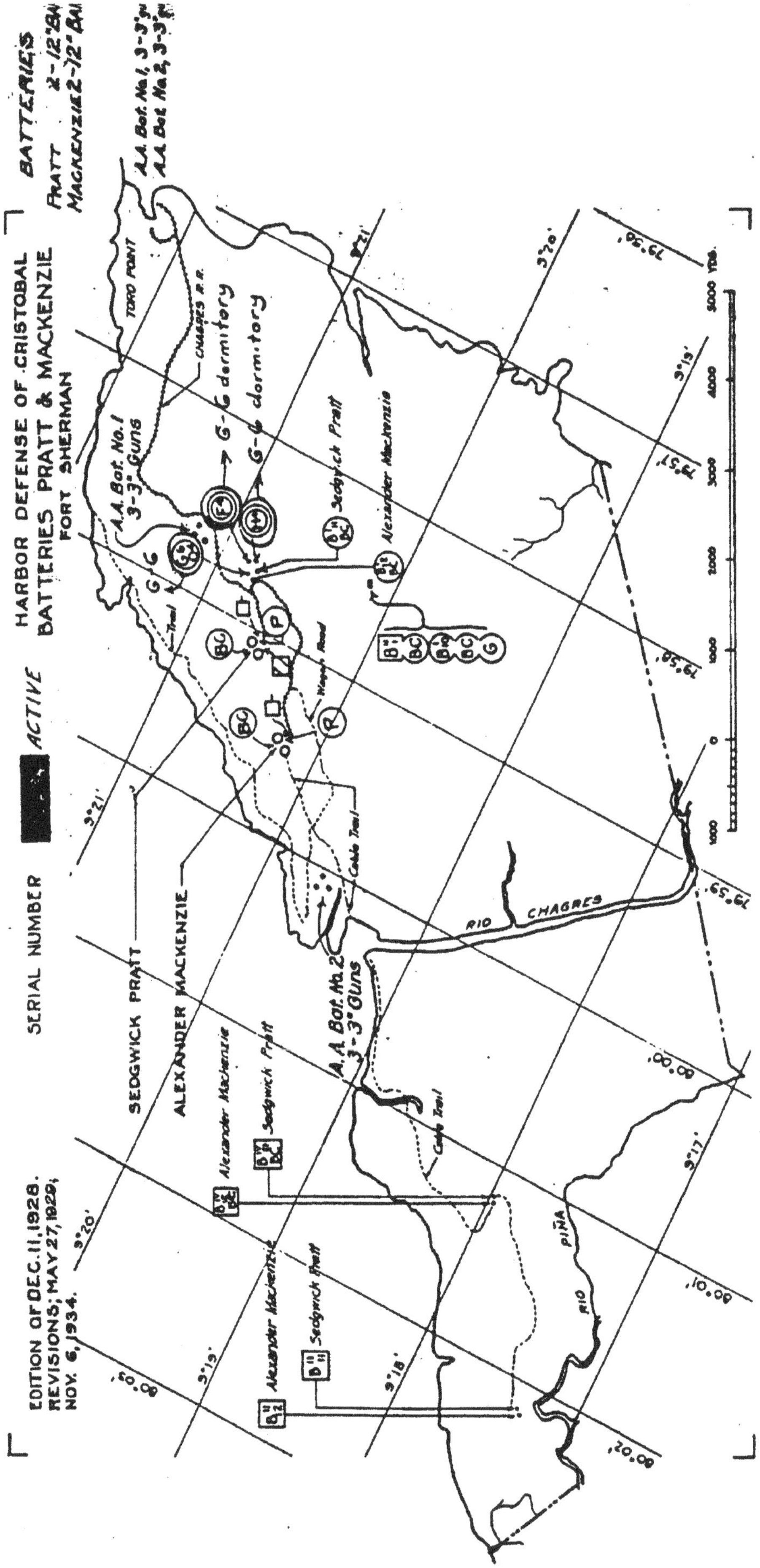

HARBOR DEFENSE OF CRISTOBAL
BATTERIES PRATT & MACKENZIE
FORT SHERMAN
BATTERIES
SERIAL NUMBER
ACTIVE
SEDGWICK PRATT
ALEXANDER MACKENZIE
EDITION OF DEC.11,1928.
REVISIONS; MAY 27,1929;
NOV. 6,1934.
TORO POINT
A.A. Bat. No.1
3-3" Guns
A.A. Bat. No.2
3-3" Guns
G-G dormitory
G-G dormitory
Sedgwick Pratt
Alexander Mackenzie
Wagon Road
Cable Trail
RIO CHAGRES
RIO PIÑA

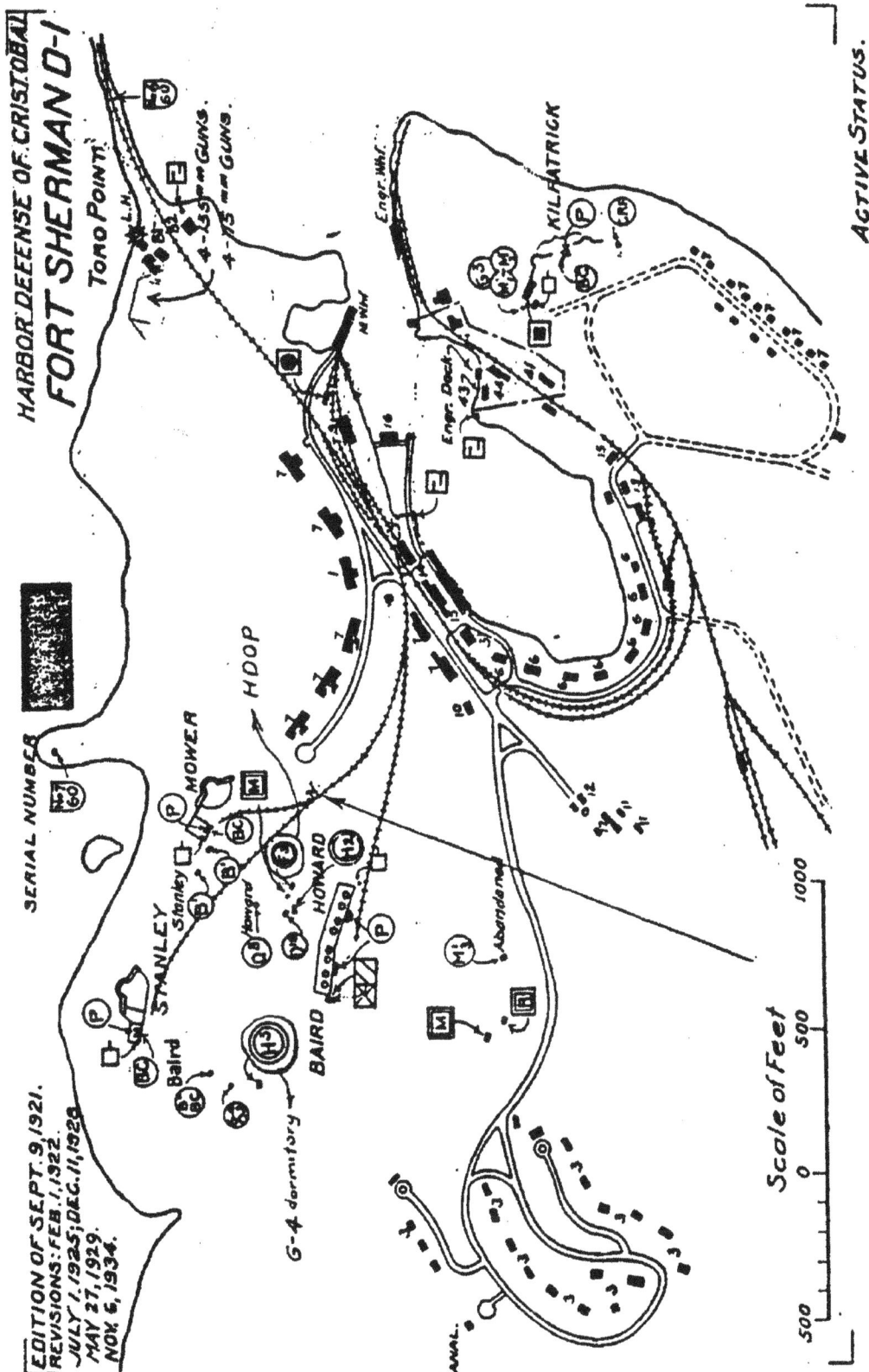

HARBOR DEFENSE OF CRISTOBAL
FORT SHERMAN D-1
Toro Point

SERIAL NUMBER

EDITION OF SEPT. 9, 1921.
REVISIONS: FEB. 1, 1922.
JULY 1, 1925; DEC. 11, 1929.
MAY 27, 1929.
NOV. 6, 1934.

ACTIVE STATUS.

Scale of Feet

LEGEND
1. ADMINISTRATION BLDG.
2.
3. OFFICERS QRS.
4.
5.
6. N.C. OFFICERS QRS.
7. BARRACKS.
8.
9.
10. BAND BARRACKS.
11. STABLE.
12. WAGON SHED.
13. GARAGE.
14. STORE HOUSE.
15. SHOP.
16. BOAT HOUSE.
17. ORDNANCE ST. HO.
18. STOREHOUSE.
19. " ELECTRICAL SHOP.
20. " BLACKSMITH SHOP.
21. " CARPENTER SHOP.
22. LIGHT HOUSE DIV, PANAMA CANAL.
23. U.S. NAVY.

BATTERIES
Baird ——— 1-12"
Howard —— 4-12"
Stanley —— 3-14"
Mower —— 1-6"
Kilpatrick. 2-10"
4 — 155 mm 6 GUNS
4 — 75 mm GUNS

The battery was equipped with two 12-inch rifles on barbette mounts. The guns of Battery Pratt not only outranged the heavier weapons of Batteries Mower and Stanley but also covered areas west of the limits of the 14-inch rifles. In September 1942, the casemating of Battery Pratt was undertaken to provide overhead protection from aerial and naval bombardment. To the southwest of Battery Pratt, Battery Mackenzie was erected, and it was a twin to Battery Pratt, equipped with two 12-inch rifles of the same model and mount. Battery Mackenzie was not casemated as it was decided that the battery's all-round fire was more important that adding overhead protection. A proposed Battery 151 (casemated 16-inch) was never built, although several gun tubes were delivered to Fort Sherman. Battery Pratt was later converted as the alternate command post for Headquarters, U.S. Southern Command and later became an important telecommunication switching point for undersea cables for Panama. When Fort Sherman no longer had a harbor defense role, the reservation became the U.S. Army's Jungle Operations Training Center after 1953. Also, a HAWK air defense missiles were emplaced on the Fort Sherman reservation from October 1960-1968. In 1999, the U.S. Army transferred control of Fort Sherman to the Government of Panama. The reservation now has several uses including barracks for the Panamanian Public Forces, a hotel & marina, tourism at Fort San Lorenzo (Parque Nacional San Lorenzo), and several other businesses. All the fortifications except for Battery Pratt are abandoned and lost in the jungle. Public access is available but keep in mind the structures are in poor condition.

Fort Sherman Gun Batteries

- **MOWER**: A battery for a single 14-inch disappearing gun built as a sister unit with Battery Stanley on the central part of the Toro Point Fort Sherman reservation. Engineer E.E. Winslow submitted a preliminary design on August 31, 1911, and final plans on March 16, 1912. It was of conventional late Taft design, with an adjacent, same-level magazine (withdrawn on the left side) for horizontal ammunition service. The battery had its own BC, plotting room, small power plant with two 25-kw generators and gasoline supply, and mechanical range indicators. Work was done from 1912 to 1914 for transfer on July 11, 1916 at a construction cost of $128,174.14. It was named on December 12, 1911 in General Orders No. 153 for Major General Joseph Anthony Mower of Civil War service. The battery was armed with one 14-inch Model 1910 gun on a Model 1907M1 disappearing carriage (#3/#8). The emplacement was served by a supply rail line from Shelter Cove. It served until late during World War II, being recommended for abandonment and scrapping in 1946. The emplacement still exists, though badly overgrown. The battery is open to the public.

- **STANLEY**: A second battery for a single 14-inch disappearing gun built on the Toro Point reservation to the west of sister Battery Mower. Engineer E.E. Winslow submitted preliminary design on August 31, 1911, and final plans on March 16, 1912. It was of conventional late Taft design, with an adjacent, same-level magazine (withdrawn on the left side) for horizontal ammunition service. The battery had its own BC, plotting room, small power plant with two 25-kw generators and gasoline supply, and mechanical range indicators. Unlike Mower it had just a single entry into the magazine. Work was done from 1912 to 1913. It was transferred on July 11, 1916 for a cost of $176,483.42. The battery was named on December 12, 1911 in General Orders No. 153 for Major General David Sloane Stanley, commander of the Fourth Corps during the Civil War. It was armed with one 14-inch Model 1910 gun on a Model 1907M1 disappearing carriage (#5/#9). The battery was served by an ammunition rail line which extended from the docks at Shelter Cove. The gun was last fired on December 13, 1943 and manning ceased after 1944. The gun was scrapped between 1946 and 1948. The emplacement still exists but is heavily overgrown. The battery is open to the public.

- **BAIRD – HOWARD:** The mortar battery for Fort Sherman, consisting of four dual mortar pits arranged in line. It was in the center of the reservation, just inland and to the south of the two 14-inch batteries. Engineer E.E. Winslow submitted preliminary plans of August 31, 1911 and complete plans on January 15, 1912. It had four similar pits of two Model 1912 mortars each, with adjacent magazines and on both flanks galleries with numerous quarters and support rooms. The battery entrance galleries connected to the ammunition railway that led to the fort pier. Work was done between 1913 and 1915, and transfer made early in 1916 with a cost of $150,316.75. The position was covered to the south by Sherman's unique infantry wall. The western two pits were named in General Orders 153 of December 12, 1911 for Brigadier General Absalom Baird of Civil War service and the eastern two pits named for Major General Oliver O. Howard from the same conflict. They were armed with eight 12-inch Model 1912 mortars on Model M1896MIII carriages (Howard: #12/#10, #15/#11, #27/#12, and #28/#17 and Baird: #25/#26, #2/#27, #18/#28, and #4/#1). This armament served until declared obsolete on June 15, 1943 (though it appears that much of the armament may have been dismounted for five years between 1927 and 1932). In later years Battery Howard had an iron cover pit over the pits for rain cover and was used as a storage area. Also, in postwar years Battery Howard's plotting room was removed. Thus modified the emplacement still exists but is overgrown. The battery is open to the public.

- **KILPATRICK:** A battery for two 6-inch disappearing guns emplaced on the northern tip of Shelter Point, covering the northern side of the Toro Point breakwater and the minefield. Plans by army engineer E.E. Winslow were submitted on September 9, 1912. The plan was a modification of recent mimeograph No. 59, but had raised magazines to allow adjacent-level ammunition service and concrete side walls replaces with gentle earthen slopes. To conceal the battery from small boats, the parapet was extended rearward on the right flank. A BC station was on the top of the traverse between the guns with plotting room immediately underneath. No power room was installed, electricity coming from the main batteries instead. Gun centers were at 186-feet. The Atlantic mine casemate was constructed on the immediate left flank of the battery. Work was done in 1915-16 and transfer made on June 23, 1916 for a cost of $149,193.39. The unit was named on December 12, 1911 in General Orders No. 153 for Major General Judson Kilpatrick of Civil War service. It was armed with two 6-inch Model 1908 guns on Model 1905M1 disappearing carriages (#11/#25 and #10/#24). It served throughout World War II, not being recommended for abandonment until early 1946. The guns and carriages were dumped at sea on August 27, 1946. In later years the site was used for the Jungle Warfare School and zoo. The emplacement still exists. The battery is open to the public.

- **MACKENZIE:** A dual 12-inch long-range battery built as part of the 1915 Board of Review Program to update the Atlantic defenses of the Panama Canal. MacKenzie and Battery Pratt were built at the same time to the southwest of the Taft batteries at a considerable distance. Specifically, they were required to provide cover over the Gatun spillway and the cover a fleet debouching into the Atlantic through the canal. Initial funding of $400,000 for both batteries (or rather for emplacements for four 12-inch howitzers) was obtained in the Sundry Civil Act of July 1, 1916. Preliminary plans were submitted on August 2, 1916. Site selection was approved on August 11. Final plans, after some discussion and changes, were submitted on November 10, 1916. Work was done from October 1916 to May 1923, for final transfer on October 10, 1923 at a cost of some $401,796.78. It was of conventional open-back design. The battery was named in General Orders No. 13 of March 27, 1922 for Major General Alexander MacKenzie who died in 1921. It was armed with two 12-inch Model 1895M1A4 rifles on Model 1917 barbette carriages (#74/#18 and #54/#16). Gun trunnions elevation was at 174-feet. It served actively between the wars, and unlike Battery Pratt, was never given overhead casemated protection, probably in a desire to retain one battery with a full 360-degree potential firing coverage. After being deleted in 1948 it served for many years as a command post of military forces in the area and then special forces training area. The emplacement still exists in abandoned and very overgrown condition. The battery is open to the public.

- **PRATT**: The sister 12-inch long-range battery to Battery Mackenzie at Fort Sherman. It was built close by on the reservation and also fired to the northwest. Initial funding of $400,000 for both batteries (or rather for emplacements for four 12-inch howitzers) was obtained in the Sundry Civil Act of July 1, 1916. Preliminary plans were submitted on August 2, 1916. Site selection was approved on August 11. Final plans, after some discussion and changes, were submitted on November 10, 1916. Work was done from October 1916 to May 1923, for final transfer on October 10, 1923 at a cost of some $428,702.55. It was named in General Orders No. 13 of March 27, 1922 for Brigadier General Sedgwick Pratt who died in 1920. The armament consisted of two 12-inch Model 1895M1A4 guns on Model 1917 barbette carriages (#43/#20 and #55/#17). Gun trunnion elevation was at 231-feet. It carried this armament throughout its service life, until disarmed and disposed of by authority of May 10, 1948. The battery was provided with a concrete casemate during 1942-43. After the war it served various command functions and eventually was a major telecommunications facility. The casemated work still exists and continues to be used as undersea cable transmission point. The battery to closed to the public.

- **AMTB-3C**: An anti-motor torpedo boat battery of four fixed 90mm guns with M3 barbette carriage erected at Fort Sherman. It was located directly on Shelter Point in front of the 6-inch disappearing gun battery. It was built from late 1942 to mid-1943. It consisted of simple concrete gun blocks. It was disarmed in 1948. The gun blocks remain today. The battery is open to the public.

(0569-13486-12)(9-15-32-9A)(12-1200) *Fort Sherman, C.Z.*   17958 A.C.

Fort Sherman 1932 (NARA)

Battery MacKenzie Fort Sherman, Panama Canal Zone (Terry McGovern)

Battery Pratt Fort Sherman, Panama Canal Zone (Terry McGovern)

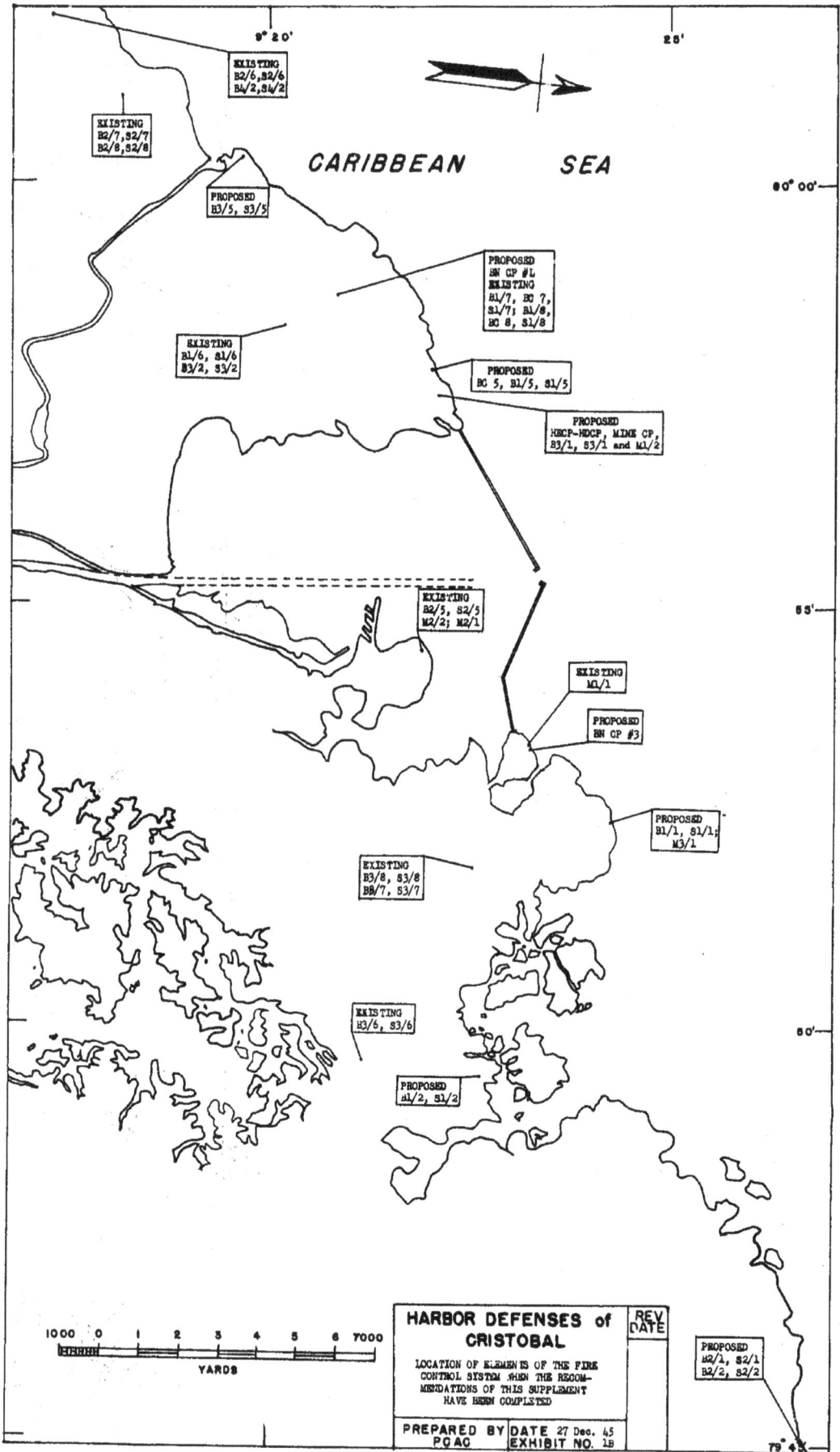

CARIBBEAN SEA

EXISTING
B2/6, S2/6
B4/2, S4/2

EXISTING
B2/7, S2/7
B2/8, S2/8

PROPOSED
B3/5, S3/5

PROPOSED
BN CP #L
EXISTING
B1/7, BC 7,
S1/7; B1/8,
BC 8, S1/8

EXISTING
B1/6, S1/6
B3/2, S3/2

PROPOSED
BC 5, B1/5, S1/5

PROPOSED
HECP-HDCP, MINE CP,
B3/1, S3/1 and M1/2

EXISTING
B2/5, S2/5
M2/2; M2/1

EXISTING
M1/1

PROPOSED
BN CP #3

PROPOSED
B1/1, S1/1;
M3/1

EXISTING
B3/8, S3/8
BB/7, S3/7

EXISTING
B3/6, S3/6

PROPOSED
M1/2, S1/2

PROPOSED
B2/1, S2/1
B2/2, S2/2

1000 0   1   2   3   4   5   6 7000
YARDS

HARBOR DEFENSES of
CRISTOBAL

REV
DATE

LOCATION OF ELEMENTS OF THE FIRE
CONTROL SYSTEM WHEN THE RECOM-
MENDATIONS OF THIS SUPPLEMENT
HAVE BEEN COMPLETED

PREPARED BY
PCAC
DATE 27 Dec. 45
EXHIBIT NO. 1B

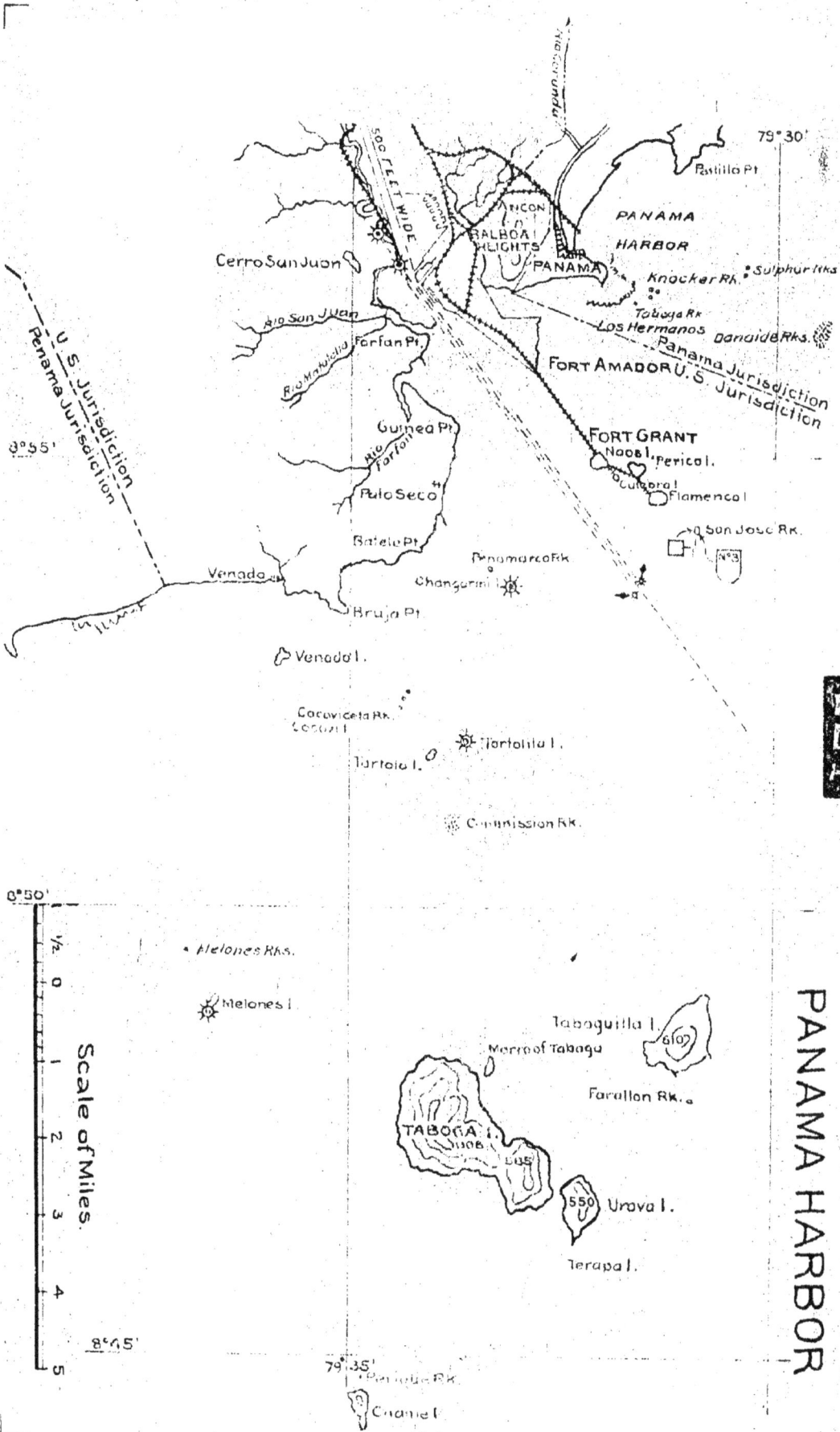

EDITION OF NOV. 8, 1916.

SERIAL NUMBER

424

DEFENSES OF THE CANAL ZONE

PANAMA HARBOR

Scale of Miles.

THE HARBOR DEFENSES OF BALBOA – PANAMA CANAL ZONE (PACIFIC ENTRANCE)

Fort Amador (1911-1979/1996) is located in Balboa between the Panama Canal and the City of Panama City on Panama Bay. In August 1911, when the first ground was broken in preparation for the construction of the Pacific fortifications, the area now known as Fort Amador was little more than a broad stretch of mud flats and swamp. Used as a dump for material removed from the canal, the area was appropriately referred to as "Balboa Dumps". The Isthmian Canal Commission, however, had started building a breakwater from the mainland, in the vicinity of Balboa, to Naos Island to protect the canal channel from cross currents. Fort Amador was originally considered the mainland portion of Fort Grant until 1917. Fort Amador is named in General Orders 153 of 1911 for Manuel Amador Guerrero, the first president of Panama. Although Fort Amador was built and used primarily for housing the Coast Artillery units that manned the batteries of Fort Grant, some armament was installed. On the southern extremity of the post, on a little rise just north of the Officers' Open Mess, were Battery Birney (6-inch DC) and Battery Smith (6-inch DC), and also AMTB Battery 7A (90mm). Fort Amador was the primary garrison area for the harbor defenses of Balboa until 1948. The post then served several roles including the Headquarters U.S. Army, Southern Command and U.S. Naval Forces, Southern Command. Most of the post was turned over to Panama in 1979 and became headquarters of the Panamanian Defense Force until 1989. Portions of the post remained under American control until 1996. Most original military buildings were demolished by 2000 for new commercial development, such as large convention center and bio-museum. Nearby at Balboa Heights was Quarry Heights Military Reservation (1916-1997), with an underground joint Army/Navy command bunker/tunnel complex on Ancon Hill.

Fort Amador Gun Batteries

- **SMITH:** A battery for two 6-inch disappearing guns built at the very southern tip of the Amador reservation, near the spit of land starting the causeway to the fortified islands. The land had been formed with the excavation of the canal and was termed at the time the Balboa Dump. Plans were drawn up by army engineer E.E. Winslow and submitted on December 7, 1912. The plan was a modification of recent mimeograph No. 59, but had raised magazines to allow adjacent-level ammunition service and concrete side walls replaces with gentle earthen slopes. Gun centers were at 170-feet. An officer's room and latrine were placed on the right flank of the emplacement. A BC station was on the top of the traverse between the guns with plotting room immediately underneath. No power room was installed internally, but a small station on the battery's left flank supplied both it and nearby Battery Birney. Work being done from 1913 to 1915. The battery was transferred on March 5, 1917 for a construction cost of $154,222.03. It was named in advance on December 12, 1911 in General Orders No. 153 for Major General Charles F. Smith who died in 1862 during the Civil War. The battery was armed with two 6-inch Model 1908 guns on Model 1905MII disappearing carriages (#14/#29 and #16/#28). These guns were scrapped by a salvage order dated June 5, 1943. In postwar years the emplacement was filled-in, and new army housing erected on top. Nothing now remains of this battery.

- **BIRNEY:** A second dual 6-inch battery at the Fort Amador Balboa Dump, it was emplaced just to the east of Batty Smith. Like that battery it had a field of fire to both sides of the causeway to the fortified islands. Plans were drawn up by army engineer E.E. Winslow and submitted on December 7, 1912. The plan was a modification of recent mimeograph No. 59, but had raised magazines to allow adjacent-level ammunition service and concrete side walls replaces with gentle earthen slopes.

DEFENSES OF THE PANAMA CANAL

FORT AMADOR

NORTHERN PART

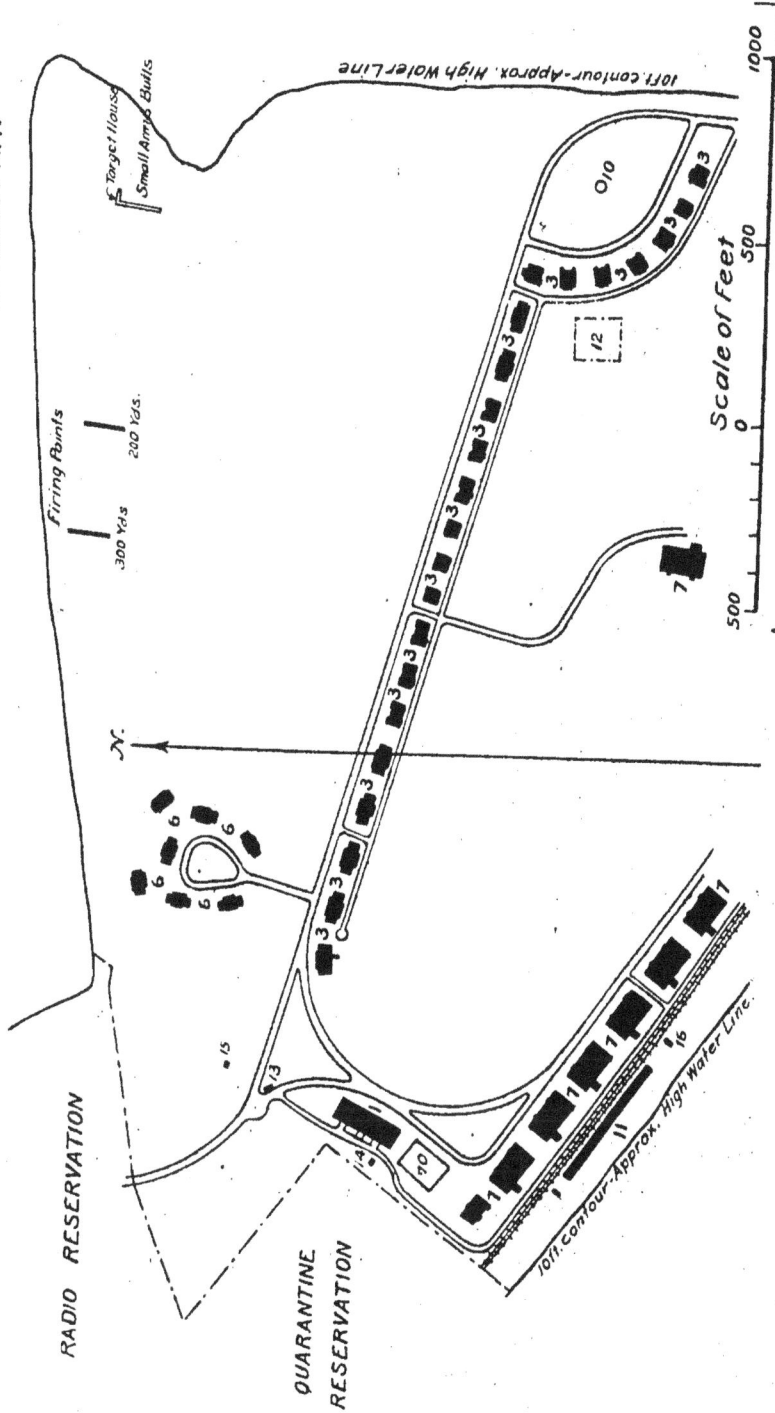

SERIAL NUMBER 124

EDITION OF SEPT. 9, 1921.
REVISIONS: MAY 10, 1921.

LEGEND

1. ADMINISTRATION BLDG.

3. OFFICERS QRS.

6. N.C. OFFICERS QRS.
7. BARRACKS.

10. BAND STAND
11. VOCATIONAL TRAINING.
12. TENNIS COURT.
13. MAIN GATE.
14. SUB-POWER HUT.
15. WATER METER HUT.
16. CO. MECHANIC SHOP.
70. OPEN AIR THEATER.

RADIO RESERVATION

QUARANTINE RESERVATION

Firing Points

300 Yds.

200 Yds.

Target House

Small Arms Butts

10ft. contour-Approx. High Water Line

Scale of Feet

1000   500   0   500

10ft. contour-Approx. High Water Line.

N.

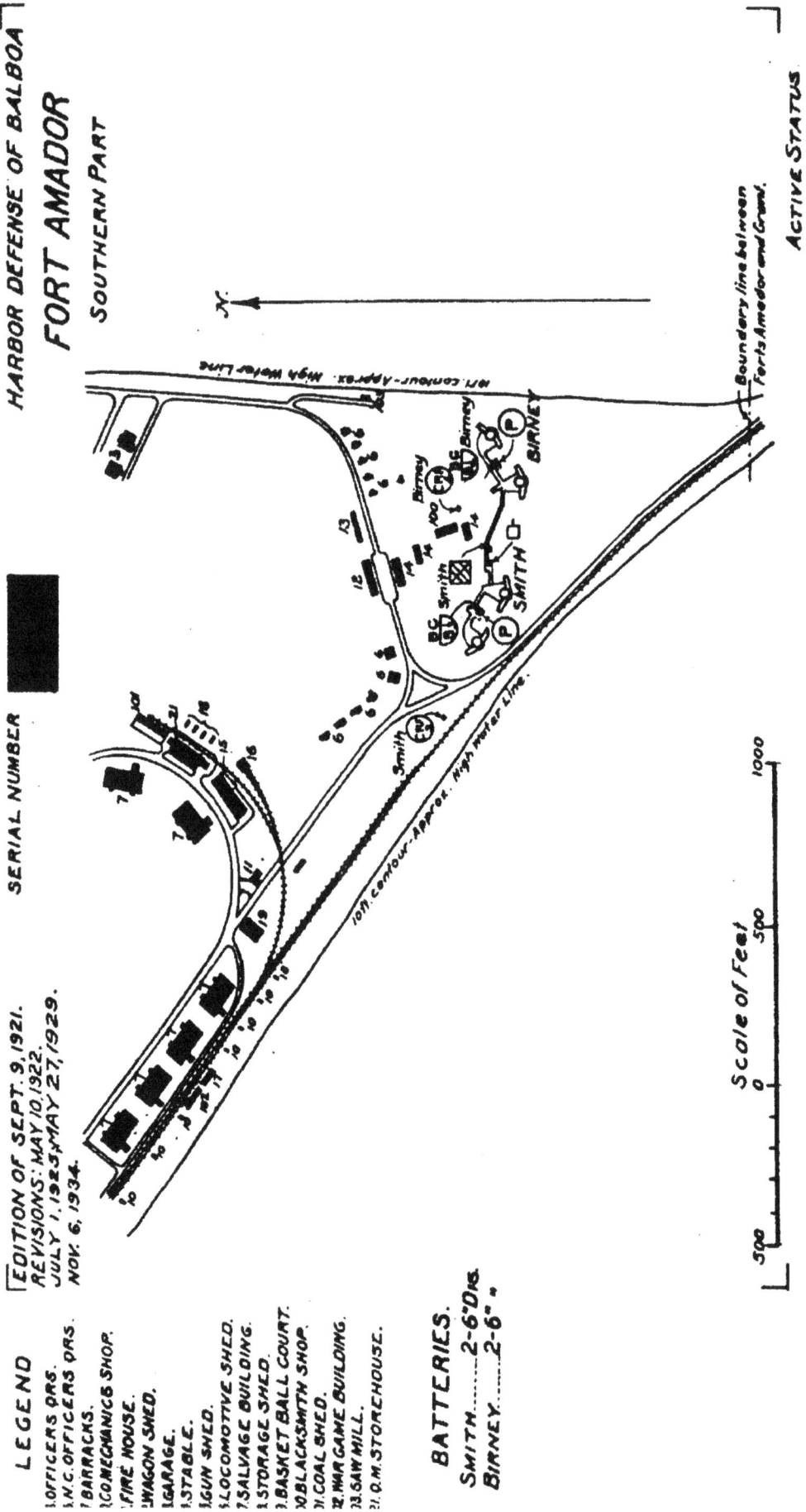

HARBOR DEFENSE OF BALBOA

FORT AMADOR

SOUTHERN PART

ACTIVE STATUS

SERIAL NUMBER

EDITION OF SEPT. 9, 1921.
REVISIONS: MAY 10, 1922.
JULY 1, 1923-MAY 27, 1929.
NOV. 6, 1934.

Boundary line between
Forts Amador and Grant.

Ref. contour - Approx. High Water Line

10 ft. contour - Approx. High Water Line

Scale of Feet

500   0   500   1000

LEGEND

1. OFFICERS QRS.
2. N.C. OFFICERS QRS.
3. BARRACKS.
4. CO. MECHANICS SHOP.
5. FIRE HOUSE.
6. WAGON SHED.
7. GARAGE.
8. STABLE.
9. GUN SHED.
10. LOCOMOTIVE SHED.
11. SALVAGE BUILDING.
12. STORAGE SHED.
13. BASKET BALL COURT.
14. BLACKSMITH SHOP.
15. COAL SHED.
16. WAR GAME BUILDING.
17. SAW MILL.
18. Q.M. STOREHOUSE.

BATTERIES.

SMITH ...... 2-6"Drs.
BIRNEY ..... 2-6" "

(0330~1349E-12)(9-14-32... (12-2000), Fort Amador, C.Z.   17952A.C.

Fort Amador, Panama Canal Zone 1932 (NARA)

The Fortified Islands of Fort Grant Planama Canal Zone (NARA)

Fort Amador (Terry McGovern)

The fortified island of Fort Grant (Terry McGovern)

A BC station was on the top of the traverse between the guns with plotting room immediately underneath. No power room was installed, that coming from a small one near Battery Smith. It was built from 1913 to 1915 and transferred on March 5, 1917 for a cost of $154,222.03. It had been named in late 1911 for Major General David B. Birney of Civil War service. The battery was armed with two 6-inch Model 1908 guns on Model 1905MII disappearing carriages (#17/#30 and #18/#31). These served actively until directed salvaged on June 15, 1943. Like Smith after the war the battery was buried for an adjacent military housing project. Nothing remains of it today.

- **AMTB-7A**: An anti-motor torpedo boat battery of two fixed 90mm guns with M3 barbette carriage erected at Fort Amador. It was built from late 1942 to July 1943. It consisted of simple concrete gun blocks and adjacent magazine. It was disarmed in 1948. No remains of the battery exist today.

Fort Grant (1911-1979) is located on several islands in Panama Bay, just offshore from the Pacific entrance to the Panama Canal. Fort Grant and Fort Amador were designated by the War Department General Orders No 153 (25 November 1911) as a single military reservation in honor of Lt. Gen. Ulysses S. Grant, USA, 18th President of the U.S. Fort Grant was used primarily for coastal defense and included a number of large batteries on the various islands. To supply them, the causeway was extended to connect from Naos to the other nearby islands, Culebra, Perico, and Flamenco, all of which had batteries of various sizes. Fort Grant also included the nearby unconnected islands of San Jose, Panamarca, Changarmi, Tortolita, Torola, Taboga, Cocovieceta, Cocovi, and Venado. The four connected island became known at the "Fortified Islands", connected by a causeway to Fort Amador. They were in an ideal location for defensive fortifications as they were 12,000 to 14,000 yards south of Miraflores Locks, making them well placed to engage a hostile naval force before it could come within range of the Panama Canal vital installations. On the summit of Flamenco Island, the most seaward of the group, was erected Battery Warren with two 14-inch guns on disappearing carriages commanded the entire area of seaward approach except for a small dead space behind Taboga Island. During construction of Battery Warren, an elevator was installed in a vertical shaft which was sunk 200 feet from the summit to connect with a horizontal tunnel which entered from the mortar batteries on the north side of the island. The battery was later used as a HAWK air defense missile platform (1960-1970). In addition to the 14-inch rifles atop Flamenco Island, a series of three mortar batteries were constructed on the landward side of the island just above the shoreline. The three mortar batteries were similar, each having four 12-inch mortars permitting them to cover all approaches to the canal. At extreme range, rounds could be dropped about one mile beyond Taboga Island. The area formerly used for accommodations in these three batteries was later used for barracks, mess hall, and administrative buildings for the HAWK missile unit. Battery Newton was erected on top of Perico Island at an elevation of some 230 feet. The battery accommodated one 16-inch gun mounted on a disappearing carriage. Located nearby were two 105mm AA guns. Naos Island was also heavily fortified. Along the top of the 100-foot ridge that forms the island were Batteries Buell and Burnside. Each of the two batteries mounted two 14-inch rifles on disappearing carriages allowing the guns covered virtually every possible approach to the Pacific entrance of the canal. Although each battery was an independent installation with its own ammunition bunkers, plotting rooms, and communications systems, an underground passage connected all four guns. Entrances to the batteries are from the causeway road through long staircases. In the vicinity of the entrances to the two batteries stand a series of three Ordnance buildings which were used as storage and repair facilities for the guns of Fort Grant as well as a controlled submarine mine complex. The harbor minefield consisted of 16 groups of 19 buoyant mines each during World War I, and 26 groups of 13 ground mines each, with hydrophones, during World War II, controlled by the mine casemates on Naos and Flamenco Islands. Located on Naos at sea-level in front the 14-inch batteries was Battery Parke with two 6-inch disappearing guns. Naos Island also had three 3-inch AA guns on top of the island. In 1928 the first of two 14-inch railway guns was brought to the Canal Zone. The maximum range of these guns was 48,000 yards, double that of the 14-inch rifles at Batteries Buell and Burnside. Two firing positions were built on Culebra Island for Battery 8 (1929-1946) to allow the railway guns to track the movement of warships. Two underground magazines with overhead trolleys which were tunneled into the island adjoining the railway track. Fort Grant was turned over to Panama in 1979. Flamenco Island was used as a prison until 1989, then by the Panamanian Coast Guard until 2008, when Battery Warren was partly destroyed for a hotel (that was not built) and the mortar batteries were converted to a two-level shopping mall. Perico Island has been greatly enlarged to accommodate a marina, cruise ship terminal, high-rise condominiums, and other entertainment businesses. Naos Island was used by the Panamanian military until 1989, then sold to developers that destroyed Battery Parke for a high-rise condominium, while a large water tank was built on part of Battery Buell, and a restaurant is now located on Battery Burnside. The Smithsonian Institute uses former submarine mine complex on Naos Island and Culebra Island for their Tropical Research Institute. Most of the former fortifications are closed to the public and permission is required to access what remains.

Fort Grant Gun Batteries

- **BUELL**: An emplacement for two 14-inch disappearing guns built on the hill crest of Naos Island of Fort Grant reservation. It was contiguous to sister Battery Burnside just to the northwest. All four emplacements were planned and submitted by engineer E.E. Winslow, provisionally on September 14, 1911 and finally on April 12, 1912. In common with the other contemporary batteries in Panama, it featured horizontal ammunition service, with the magazines on the same level as the loading platform to the left. Each gun had its own BC and plot. Gun centers were 320-ft. apart. A rock or earth parados was behind the battery. The magazines were connected by a covered tunnel gallery, which also included a large power room with four 25-kw generators and gasoline storage nearest Battery Burnside. It was built between 1912 and 1914 for final transfer to service troops on December 1, 1916. Its transfer construction cost was $425,096.73. On December 12, 1911 in General Orders No. 153, it was named for Major General Don Carlos Buell, Civil War officer and Assistant Adjutant General. The number two position was splayed out at almost 90-degrees from its partner, with its magazine almost directly behind the gun position. The battery was armed with two 14-inch Model 1910 guns on Model 1907M1 disappearing carriages (#7/#12 and #6/#11). Gun No. 1 had a trunnion height of 97.4-feet, gun No. 2 was at 92.4-feet. It retained this armament until disarmed in early 1946. The emplacement was used for a variety of purposes postwar but was abandoned until recently. The battery's two emplacements have been converted into a restaurant and music venue. The battery's magazines are to become a museum and support areas for the restaurant The battery is closed to the public.

- **BURNSIDE**: A second emplacement for two 14-inch disappearing guns built on Naos Island contiguous with Battery Buell. All four emplacements were planned and submitted by army engineer E.E. Winslow, provisionally on September 14, 1911 and finally on April 12, 1912. In common with the other contemporary batteries in Panama, it featured horizontal ammunition service, with the magazines on the same level as the loading platform to the left. Each gun had its own BC and plot. Gun centers were 320-ft. apart. A rock or earth parados was behind the battery. The magazines were connected by a covered tunnel gallery, which also included a large power room with four 25-kw generators and gasoline storage as part of this emplacement. It was constructed in 1912-13. It was located on the hilltop of the island, firing to the southwest. On the north side of the battery was a tunnel containing a cableway to bring ammunition up from road level to the entire gun line. The service tunnel connected the battery to a large fire control complex at the top

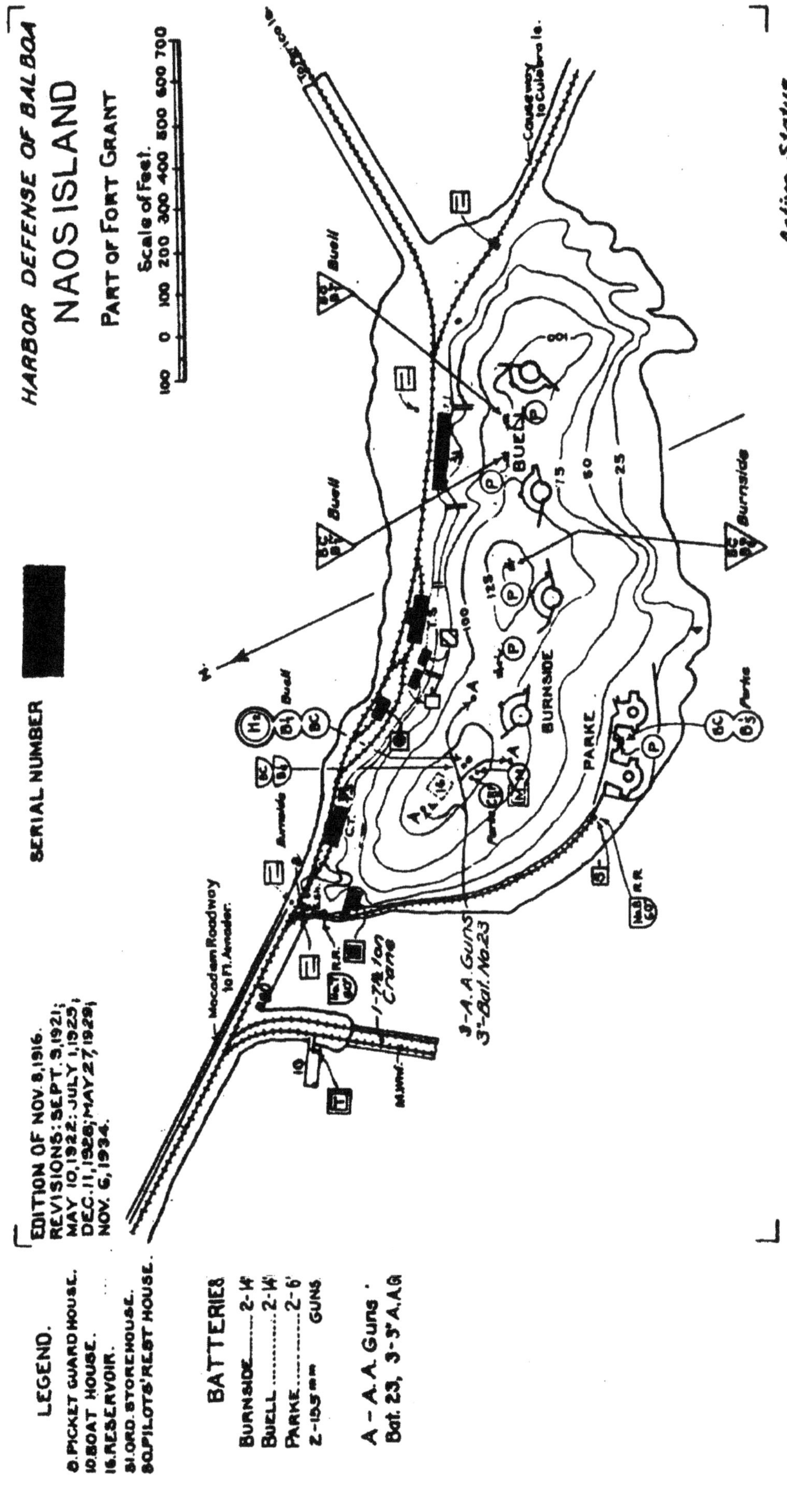

HARBOR DEFENSE OF BALBOA
NAOS ISLAND
PART of FORT GRANT

Scale of Feet.
100 0 100 200 300 400 500 600 700

Active Status.

SERIAL NUMBER

EDITION OF NOV. 8, 1916.
REVISIONS: SEPT. 3, 1921;
MAY 10, 1922; JULY 1, 1925;
DEC. 11, 1928; MAY 27, 1929;
NOV. 6, 1934.

LEGEND.
8. PICKET GUARD HOUSE.
10. BOAT HOUSE.
16. RESERVOIR.
31. ORD. STOREHOUSE.
80. PILOTS' REST HOUSE.

BATTERIES
BURNSIDE .... 2-14"
BUELL .... 2-14"
PARKE .... 2-6"
2-10.5 mm GUNS

A - A.A. Guns
Bat. 23, 3-3" A.A.G

Naos, Perico, Culebra and Flamenco Islands (Terry McGovern)

Perico, Culebra and Naos Islands (Terry McGovern)

of island (now destroyed by a large water tank). It was armed with two 14-inch Model 1910 guns on Model 1907M1 disappearing carriages (#2/#7 and #4/#10). Gun No.1 had a trunnion height of 110-feet, gun No, 2 of 102-feet. On December 12, 1911 in General Orders No. 153, it was named for Major General Ambrose Burnside who died in 1881. The battery kept its original armament through its service life, not being eliminated and disarmed in 1946. For a while postwar it was used for civil defense storage. In recent years the island has been commercially developed and the surface emplacements filled in, though access tunnels and magazines underneath still partially exist. The battery is closed to the public.

- **PARKE:** A battery for two 6-inch disappearing guns also built on Naos Island, just to the east of Burnside down the slope of the hill on the northern shoreline. Plans were drawn up by army engineer E.E. Winslow and submitted on July 11, 1912. The plan was a modification of recent mimeograph No. 59 but had raised magazines to allow adjacent-level ammunition service and concrete side walls replaced with gentle earthen slopes. Unlike the other Panama 6-inch batteries with shared magazines in the central traverse, with the danger of heavy fire from the left, separate shot and powder rooms were built for each emplacement. Also, the guns were given elevation capable of 5-degree depression to cover waters in close. To conceal the battery from small boats, the parapet was extended rearward on the left flank. A BC station was on the top of the traverse between the guns with plotting room immediately underneath. No power room was installed, that coming from the one in adjacent Battery Burnside instead. Work was done from 1913 to 1914 for transfer made on September 7, 1916 at a cost of $143,111.89. It was named on December 12, 1911 in General Orders No. 153 for General John G. Parke, U.S. Volunteers and Civil War officer and later surveyor. It was armed with two 6-inch Model 1908 guns on Model 1905MII disappearing carriages (#8/#22 and 9/#23). After the war it was abandoned, the armament being dumped at sea on August 27. 1946. The emplacement remained for a good number of years but was finally destroyed and removed for commercial developments in the 2010s.

- **NEWTON:** A battery for a single 16-inch gun at the crest of Perico Island. While guns of 16-inch size were contemplated for a number of years for the Endicott defenses, early on just a single prototype example was ever produced as the Model 1895 35-caliber gun. For years is hung around arsenals and proving grounds, A disappearing carriage was also developed, though somewhat later and finally emerged as the Model 1912 disappearing carriage. In the Taft period it was decided to mount this single gun and carriage on Perico Island as part of the defenses at Fort Grant. On July 10, 1913 a plan by E.E. Winslow was submitted for a one-story, horizontal ammunition service emplacement, similar to the 14-inch emplacements already adopted in these defenses. Magazine was withdrawn on the left side. Entrance was from the rear, and the main rear corridor contained the storerooms, latrines, and power room. It was mounted on the very top of the high hill that was Perico, the trunnions having an elevation of 236-feet. Access was by the railroad that spiraled several times around the island to the summit. There was also a broad stairway leading to the battery from sea level on the rear of the island. The emplacement was commenced in 1915 and completed in March 1917, being transferred for a cost of $229,126,91. It was armed with one 16-inch gun (Model 1895 No. 1) on the Model 1912 disappearing carriage (also No. 1). It was named by General Orders on December 12, 1911 (in advance of actual construction) for Major General John Newton, of Civil War service anda former Chief of Engineers. This older gun did not prove particularly successful in service, in May 1929 it was declared inactive, though still retained. Salvage was finally directed on June 15, 1943 and it was scrapped soon thereafter. Eventually in postwar years the site was used for air traffic control radar station by the Panamanians, and as such still exits. The battery is not open to the public.

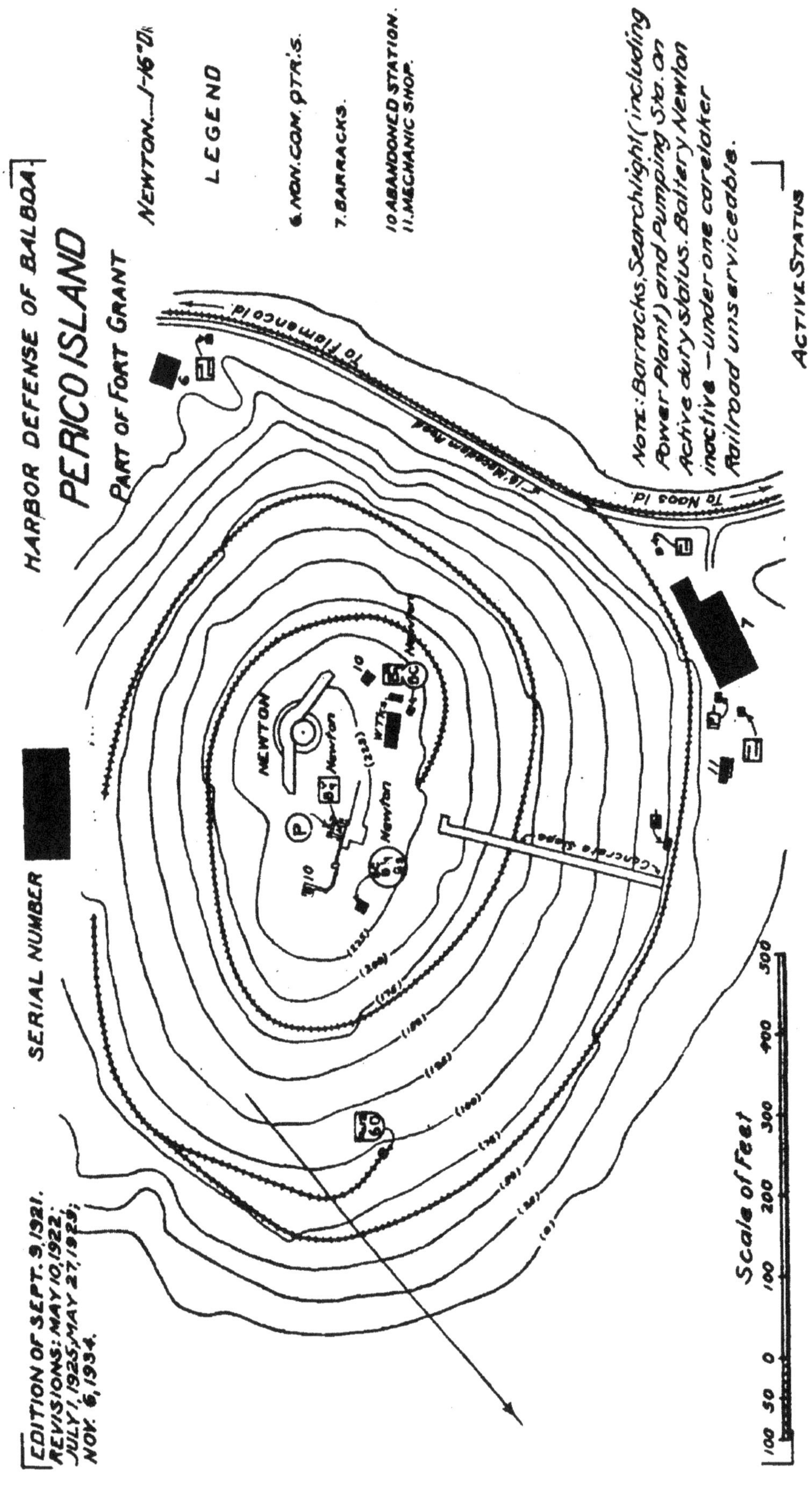

HARBOR DEFENSE OF BALBOA
PERICO ISLAND
PART OF FORT GRANT
SERIAL NUMBER
NEWTON....1-16"DR
LEGEND
6. NON.COM. QTR'S.
7. BARRACKS.
10 ABANDONED STATION.
11. MECHANIC SHOP.
NOTE: Barracks, Searchlight (including Power Plant) and Pumping Sta. on Active duty status. Battery Newton inactive - under one caretaker Railroad unserviceable.
ACTIVE STATUS
EDITION OF SEPT. 9, 1921.
REVISIONS: MAY 10, 1922;
JULY 1, 1925; MAY 27, 1929;
NOV. 6, 1934.
To Flamenco Id
To Naos Id
NEWTON
Newton
Concrete Slope
Scale of Feet
100 50 0 100 200 300 400 500

Flamenco Island (Terry McGovern)

Perico and Flamenco Islands (Terry McGovern)

- **WARREN:** A battery for two 14-inch disappearing guns emplaced on the hilltop of Flamenco Island. Original plans had called for four guns on this island, but it was simply too small to practically accommodate this many. In June 1911 E.E. Winslow recommended a change to a battery for two guns, and even those would have a two-story magazine to conserve space on top. Detailed plans were approved on July 1, 1911, and the final submission prior to construction was made on February 3, 1912. It was of unusual design, with a double story structure. Both shell and powder were available from magazines at the gun level, with the reserve ammunition magazine on the lower level, along with the battery's power room. Both guns shared the central magazine. Also, there was incorporated an elevator shaft connecting to a tunnel below running out at the level of the mortar battery on the reverse side of the island. An 8-foot square elevator rated for a loading of 4500-lb. was available for both personnel and supply transfer. A stairway also ran the backside of the island from the mortar level to the battery site. The two guns were spaced with 320-foot centers, trunnion height was 281-feet. Work was done from 1912 to 1915 with transfer being made on July 17, 1917 for a construction cost of $566,341.52. It was named in General Orders No. 153 of December 12, 1911 for Major General Governor K. Warren, chief engineer for the Army of the Potomac during the Civil War. It was armed with two 14-inch Model 1910 guns on Model 1907M1 carriages (#11/#15 and #12/#16). Some armament reports mistakenly report gun No.1 vs. 11 here, but that is believed to be a repeated error. The battery served until late in the war, the guns being scrapped about 1946. The emplacement, somewhat modified for a later Hawk Missile battery by filling in part of the emplacements, still exists on undeveloped commercial property. Efforts were made to develop a hotel on site, resulting in several fire control structures at the center of the emplacement being destroyed. The hotel development did not move forward so the battery is currently abandoned. The battery is not open to the public.

- **CARR – MERRITT – PRINCE:** A series of seacoast mortar emplacements built on Flamenco Island as part of the Taft defenses of Fort Grant. Originally projected for eight mortars in four pits, preliminary plans were submitted by Corps of Engineer officer E.E. Winslow on July 7, 1911. By December 27th of the same year the plans had changed to twelve mortars in six pits. It was built as an inline series of six dual mortar pits, in a line on a shelf on the rearward side of Flamenco Island. Magazines were in front of the pits to help reduce lateral dimensions. Behind the line of pits were a substantial casemated series of rooms in a parados for support, quarters, and storage. BC and plotting rooms were located between each pair of emplacements on the back of the traverses; mechanical range transmission devices were utilized. Trunnion height was calculated at just 24.2-feet for all the mortars. A large power room was worked into the plan in the western mortar pits (Battery Merritt). When completed it was organizationally split into three separate batteries each with four guns. The two western pits were named by General Orders No. 153 of December 12, 1911 Merritt for Major General Wesley Merritt, the two center pair Prince for Brigadier General Henry Prince, and the two eastern ones Carr for Major General Joseph B. Carr. The combined emplacement was built from 1912 to 1914 and transferred on September 7, 1916 for a total construction cost of $622,718.94. A unique, long service tunnel ran from the battery's No. 1 traverse directly under Flamenco Island to an elevator shaft which ascended to Battery Warren. It was armed in 1916 with twelve 12-inch Model 1912 mortars on Model 1896MIII carriages (Merritt: #13/#9, #16/#8, #14/#7, and #10/#6, Carr: #7/#37, #12/#5, #6/#18, and #5/#13, and Prince: #11/#3, #3/#2, #8/#4 and #9/#15). By the middle of World War II, the battery was deemed obsolete and was authorized for removal in 1943. After this the emplacements were not used for armament, but during the 1960s was conveniently used as a support structure for nearby HAWK anti-aircraft missile batteries. Later this portion of the island was sold for commercial development, and shops and restaurants were built into the old

HARBOR DEFENSE OF BALBOA
FLAMENCO ISLAND
PART OF FORT GRANT.

ACTIVE STATUS

SERIAL NUMBER

EDITION OF: DEC.7,1915,
REVISIONS NOV. 8,1916.
SEPT.9,1921.
MAY 10,1922.
JULY 1,1925.
DEC.11,1925.
MAY 27,1929.
NOV. 6, 1934.

BATTERIES.
CARR _____ 4-12"M.
MERRITT ____ 4-12"M.
PRINCE _____ 4-12"M.
WARREN ____ 2-14"Dis.
4-155 mm GUNS.

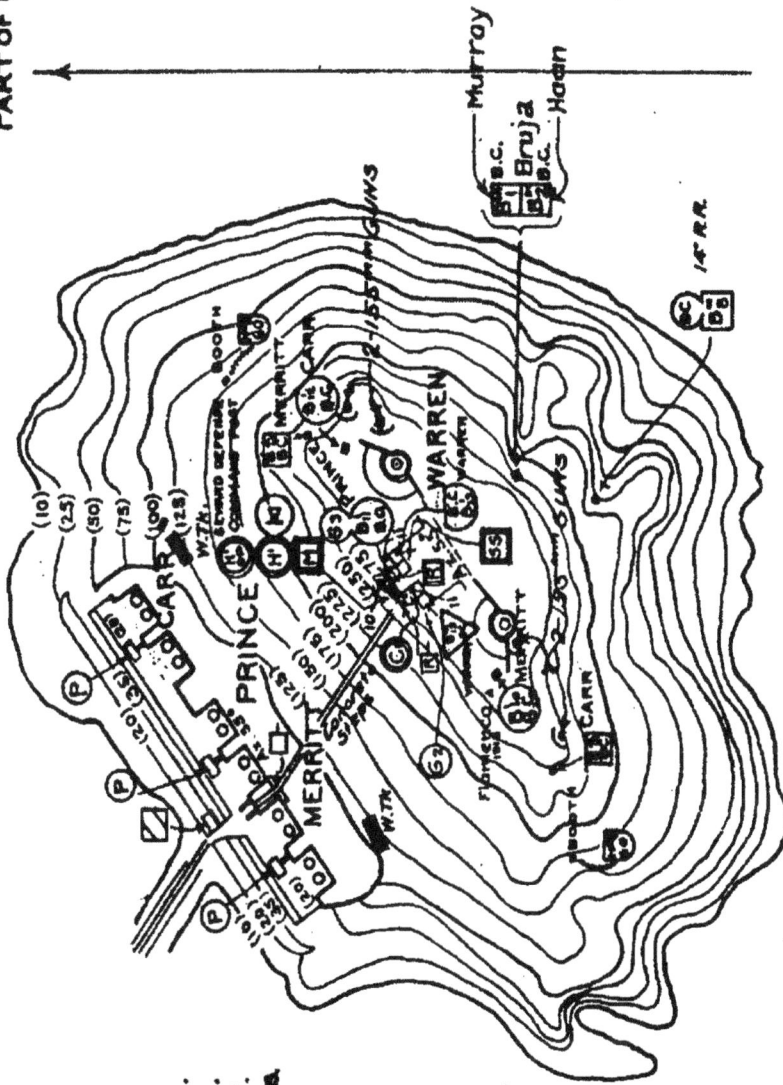

Scale of Feet.

mortar emplacements of Batteries Prince and Carr. Battery Merritt was not built-on by the mall. Access to elevator tunnel and power rooms remain. As such much of it still exists. The batteries are not open to the public.

- **Railway Battery:** On Culebra Island emplacements for two railway 14-inch guns were built in the early 1930s. Culebra Island was in between and connected by causeway to Naos and Perico Islands. These emplacements were in conjunction with similar gun blocks built on the Atlantic side at Fort Randolph. Basically, it involved two reinforced turning blocks, adjacent ammunition magazines and service sidings, and support facilities. This was a low-level site, gun trunnion heights were just 38.25-feet (gun No, 1) and 32.2-feet (gun No. 2). The two blocks were originally built from 1929 to 1930 for transfer on August 14, 1930 at a cost of $50,000. The intended guns were Model 1920MII guns on Model 1920 carriages (#9/#2 and #8/#3). Discarded after World War II, parts of the connecting rail network and protected magazines still exist on this island. The battery site is open to the public.

14-inch Railway gun and carriage M1920 at Culebra Island, Fort Grant (NARA)

HARBOR DEFENSE OF BALBOA
CULEBRA ISLAND
PART OF FORT GRANT.

SERIAL NUMBER

EDITION OF NOV. 8, 1916.
REVISIONS: SEPT. 3, 1921.
MAY 10, 1922; JULY 1, 1925;
DEC. 11, 1926; MAY 27, 1929;
NOV. 6, 1934.

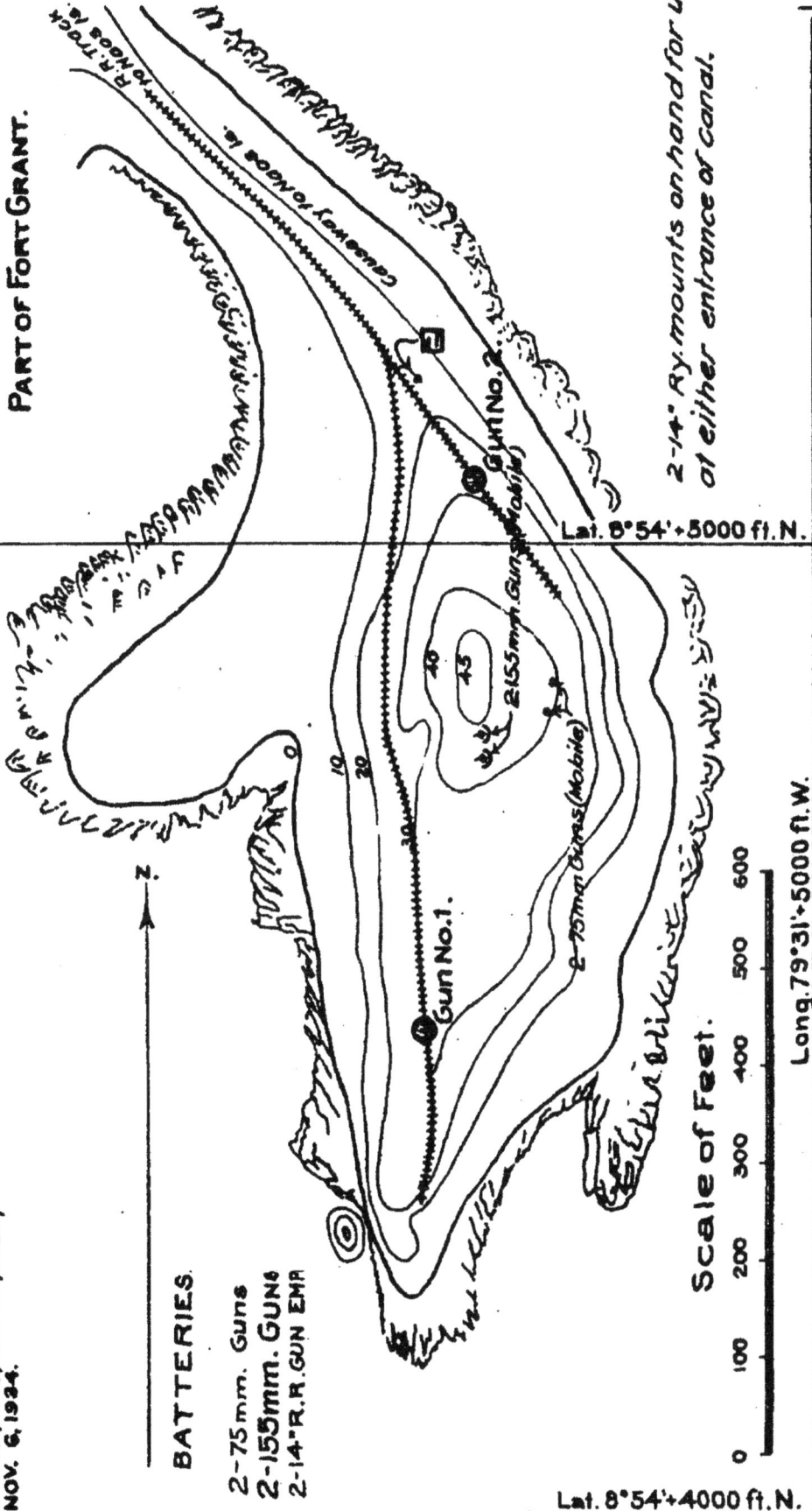

BATTERIES.
2 - 75 mm. Guns
2 - 155 mm. Guns
2 - 14" R.R. Gun EMP.

Gun No. 1.
Gun No. 2.
2 - 155 mm. Guns (Mobile)
2 - 75 mm. Guns (Mobile)

R. Track
Phasos Ia
Causeway to

2 - 14" Ry. mounts on hand for use
at either entrance of canal.

Lat. 8°54' + 5000 ft. N.
Lat. 8°54' + 4000 ft. N.
Long. 79°31' + 5000 ft. W.

ACTIVE STATUS.

N.

Scale of Feet.
0   100   200   300   400   500   600

Fort Kobbe (1918-1999) is located at Bruja Point to the west side of the entrance to the Panama Canal which is on the opposite side of the canal entrance from Fort Amador. Known as Bruja Point Military Reservation until 1928, then named Fort Bruja until 1932. On 15 April 1932 in General Orders 4, it was renamed to honor Major General William A. Kobbe. It was a sub post of Fort Amador until 1941, and again after 1946. Battery Murray was near Bruja Point, while Battery Haan was on Batele Point. Two 16-inch naval-type rifles on barbette mounts were each installed at Battery Haan and Battery Murray. The barbette mounts permitted a 360° transverse, covering all approaches to the canal entrance for 25 miles. In March 1942 work was started on the casemating of Battery Murray. The railroad track in back of each gun emplacement was covered by a concrete tunnel, which intersected another tunnel leading from the gun to a rear entrance. The tunnels, and magazines opening onto them, were equipped with an overhead trolley to handle ammunition and powder, and the whole system was covered with a mound of earth about thirty feet high. A thick semi-circular slab of concrete was cantilevered over the gun breech for additional protection. The four 16-inch guns were placed on secondary manning status in 1944 and were dismantled and scrapped in 1948. Today, Battery Murray's casemated gun houses are abandoned while Battery Haan two emplacements are buried. AMTB Battery 6 (90mm) was on Batele Point, Battery 3 and Battery Z (3A) (four 155mm GPF on Panama mounts) were on Bruja Point, and Battery AZ (1917-1946) (two fixed 4.7-inch, replaced by two 75mm in 1919) was on Bruja Point. Howard Army Airfield was established in 1937 on the northern portion of the reservation, originally known as Bruja Point Field. It became a separate post in 1946, becoming Howard Air Force Base in 1947. The former military post is now the Panamanian international business park and mixed-use residential community of "Panama Pacifico", with the former garrison area now the "Town Center". There is open public access to the former fort, while the 16-inch batteries are fenced in.

Fort Kobbe Gun Batteries

- **MURRAY**: Almost from the completion of the Panama Canal's initial defenses, plans were made for its modernization. This became more urgent during World War I when longer ranged guns from the newest battleships came into common use. Plans for bigger guns were soon initiated. These initially called for new 16-inch gun batteries on Taboga Island on the Pacific side, and formal recommendations were made on October 13, 1920 for four such guns to go into barbette batteries there. But in 1922 a substitute site on the mainland at Bruja Point was selected, primarily on the basis that Taboga was too isolated for economic maintenance. This location eventually received batteries Murray and Haan on the new reservation named Fort Kobbe. Also, recommendations were that the mounts be close together to economize on service facilities. Murray was begun first, in the spring of 1925 and done for provisional transfer in July 1926. Full transfer was made on March 9, 1929 for a construction cost of $521,160. It was the western two guns of the set and consisted of open gun blocks and connecting magazines reached by rail spurs. The battery had its own small reserve power station and fuel supply. Gun trunnion height was just 48.7-feet. It was the first battery armed with the new naval 16-inch guns made available by the scrapping of battleships by the Washington Treaty. It was armed with two 16-inch navy MkIIM1 guns on Model 1919M1 barbette carriages (#57/#9 and #58/#11). It was named for Major General Arthur Murray, former Chief of Coast Artillery. During World War II (1943-43) each of the two gun blocks was casemated with heavy overhead protection, though this did restrict the field of fire for the weapons. The battery was disarmed in 1948 but continued to serve as ammunition storage well into the 1970s. The casemated gun houses still exist along with several of the dispersed magazines as well as PSR and Power Plant. Parts of the battery are open to the public, while the gun casemates are fenced in.

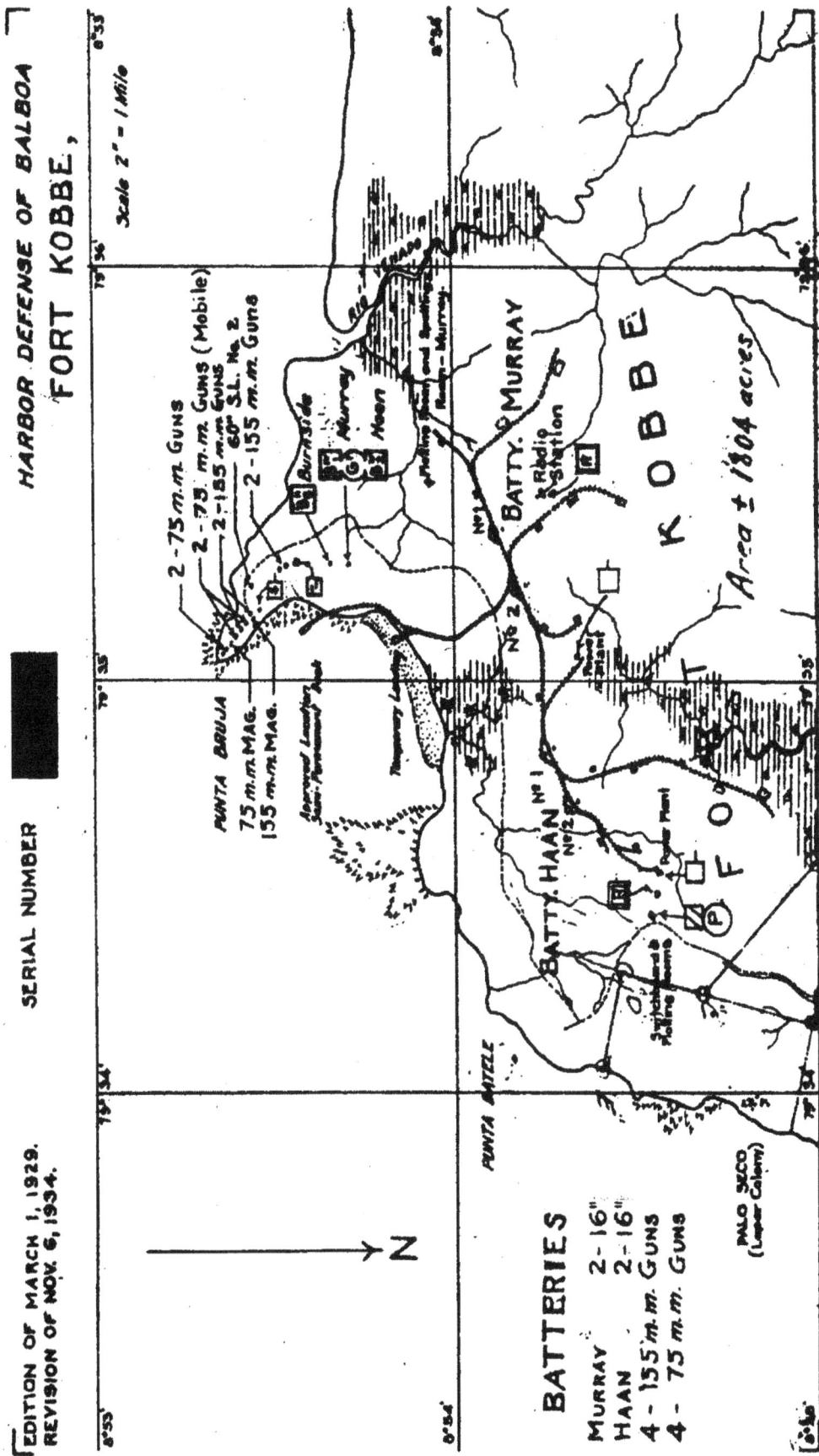

HARBOR DEFENSE OF BALBOA,
FORT KOBBE,

Scale 2" = 1 Mile

EDITION OF MARCH 1, 1929.
REVISION OF NOV. 6, 1934.

SERIAL NUMBER

DESIGNATION OF FORT BRUJA CHANGED TO
FORT KOBBE, W.D., G.O. No. 4, APRIL 15, 1932.

BATTERIES
MURRAY   2 - 16"
HAAN   2 - 16"
          4 - 155 m.m. GUNS
          4 - 75 m.m. GUNS

EXISTING CONSTRUCTION
PROPOSED CONSTRUCTION

2 - 75 m.m. GUNS
2 - 75 m.m. GUNS (Mobile)
2 - 155 m.m. GUNS
60" S.L. No. 2
2 - 155 m.m. GUNS

PUNTA BRUJA
75 m.m. MAG.
155 m.m. MAG.

BATTY. MURRAY

KOBBE

Area ± 1804 acres

BATTY. HAAN Nº 1
         Nº 2

PUNTA BATELE

PALO SECO
(Leper Colony)

Battery Murray Fort Kobbe, Panama Canal Zone (Terry McGovern)

Battery Murray Fort Kobbe, Panama Canal Zone (Terry McGovern)

- **HAAN:** The second set of 16-inch guns for the Kobbe reservation, planned and built in close conjunction with Battery Murray. Haan was the eastern two-gun blocks, also consisting of completely open sites connected by rail to separate ammunition storehouses. It had its own power room. The ammunition supply railway connected both batteries and all four gun blocks. Trunnion height was 37.7-feet. Work was slightly later than Battery Murray, taking place in February 1928 until initial transfer on November 1928. Official transfer was made on March 10, 1929 for a cost of $538,041. It was named for Major General William G. Haan of Spanish and World War I service. It was armed with two 16-inch navy MkIIM1 guns on Model 1919M1 barbette carriages (#59/#12 and #61/#13). Unlike Murray it was in a less exposed position and was not given casemated protection. No doubt the desire to retain at least a portion of the armament's all-round fire capability was also important to this decision. It was disarmed in 1948; the area being used postwar as part of the bayonet course. Today many of the dispersed magazines, PSR, Power Plant, and the covered-over blocks still exist. The battery is not open to the public.

- **AMTB-6:** An anti-motor torpedo boat battery of two fixed 90mm guns with M3 barbette carriage erected at Fort Kobbe at Batele Point. It was built from late 1942 to July 1943. It consisted of concrete gun blocks and adjacent magazine. It was disarmed in 1948. The battery is open to the public, but difficult to locate in the jungle.

Taboga Island Gun Batteries

- Taboga Island was actively considered for battery sites for projects over the history of the canal's defenses. The 1915 Board of Review Project called for two 16-inch and 12 12-inch mortars to be built here. But this was not implemented. In 1918 the plans changed to two 16-inch guns and eight 16-inch howitzers. Land acquisition actually began in 1919, but with some difficulty and considerable expense. In 1920 four 16-inch guns on barbette carriages were recommended here, but soon dropped in favor of the Bruja Point location that would become Fort Kobbe. During the 1940 project, a casemated dual 16-inch battery (Battery Construction No. 151) was slated for Taboga Island. Like the other projects it was soon cancelled. While Taboga received emplacements for searchlights and temporary emplacements for 155mm guns, no permanent fortification batteries were ever emplaced here.

Culebra Island (Terry McGovern)

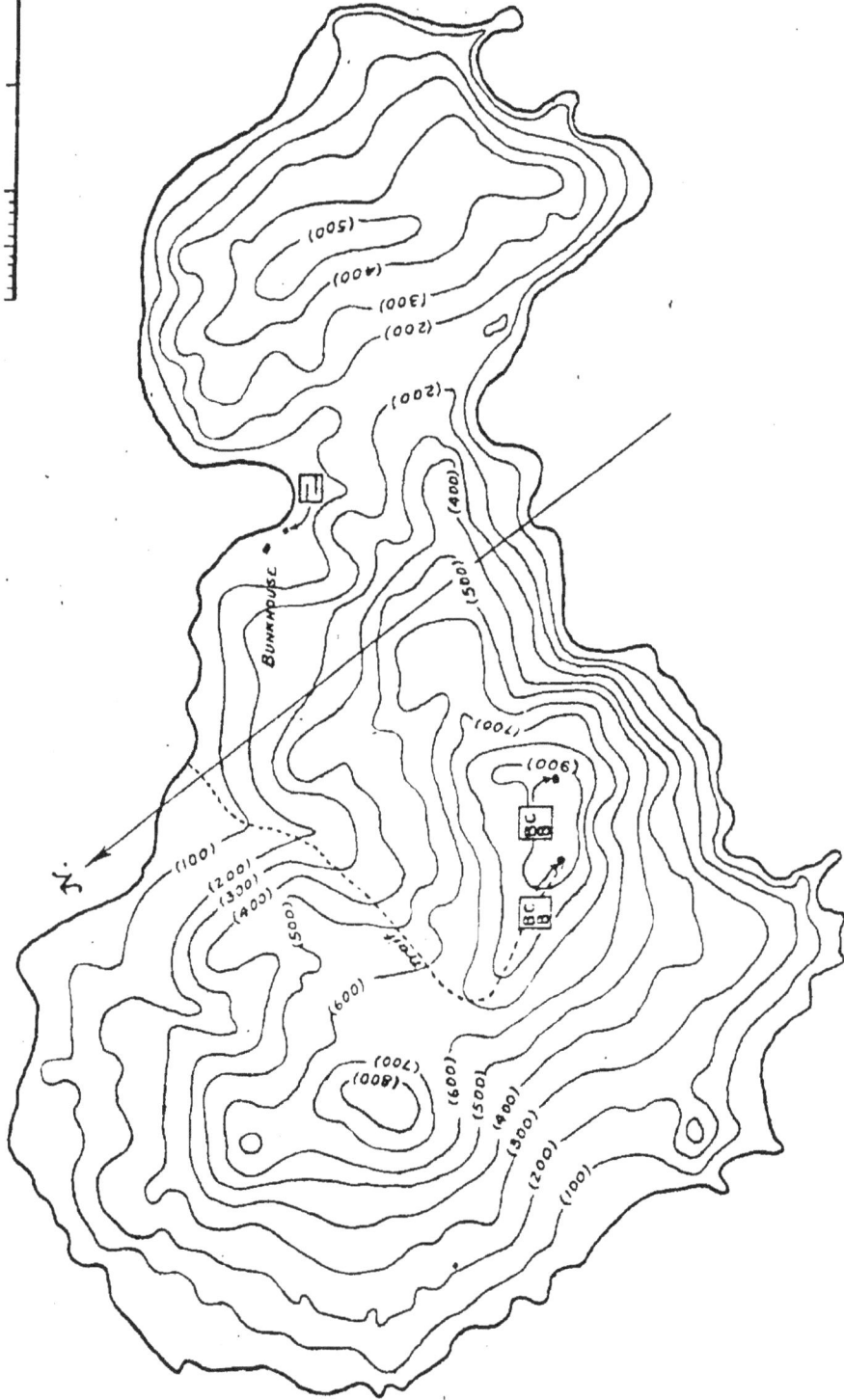

DEFENSES OF THE PANAMA CANAL

TABOGA ISLAND

Scale of Feet

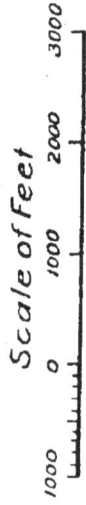

SERIAL NUMBER 12 Y

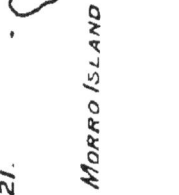

MORRO ISLAND

EDITION OF SEPT. 9, 1921.
REVISIONS : MAY 10, 1922.

BUNKHOUSE

(500)
(400)
(300)
(200)
(100)

(500)
(400)
(300)
(200)

(200)

(400)
(500)

(700)
(900)
BC

BC

N

(100)
(200)
(300)
(400)
(500)
(600)
trail

(700)
(800)
(600)
(500)
(400)
(300)
(200)
(100)

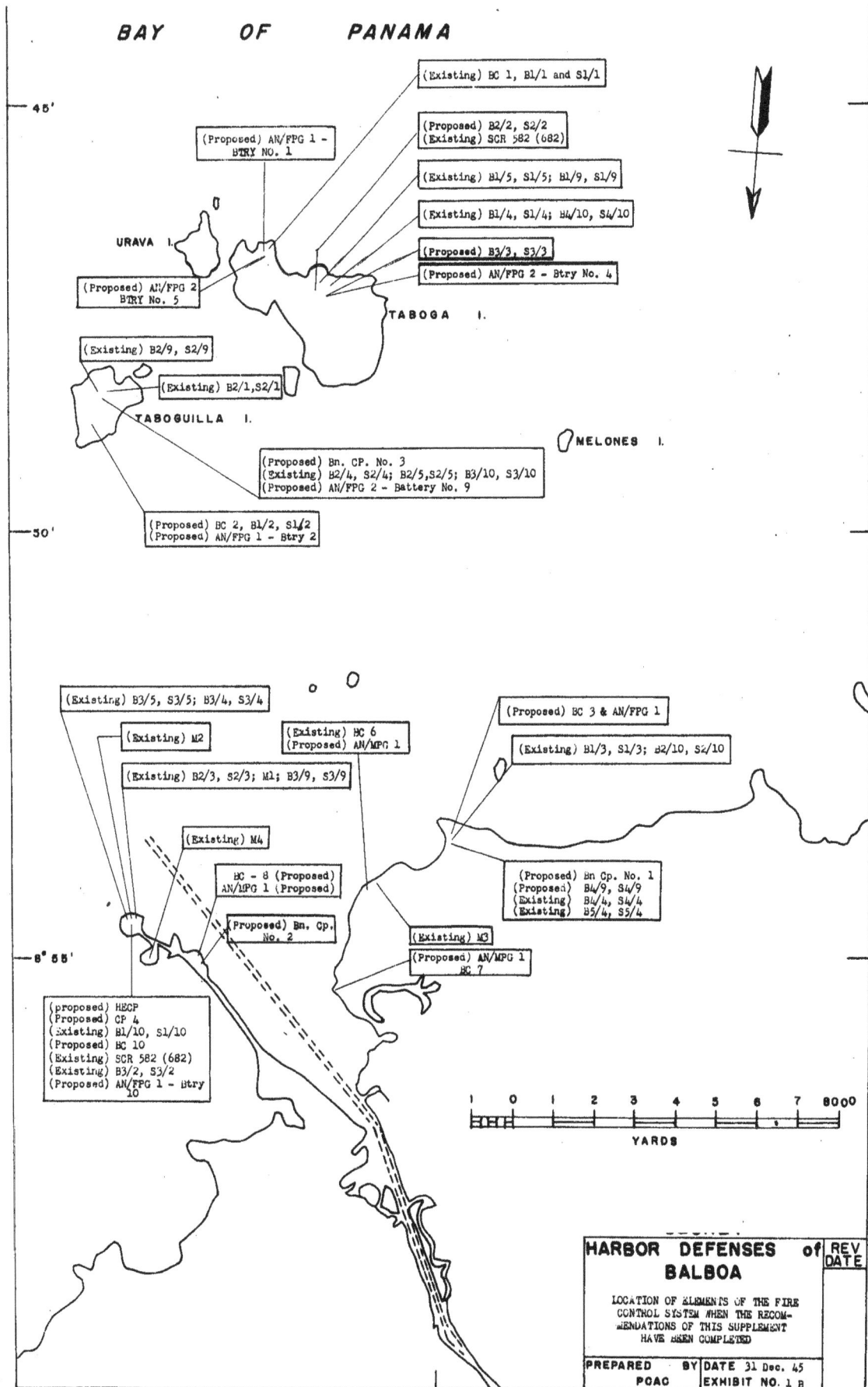

BAY OF PANAMA

(Existing) BC 1, B1/1 and S1/1

(Proposed) AN/FPG 1 - BTRY NO. 1

(Proposed) B2/2, S2/2
(Existing) SCR 582 (682)

(Existing) B1/5, S1/5; B1/9, S1/9

(Existing) B1/4, S1/4; B4/10, S4/10

(Proposed) B3/3, S3/3
(Proposed) AN/FPG 2 - Btry No. 4

URAVA I.

(Proposed) AN/FPG 2 BTRY No. 5

TABOGA I.

(Existing) B2/9, S2/9

(Existing) B2/1, S2/1

TABOGUILLA I.

MELONES I.

(Proposed) Bn. CP. No. 3
(Existing) B2/4, S2/4; B2/5, S2/5; B3/10, S3/10
(Proposed) AN/FPG 2 - Battery No. 9

(Proposed) BC 2, B1/2, S1/2
(Proposed) AN/FPG 1 - Btry 2

(Existing) B3/5, S3/5; B3/4, S3/4

(Proposed) BC 3 & AN/FPG 1

(Existing) M2

(Existing) BC 6
(Proposed) AN/MPG 1

(Existing) B1/3, S1/3; B2/10, S2/10

(Existing) B2/3, S2/3; M1; B3/9, S3/9

(Existing) M4

BC - 8 (Proposed)
AN/MPG 1 (Proposed)

(Proposed) Bn Cp. No. 1
(Proposed) B4/9, S4/9
(Existing) B4/4, S4/4
(Existing) B5/4, S5/4

(Proposed) Bn. Cp. No. 2

(Existing) M3

(Proposed) AN/MPG 1
BC 7

(proposed) HECP
(Proposed) CP 4
(Existing) B1/10, S1/10
(Proposed) BC 10
(Existing) SCR 582 (682)
(Existing) B3/2, S3/2
(Proposed) AN/FPG 1 - Btry 10

YARDS

HARBOR DEFENSES of BALBOA

REV DATE

LOCATION OF ELEMENTS OF THE FIRE CONTROL SYSTEM WHEN THE RECOMMENDATIONS OF THIS SUPPLEMENT HAVE BEEN COMPLETED

PREPARED BY POAC | DATE 31 Dec. 45
EXHIBIT NO. 1 B

OVERSEAS BASES – CARIBBEAN

The Caribbean defenses were located on a combination of islands, some with low, tropic seashores and other with volcanic hills and cliffs. The climate is tropical and access to these coast defense sites can be difficult. These harbor defenses were mainly developed during World War II using the designs of the 1940 Program. Included in this region were the defenses of Puerto Rico, Virgin Islands, and Guantanamo Bay, Cuba. Temporary U.S. defenses during World War II were installed on Trinidad, Aruba, and Curacao.

THE HARBOR DEFENSES OF VIEQUES SOUND (ROOSEVELT ROADS) – PUERTO RICO AND VIRGIN ISLANDS

Fort Segarra (1942-1948) is located primarily on Water Island, St. Thomas, U.S. Virgin Islands. This military reservation was established as part of the major effort to defend the Vieques Sound, also known as Roosevelt Roads, which stretches from eastern Puerto Rico to St. Thomas. Fort Segarra was named in General Orders 44 of 1944 in honor of Lieutenant Colonel Rafael Angel Segarra (from Puerto Rico) a highly decorated World War I veteran. Using the similar designs as the 1940 Program, a #400 Series battery with two 8-inch naval guns on long-range barbette carriages (Battery Construction #401) was planned as part of the permanent defense. Located on Fortuna Hill (900 feet elevation) this battery only had site cleared and gun blocks constructed before the project was canceled. Also, using the same designs at the 1940 Program, a #200 Series battery with two 6-inch shielded barbette guns (Battery Construction #314) was started as part of the permanent defense. This battery, located at Flamingo Point on Water Island, was never completed as construction was abandoned (97% complete) in 1944 due to the realization that the enemy was no longer a threat to St. Thomas. Also located on Water Island were two 90mm AMTB batteries, one at Druif Point and one at Providence Point. Several fire control and searchlight stations were also constructed for this fort. The uncompleted post was transferred to the Army's Chemical Warfare Division in 1948 for testing poison gas on goats and pigeons. The island was leased by the Federal government to private developers from 1952 until 1996 and then sold to private interest and remains so today. Access to the island is by ferry.

Fort Segarra Gun Batteries

- **Battery #314**: A dual 6-inch barbette battery emplaced at Flamingo Point which is at the southern extremity of Water Island, which in turn is just to the south of St. Thomas Island. Work was done from October 30, 1942 until December 31, 1943. At that time, it was reported as 97% complete, basically just missing its armament and some fire control equipment which had already been deferred. It was to have carried two 6-inch guns Model M1(T2) on Model M4 barbette carriages (no tubes, but carriages #67 and #68 were delivered to the site). Work was never completed, and the suspended battery remained unfinished. It was abandoned after the war, but the emplacement still exists. The battery is open to the public, but a ferry or boat is required to get to Water Island.

- **AMTB Providence Point**: An AMTB battery for two 90mm fixed and two 90mm mobile guns emplaced at Providence Point on the northern side of Flamingo Bay on the western side of Water Island. Apparently built in 1943 or 1944, it consisted of two gun blocks with surrounding parapets for the fixed guns and adjacent concrete ready magazines. It was abandoned after the war when the armament was removed. The gun blocks still exist. The site is on private property and is not open to the public.

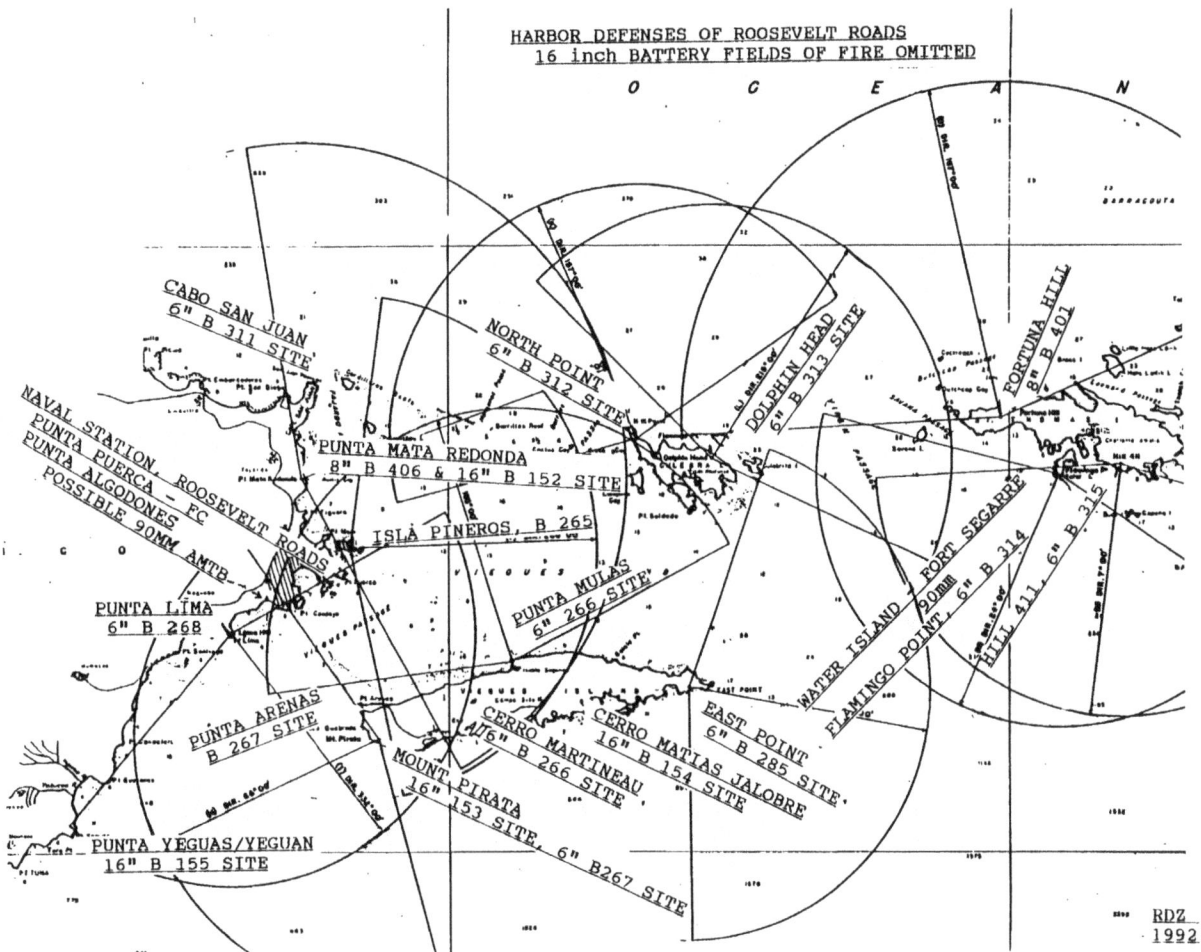

Map by Robert D. Zink, 1992

- **AMTB Druif Point:** An AMTB battery for two 90mm fixed and two 90mm mobile guns emplaced at Druif Point on the southern side of Flamingo Bay on the western side of Water Island. Apparently built in 1943 or 1944, it consisted of two gun blocks with surrounding parapets for the fixed guns and adjacent concrete ready magazines. It was abandoned after the war when the armament was removed. The gun blocks still exist. The site is on private property and is not open to the public.

- **AMTB Muhlenfels Point:** An AMTB battery for two 90mm fixed and two 90mm mobile guns emplaced at Muhlenfels Point which is three miles south of Charlotte Amalie and across from eastern side of Water Island. Apparently built in 1943 or 1944, it consisted of two-gun blocks with surrounding parapets for the fixed guns and adjacent concrete ready magazines. It was abandoned after the war when the armament was removed. The gun blocks still exist as part of the Marriott Frenchman's Reef Hotel. The battery is not open to the public.

Fortuna Hill Military Reservation (1944-1948) is located near Fortuna, St. Thomas on Fortuna Hill (900 feet elevation) on the western slope of the main island.

- **Battery #401** A dual 8-inch barbette batter emplaced here on Fortuna Hill—the highest point on the western side of St. Thomas Island as part of the Roosevelt Roads defenses. It was planned to follow the design of other expedient World War II 8-inch batteries with open gun blocks separated by 240-feet and a heavily protected traverse magazine, power, and plotting room in between. Work was done from November 30, 1942 until January 30, 1943. Only the blocks were completed, the

Battery #314 on Water Island, St. Thomas Virgin Islands (Terry McGovern)

Battery #314 on Water Island, St. Thomas Virgin Islands (Terry McGovern)

magazine was not built prior to structural work being suspended and then cancelled. Even though suspended, the guns were placed on the blocks in early 1944. They consisted of two 8-inch navy guns MkVIM3A2 on Model M1 barbettes (tube numbers #130 and #204, carriages #9 and #10). This armament had been shipped to the island on December 12, 1942. After the war these were removed, probably in 1946 and the reservation sold into private ownership. The two-gun blocks still exist in front of a private residence. The battery site is not open to the public.

Fort Charles W. Bundy (1942-1948) is located primarily on Vieques Island, near Ceiba on the eastern end Puerto Rico. This military reservation was established as part of the major effort to defend the Vieques Sound, also known as Roosevelt Roads, which stretches from eastern Puerto Rico to St. Thomas. The Roosevelt Roads Naval Operating Base was established in 1943, intended to be the major American naval base in the Caribbean region (the Atlantic's "Pearl Harbor"). Construction was suspended in 1944. The former U.S. Army post became the southern section of the Naval Station when it was reactivated in 1957. The Naval Base closed in 2004. A temporary battery was built in 1941 using two 155mm GPF guns on Panama mounts at Punta Yeguan. While many plans were made to construct as many as a dozen permanent batteries using up to 16-inch guns, only four batteries were ever undertaken. Using the similar designs as the 1940 Program, Battery #406 for two 8-inch naval guns on long-range barbette carriages was planned as part of the permanent defense. Located on Punta Mata Redonda this battery only had the site cleared and gun blocks constructed before the project was canceled. Also, using the same designs at the 1940 Program, Battery #265 and Battery #268 for two 6-inch shielded barbette guns each were built as part of the permanent defense. Battery #265, located on Isla Pineros, and Battery #268, located at Punta Lima, were completed, but only Battery #265 was armed. Also located on Punta Algodones was a 90mm AMTB battery and the HECP/HDCP for the harbor defenses. Several fire control and searchlight stations were also constructed for this fort. After World War II, Fort Bundy's part of its reservation became part of the U.S. Naval Station – Roosevelt Road while the rest was return to its former owners or sold to private interest and remains so today.

Fort Bundy Gun Batteries

- **Battery #265**: One of the dual 6-inch barbette batteries intended for the defenses of Roosevelt Roads. It was emplaced on Pineros Island, just offshore of Puerto Rico near the new Roosevelt Roads Navy Base. Work was begun on December 11, 1942 and proceeded through most of 1943. By May 31, 1944 it was listed at 97% complete. The armament had been received in March 1943 and consisted of two 6-inch guns Model 1903A2 on Model M1 barbette carriages (#26/#50 and #11/#31). It was disarmed shortly postwar, broken parts of the carriage around the emplacement suggest the armament may have been demolished or scrapped in place. The emplacement still exists and has recently been transferred from U.S. Navy hands to private development. The battery site is not open to the public.

- **AMTB Punta Algodones**: An AMTB Battery of two 90mm fixed and two 90mm mobile guns erected on the south side of the new naval base at Punta Algodones. It was probably built (mainly just concrete blocks for the two fixed guns) in 1943, being armed and serving until removal postwar in 1946. No information on any surviving elements has been obtained.

- **Battery #268**: A dual 6-inch barbette battery located on southeastern Puerto Rico as part of the Roosevelt Roads defenses. It was built into the side of a hill at Punta Lima, necessitating the adoption of the "bent" or side-entry plan for the battery, where the access entry to the power and plotting room was on the side of the battery rather than directly in the rear. Work was done from December

31, 1942, and likely the major structural parts were completed in late 1943. Further completion was suspended on October 15, 1943 when reported at 97% complete. The armament was never mounted or probably even received at site, but 6-inch carriages M4 #65 and #66 were allocated to the battery. It was never named. For many years the reservation was a Puerto Rican prison, but it has relatively recently been sold into private ownership where the emplacement still exists. The battery site is not open to the public.

- **Battery #406**: A dual 8-inch barbette battery located in northeastern Puerto Rico Island at a location known as Mata Redondo. It was designed similar to Battery #401 at St. Thomas—two open gun platforms for unshielded barbette carriages and a single, shared protected traverse magazine. Work was done from October 30, 1942 until December 30th of the same year. The armament was received in February 1943 and consisted of two navy 8-inch guns Model MkVIM3A2 on Model M1 barbette carriages (tube numbers #158 and #236, barbettes were serial numbers #7 and #8). Little physical work had been done at the site beyond the concrete pouring of the gun blocks, no work had been started on the magazine besides excavation of the foundation, when further construction was suspended. Orders were given to mount the guns and carriages even in this incomplete state, and that was apparently done. They were removed early postwar and the site abandoned without any further work. Within the property of a quarry site, the blocks survived for many years, but their current status is unknown.

Vieques Island Batteries	*#153, #154, #266, #267, #285* (all planned)
Culebra Island Batteries	*#312, #313* (all planned)
Punta Mata Redonda	*#152* (planned)
Punta Yeguas	*#155* (planned)
Cabo San Juan	*#311* (planned)

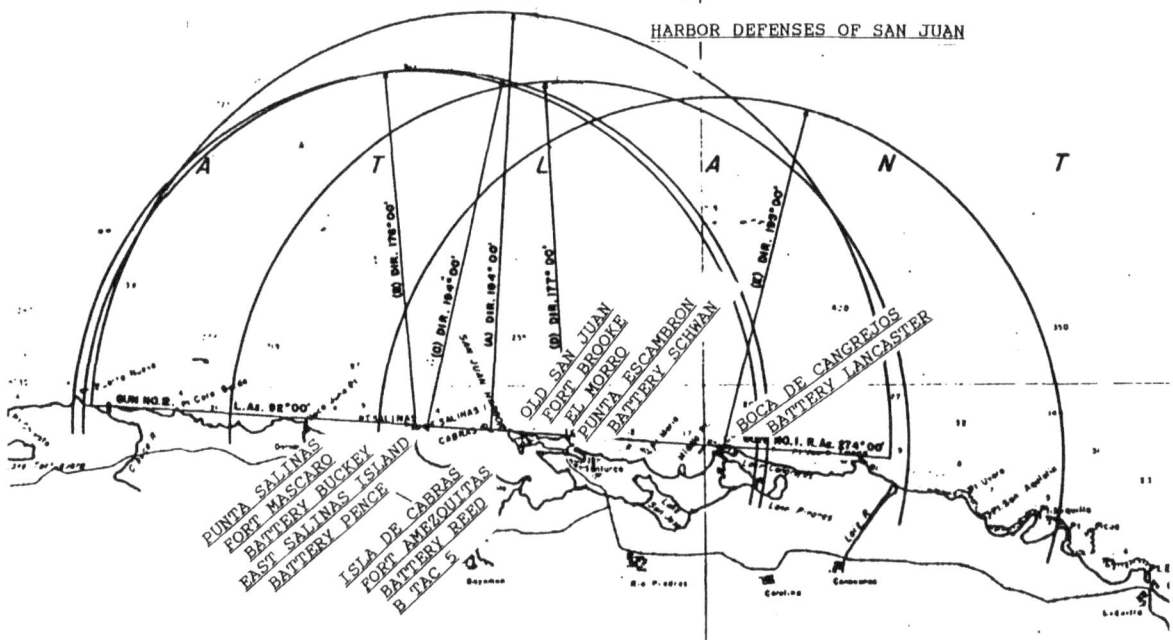

Map by Robert D. Zink, 1992

THE HARBOR DEFENSES OF SAN JUAN – PUERTO RICO

Fort Brooke (1941-1949/1966) is located primarily in the former Spanish defenses of Castillo de San Felipe del Morro and Castillo de San Cristobal that defend the City of San Juan, Puerto Rico. This was the U.S. Army's main garrison post in Puerto Rico, centered mainly around the historic El Morro and San Cristóbal Castles. Originally known as the San Juan Military Reservation. Renamed under General Orders 10 of 1943 in memory of United States Army Major General John Rutter Brooke from Pennsylvania, better known as "John Ruller" during the American Civil War and Spanish-American War. This military reservation was taken over the United States during the Spanish-American War in 1898. The Americans installed two 4.7-inch guns during World War I at El Morro. Numerous barracks and quarters covered the open plain below El Morro ("El Campo del Morro"). The old Ballajá Barracks became the Fort Brooke Army Hospital (aka Rodriguez Army Hospital) in 1943. The Convento de los Dominicos, originally built in 1523, was used in World War II as the administrative headquarters of Fort Brooke and the U.S. Army Caribbean (Antilles) Command. During World War II, the 4.7-inch emplacement at the very tip of El Morro had a 3-inch rapid-fire gun installed to serve as an exam battery for ship entering San Juan harbor. A temporary battery was built in 1941 using two 155mm GPF guns on Panama mounts at Princesa Battery, part of San Cristobal. Several fire control stations were built into the battlements of El Morro and San Cristobal. A large HECP/HDCP was also built in the moat of San Cristobal. Also, using the same designs at the 1940 Program, two #200 Series batteries with two 6-inch shielded barbette guns (Battery Construction #263 and #264) were built as part of the permanent defense. Battery Schwan (#263 – 1942-1949) was located on Punta Escambron, near Fort St. Geronimo. This battery was destroyed in 1965 to make room for a swimming pool. Battery Lancaster (#264-1942-1946) was located at Punta Cangrejos, near the eastern end of San Juan's international airport. This battery was later turned into an aquarium, but today it privately owned. Several fire control and searchlight stations were also constructed for this fort. After World War II, Fort Brooke continued to serve as the U.S. Army's headquarters until 1961 when the U.S. Army departed, and these historic defenses were turned over to the U.S. National Park Service.

El Morro, San Juan (Terry McGovern)

Fort Brooke Gun Batteries

- Three relocated Armstrong 4.7-inch guns were emplaced at various locations around the old fortifications of San Juan just before World War I . These weapons originated with those mounted at Battery Gatewood, Fort Monroe. They were removed sometime between 1910 and 1914 and sent to the Augusta Arsenal. From there they went to San Juan—where three were mounted, the fourth presumably becoming a spare barrel. The three site locations were the point of El Morro fortress, at Princesa battery just to the east of San Cristobal fortress and at the St. Elena bastion to the southeast of El Morro on the inside of the harbor entrance. Each emplacement had one Armstrong 4.7-inch gun with shield on pedestal mount. It appears they may have been emplaced between 1915 and about 1920 when this type of gun was declared obsolete and removed from service. The El Morro site was subsequently used for a 3-inch battery. The emplacement for the St. Elena bastion position still exists at the San Juan National Historic Site. The battery is open to the public.

- A 3-inch gun was located at the point position of San Felipe del Morro in San Juan, where previously a single 4.7-inch gun had been located in the 1910s. It consisted of a single 3-inch Model 1903 gun and pedestal (#85/#47). The emplacement was little more than a platform with base bolts and an adjacent ready ammunition box. Dates of service are unknown but probably would have been from 1942 to 1946. Photographs show it in place in 1944. The emplacement remains today and is open to the public.

Fort Amezquita (1941-1948) is located near the former Spanish defenses of Fort Amezquita on Isla Las Cabras that defend the City of San Juan, Puerto Rico. This military reservation was taken over the United States during the Spanish-American War in 1898. Named for Juan de Amézqueta of the Spanish Puerto Rican Militia in General Orders 10 in 1943. A temporary battery was built in 1941 using two 155mm GPF guns on Panama mounts on this island that western side of the entrance to Sun Juan Harbor. Started in 1941 was the construction of Battery Reed (1941-1948), a 12-inch casemated battery. This battery used two 12-inch Model 1895 guns on Model 1917 barbette carriages. Using a design very similar to one used for a #100 Series battery, these two 12-inch guns were emplaced in protected gun houses and service gallery connected them. Today the battery is used by San Juan police as a training area. Several fire control and searchlight stations were also constructed for this fort. A secondary harbor entrance control post (HECP) with an SCR-582 radar was also located here. The fort also had an 90mm AMTB battery. After World War II, Fort Amezquita was transferred to the city of San Juan to be used as park.

Fort Amezquita Gun Batteries

- **REED:** A battery for two 12-inch guns in a casemated emplacement built as the heaviest battery of the new San Juan defense project of 1939. San Juan was not included in the 1940 Project, having its own approved project dating from slightly earlier, though incorporating many of the characteristics and weapon types of the national program. This battery was first proposed on September 1, 1939 under a plan to provide a 12-inch battery for San Juan, at a suggested cost of $847,000. Approval of the battery came from the Secretary of War on October 6, 1939. It was to be emplaced at Cabras Island on the western side of the harbor entry. The emplacement design resembled the later large casemated types, but with a simplified magazine arrangement. It was of unique design and layout never precisely duplicated elsewhere. The 12-inch guns were removed from old Battery Torbert at Fort Delaware (two guns to be mounted, a third as a ready spare). The two Model 1917 long-range carriages were spares that had been at government arsenals for a number of years. Work was done

Battery Reed, Fort Amezquita (Terry McGovern)

from late 1939 to 1941. A transfer date and final cost have not been located. The gun tubes were sent here on October 25, 1941. It was armed with two 12-inch Model 1895M1A4 on Model M1917 barbette carriages (#15/#32 and #18/#1, with tube #21 as the spare stored near the battery). It was never given a construction number but eventually named in General Orders No. 13 of November 24, 1941 for Brigadier General Henry A. Reed. It served through the war, being removed in 1947 and abandoned as a gun site. In later years the site was occupied by the San Juan Police Academy. The emplacement still exists on police academy property. The battery is closed to the public.

- **AMTB Cabras**: An AMTB battery located just in front of and to the eastern flank of Battery Reed. It covered the main entry into the harbor. It consisted of four fixed emplacements for 90mm guns on M3 coast defense shielded carriages. It probably served between 1943 and postwar, being removed in 1946 or 1947. The gun pads still exist on the property of the San Juan Police Academy. The battery is closed to the public.

Fort Mascaro (1941-1948) is located Punta Salinas to west of the City of San Juan, Puerto Rico. Using the same designs at the 1940 Program, two #200 Series batteries with two 6-inch shielded barbette guns (Battery Construction #261 and #262) were built as part of the permanent defense. Battery Buckey (#261 – 1941-1948) was located on Punta Salinas proper, while Battery Pence (#262 – 1941-1948) was located at Punta Salinas Island, about a half of mile away from #261. These batteries were later turned into support structures for a radar facility in 1964. Several fire control and searchlight stations were also constructed for this fort. After World War II, Fort Mascaro served as the U.S. Air Force radar station until it was turned over the Puerto Rico National Guard for continued use as a radar site. No public access.

Fort Mascaro Gun Batteries

- **BUCKEY (#261)**: A battery for two modern 6-inch barbette guns erected as half of a pair of new batteries for this reservation at Point Salinas. It was located on the headland of the point and followed standard 200-series design. It was constructed under the 1939 San Juan project but later brought under the auspices of the 1940 Program as Battery #261. Work was begun on September 24, 1941 and essentially completed by August 1943. While the work was completed, the armament delivery was deferred. In 1944 ordnance returns it had been assigned M4 barbette carriages #52 and #24 (both designed to take the new M1(T2) 6-inch gun). The battery was never fully completed or armed. It was named in General Orders No. 15 of March 31, 1942 for Colonel Mervyn Chandos Buckey, US Army, who received the Distinguished Service Medal in World War 1. The emplacement was abandoned postwar. In 1964 the emplacement was converted for use as an Air Force radar and communication site. The emplacement remains in use to support the radar station. It still exists on property of the Puerto Rico Air National Guard. The battery is closed to the public.

- **PENCE (#262):** The second modern, 6-inch barbette gun battery for East Point Salinas (Island), near the point. Work was done from September 24, 1941 until June 30, 1943 and referred to at the time as Battery #262. It was of conventional design, and unlike its sister battery was armed fairly promptly. This armament was received in March 1943 and consisted of two 6-inch guns Model 1903A2 on Model M1 barbette carriages (#19/#13 and #39/#14, shipped here February 25, 1943). It was named in General Orders No. 15 of March 31, 1942 for Major William Perry Pence. It served during the latter stages of the war, not being eliminated until early postwar, probably in 1946 or 1947. The emplacement still exists on Puerto Rico Air National Guard property in abandoned condition. The battery is closed to the public.

Punta Escambrón Military Reservation (1941-1949) is located in the Puerta de Tierra area of San Juan. Located here was Battery Schwan/#263 near Fort San Gerónimo. The battery was destroyed in 1965 for an Olympic pool and stadium. That complex was later destroyed as well and the site is now a public park.

Punta Escambrón M.R. Gun Battery

- **SCHWAN (#263):** A standard dual 6-inch barbette battery built in compliance with the 1939 approved program for the defenses of San Juan at Escambron Point. It was constructed as Battery #263 between September 24, 1941 and March 31, 1943. It was of conventional 200-series design type. It received its armament in March 1943. This was two 6-inch guns Model 1903A2 and model M1 barbette carriages (#70/#46 and #69/#47). It fired to the northeast. It was named in General Orders No. 15 of March 31, 1942 for Major General Theodore Schwan. The battery served until just postwar, probably being disarmed in 1946 or 1947. The emplacement itself was demolished in the mid-1960s to make room for the Olympic swimming pool and stadium built at the site. Some concrete remains exist on the beach.

Punta Cangrejos Military Reservation (1942-1946) is located in Boca de Cangrejos, near Loíza. Located here was Battery Lancaster / 264 on Punta Cangrejos, north of the east-end of the runway at Luis Muñoz International Airport. The abandoned battery was later used as an aquarium from 1970-1975. The battery still remains, now overgrown in an area of private businesses and restaurants.

Punta Cangreios M.R. Gun Battery

- **LANCASTER (#264):** A standard dual 6-inch barbette battery built in compliance with the 1939 approved program for the defenses of San Juan at Point Cangrejos. During construction it was known as Battery #264. Work was done from January 2, 1942 and March 30, 1943. The armament was received in March 1943. It consisted of two 6-inch guns Model 1903A2 on Model M1 barbette carriages (#29/#88 and #60/#89). It was named in General Orders No. 15 of March 31. 1942 for Lieutenant Colonel James W. Lancaster. It served out the war and was disarmed probably in 1946 or 1947. In the mid-1970s the property was acquired. It was modified to become part of the Ocean Life Park Aquarium. The aquarium has been abandoned for many years and battery is very overgrown. The battery is open to the public.

THE HARBOR DEFENSES OF GUANTANAMO BAY – CUBA

Conde Bluff and Cuzco Hill Military Reservations (1906-1945) are located at Conde Bluff and Cuzco Hill at the Guantanamo Bay Naval Station, Cuba. Constructed in 1906 as part of the Taft Program each reservation received one battery. Conde Bluff (aka Fort Conde) had an emplacement for four 6-inch disappearing guns, while Cuzco Hill (aka Fort McCalla) received two emplacements for four 3-inch rapid-fire guns. The batteries were never armed. Also built were two fire-control stations for the 6-inch guns; a power plant at Fort Conde; a concrete shelter for a portable 30-inch searchlight at Fort McCalla; a mine casemate, torpedo storehouse, cable tank, and loading room for the harbor controlled mine defense; an ice plant; four sets of quarters; and a mess hall. The U.S. Army left in 1908 due to lack of funds but still retained title to the parcels until 1928 and 1940 respectively. The Leeward Point Reservation (1906-1912) was never used. The Marine Barracks was relocated to Fisherman's Point after 1908. The Marine Barracks on Fisherman's Point were then relocated to Casa Point (Marine Site 1) and Defense Point (Marine Site 2). An underground command bunker was built in 1942 near the Naval Station administration building. The harbor entrance and the north boundary channel were protected by anti-sub nets and booms in World War II. A Marine Defense Battalion was sent here in February 1941, with several 5-inch naval guns and 3-inch AA guns emplaced in the Cuzco Hills above the Naval Air Station complex; on Conde Bluff; and at Leeward Point. The Marines also emplaced an SCR-268 surface search radar and an SCR-270 early warning AA radar. The defenses were deactivated in 1945. These unarmed batteries remain today as part of the U.S. Naval Station – Guantanamo Bay, Cuba.

Guantanamo Bay Gun Batteries

- **Conde Bluff Battery:** This was the heaviest battery emplacement actually authorized for the defenses of Guantanamo. The site selected was a coral outcropping bluff above the beach west of Hicacal Beach, due north of Fisherman's Point. Plans were submitted on March 30, 1905. Plans were for a conventional, in-line, battery emplacement for four 6-inch guns on disappearing carriages, firing to the south. It closely followed the recommended plans in mimeograph No. 59. There were gun centers of 125-feet. Magazines were shared under the traverses between the first two and then the last two guns. Work was done in 1905-1906 and almost finished except for some placement of sand in the parapets and flanks on June 1, 1907. While it was not transferred, estimated cost to this point was $140,000. The battery was to have carried four 6-inch guns Model 1905 on Model 1905 disappearing carriages. In fact, the carriages were supplied and actually mounted (#2, #3, #4, and #5). The tubes were never received. When the U.S. Navy decided not to develop Guantanamo as a naval station, the U.S. Army suspended defensive works, and the incomplete 6-inch battery was caught in the midst of completing. By late 1908 it had only a caretaker assigned, though the carriages remained here for several years. Eventually they were dismounted. In World War I temporary 5-inch ex-naval guns were emplaced in at least two of the old disappearing platforms and manned by a Marine Corps detachment. The old, abandoned emplacement still exists on the American Guantanamo base.

EDITION OF JAN.24,1911. SERIAL NUMBER **117**   GUANTANAMO BAY, CUBA

CONDE BLUFF

Conde Beach

Leeward Pt.

CUZCO HILLS.

Nautical Miles

Statute Miles

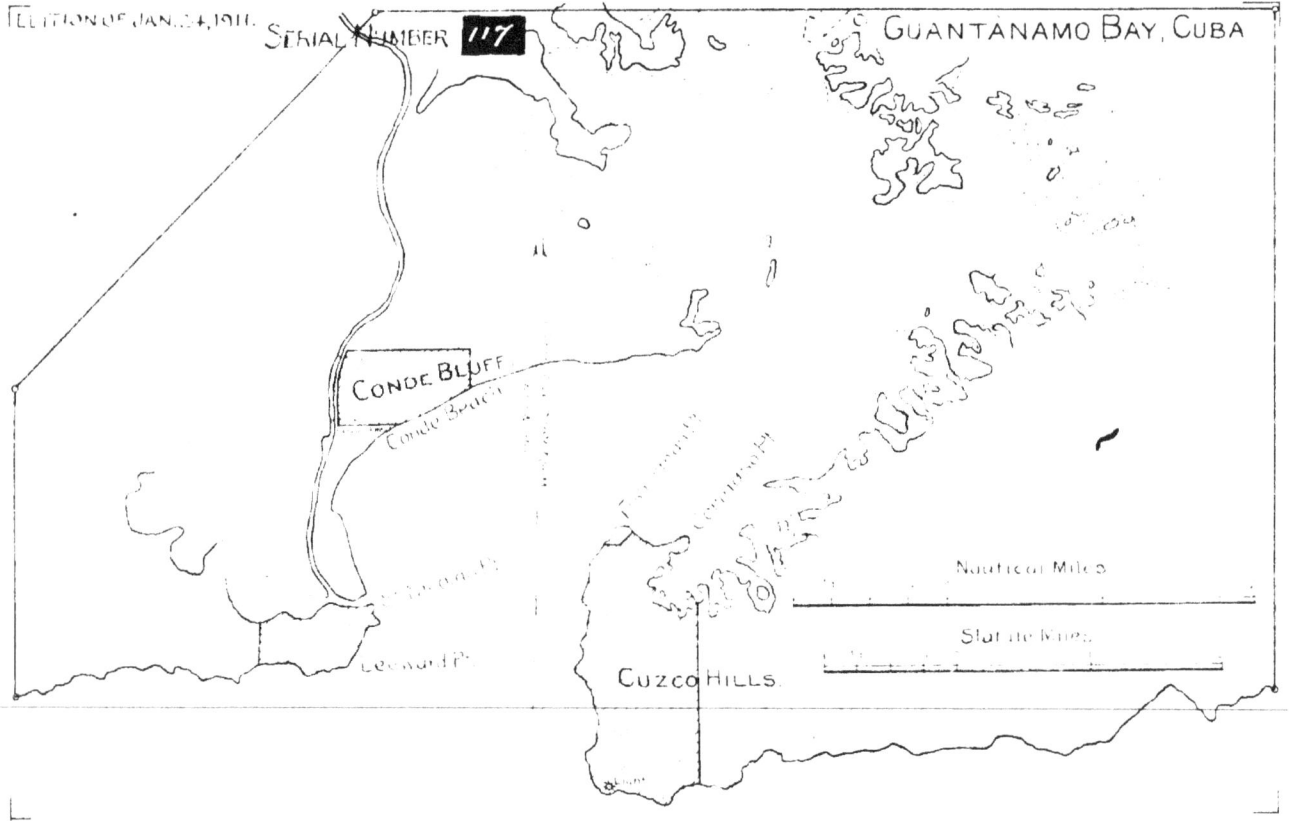

EDITION OF OCT.19, 1905.   SERIAL NUMBER **126**   GUANTANAMO BAY, CUBA.

CUZCO HILLS.

1000'   0   1000'   2000'   3000'
Scale.

Proposed Purchase

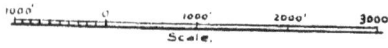

LEGEND.
1. Cn.Assist's Qrs.
2. Office.
3. Quarters.
4. Mess.
5. B.S. Stores.
6. Labor.

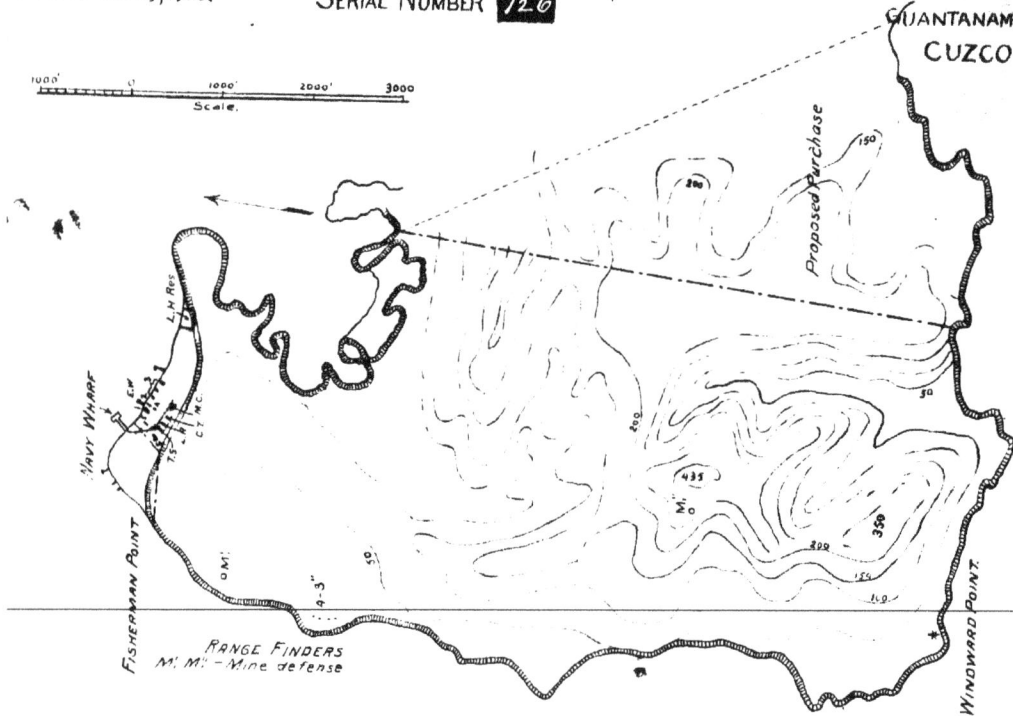

MINE DEFENSE.

M.C. Mining Casemate.
T.S. Torpedo Storehouse.
L.R. " Loading Room.
C.T. Cable Tank.

NAVY WHARF

FISHERMAN POINT

WINDWARD POINT

RANGE FINDERS
M' M'' - Mine defense

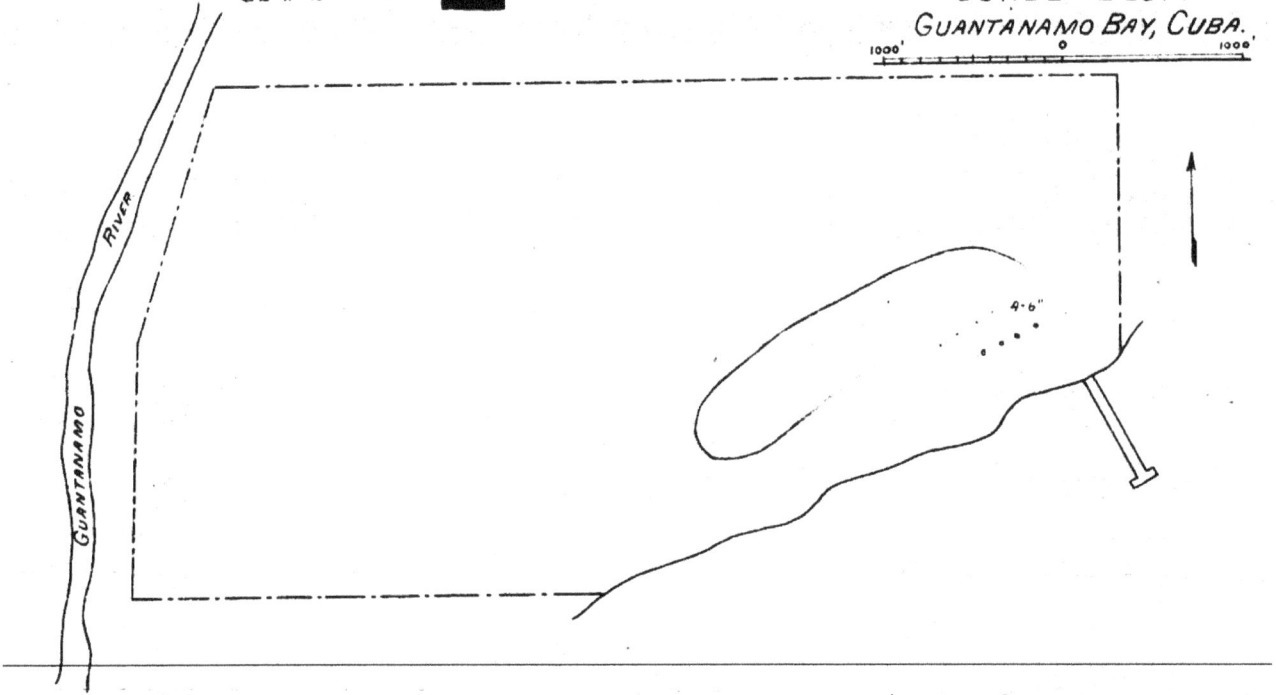

EDITION OF OCT. 19, 1905 SERIAL NUMBER 132

CONDE BLUFF
GUANTANAMO BAY, CUBA.
1000'   0   1000'

RIVER

GUANTANAMO

4·6"

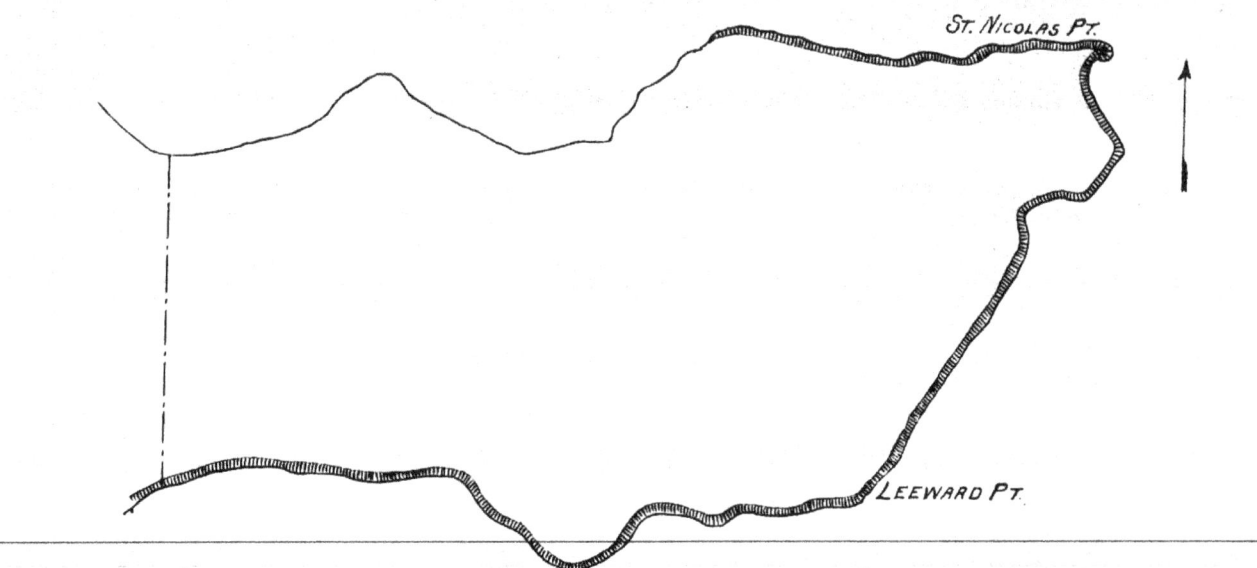

EDITION OF OCT. 19, 1905. SERIAL NUMBER 124

ST. NICOLAS POINT.
GUANTANAMO BAY, CUBA.
1000'   0   1000'

ST. NICOLAS PT.

LEEWARD PT.

- **Cuzco Hill Batteries:** One of two dual 3-inch pedestal gun emplacements put on the Cuzco Hill reservation near Fisherman's Point at Guantanamo Bay. These two emplacements (in close proximity but not immediately adjacent) were at the western tip of the main southern peninsula, protecting the entrance to the bay and the projected site for the main mine field. Plans for both batteries were submitted on March 30, 1905. They were positioned on the hard coral rock, above the steep bluff overlooking the point. Plans were strictly in accord with standard Mimeograph 30 Supplement 6-9. It consisted of two platforms separated by 62-feet with traverse covering two magazines and a storeroom. Work was done from late 1905 to 1906, being virtually complete by June 1, 1907 except for some electrical wiring. It was not transferred but had an estimated cost at that time of $20,000. Further work was suspended with the cancellation of plans to construct the naval base

6-inch carriage in the unfinished Conde Bluff battery at Guantanamo Bay (NARA)

THE HARBOR DEFENSES OF GULF OF PARIA — TRINIDAD AND TOBAGO

Fort Read (1941-1949/1960) is located on northwest tip of Trinidad on the Gulf of Paria, not far from Port-of-Spain. The fort's primary purpose was to help defend the U.S. Navy's Naval Base Trinidad. Naval Base Trinidad was a US Naval Advance Base built to protect the shipping lanes to and from the Panama Canal from U-boat attacks, by sea and air. The base provided fuel for shipping, loading, and unloading of cargo ships. The base also became a repair depot, with auxiliary floating drydocks that were able to repair boats and ships in the field. Naval Base Trinidad was commissioned on June 1, 1941. Naval Base Trinidad and seven other bases in the Caribbean became known as Destroyer Bases. This name came from the U.S.-British Destroyers for Bases agreement which exchanged older US destroyers for U.S. rights to operate Advance Bases in the Atlantic. This was done so the US could have tactical bases, patrol aircraft and ships to control the Caribbean Sea. Trinidad, Bermuda, Santo Domingo and Argentia became major bases. The Naval Base was built on the northwest tip of the island on 7,940 acres, this included the land on five small islands in the Gulf of Paria. Later 3,800 more acres were added to the base, but only 1,200 acres were built up. Four bays were used for Naval activities: Carenage, Chaguaramus, Teteron, and Scotland. Two major land bases were built at Chaguaramus (Chaguaramas Naval Base) and Tucker (Tucker Naval Base). Fort Read was the U.S. Army Headquarters, and Waller Field was the main Army Air Corps field at Chaguaramas. An SCR-270 early warning radar was also located here. Carlson Field was also built. The second task after the port was built, was building a naval air station and a seaplane base at Carenage Bay. The Gulf of Paria was used for major fleet anchorage. Six modern batteries were planned to protect the anchorage from sites on the northwestern peninsula of the island, and one battery for the passage on the southwestern side. Extensive planning work was conducted, and active surveying and clearing of sites was undertaken in 1942. However, before actual concrete construction or delivery of guns could be undertaken the projects were deferred. The lack of adequate shipping resources to bring the construction equipment and materials to the island was the stated reason. Then in late 1942 the large gun project was cancelled and the smaller batteries moved to indefinite deferral due to the improving strategic situation and the realization that with their current priority nothing could be finished until way into 1944. The batteries planned for the Trinidad Defenses were to have been two casemated 12-inch gun batteries (500-series) at Corozal Point and on Chacachacare Island, Battery 271 on Corozal Point, Battery 272 and Battery 273 on Monos Island, Battery 274 on Chacachacare Island, and Battery 275 on Green Hill. None of these were built. What were actually emplaced by the U.S. were eight 155mm guns on Chacachacare Island (Battery U at Boca Grande, and Battery V at Boca de Navos); four 155mm guns in two positions on Monos Island; four 155mm guns on Green Hill (Battery W) (two guns transferred to Venezuela in 1942); 90mm AMTB/AA batteries at Fort Read, Cumuto, Port of Spain, Piarco Field, Edinburg Field, and Nelson Island; and 37mm AMTB/AA batteries at Mucurapo Point, Laventille Hill, Port of Spain, and Sangre. The HECP (with an SCR-582 radar) was located on Chacachacare Island. The Americans used contact mines and anti-submarine nets in the harbor. American coast defenses were withdrawn in 1945. Ruins of the U.S. Naval Fuel Depot still remain in the Lumber Lane area. An ammo bunker still remains on the old Huggins plantation. Today the base headquarters are a hotel and convention center.

—World War II Program sites

- *Corozol Point 12-inch Battery* (planned): A battery for two 12-inch barbette guns intended for Corozol Point. This was planned as a large, casemated battery allowing a 145-degree field of fire, perhaps similar in layout to Battery Reed in Puerto Rico. On April 15, 1942 it was approved to source the guns for the two 12-inch batteries in Trinidad from continental Batteries Haslet (Fort Saulsbury) and Kimble (Fort Travis) that were not approved for casemating under the new defense program. It was never assigned a project number. There does appear to have been some site preparation for construction, but little or probably no physical building was done when the project was cancelled and the battery eliminated on November 12, 1942.

- *Battery #271*(planned): A 1941 planned standard dual 6-inch barbette battery intended for Corozol Point. No structural work had been done when the emplacement was deferred on July 5, 1942 and then finally permanently eliminated on December 17, 1942.

- *Chacachaca Island 12-inch Battery* (planned): A battery for two 12-inch barbette guns intended for Chacachaca Island. This was planned as a large, casemated battery allowing a 145-degree field of fire, perhaps similar in layout to Battery Reed in Puerto Rico. On April 15, 1942 it was approved to source the guns for the two Trinidad 12-inch batteries from continental Batteries Haslett (Fort Saulsbury) and Kimble (Fort Travis) that were not approved for casemating under the new defense program. It was never assigned a project number. There does appear to have been some site preparation for construction, but little or probably no physical building was done when the project was cancelled and the battery eliminated on November 12, 1942

- *Battery #274* (planned): A 1941 planned standard dual 6-inch barbette battery intended for North Monos Island. No structural work had been done when the emplacement was deferred on July 5, 1942 and then finally permanently eliminated on December 17, 1942.

- *Battery #272* (planned): A 1941 planned standard dual 6-inch barbette battery intended for South Monos Isalnd. No structural work had been done when the emplacement was deferred on July 5, 1942 and then finally permanently eliminated on December 17, 1942.

- *Battery #273* (planned): A 1941 planned standard dual 6-inch barbette battery intended for Corozol Point. No structural work had been done when the emplacement was deferred on July 5, 1942 and then finally permanently eliminated on December 17, 1942.

- *Battery #275* (planned): A 1941 planned standard dual 6-inch barbette battery intended for Green Hill at the southwestern extremity of Trinidad. No structural work had been done when the emplacement was deferred on July 5, 1942 and then finally permanently eliminated on December 17, 1942.

THE HARBOR DEFENSES OF PORTLAND BIGHT – JAMAICA

Fort Simonds (1941-1949) is located near Old Harbour Bay, Jamaica. The U.S. Navy leased the British Navy's Port Royal Dockyard and used Old Harbour Bay (Portland Bight), west of Kingston, as a fleet anchorage. The U.S. Naval Air Station was located on Little Goat Island until 1944. Fort Simonds was the U.S. Army Headquarters. The U.S. Army Coast Artillery emplaced four 155mm guns on Little Goat Island from 1942-1944. Battery 285 was planned here but it was never built (this battery number was then assigned to Vieques Island, Puerto Rico). The U.S. Army Air Corps built Vernam Field (vacated 1949). An SCR-270 early warning radar was also located here.

Fort Simonds Gun Battery

- *Battery #285* (planned): Plan for the battery was to be a standard dual, 6-inch barbette battery for Jamaica. The United States gained access to a Jamaican base as a result of the September 1940 Destroyers for Bases deal. In April 1941 a program expanding the fixed defenses to this location was approved. Just a single new, fixed battery was approved to be built in Jamaica—at Portland Bight in Heathshire. Apparently a standard 200-series type. It was never built and probably never got beyond just conjecture. During planning it was known as Battery Construction No. 285.

OVERSEAS BASES – NEWFOUNDLAND AND BERMUDA

The Overseas Bases – Newfoundland, Bermuda, and Iceland are characterized by isolated islands, rugged landscapes, and naval operating bases (both airfields and docks). The destroyers-for-bases deal was an agreement between the United States and the United Kingdom on September 2, 1940, according to which 50 Caldwell, Wickes, and Clemson-class U.S. Navy destroyers were transferred to the Royal Navy from the U.S. Navy in exchange for land rights on British possessions. These harbor defenses were developed under the 1940 Program and during World War II as America acquired access to additional oversea bases. Temporary U.S. defenses during World War II were also installed in Greenland (Bluie West #1 two 90mm 1942-1944 at Narsarsuaq and Bluie West #7 two 90mm 1942-1944 at Kangilinnguit), Ascension Island, and Iceland.

THE HARBOR DEFENSES OF ARGENTIA – NEWFOUNDLAND

Fort McAndrew (1940-1946/1994) is located at Argentia, Newfoundland (a British crown colony until 1949) to defend the U.S. naval operating base (naval docks and airfield) that was established by the U.S. military after exchanging basing rights with Britain for 50 destroyers in 1940. Originally called the Marquise Military Reservation until March 1942. Named in General Orders 4 of 1942 for Maj. Gen. James W. McAndrew, USA. The primary coast defenses were based on the same designs at the 1940 Program as two #200 Series batteries each with two 6-inch shielded barbette guns (Battery Construction #281 and #282) were built as part of the fort's permanent defense. Battery #281 was located at Argentia North (on hills overlooking the peninsula on which the airfield was built), while Battery #282 was located on Hill 195. Significantly, both these batteries retained their 6-inch shielded barbette guns until 1994. In that year, the two 6-inch guns of Battery #281 were removed and shipped to Fort Columbia State Park in Washington State where they were installed in Battery #246. The 6-inch guns of Battery #282 remain on site today. Other coast defenses included a 3-inch rapid-fire battery at Isaac's Point, two unnamed two-gun 6-inch naval gun batteries (temporary), two 155mm guns on Panama mounts, a 90mm AMTB battery at Roche Point, and another 90mm AMTB battery at Ship Harbor Point. Another 90mm AMTB battery was located south of Placentia at Black Point. Also located at Argentia were a SCR-582 radar station and a HECP/HDCP for the harbor defenses. A U.S. Navy-manned indicator loop station (Station 1X) was at Argentia, this detected submarines via their magnetic signature. Several fire control and searchlight stations were also constructed for this fort. After World War II, Fort McAndrew was transferred to the U.S. Navy. The U.S. Naval Station Argentia remained in operation until the base was closed in 1994 and return to Government of Newfoundland. Several of the military housing barracks were destroyed in 1999, and the large Officers' Quarters building was destroyed in 2000. The site is now being used as an industrial development park and many of structures have been removed. Argentia Management Authority Inc. was established for the economic redevelopment of Argentia for the benefit of businesses, residents and other stakeholders in the Placentia region of Newfoundland & Labrador. Now called the Port of Argentia. The Battery 282 site is now designated a Municipal Heritage Site, located on the west side of Charter Ave., west of the "Pinnacles" and south of Shag Ponds. The AMA Backland Trail in Southside leads to the old U.S. Naval Ammunition Magazines and a naval lookout post. The former fort is open to public and former battery sites are abandoned.

Fort McAndrew Gun Batteries

- **Battery #281:** A World War II program battery for two 6-inch barbette guns emplaced to protect the new U.S. base at Argentia. It was the eastern unit of the pair emplaced adjacent to each other. Construction was accomplished from January 10, 1942 until November 10th of the same year. It was of standard 200-series plan. Armament was shipped here on March 8, 1943. The battery was

Battery #281 Argentia

One of the 6-inch guns of Battery #282 (Terry McGovern)

armed with two 6-inch Model 1903M2 guns on Model M1 barbettes (#30/#9 and #61/#10). The battery served until early postwar, probably being taken out of service in 1946. However, the guns and shielded carriages were not scrapped, but left in place, though abandoned for many years. The site operated as McAndrew Air Force Base until 1955, when it became the Argentia Naval Air Station, which in turn was closed in 1975. Eventually the fort reservation became an industrial park. In 1993 the guns and carriages were removed and sent for display purposes to Fort Columbia, Washington. The emplacement still exists at the industrial park. The battery is open to the public.

- **Battery #282:** A second modern World War II program battery for two 6-inch barbette guns emplaced to protect the new U.S. base at Argentia. It was the western unit of the pair emplaced at a site known locally as Hill 195. Construction was done from December 20, 1941 until November 1, 1942. It was of conventional 200-series design plan. The battery was armed with two 6-inch Model 1903M2 guns on Model M1 barbettes (#13/#44 and #8/#45). The battery served until early postwar, probably being taken out of service in 1946. However, the guns and shielded carriages were not scrapped, but left in place, though abandoned for many years. The site operated as McAndrew Air Force Base until 1955, when it became the Argentia Naval Air Station, which in turn was closed in 1975. Eventually the fort reservation became an industrial park. The emplacement with its original armament still in place, exists on this facility. The battery is open to the public.

- **AMTB Roche Point:** An AMTB battery for two fixed 90mm guns emplaced at Roche Point. It was probably built in late 1942 and served from 1943 to early postwar, probably being removed in 1946.

- **AMTB Ship Harbor:** An AMTB battery for two fixed 90mm guns emplaced at Ship Harbor Point. It was probably built in late 1942 and served from 1943 to early postwar, probably being removed in 1946.

- A battery for two 3-inch Model 1902 pedestal guns emplaced at this location overlooking Isaac Head. This work was done in 1942, the guns served until probably 1945 or 1946 before being removed. Remains of the emplacement still exist.

THE HARBOR DEFENSES OF SAINT JOHN'S — NEWFOUNDLAND

Fort Pepperrell (1941-1947/1961) is located along the north shore of Quidi Vidi Lake in Pleasantville, St. John's. Camp Alexander was the cantonment area before the fort was finished in 1941. Named in honor of Sir William Pepperrell (1696–1759) of Kittery, Maine, commander of a force of 4,200 soldiers and sailors aboard 90 ships, who captured the French seaport at Louisbourg after a 46-day siege on June 16, 1745. The headquarters post of the U.S. Army's Newfoundland Base Command during World War II. Became a U.S. Air Force base in 1947. The Canadian Military Headquarters for Newfoundland was transferred here from Buckmasters Field in 1961 after the Americans left. The former post is now mostly used by the RCMP and the Fisheries Canada Centre. The Canadian Forces Station St. John's still uses a portion of the former base. Several original buildings still exist, now known as the Pleasantville community, although many have recently been torn down. Several magazines are still extant in the White Hills area. The U.S. Army built a two-gun coastal defense battery on Signal Hill, North Head, St. John's in 1941, consisting of two 8-inch M1888 guns on M1918 barbette carriages, located at the present-day parking lot for Cabot Tower. The two guns were removed and relocated to Red Cliff in 1942. A battery of four 155mm guns on Panama mounts was also located here in 1942, until moved to Middle Cove and Manuels. This site was considered a sub post of Fort Pepperrell. A grass-covered mound near the parking lot is one of the gun sites. Parking lot improvements in the fall of 2016 exposed at least one of the 155mm gun's Panama mounts. Both Signal Hill and Red Cliff are open to the public.

THE HARBOR DEFENSES OF BERMUDA

Fort Bell (1940-1946/1957/1995) is located at several locations on the islands of Bermuda (a British crown colony) to defend the U.S. naval operating base (naval docks and airfield) that was established by the U.S. military after exchanging basing rights with Britain for 50 destroyers in 1940. Burrow's Hill, St. David's Island, was the U.S. Army headquarters, and Kindley Field on Long Bird and St. David's Islands was the U.S. Army Air Corps base (now the Bermuda International Airport). The airfield was begun in July 1941 and was completed and operational by December 1941. The primary coast defenses were based on the same designs at the mainland 1940 Program. Two #200 Series batteries each with two 6-inch shielded barbette guns (Battery Construction #283 and #284) were built as part of the islands' permanent defenses. Battery #283 (1943-1945) was located on Tudor (Stone) Hill in Southampton Parish (overlooking U.S. Naval Base on Morgan Island), while Battery #284 (1943-1945) was located at Fort Victoria on St. Georges Parish. Both 6-inch batteries remain though impacted by later construction around them. Prior to completion of these two batteries, the American provided a temporary defense using four 8-inch Model 1888 guns on Model 1918 railway carriages. Two of the guns were emplaced on a short section of railway track at Fort Victoria, St. Georges, while the other two were similar emplaced at Scaur Hill Fort in Sandy's Parish. Other U.S. coast defenses included four 155mm guns on Panama mounts (Well Bay Hill on Cooper's Island and Turtle Hill on Southampton Parish), and 90mm AMTB battery (two fixed mounts) at Fort Victoria. There were 13 listed fire-control base-end stations for the large caliber guns, located east to west at Fort Victoria, Mount Hill (two) (still extant), Cooper's Island, Paynter's Hill (still extant), Cataract Hill, Gibbs Hill (two) (still extant), Tudor Hill, Wreck Hill, Daniel's Head (still extant), and the Royal Naval Dockyard. A FC station listed for Beek Hill in Southampton Parish is apparently unknown on modern maps. An SCR-296A radar was located at Skinner's Hill and another SCR-296A radar was located at High Point. SCR-582 radars were located at Gibbs Hill and Mount Hill. Coast artillery searchlight stations were located at Cemetery Hill, Fort Victoria (two), St. David's Battery (three), Surf Bay (two), Elbow Bay, Warwick Long Bay (two), Church Bay (two), and Wreck Hill (two). After World War II, Fort Bell was transferred to the U.S. Navy. The U.S. Naval Station Bermuda remained in operation until the base was closed in 1995 and return to Government of Bermuda. Most of Fort Bell has since been demolished and redeveloped. None of the railway gun battery sites remain. The Cooper's Island 155mm battery site was destroyed in the early 1960s for a NASA tracking station, although the underground magazines may still exist. One Panama mount still remains at the Turtle Hill 155mm battery site, adjacent to the Fairmont Southampton Hotel. The U.S. Navy operated a Magnetic Indicator Loop Station at British Fort St. Catherine. An American submarine base was built on Ordnance Island in St. George's Harbour. The main American Naval Base was on Morgan's Island in Southampton Parish, with subposts on Tucker's Island and King's Point. The U.S. Navy took over Fort Bell/Kindley Field (Kindley Air Force Base after 1948) from the Air Force in 1970 (Naval Air Station Bermuda). The last American and Canadian forces left the island in September 1995. Most of the former coast defense sites are now being used for a variety of residential and commercial uses.

Fort Bell Gun Batteries

- **Battery #284:** The plan for American seacoast defenses for the defense of Bermuda was approved and issued on April 16, 1941. In addition to interim batteries of 8-inch Model 1918 railway batteries (4 mounts total) and 155mm Panama Mounts, the project called for the construction of two permanent, dual 6-inch barbette batteries. These were intended to be built and armed just like the 200-series batteries of the domestic 1940 Seacoast Defense Project. This one was located at the former British fort site of Fort Victoria, on St. George's Island to the northeast in the archipelago. The battery was built to the east of the old work, primarily on the outer earthwork cover, but the rear central

Scaur Hill Fort, Sandy Parish, Bermuda (terry McGovern)

Fort Victoria, St. George Parish, Bermuda (terry McGovern)

entry into the power room was actually built into the outer ditch wall of the older fort. Otherwise, it followed the dimensions and layout plan of standard 200-series batteries. Work was done from February 2, 1942 to May 1, 1943. It was turned over on June 9, 1943 at a cost of $488,021. It was reported armed on August 15, 1943, with the guns shipped to Bermuda on February 24, 1943. The battery was armed with two 6-inch guns Model M1903A1 on barbette carries M2 (#41/#11 and #43/#12). The battery was never named, just known as Battery Construction No. 284 during construction and service. It served until early in the postwar period, probably being disarmed in 1945-1946. Most of the battery has been covered over with new development, but the entry into the power room in the ditch of old Fort Victoria still persists. The battery is not open to the public.

- **Battery #283**: The second battery for two 6-inch dual barbette guns built by the Americans to augment the fixed defenses of Bermuda. Its location and authorization also date from the April 16, 1941 Project. This battery was located on the southwestern end of Bermuda Island, at a location generally known as Stone Hill (although also referred to as Tudor Hill). The battery plan is consistent with the most common 200-series types, with separate open gun platforms with shielded guns and a traverse magazine, power, radio, and plotting room between. Guns were at a trunnion elevation of 151-feet. Work was done from April 15, 1942 to May 1, 1943. Transfer was made on June 9, 1943 for a cost of $376,478. It was armed on August 15, 1943. The battery carried two 6-inch guns Model M1903A2 on carriages M1 (#49/#49 and #50/#49). The battery was never named, just known as Battery Construction No. 283 during construction and service. It was initially manned by the 27th Coast Artillery Regiment, later that unit was reorganized as separate battalions, this battery being operated by the 854th Coast Artillery Battalion. It served until the early postwar period, probably being disarmed in late 1945 or 1946. The emplacement still exists on public park land in Bermuda, though one of its magazine entries is buried. The battery is closed to the public.

- **AMTB Fort Victoria**: An AMTB Battery of two 90mm fixed and two 90mm mobile guns erected on the east side of Fort Victoria in St. George Island. It was probably built (mainly just concrete blocks for the two fixed guns) in 1943, being armed and serving until removal postwar in 1946. No information on any surviving elements has been obtained.

Installing a 6-inch gun and carriage for Battery #284 at Fort Victoria, Bermuda

www.ingramcontent.com/pod-product-compliance
Lightning Source LLC
Chambersburg PA
CBHW040259100426
42811CB00011B/1312